Safety by Design

Safety-critical systems with more digitalization, automation, and AI require integration of human factors to ensure safety, efficiency, and meaningful human control. Across sectors such as aviation, maritime, energy, transportation, and remote operations, accident investigations repeatedly show that failures often stem not from technology alone but from inadequate scoping, poor human-centered design, alarm overload, fragmented automation, and organizational conditions that leave operators unprepared during critical situations. Disasters such as the Boeing 737 MAX accidents demonstrate how safety can be compromised when humans and systems are misaligned, while successful recoveries (such as the US Airways Flight 1549 landing on the Hudson River) show the vital role of design-supported human performance in demanding conditions.

Safety by Design: Human-Centered Approaches to AI, Automation, and Remote Operations explores the relationship between human capabilities, organizational readiness, and complex technical systems. It offers best practices for designing safety-critical systems that integrate human factors from concept to operation. The book introduces Meaningful Human Control, a framework ensuring humans retain authority, awareness, and intervention ability, aligning with emerging EU AI regulations. Readers will learn practical methods to design systems with usability, transparency, and oversight, alongside insights into automation, accident analysis, and supporting operators under stress. These principles help create safer, more resilient, and trustworthy systems.

This book has been written for human factors students and specialists, system designers, engineers, safety professionals, regulators, and project managers working on safety-critical applications involving automation, AI, or remote operations.

Safety by Design

Human-Centered Approaches to AI, Automation, and Remote Operations

Edited by

Frøy Birte Bjørneseth
Stig Ole Johnsen
Ole Andreas Alsos
Vidar Hepsø
Gunhild Birgitte Sætren

CRC Press
Taylor & Francis Group
Boca Raton London New York

CRC Press is an imprint of the
Taylor & Francis Group, an **informa** business

Designed cover image: NTNU/SCL-Shore Control Lab

First edition published 2026
by CRC Press
2385 NW Executive Center Drive, Suite 320, Boca Raton FL 33431

and by CRC Press
4 Park Square, Milton Park, Abingdon, Oxon, OX14 4RN

CRC Press is an imprint of Taylor & Francis Group, LLC

Library of Congress Cataloging-in-Publication Data
Names: Bjørneseth, Frøy Birte editor | Johnsen, Stig Ole editor | Alsos, Ole Andreas editor | Hepsø, Vidar editor | Sætren, Gunhild B. editor
Title: Safety by design : human-centered approaches to AI, automation, and remote operations / edited by Frøy Birte Bjørneseth, Stig Ole Johnsen, Ole Andreas Alsos, Vidar Hepsø, and Gunhild B. Sætren.
Description: First edition. | Boca Raton, FL : CRC Press, 2026. | Includes bibliographical references and index.
Identifiers: LCCN 2026013506 | ISBN 9781041229728 hardback | ISBN 9781041244141 paperback | ISBN 9781003741824 ebook
Subjects: LCSH: Automation--Safety measures | Industrial safety | Automation--Human factors | Remote control | Artificial intelligence--Industrial applications
Classification: LCC T59.5 .S24 2026
LC record available at https://lccn.loc.gov/2026013506

ISBN: 978-1-041-22972-8 (hbk)
ISBN: 978-1-041-24414-1 (pbk)
ISBN: 978-1-003-74182-4 (ebk)

DOI: 10.1201/9781003741824

Typeset in Times
by SPi Technologies India Pvt Ltd (Straive)

Contents

PART 3 Key Techniques for Designing for Safety and HF in Emerging Environments

PART 4 Workload Assessment and Training

PART 5 Learning from Practice and Accidents

About the Editors

Frøy Birte Bjørneseth is an Associate Professor in Human Factors in the Department of Ocean Operations and Civil Engineering at the Norwegian University of Science and Technology (NTNU) in Ålesund, Norway. She holds a PhD in Computer Science, with a focus on human–machine interaction and human factors in the maritime environment. She is also a Principal Engineer in Human Factors with 18 years of experience at Rolls-Royce Marine (now Kongsberg Maritime AS), a worldwide supplier of maritime equipment. Dr. Bjørneseth's research focuses on human factors in maritime operations, with a primary focus on safety-critical applications and a holistic, human-centered design approach in operational environments. Her approach emphasises usability, safety, and operational insight, contributing to the development of human-centered maritime technologies.

Stig Ole Johnsen is a Senior Safety Scientist and Lecturer at NTNU (Norwegian University of Science and Technology) and NORD University in Norway. He holds a PhD and a Master's degree in Technology Management and Science. His research focuses on safety science, human factors, and design in critical systems, with expertise in remote operations, automation, and AI in safety-critical environments. Dr. Johnsen lead the Human Factors in Control network in Norway (from conception 2004 to 2025) and has contributed to numerous accident investigations (e.g., Deepwater Horizon). His work spans multiple industries, including oil and gas, maritime operations, and autonomous systems. He has published extensively on meaningful human control in automation, the verification and validation of safety-critical systems, and human factors in remote operations. His expertise bridges theoretical safety science with practical applications in increasingly automated environments.

Ole Andreas Alsos is a Professor of Interaction Design at the Norwegian University of Science and Technology (NTNU) and serves as Head of the Shore Control Lab. He holds an MSc in Computer Science and a PhD in Computer and Information Science from NTNU. Dr. Alsos specializes in human–computer interaction, with a focus on maritime technology, autonomous ships, and the design of shore control centres. His research explores how to enhance situational awareness and safety in maritime operations through innovative interface design. As the leader of the Shore Control Lab, he oversees research on autonomous vessel monitoring and control systems, including work with the autonomous ferry prototypes milliAmpere1 and milliAmpere2.

Vidar Hepsø is a Principal Researcher at Equinor Research and Technology and an Adjunct Professor at the Norwegian University of Science and Technology (NTNU). He earned his PhD in Social Anthropology from NTNU and specializes in integrated and remote operations, as well as human factors in technological environments. At Equinor, Dr. Hepsø serves as Human Factors Integration Lead, focusing on scaling control room solutions and remote operations for offshore wind across Norway, the UK, the US, and Poland. His research examines how new information and communication technologies enable innovative collaborative practices in industrial settings. Dr. Hepsø has published extensively on socio-material practices, information infrastructure development, and digitalization challenges in the energy sector.

Gunhild Birgitte Sætren is a Professor and Manager at the Arctic Safety Centre at the University Centre in Svalbard (UNIS), Norway. She specializes in safety psychology and human factors, with extensive experience in high-risk contexts and a focus on human–machine interaction. She manages the Arctic Safety Centre at the University Centre in Svalbard, overseeing research projects on human performance in extreme environments. She has also been awarded the title of Excellent Teaching Practitioner for her research and teaching in safety psychology, safety management, and human factors over many years. Consequently, she is a highly sought-after lecturer at several universities.

Contributors

Dr. Ole Andreas Alsos leads NTNU's Shore Control Lab, specializing in interaction design, prototyping, and digital innovation. He holds an MSc and a PhD in computer science from NTNU and serves as Vice Dean of Innovation. His work bridges design research, technology development, and user-centered innovation.

Dr. Sonja Biede-Straussberger specializes in cognitive psychology and human factors in aviation. She led Airbus' contributions to SESAR and focuses on integrating scientific research with industrial practices. She actively participates in international committees and contributes to aviation standards and human performance assessment.

Dr. Frøy Birte Bjørneseth combines research and industry expertise in human-centered maritime system design. With 18+ years in the maritime sector, she focuses on usability, operator workload, and safety-critical environments, bridging theory and practice in operational maritime management.

Dr. Clark Borst specializes in human-centered aviation automation, applying Cognitive Systems Engineering and Ecological Interface Design principles. His research advances human–automation collaboration, system transparency, and adaptive automation, enhancing pilot interaction with automated systems.

Dr. Houda Briwa develops explainable AI methods for modeling human performance in safety-critical systems. Her research integrates human factors, alarm system design, and Bayesian networks to improve alarm management and human reliability assessment in oil and gas operations.

Piotr Cichowicz specializes in safety transformation, human performance, and organizational change. With a background in drilling engineering and HSE leadership, he collaborates on improving safety programs and reducing risks across high-risk industries.

Dr. Jonathan Earthy leads human-centered design assurance at Lloyd's Register, focusing on socio-technical systems, resilience, and human-centered AI. He contributes to international maritime projects, chairs ISO committees, and has authored numerous standards and publications.

Dr. Maren Helene Rø Eitrheim integrates human and technical perspectives to improve system safety in complex operations. With more than 15 years of experience, her research spans automation, staffing strategies, and human factors across petroleum, nuclear, and healthcare sectors.

Dr. Rune Elvik has researched road safety since 1980, publishing over 190 papers. He served as an editor for major journals and focuses on accident prevention and transport safety, aiming to surpass 200 publications.

Marius Fernander applies Human Factors methods like Petro-HRA to improve safety and performance in complex industries. His work bridges theory and practice, focusing on practical solutions for safety-critical environments.

Nick Grigg specializes in design research and analysis, using qualitative and quantitative methods to support user-centered decisions in digital product development and market research.

Stein Helgar advises on safety-critical design for high-risk sectors, focusing on user experience, automation, and system resilience. His work integrates human factors principles into industrial innovation.

Dr. Vidar Hepsø specializes in digitalization and socio-technical systems, contributing to renewable energy control room projects. With 30+ years in the energy industry, he advances human-centered approaches in technology-intensive operations.

Dual Herzog researches cybersecurity and AI collaboration in threat detection. With a background in psychology and organizational operations, he focuses on training humans to work effectively with AI systems.

Lars Hurlen specializes in control room design and human–system interaction across safety-critical domains. His work focuses on operator support tools and the impact of digitalization on human performance.

Jon Bernhard Høstmark integrates human factors into maritime cybernetics and automation. His work focuses on user-centered design, human–system integration, and safety in advanced digital systems.

Dr. Stig Ole Johnsen chairs the Human Factors in Control network and serves as ESREL Chair of Accident and Incident Modeling. With a PhD from NTNU and a Master of Technology Management from MIT, his work spans safety management, accident investigation, and human–technology interaction across multiple sectors.

Stine Skaufel Kilskar researches safety management, resilience, and Human Factors at SINTEF Digital. She holds an MSc in Industrial Economics from NTNU and is pursuing a PhD. Her work spans oil and gas, construction, maritime, energy, transportation, and healthcare sectors, focusing on learning from incidents and Man, Technology, Organization (MTO) analysis.

Dr. Gernot Konrad leads Human Factors initiatives at Honeywell Aerospace, focusing on avionics systems for diverse aviation markets. He holds advanced degrees in Aerospace Engineering and is an accredited HF Specialist. He chairs European Union Aviation Safety Agency (EASA) HF committees and lectures on flight deck design and certification.

Dr. Maria Chiara Leva leads TU Dublin's Human Factors in Safety and Sustainability Research Group. She co-chairs the Occupational Health and Safety Master's program and has authored 80+ publications. Her work focuses on operational risk assessment, safety management, and promoting female researchers in safety-critical fields.

Andreas Myskja Lien is a design engineer specializing in safety-critical solutions for defence, maritime, and advanced industrial operations. He focuses on integrating human factors with emerging technologies (AI, machine learning, and automation).

Dr. Jinchao Lin specializes in human–autonomy collaboration, human performance evaluation, and psychophysiological assessment. His research focuses on automation impacts in nuclear power plants, stress, workload, and fatigue in safety-critical systems.

Dr. Anders L. Madsen develops Bayesian network algorithms for safety-critical systems, environmental risk modeling, and fault detection. He is the CEO of HUGIN EXPERT and a professor at Aalborg University, focusing on probabilistic AI applications.

Dr. Gerald Matthews researches human factors, stress, and cognitive performance in demanding settings. He co-developed the Dundee Stress State Questionnaire and advanced multivariate models of sustained performance. He is a Fellow of IAAP and Specialty Chief Editor for *Frontiers in Psychology*.

Dr. Max Mulder specializes in human-centered aviation systems, focusing on manual control cybernetics, ecological interface design, and pilot-automation cooperation. His research advances higher level automation in aviation.

Dr. Marcin Nazaruk transforms safety through proactive learning and human performance strategies. He authored industry guides on "Learning from Normal Work" and develops tools for frontline staff and leaders, combining safety psychology with 20 years of industry experience.

Dr. Salman Nazir leads the Human Factors Research Group at USN, focusing on maritime operations, safety management, and training in complex systems. His work includes simulator-based training, offshore safety, and human–technology interaction.

Dr. Tor-Olav Nævestad researches safety culture, resilience, and disaster risk management in transport and high-risk sectors. His work applies safety culture concepts to improve road safety for private and professional users.

Dr. Marinus M. (René) van Paassen specializes in human–machine interaction, haptics, and simulation. His research applies ecological interface design principles to vehicle control and cockpit systems, enhancing human-centered aviation technologies.

Jooyoung Park researches human-centered design and maritime human factors, focusing on digital navigational tools and emerging technologies to improve situational awareness, decision-making, and safety in autonomous and remote maritime operations.

Dr. Ross O. Phillips studies human factors, safety culture, and decision-making in transportation. His work focuses on human-centered system design for automated freight vehicles and improving safety and sustainability in logistics.

Dr. Martin Rasmussen Skogstad coordinates the SAFE – Centre for Civil Safety and Security Research. His work explores the cognitive and organizational aspects of decision-making and safety outcomes, with a focus on AI integration in safety-critical systems.

Dr. Gunhild Birgitte Sætren specializes in safety psychology and human factors in extreme environments. As the manager of the Arctic Safety Centre, she oversees research on human performance and safety management in high-risk contexts.

Dr. Stig Winge's research focuses on safety management, barrier analysis, and learning from incidents in the oil and gas and construction industries. His work improves safety governance and operational reliability.

Preface

Safety by Design addresses a key modern challenge: ensuring that increasingly digitalized, automated, and AI-enabled systems stay safe, efficient, and under meaningful human control.

Across sectors such as maritime, energy, transportation, and remote operations, accident investigations reveal recurring systemic weaknesses: inadequate scoping, poor design, alarm overload, fragmented automation or AI behavior, poor systematic assessment of workload, insufficient training, and organizational conditions that leave humans unprepared for critical moments. The opening chapters show how design flaws on modern ship bridges diminish situational awareness and create cognitive traps – transforming advanced technology into a safety hazard when Human Factors are not integrated early in the design process.

At the same time, the book demonstrates that when Human Factors are applied effectively, safety and efficiency significantly improve. Aviation's outstanding safety record, Norway's world-leading road safety outcomes, best practices in remote operation of automated ferries, and the impressive usability of the Unified Bridge concept all show what can be achieved when technology, human capabilities, and organizational learning are aligned.

At the heart of this book is the practice of Meaningful Human Control (MHC). This concept describes what it means for humans to retain authority, oversight, and the capacity to intervene when systems fail or behave unexpectedly. MHC ties together design, operations, and learning within a systemic, non-blaming improvement loop. The chapters present cases within a unifying framework for developing safe autonomy, remote supervision, and AI-enabled decision-making systems. They show that safety cannot be added afterward; it must be designed in from the start through comprehensive, human-centered engineering. They also highlight how organizational readiness, psychological safety, and learning from work-as-done influence whether control is maintained or lost.

The book concludes with a synthesis of cross-industry lessons and a call to action: Safety by Design is no longer optional. It forms the basis for safe, trustworthy automation and AI. Achieving this demands a lifecycle approach in which design integrity, operational vigilance, and organizational learning support one another to prevent failures and build resilient socio-technical systems.

Part 1

Defining Human-Centred Approaches in an Environment of Automation, AI, and Remote

1 Designing for the Night Watch

Human Factors Challenges on Modern Ship Bridges

Ole Andreas Alsos and Frøy Birte Bjørneseth

1.1 INTRODUCTION

The starting point and inspiration for this chapter was a field visit to a brand-new passenger vessel. At first sight, the bridge seemed impressive – large, clean, and seemingly well-organised with modern digital consoles, integrated sensors, and comfortable workstations. However, to those with operational experience, such as the highly experienced captain who guided us, it quickly became clear that the bridge design was sometimes misaligned with the realities of seafaring work. One obvious flaw was the inability to dim many of the displays and control panels. This is not merely a convenience issue; it directly compromises night vision (Hess et al., 1990), a crucial factor for safe navigation in low-light conditions. In response, the crew resorted to improvisational solutions, taping paper and posters over displays, potentially covering critical information.

Modern ship bridges, at first glance, appear to be marvels of technological progress. Well equipped with numerous displays and panels, spacious and well-lit, and sometimes even aesthetically appealing, they evoke the image of a highly efficient, future-ready operational environment. Yet, as the example above shows, beneath this polished surface lies a persistent and serious problem: the operational needs of seafarers, particularly during critical tasks like night sailing or navigation in complex and congested waters, have often been neglected in design processes.

As you will see throughout this chapter, the observations above were not isolated. Despite over a decade of research emphasising the importance of Human Factors (HF) and user-centred design (UCD) in maritime systems (Bjørneseth, 2021; Kirwan, 2025), economic pressures following the offshore market downturn stalled progress (Razmjooei et al., 2023). As a result, bridges today often represent a regression rather than an evolution in ergonomic and cognitive design: interfaces remain inconsistent across vendors, system complexity has increased, alarm management is frequently inadequate, and integration across subsystems is still weak.

The consequences of these shortcomings are well documented. Poorly designed ship bridges pose a significant risk to maritime safety by increasing cognitive workload, degrading situational awareness, and heightening the risk of operator error (Danielsen, 2022). This is tragically highlighted by numerous incidents, some of which are revisited later in this chapter. These incidents are followed by a description of the various standards and design processes deemed necessary for the design of bridge equipment and systems. In addition, the chapter discusses the systemic HF deficiencies in modern ship bridge designs, critiques the industry's detachment from user-centred principles, and offers strategies to improve integration between human operators and complex maritime systems. Finally, some examples of better bridge equipment and system design are outlined.

1.2 HUMAN FACTORS FOUNDATIONS FOR SHIP BRIDGES

Maritime navigation is a cognitively demanding activity conducted in dynamic, high-risk environments. Understanding how ship bridge design influences safety requires first examining the human

DOI: 10.1201/9781003741824-2

strengths and limitations that shape human performance at sea, also known as HF. This section introduces four foundational concepts – situation awareness (SA), cognitive workload, alarm management, and human–system integration (HSI) – which form the analytical basis for the accident cases discussed later.

The HF principles outlined next provide a foundation for understanding the systemic issues often seen in maritime accidents.

1.2.1 Situation Awareness

SA is a cornerstone of safe navigation. Endsley (1995) defines SA in three levels: (1) *perception* of relevant information (e.g., traffic, alarms, environmental cues), (2) *comprehension* of what this information means in context, and (3) *projection* of future states (e.g., collision risks or environmental changes).

On the ship bridge, SA depends on how well system layouts, information presentation, and workstation arrangements support an uninterrupted link between the operator and the external environment. When information is scattered across poorly integrated systems – such as radar, ECDIS (Electronic Chart Display and Information System), communication equipment, and automation displays – operators must mentally assemble a coherent picture from disparate sources. This increases cognitive effort and reduces the ability to form a clear 'situation at a glance'.

Research consistently identifies the loss of SA as a primary causal factor in maritime incidents. Grech et al. (2002) found that SA-related failures accounted for 71% of human errors in maritime accidents, a statistic further supported by Stratmann and Boll (2016).

As shown in Section 1.3, design choices influence all three SA levels, from misperception to misjudgement and failure to anticipate hazards.

Achieving sufficient SA requires human-centred design (HCD) processes, including contextual inquiry, task analysis, and human–machine interface (HMI) design aligned with standards such as ISO 9241-210 (2019) and IEC 63303 (2024).

1.2.2 Cognitive Workload

Cognitive workload is the mental effort required to perform a task. In maritime operations, workload varies with traffic density, weather, system complexity, proximity to hazards, and operational speed.

Poor ship bridge design frequently imposes unnecessary workload on the operators by requiring them to manage overly complex or unintuitive user interfaces. This is often caused by multi-vendor interfaces with inconsistent designs, information spread across distant displays, and poor system feedback.

These demands divert attention from essential watchkeeping tasks, such as monitoring the environment, anticipating developing situations, and coordinating within the ship's bridge team. High workload reduces the cognitive resources available for monitoring, prediction, and coordinated team action.

Workload can be assessed using standard HF methods, such as observational studies and task analysis, as well as tools like the NASA-TLX. An example of this is Gould et al. (2009), who investigated sleep deprivation among operators navigating high-speed vessels. Such analyses are rarely conducted in maritime design projects, despite their value in predicting – and preventing – overload. This gap is supported by Lützhöft (2004) and Costa (2018).

1.2.3 Alarm Fatigue

According to IEC 62923-1 (International Electrotechnical Commission, 2018), an alert is a generic term that encompasses all notifications issued by shipboard systems to draw attention to abnormal situations. They are categorised into three levels: *alarm* (requiring immediate action), *warning* (requiring awareness and possible action), and *caution* (requiring monitoring). Hence, 'alert' is the umbrella term, and 'alarm' is a subset of alerts.

Alarms are intended to highlight abnormal or hazardous conditions, but they only support operators effectively when well-designed and well-managed. Alarm fatigue arises when alarms are too frequent, poorly prioritised, repetitive, lack actionable information, and are triggered by predictable, non-critical events.

Under such conditions, operators quickly lose trust in the alerting system, reducing the likelihood that genuine hazards will be noticed in time. In safety-critical contexts, this can cause the operator to ignore or delay responses when real danger occurs.

International standards – such as IEC 62682, IEC 62923-1, and IMO MSC.302(87) – provide clear guidance on alert behaviour, grouping, and prioritisation. Nonetheless, real-world ship bridge systems often produce alarm cascades that obscure key information and overwhelm operators. As later accident cases show, poor alarm management can substantially degrade situational awareness and hinder timely decision-making.

Effective alarm design requires systematic HF work, including task analysis, workload assessment, HMI design, SA analysis, prioritisation workshops, prototyping, and operator-in-the-loop testing.

1.2.4 Human–System Integration

HSI refers to the coordinated design of technology, tasks, roles, and environments to support human performance. In maritime contexts, effective HSI ensures that the designers of ships, equipment, and software align system behaviour with operators' capabilities and the context of use. Key HSI considerations include (1) ensuring consistent interaction principles across systems; (2) designing spatial layouts that support visual scanning and communication; (3) providing precise, meaningful system feedback; (4) supporting teamwork and shared situational awareness, and (5) making automation behaviour transparent and predictable.

In practice, many ship bridge systems still reflect a technology-driven rather than a human-centred approach. In other words, engineering constraints dominate user needs. As a result, equipment may be technically functional yet operationally complex, requiring operators to adapt to system limitations rather than for the system to adapt to the operators and the context of use.

Failures in HSI manifest not only as isolated ergonomic issues but as system-level breakdowns in perception, decision-making, coordination, and response. The accident cases presented next show how these weaknesses increase risk during time-critical operations.

1.3 MARITIME ACCIDENTS CAUSED BY HUMAN FACTORS AND POOR BRIDGE DESIGN

A series of high-profile maritime accidents, presented next, highlight critical vulnerabilities in the intersection between human performance and ship bridge design. These incidents reflect how inadequate HSI and poor situational awareness can lead to catastrophic outcomes.

Viewing accidents through the HF lens presented in the previous section shifts the focus from blaming individuals to understanding how design, environment, and organisational factors shape performance.

1.3.1 *City of Rotterdam* (2015)

Danielsen and Petersen (2022) show that ship design can contribute to 'human error', as seen in the collision between *City of Rotterdam* and *Primula Seaways* in the river Humber in 2015 (Marine Accident Investigation Branch, 2017). The *City of Rotterdam* adopted an unconventional design featuring a round bow to achieve a more aerodynamic shape and improved fuel economy. The combination of the unconventional design and the placement of the communication equipment created a relative motion illusion, confusing the pilot and leading to inadequate changes of heading, ultimately resulting in a collision with *Primula Seaways*.

In the accident investigation report, the investigation noted that although innovations in ship design can contribute to safety, stricter adherence to the ergonomic principles outlined in SOLAS V/15 (IMO, 2002) should have been applied in this case (Marine Accident Investigation Branch, 2017). The accident investigation revealed that several pilots found piloting the vessel 'disconcerting' or 'uncomfortable' due to the bridge's design. The report states that operating this ship for several years without navigational accidents was primarily due to the crew and pilots' adaptations and coping strategies.

1.3.2 *KNM Helge Ingstad* (2018)

A similar pattern was one of the factors contributing to the accident in which the Norwegian Royal Marine's frigate, *KNM Helge Ingstad*, collided with the tanker *Sola TS* and sank in the Norwegian archipelago (Norwegian Safety Investigation Authority, 2019). The accident stemmed from the crew's failure to correctly identify the tanker, misinterpreting its incorrect use of deck lights as shore lights from the nearby Sture oil Terminal. The frigate sank and was condemned (Figure 1.1).

Investigations pointed to cognitive overload, poor team coordination, and an overly cluttered radar and visual display environment (Norwegian Safety Investigation Authority, 2019). The bridge team suffered from confirmation bias and lacked effective cross-checking mechanisms.

Alsos et al. (2023) analysed the accident from a systems perspective. They identified several underlying causes, including poor design of critical support systems, systemic weaknesses within the organisation, and inadequate design of the bridge crew's workspace and bridge equipment.

- The ship bridge layout and work environment did not support coordinated team communication.
- Poor equipment placement forced operators to split attention between communication and navigation tasks.
- The visual and radar display arrangements did not promote or support shared situational awareness on the bridge.

These findings directly illustrate how poor bridge design interacts with team cognition, amplifying risk during periods of high workload.

FIGURE 1.1 *KNM Helge Ingstad.*

Source: Forsvaret.

1.3.3 *Viking Sky* (2019)

The incident involving the cruise ship *Viking Sky* in March 2019 is a well-documented example of how automation systems can contribute to critical failures under certain conditions. According to Porathe (2023) and supporting reports (Norwegian Safety Investigation Authority, 2024a), the *Viking Sky* experienced a complete loss of propulsion while navigating the hazardous waters of Hustadvika on the Norwegian coast. The root cause was traced to low lubricating oil levels in all four diesel generators, with sensors detecting dangerously low oil pressure likely due to sloshing caused by rough seas. The automation system interpreted this as a critical fault and shut down all four generators to prevent engine damage. This resulted in a total blackout and loss of propulsion, leaving the vessel adrift in high winds and shallow waters with nearly 1,400 passengers on board. The key issue highlighted in the article is that the automation system lacked contextual awareness and could not distinguish between a real oil shortage and a temporary sensor anomaly caused by ship motion. This led to a safety-driven shutdown, followed by an alarm avalanche, which made it difficult for the crew to identify and address the root causes. The shutdown underscores the need for improved human-centred design of automation behaviour that allow for operator override or contextual decision-making.

1.3.4 *PSV Sjøborg* (2019)

Poor decision-making and loss of situational awareness are recurring patterns that also led to the 2019 accident in which the *PSV Sjøborg* collided with the oil rig *Statfjord A* in the North Sea. The investigation report by the Petroleum Safety Authority Norway (PSA) (2019) concerning the collision mentions alarm overload as a contributing factor. During this accident, the bridge crew on *Sjøborg* experienced a high number of alarms, more than 20 alarms in 45 minutes before and during the incident. This alarm overload contributed to reduced situational awareness and delayed the crew's ability to respond effectively to the developing loss of position. The report highlights that the alarm system was not optimally configured and that the crew had difficulty distinguishing between critical and non-critical alarms in the heat of the moment. This is a typical example of what is often referred to in HF literature as 'alarm fatigue', where too many simultaneous or frequent alarms can overwhelm operators, leading to slower or incorrect responses (Nygård, 2019).

1.3.5 *USS John S. McCain* (2017)

In 2017, the *USS John S. McCain*, a U.S. Navy guided-missile destroyer, collided with the oil tanker *Alnic MC* in the Singapore Strait, one of the world's busiest shipping lanes. The collision resulted in the deaths of 10 sailors and injuries to 48 others. Investigations identified several contributing factors, including inadequate training, poor communication among the bridge crew, and a loss of situational awareness (NTSB, 2019). A key factor in the accident was confusion caused by the ship's touchscreen-based steering and throttle control system. During a routine transfer of control between consoles, operators unintentionally changed the steering control settings, leading to a perceived loss of steering. The touchscreen interface was overly complex, lacked clear feedback, and had inconsistent configurations, which contributed to the crew's loss of situational awareness. Following the accident, the U.S. Navy decided to replace touchscreen controls with traditional physical throttles and steering systems to improve usability and safety.

1.3.6 *USS Fitzgerald* (2017)

In a similar collision off the coast of Japan in 2017, the *USS Fitzgerald* collided with the container ship *ACX Crystal*, which resulted in the deaths of seven sailors. The probable cause was the *Fitzgerald's* bridge team's failure to take early and substantial action to avoid collision.

Contributing factors included poor communication and cooperation within the *Fitzgerald* crew, insufficient planning by the commanding officer, inadequate training and fatigue, and the Navy's practice of not broadcasting automatic identification system (AIS) signals. Several design-related issues on the ship's bridge contributed to the failure (National Transportation Safety Board, 2020). Poor integration between the bridge and the Combat Information Centre limited shared situational awareness, while radar and navigation systems featured complex, cluttered interfaces that were difficult to interpret under pressure. The bridge lacked effective collision warning systems and decision-support tools, and the layout of displays and workstations hindered teamwork and monitoring. These design flaws compromised the crew's ability to detect and respond to the developing collision risk in a timely manner.

1.3.7 *Tyrhaug* (2023)

In 2023, the high-speed passenger ferry *Tyrhaug* collided with the aquaculture service vessel *Frøy Loke* in the Edøyfjord, Norway, under clear and calm weather conditions, good visibility, and little other traffic. The accident resulted in significant damage to both vessels and injuries to several individuals. Investigations revealed that *Tyrhaug*'s bridge was not manned in accordance with standard procedures at the time of the accident. The primary navigator was distracted by a passing sailboat, while the support navigator was occupied with administrative tasks in a separate room, leaving the bridge effectively undermanned (Norwegian Safety Investigation Authority, 2024b). The Norwegian Safety Investigation Authority highlighted the importance of adhering to established watchkeeping procedures and ensuring that administrative tasks do not compromise navigational duties. The incident underscored the critical need for vigilance and proper manning on the bridge to prevent similar accidents in the future.

1.3.8 *NCL Salten* (2025)

In a recent accident, the container ship *NCL Salten* grounded in Trondheimsfjorden, Norway, narrowly missing a residential home (Figure 1.2). No accident report has been published yet, but investigations revealed that the second officer, who was alone on watch, fell asleep, leading to the incident (Norwegian Maritime Authority, 2025).

A subsequent inspection by the Norwegian Maritime Authority uncovered that the ship's Bridge Navigational Watch Alarm System (BNWAS), designed to alert the crew if the watch officer becomes incapacitated, had been switched off.

FIGURE 1.2 *NCL Salten.*

Source: Politiet.

The list could be extended with several additional examples. Still, across these incidents, bridge design shortcomings – including suboptimal layout, unintuitive interfaces, poor visibility, or high cognitive load – were compounded by HF failures, such as miscommunication, fatigue, and training gaps. They collectively underscore the need for UCD in maritime interfaces and a systems approach to bridge operations, where human capabilities and limitations are integral to design and procedure development. Several standards, frameworks, processes, and methods can help naval architects and system developers design ships that support crew navigation to their destinations safely, as we will see in the next section.

1.3.9 Patterns across Accidents

When analysing the accidents, we see that some stem from the way ship bridge design interferes with visual perception, spatial awareness, and the ability to gather and integrate visual cues to sustain SA (e.g., the *City of Rotterdam* and *the KNM Helge Ingstad*). Other accidents demonstrate how poorly designed automation and alert systems can introduce new vulnerabilities, especially when automation lacks contextual awareness or when alarms overwhelm the operator (e.g., *Viking Sky* and *PSV Sjøborg*). Several accidents demonstrate how interface complexity, inconsistent layouts, and ambiguous control modes can overwhelm bridge crews (*USS John S. McCain* and *USS Fitzgerald*). Other accidents illustrate how bridge design interacts with human limitations such as fatigue, monotony, and divided attention – particularly during night watches or periods of low workload.

1.4 HUMAN FACTORS AND HUMAN-CENTRED DESIGN

In many accidents, the officer on watch is blamed and prosecuted for 'human errors' deemed to have caused the accident (Marine Insight, 2025). However, as several of the accidents in the previous section suggest, many human errors are, directly or indirectly, caused by design flaws in the bridge layout or navigation systems. Many of these design problems could have been easily avoided early in the ship's and its systems' design by using well-established design standards, frameworks, processes, and methods. Some of them are presented next.

1.4.1 Human-Centred Design

ISO 9241, Ergonomics of human-system interaction (ISO, 2019), focuses on creating systems that are usable, efficient, and satisfying for users. At its core, the standard emphasises a human-centred approach to design, ensuring that interactive systems align with users' needs, capabilities, and limitations. Key principles include designing systems that are suitable for users' tasks and efficient to use, ensuring interactions are self-explanatory, and aligning with user expectations. It also stresses the importance of learnability, giving users control over their interactions, minimising errors, and fostering a positive and engaging user experience. Together, these principles guide designers and developers in building systems that not only function well but also help users achieve their goals effectively and safely.

ISO 9241 contains Part 210, Human-centred design for interactive systems *(HCD)* (ISO, 2019). The approach is described as iterative and problem-solving, placing the needs, limitations, and experiences of people, as well as the context of use, at the core of the design process. Rather than starting with technology or system constraints, HCD begins by gaining a deep understanding of users through observation, interviews, and other methods. Iterations of prototyping and usability testing follow. This is done to ensure that the resulting products, interfaces, or systems are designed and built to be intuitive, safe, and effective.

The HCD process, first illustrated in the ISO 13407 standard (1999), remains relevant and is illustrated in Figure 1.3. Today, variations of the HCD process have emerged, but with roughly the same steps.

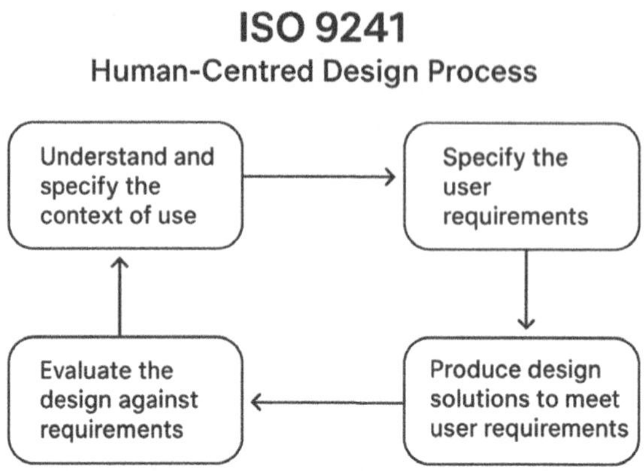

FIGURE 1.3 Simplified version of the human-centred design process.

Source: ISO 9241-210.

In complex, safety-critical domains such as maritime operations, HCD helps bridge the gap between human capabilities and technological complexity by designing tools that enhance situational awareness, decision-making, and collaboration. By involving users throughout the development process, HCD not only enhances usability and performance but also reduces the likelihood of errors and improves overall system resilience. A natural extension of the HCD approach is systemic design thinking, which is presented later.

1.4.2 CRIOP: Scenario-Based HF Verification

CRIOP (Aas et al., 2009; Johnsen et al., 2025) is a methodology initially developed for the offshore oil and gas industry to verify HF and operability in complex control centres. It combines (1) structured checklists with best practices to document 'Work as Imagined'; (2) scenario walk-throughs to explore 'Work as Done' with an experienced workforce; and (3) evaluation of team communication and coordination, with assessment of system support in normal, degraded, and emergency operations.

The method supports organisational learning and development by conducting the CRIOP workshop with management, the workforce, and expert designers. Identified issues are prioritised collaboratively, and mitigating actions are followed up on. The method yields the greatest value when applied early, in concept and design, before architectural or layout decisions become fixed. But the technique is also suited for use in detailed design, installation, and after implementation to check that the HF challenges are mitigated in 'work as done'.

The CRIOP method is highly relevant for maritime bridge design because it is relatively inexpensive and reveals interaction problems early. Further, it identifies missing safety barriers, highlights cognitive and organisational vulnerabilities, and enables multidisciplinary teams to evaluate design alternatives collectively.

Recent studies (Park et al., 2025) demonstrate that CRIOP can be adapted for autonomous ferry operations, underscoring its applicability across new and emerging operational contexts.

1.4.3 Bridge and Control Centre Design Standards

For the arrangement of equipment and layout of ship bridges, ISO 8468:2007 on Ship's bridge layout and associated equipment (ISO, 2007) provides standardised guidelines and requirements.

Its goal is to enhance navigational safety, usability, and operational efficiency by ensuring that bridge design supports the safe and effective performance of navigational tasks. The standard addresses the ergonomic placement of instruments, visibility from the bridge, access to controls, and the integration of systems such as radar, ECDIS, autopilot, and communication devices. It is intended for designers and shipbuilders.

Similarly, ISO 11064-4:2004 on Ergonomic design of control centres (ISO, 2013) provides comprehensive guidelines for the ergonomic design of control centres, where critical monitoring and decision-making occur, such as in industrial plants, transportation hubs, energy facilities, and maritime shore control centres. The standard covers the entire design lifecycle, from requirements analysis and layout planning to workstation design, display and control arrangement, and environmental considerations (e.g., lighting, acoustics, thermal comfort). Its primary goal is to enhance situational awareness, reduce operator workload, and improve safety and performance through HCD.

International standards describing best practice provide a structured way to involve users. For example, ISO 9241-210 (2019) outlines a human-centred design process. In addition, human factors engineering includes methods such as observation studies, eye tracking, task analysis, HMI design standards, and alarm standards.

In addition, (1) *SOLAS* (Safety of Life at Sea) includes comprehensive regulations for alarm systems on ships to ensure safety and compliance (International Maritime Organization, 2020). (2) *IMO Resolution MSC.302(87)* provides performance standards for bridge alert management, aiming to reduce the number of unnecessary alarms and improve the handling of critical alerts (International Maritime Organization, 2010). (3) *IMO Resolution MSC.252(83)* covers the performance standards for integrated navigation systems (INS), including alert management (International Maritime Organization, 2007). (4) *ISO 17894:2005* specifies requirements for the design, installation, and testing of alarm systems on ships (International Organization for Standardization, 2005), and (5) *IEC 62616:2010* provides guidelines for maritime navigation and radiocommunication equipment and systems. In addition to these, regulatory institutions such as Lloyd's Register, DNV, and the American Bureau of Shipping offer guidelines and reports on effective alarm management to ensure safety and crew well-being on board ships (Lloyd's Register, 2013; DNV, 2021; American Bureau of Shipping, 2010).

1.4.4 Systemic Design Thinking

Systemic design thinking takes HCD one step further, incorporating more stakeholders into the very beginning of the development process (Blokland and Reniers, 2020). This approach combines the holistic perspective of systems thinking with the more creative and exploratory, human-centred methods of design thinking, integrating the user-focused mindset of design thinking with the broad, interconnected perspective of systems thinking. It emphasises a holistic understanding of problems, recognising that changes in one part of a system can affect the whole system.

Central to this approach is the concept of interconnectedness, which emphasises the dynamic relationships between social, technological, environmental, and economic elements. Through co-creation and stakeholder engagement, it brings all relevant aspects into the design process early. The approach values iteration and prototyping to refine ideas through testing and feedback. Thinking holistically is especially useful when there are many interconnected problems in a complex, and often safety-critical, environment.

This approach aligns closely with modern safety science (e.g., Safety-II by Hollnagel (2014)), which incorporates successes and views operator performance as emerging from complex socio-technical interactions.

1.4.5 Design Systems and Standards

One way to improve the user interfaces of bridge equipment is to standardise them across ships and across different vendors using design systems.

The OpenBridge Design System is an open-source framework developed to enhance the HCD of ship bridges and maritime equipment (OpenBridge, 2025). It offers a structured set of user interface components, interaction principles, and reference layouts that align with international standards and established best practices in usability and safety. By promoting consistency and interoperability across systems, OpenBridge facilitates the development of bridge solutions that enhance situational awareness, reduce cognitive workload, and minimise the risk of human error (Nordby et al., 2019). The design system is intended for use by shipbuilders, equipment manufacturers, and software developers, enabling more efficient design processes and improved user experience in complex maritime operations.

A similar design system, OpenCrane, has been developed for crane user interfaces (Sitompul et al., 2023a, b), and studies have shown that such design systems make it easier for developers and technologists without design competence to create human-centred interfaces (Vehus, 2024).

Applying the aforementioned standards, design processes, and design systems in the design and development of ship bridges and bridge systems can make ship navigation more user-friendly and safer. Despite the availability of all these aids, accidents still occur due to design failures, as discussed in the previous section. Next, we will dive deeper into common design failures on ship bridges.

1.5 RECURRING DESIGN FAILURES AND THEIR CONSEQUENCES

The issues we identified on board the new passenger ship echo challenges that have persisted for 10 to 15 years (Danielsen, 2023), where very little has changed concerning alert and alarm handling.

Firstly, the number of alarms and warnings has not decreased, and alarm fatigue (IEC 62923-1, 2018) remains a significant cognitive hazard. Secondly, the lack of system integration forces navigators to mentally reorient and combine information from disparate sources, increasing the cognitive workload and the risk of errors. In addition, display layout and visibility are poor. Critical displays are often placed too far from the operator, requiring physical movement that disrupts situational awareness. Finally, the non-intuitive user interfaces of the systems frequently behave more like office IT equipment than tools designed for dynamic, high-risk, and rough environments, requiring fine motor skills and coordination to navigate a mouse pointer between small tool menus.

These flaws are not merely inconvenient and annoying; they undermine safety. They increase the risk of slips, lapses, and mistakes, particularly during night watches when cognitive performance is naturally impaired due to circadian rhythms, or during periods of increased cognitive workload and high stress.

1.5.1 Cognitive Workload and Situation Awareness

The workload on many bridges is often unnecessarily high due to poorly designed user interfaces across the various systems onboard, each of which is operated differently. The captain we accompanied illustrated this vividly. To dim displays for night navigation, he sometimes needed to navigate through up to 30 touch commands – a tedious and mentally demanding task during critical operations. At dawn, when he wanted them brighter again, it was nearly impossible to see anything on the dark, dimmed displays, and it wasn't very easy to adjust them back to normal. Moreover, on the passenger ship, critical navigational screens were placed several metres from the steering controls, forcing the navigator to move physically, disrupting their 'head-up' time – the period when they can visually scan the environment. This degraded both Level 1 (perception) and Level 2 (comprehension) of SA, as outlined by Endsley (1995).

Unfortunately, interaction design flaws are not uncommon. They were also a significant contributing factor to the USS John S. McCain accident, where the complex and inconsistent touchscreen interface contributed to the crew's loss of situational awareness (Mallam et al., 2020).

1.5.2 Bridge Layout

The bridge's layout can sometimes lead to a higher workload. In restricted waters and under bridges, our captain on the passenger vessel was forced to stand between steering positions to safely pass directly in the centre. However, he then lost direct line of sight to the fairway due to the thick window beam in front of him. This physical misalignment required constant body adjustments, forcing the captain into a solo tango dance between the window beam, increasing workload, and the risk of navigational errors. In the *City of Rotterdam* accident mentioned earlier, the structural design flaws of the bridge were the primary contributing cause (Marine Accident Investigation Branch, 2017), highlighting the importance of the bridge's design layout. On board *KNM Helge Ingstad*, an awkward bridge layout made it impossible for the officer to talk on the radio and use the radar/ ECDIS simultaneously, increasing strain and reliance on his short-term memory (Norwegian Safety Investigation Authority, 2021).

1.5.3 Multitasking

Humans are generally poor at multitasking and tend to struggle when tasks exceed their cognitive limits. One example is the aforementioned *Tyrhaug* accident, in which one of the officers on watch attempted to perform administrative tasks while also serving as lookout (Norwegian Safety Investigation Authority, 2024b). This accident illustrates the need to conduct task analysis and assess the physical and cognitive workload (Stanton, 2006). This can help develop operational procedures that put the crew's workload in the 'sweet spot' of their capabilities, neither overloading them nor making the work boring and unengaging.

1.5.4 Alarm Management Deficiencies

During our visit to the passenger ship, an alarm sounded every five minutes, forcing the first mate to repeatedly leave his operator position, turn his back to the bridge windows and his lookout duties, and walk to the back of the bridge to deactivate it. He did not even look at the alarm display. In HF terms, this represents a classic case of alarm mismanagement, where frequent and irrelevant alarms desensitise operators, delay critical responses, and contribute to accidents. The *Helge Ingstad* accident, in which 11 non-critical alarms sounded minutes before the collision, is a prime example (Alsos et al., 2023).

Proper alarm management – including prioritisation, suppression of nuisance alarms, and meaningful differentiation – is a well-established requirement in safety-critical industries such as aviation and nuclear energy. Yet maritime bridge systems often fail in this regard.

1.5.5 Organisational and Systemic Failures

Wróbel (2021) critically examined the widely held belief that human error accounts for approximately 80% of maritime accidents. Results indicate that although original research generally agrees that human error contributes significantly to maritime accident occurrence, the widely accepted 80% rate is unsubstantiated.

The study found that human error is a significant factor in maritime incidents, but also that attributing accidents predominantly to human error can oversimplify complex causal relationships and potentially divert attention from systemic issues within maritime operations. In other words, human error is a symptom of deeper systemic problems. The study advocates for a more nuanced understanding of accident causation that considers a broader range of contributing factors beyond human error alone.

Yet, when an incident or accident occurs, blame typically falls on the individual, as seen in the verdict following the *Helge Ingstad* accident (Hordaland District Court, 2023). Despite this, deeper system-level factors – such as poor bridge layout, inadequate integration, and cognitive overload – are rarely scrutinised or corrected.

The lack of accountability for HF engineering in ship design processes maintains unsafe conditions and enables accidents to continue. Yards, equipment manufacturers, and ship designers are rarely held responsible for the operational usability of their products. Meanwhile, seafarers are seldom meaningfully involved in design processes; at best, they are invited to 'validate' near-finished designs rather than co-develop solutions from early stages. Nevertheless, when accidents inevitably occur due to systemic errors, seafarers are the ones prosecuted, fined, or jailed.

This absence of UCD practices is stark. When we ask ship designers 'How do you include the human operator in your design process?', the most common responses are blank stares or vague statements. Actual UCD requires field observations, task analyses, iterative prototyping with user feedback, and rigorous usability testing – practices that are still rare in the maritime sector. Hsieh et al. (2024) conducted a comprehensive literature review that compiled system design and evaluation studies in the maritime domain. The study reviewed 130 English-language studies in the maritime domain published between 2010 and 2024. The studies matched the criteria: (1) conducted in the scope of maritime or vessel traffic, (2) exploring the design or evaluation of systems and HMI, and (3) adopting an HCD or a UCD approach.

The study concluded that the usability of maritime systems plays a key role in marine safety and in operator workload. There is a consistent upward trend in the number of studies from 2010 to 2023, indicating the growing significance and attention to HCD and usability in maritime studies. This illustrates that the knowledge needed to improve focus in the supplier industry is available; however, the focus needs to shift at a strategic level to utilise it effectively in product development.

1.5.6 Indicators of Poor Usability: Seafarer Workarounds

One clear indicator of poor bridge design is the prevalence of seafarer improvisations. Sticky notes over bright displays, magnets on active equipment, Figure 1.4, manual labels on malfunctioning panels, and handwritten cheat sheets are the mariner's desperate attempts to 'fix' human-machine interfaces ill-suited to operational needs.

FIGURE 1.4 Adaptation – Whiteboard magnet indicating active pump on aft-bridge console onboard a platform supply vessel.

Photo: Frøy Birte Bjørneseth.

Rather than dismissing these adaptations as 'crew ingenuity', system developers should treat them as critical indicators of system failure, prompting formal HF assessments and redesign efforts.

NCL Salten grounded when the officer on watch fell asleep, and the deactivated BNWAS failed to wake him (Norwegian Maritime Authority, 2025). It is not yet known whether this was done on this particular vessel. Still, our observations from several other vessels show that a common way to disable the BNWAS and prevent annoying alarms is to hang a dangling object in front of the motion sensor, effectively deactivating the system's alarms.

1.6 TOWARDS BETTER DESIGN: EMBEDDING HUMAN FACTORS EARLY

Improving bridge design demands a holistic approach that involves a systemic design-thinking approach which integrates (1) *technical systems*, such as navigation, communication, and automation; (2) *physical environments*, that is, the bridge layout, ergonomics, and visibility; and (3) *cognitive needs*, such as workload management, information processing, and situational awareness (Stanton, 2006). This integration must occur early, during requirements gathering and conceptual design phases, not retroactively. Fixing problems in later stages of the development process can cost up to 1,000 times more than fixing them in the early design phases (Boehm, 1981).

In the maritime supplier industry today, a technology-driven approach is often chosen. The technical requirements dominate the functional requirements, as research shows that small details that do not support the operator's interaction patterns can cumulatively become cognitively demanding and consume cognitive capacity (Bjørneseth, 2021). This leaves less capacity to handle safety-critical situations, as the regular operation without any significant deviations is already overloading the operator.

Another sign of technology-centricity we observed is the tendency to downplay usability testing. Significant resources are often spent testing the technology to ensure it functions correctly in accordance with the requirements specifications, without errors. This is achieved using a comprehensive test portfolio that includes unit tests, system tests, integration tests, acceptance tests, and numerous other types of tests. However, none of these tests measures the system's user-friendliness. Only usability tests with representative users in realistic environments can reveal whether end users can complete their tasks correctly, efficiently, and safely at sea.

Another problem is the vague definitions of usability requirements. As observed in other domains (Alsos, 2023), technical requirements often dominate the requirements documentation, which can span hundreds or thousands of pages. In contrast, usability requirements typically fill a few pages at most, often in vague, imprecise terms that are impossible to measure or assess, such as 'the system should be intuitive and user-friendly'. Instead, the usability requirements should be defined with the same rigour as technical requirements. An example is: '95 % of the officers with a D1 certificate should, after completing 1 hour of training, be able to put the vessel in DP mode within 30 seconds successfully'.

A usability requirement in this format is easy to assess. A failed test (e.g., 50% of the officers took more than 30 seconds to complete the task) requires the designers and developers to return to the drawing board to improve the system or redesign its user interface.

Many technologists and developers lack on-the-job experience and develop ship bridge systems from modern, comfortable offices. Spending time at sea and experiencing fatigue, alarm overload, and poor visibility firsthand would provide invaluable insights to empathise with the seafarers and understand their needs. The first two phases of the HCD process (ISO 9241-210, 2019), which include discovery and understanding users and the context of use, are vital for the subsequent phases that address creation, design, and the final product, ensuring the crew's needs are successfully met. Field-based ethnography, participatory design workshops with mariners, and operational scenario testing should become standard practices. It should even be part of developers' standard job training.

The maritime supply industry has many comprehensive standards, regulations, and guidelines covering a range of topics to navigate. However, most of the attention and financial resources are given to the rules that classify the vessel and approve it for operation. Guidelines and standards that are not included in these regulations are considered 'nice to have' and not prioritised due to cost-driving factors and prolonged development time. Safety and the operator's work environment are often far down on the priority list. Therefore, regulatory frameworks must evolve to explicitly mandate HF evaluations in bridge system certifications, similar to aviation standards like SAE ARP4761 (Society of Automotive Engineers, 1996) and EASA CS-25 (European Union Aviation Safety Agency, 2023) for cockpit systems.

As presented next, some reference projects illustrate that the knowledge and research available are not only an unattainable vision but can be applied to real-world solutions.

1.6.1 Example: Unified Bridge

The Unified Bridge (Bjørneseth, 2021) is a ship bridge concept that was developed and industrialised by Rolls-Royce Marine (acquired by Kongsberg Maritime in 2019). The idea began when the company identified the need for a ship bridge that could provide a unified user experience and address many of the issues operators encounter when controlling a vessel. It was designed with a holistic perspective. This included a redesign and rearrangement of the physical consoles, operator's chair, input devices, and software interfaces. The software interfaces were harmonised by developing a design system for graphical components with a unified look and feel across all applications situated in the ship bridge environment. The ship bridge concept was to be developed using technology available on the market, intending to realise it within a maximum of five years from idea to installation on a sailing vessel.

The Unified Bridge supported four design criteria: *safety, simplicity, performance, and proximity* to improve operator effectiveness and reduce human error:

1. **Safety**: Since up to 75–96% of maritime accidents (Allianz Global Corporate & Specialty, 2012) can involve some human error, the contemporary view is that human error is a consequence of deeper issues with the system. This can be a combination of issues such as poor design, poor training, mental overload, and fatigue (Dekker, 2004). The Unified Bridge concept aimed to reduce cognitive overload and poor interface design. A safer environment allows operators to focus on operations rather than managing complex systems, especially during critical situations. This was done by rearranging and repositioning equipment and software interfaces more efficiently, based on research (Bjørneseth et al., 2012a, 2012b).
2. **Simplicity**: Decluttering the workspace and streamlining controls reduces distractions and confusion. A more straightforward interface minimises the chance of misinterpretation and supports faster, more accurate decision-making.
3. **Proximity**: Bringing essential tools like touchscreens and controls closer to the operator enhances accessibility and efficiency. This reduces unnecessary movement and helps maintain focus during operations.
4. **Performance**: By lowering mental strain and supporting intuitive interaction, the design boosts operator performance, particularly in high-pressure scenarios. This aligns with HF research showing that reduced workload improves reaction time and decision quality.

Bjørneseth (2021) notes that research across various fields indicates that high cognitive workload reduces an individual's capacity to respond to incoming information and increases the risk of human error (Nocera et al., 2007). An example from the power industry, by Holzinger et al. (2012), illustrates how to reduce the complexity of user interfaces for safety-critical power-plant control systems. The operators need to make sense of the information presented to them; they need to perceive and interpret to make decisions (Weick, 1993). The four design criteria support this.

The main goal of redesigning the work environment on the ship bridge was, in this case, to address many of the issues operators experience. This included the following:

Alarm avalanches (including audible alarms) were addressed by adding a single physical button to silence alarms, giving the operator time to locate the cause of the alarm. The alert lists in the software applications were also redesigned, encompassing grouping and shelving, and organised in line with appropriate standards.

Glare and illumination issues were solved by placing the displays so they could be easily adjusted and tilted to avoid direct sunlight or glare from interior lights. All software interfaces could shift from day to dusk and night palettes from a single interaction point. At the same time, all ceiling lights, display brightness, and LEDs on levers, buttons, and panels could be dimmed either through a system integrator or via a physical knob at each workstation.

Crowded and cluttered consoles were addressed by reorganising all equipment, where the system mentioned earlier, which integrated functionality from many different subsystems, was key. This was achieved by moving third-party equipment, such as lantern control panels, window wipers, typhoons, and more, to an equipment station located behind the operator, and further integrating the main functionality into the remote-control system. This made it possible to save space/equipment footprint in the consoles, leaving them less cluttered and with a cleaner look, see Figure 1.5.

Issues with physical ergonomics could be addressed by incorporating variation in work position during lengthy operations, as the height of the consoles allowed the operator to push the operator chair backwards and change from seated to standing. The operator's chair itself was redesigned. The improved version had a narrow headrest and a combination of leather and fabric on the seats and the chair back, leaving the operator with an increased field of vision to both sides and a comfortable seat that helps prevent dampness and heat.

FIGURE 1.5 Unified bridge – aft bridge workstation, Photo: Kongsberg Maritime.

The very first version of the concept was an aft-bridge station, including a dynamic positioning system, on a platform supply vessel (PSV), which was showcased at the Norwegian shipping fair, NorShipping, in 2011. This developed further into a forward bridge with various console configurations, bridge wing stations, a concept for tugs, operator's chairs with levers in the armrests, cruise vessels, ro-ro vessels (ferries), mega yachts, anchor-handling tug supply (AHTS), construction vessels, and fishing vessels.

The Unified Bridge was, and still is, the only ship bridge concept developed and industrialised, implementing and scientifically documenting the HCD process throughout all phases of development. In this concept, end users (ship operators) were involved from the ideation phase through product development, culminating in a finished, industrialised product.

After five years on the market Danielsen et al. (2019) conducted a benchmarking study on the vessels on which the first prototypes of the ship bridge concept were installed. The feedback from the vessel operators was positive, with reluctance to return to a more conventional ship bridge environment.

1.6.2 Example: Design Systems

Earlier, we presented design systems, such as OpenBridge (Nordby et al., 2019), as significant contributions to more standardised, user-friendly, and safer systems. However, these design systems are not silver bullets. The use of design systems is not a substitute for HCD and HF processes and will not automatically lead to safe and user-friendly interfaces or systems.

One of the main challenges in the maritime industry today, as mentioned earlier, is that the knowledge and competence built over the past 15 years are not utilised in industry to the extent that they make a difference. This indicates a lack of strategic anchoring of the HCD approach in the organisations (Figure 1.3) and low design maturity (Danielsen et al., 2022). According to the six-level UX maturity model by Nielsen Norman Group (Pernice et al., 2021), most suppliers are at stage 2 (limited maturity); however, some are moving towards stage 3 (emergent). This indicates that the HCD approach is neither acknowledged nor given the financial resources to build a foundation that enables the production of products that can be quality-assessed against the standards, guidelines, and regulations mentioned earlier.

When organisations are at this low level of maturity, it becomes easier for them to skip the vital pre-phase in the HCD process, which, according to ISO 11064, comprises three of the five phases (A, B, and C). It is plausible that ready-made design systems, such as OpenBridge (2025), facilitate skipping pre-development work, as the design system developers have already done it. This is, however, not correct, as they have designed single graphical components using HF methodologies to some extent but have not considered the complete work environment with a systemic design-thinking perspective. Composing a whole system without the right competence in the design process – designers, interaction designers, and HF specialists with expertise in the aforementioned frameworks and standards – there is a high risk that the bridge systems may appear well-designed on the surface but conceal serious usability issues underneath that could lead to 'human error'. Using a LEGO brick analogy: the well-designed components can be carefully assembled into a wonderful and elegant whole with the right design skills, but the final creation might as well become ugly and useless if quickly thrown together by an incompetent person. A worst-case scenario for maritime safety is that the industry hands the design system to its developers and believes it can replace skilled interaction designers and HF specialists.

1.7 CONCLUSION

The analyses presented in this chapter point towards a consistent conclusion: many of the challenges observed on modern ship bridges are not the result of insufficient technology but of inadequate attention to the human operators who must use poorly designed bridges, systems, and ships under demanding conditions without consideration of their physical and cognitive strengths and limitations. The most significant barriers to safe and effective bridge operations stem from systemic design shortcomings, not from individual seafarers' shortcomings.

The path forward is therefore clear. Effective ship bridge design must be grounded in HF methods; in measurable, task-based requirements; in systematic analysis of workload to support SA and create well-organised alarm behaviour; and in involvement with those who operate vessels in demanding and dynamic environments.

Established standards and tools already provide a robust foundation for this work. What is needed now is a shift in mindset – from treating HF as a late-stage corrective measure to recognising it as a core determinant of safety and operational excellence. Prioritising HF from the start when implementing automation, AI, and remote operations will also speed up user acceptance and address key issues such as safety, efficiency, and usability.

As we were reminded during our visit to the brand-new passenger ship, shiny technology is no substitute for safe, usable HCD. The future of maritime safety depends on our willingness to listen to the experienced crew and to send engineers on board to speak with crews and hear their frustrations firsthand.

ACKNOWLEDGEMENTS

Our deep gratitude goes to the anonymous shipowner, the captain, and his crew for letting us board the ship and for explaining the challenges of the bridge systems. The writing of this book chapter was supported by the MIDAS and MAS projects, both funded by the Norwegian Research Council (project numbers 331921 and 326676).

REFERENCES

Aas, A. L., Johnsen, S. O., & Skramstad, T. (2009). CRIOP: A human factors verification and validation methodology that works in an industrial setting. In Proceedings of the 28th International Conference on Computer Safety, Reliability, and Security (SAFECOMP'09) (pp. 243–256), Hamburg, Germany, September 15–18. Springer.

Allianz Global Corporate & Specialty. (2012). *Safety and shipping 1912–2012: From Titanic to Costa Concordia*. Munich, Germany: Allianz SE. Retrieved from https://www.agcs.allianz.com/news-and-insights/reports/safety-and-shipping-review-1912-2012.html

Alsos, O. A. (2023). *Hvordan unngå fakkeltog mot Helseplattformen?* [How to avoid torchlight processions against the Health Platform?] (opinion article). Adresseavisen. Retrieved August 19, 2025, from https://www.adressa.no/debatt/i/MoVgxr/hvordan-unngaa-fakkeltog-mot-helseplattformen

Alsos, O. A., Johnsen, S. O., & Bjørneseth, F. B. (2023). *Human, technical and organizational factors related to the Helge Ingstad accident*. Norwegian University of Science and Technology (NTNU). https://hdl.handle.net/11250/3109377

American Bureau of Shipping. (2010). *Guidance notes on the application of ergonomics to marine systems*. Houston, TX: ABS.

Bjørneseth, F. B. (2021). Unified bridge – Design concepts and results. In J. R. Eid & B. Johnsen (Eds.), *Sensemaking in safety critical and complex situations* (1st ed., Ch. 9). CRC Press.

Bjørneseth, F. B., Dunlop, M. D., & Hornecker, E. (2012a). Assessing the effectiveness of direct gesture interaction for a safety-critical maritime application. *International Journal of Human-Computer Studies*, 70(10), 729–745.

Bjørneseth, F. B., Renganayagalu, S. K., Dunlop, M. D., Hornecker, E., & Komandur, S. (2012b, September). Towards an experimental design framework for evaluation of dynamic workload and situational awareness in safety-critical maritime settings. In Proceedings of the 26th BCS Conference on Human–Computer Interaction: People and Computers XXVI (pp. 309–314). *BCS Learning and Development*. https://doi.org/10.14236/ewic/HCI2012.43

Blokland, P., & Reniers, G. (2020). Safety science, a systems thinking perspective: From events to mental models and sustainable safety. *Sustainability*, 12(12), 5164. https://doi.org/10.3390/su12125164

Boehm, B. W. (1981). *Software engineering economics*. Englewood Cliffs, NJ: Prentice-Hall.

Costa, N. A. (2018). *Human-centred design for maritime technology and organizational change*. Chalmers University of Technology.

Danielsen, B.-E. (2022). The contribution of ship bridge design to maritime accidents. In K. Plant & G. Praetorius (Eds.), Proceedings of the International Conference on Human Factors in Transportation (AHFE'22) (Vol. 60). AHFE Open Access.

Danielsen, B.-E. (2023). Usability in ship bridge design – A mission impossible? A qualitative study of maritime stakeholders' perspectives on usability in ship bridge design. Doctoral Thesis, NTNU, Trondheim, Norway.

Danielsen, B.-E., Bjørneseth, F. B., & Vik, B. (2019). Chasing the end-user perspective in bridge design. Proceedings of Ergoship 2019.

Danielsen, B.-E., Lützhöft, M., Haavik, T. K., Johnsen, S. O., & Porathe, T. (2022). "Seafarers should be navigating by the stars": Barriers to usability in ship bridge design. *Cognition, Technology & Work*, 24(4), 675–691.

Danielsen, B.-E., & Petersen, E. S. (2022). Somebody else's problem? Usability in ship bridge design seen from the perspective of different maritime actors. *TransNav: International Journal on Marine Navigation and Safety of Sea Transportation*, 16(4). https://doi.org/10.12716/1001.16.04.10

Dekker, S. (2004). *Ten questions about human error: A new view of human factors and system safety*. Mahwah, NJ: Lawrence Erlbaum Associates.

DNV. (2021). *Recommended practice DNV-RP-G108: Human-centred design of alarm systems for ships*. Høvik, Norway: DNV.

Endsley, M. R. (1995). Toward a theory of situation awareness in dynamic systems. *Human Factors*, 37(1), 32–64.

European Union Aviation Safety Agency. (2023). *Certification specifications and acceptable means of compliance for large aeroplanes (CS-25), Amendment 26*. Cologne, Germany: EASA.

Gould, K. S., Hirvonen, K., Koefoed, V. F., Røed, B. K., Sallinen, M., Holm, A., ··· Moen, B. E. (2009). Effects of 60 hours of total sleep deprivation on two methods of high-speed ship navigation. *Ergonomics*, *52*(12), 1469–1486. https://doi.org/10.1080/00140130903272611

Grech, M. R., Horberry, T., & Smith, A. (2002). Human error in maritime operations: Analyses of accident reports using the Leximancer tool. *Proceedings of the Human Factors and Ergonomics Society Annual Meeting*, 46(19), 1718–1721.

Hess, R. F., Sharpe, L. T., & Nordby, K. (Eds.). (1990). *Night vision: Basic, clinical and applied aspects*. New York: Cambridge University Press.

Hollnagel, E. (2014). *Safety-I and Safety-II: The past and future of safety management*. Farnham, UK: Ashgate Publishing.

Holzinger, A., Searle, G., & Nischelwitzer, A. (2012). On complexity reduction of user interfaces for safety-critical systems. In International Conference on Availability, Reliability and Security, IFIP Cross-Domain Conference, LNCS 7465 (pp. 108–122). Springer.

Hordaland District Court. (2023, May 15). *Judgment in case T 22-2023: Criminal case against the officer of the watch on KNM Helge Ingstad* [District court judgment]. Bergen, Norway.

Hsieh, M.-H., Xia, Z., & Chen, C.-H. (2024). Human-centred design and evaluation to enhance safety of maritime systems: A systematic review. *Ocean Engineering*, 307(2), Article 118200. https://doi.org/10.1016/j.oceaneng.2024.118200

International Electrotechnical Commission. (2010). *IEC 62616:2010: Maritime navigation and radiocommunication equipment and systems — Electronic charts — Operational and performance requirements, methods of testing and required test results*. Geneva, Switzerland: IEC.

International Electrotechnical Commission. (2018). *IEC 62923-1:2018: Maritime navigation and radiocommunication equipment and systems — Bridge alert management — Part 1: Operational and performance requirements, methods of testing and required test results*. Geneva, Switzerland: IEC.

International Electrotechnical Commission. (2024). *Human machine interfaces for process automation systems* (IEC Standard No. 63303:2024). Geneva, Switzerland: IEC.

International Maritime Organization. (2002). *International convention for the safety of life at sea (SOLAS), V/15*. London, UK: IMO.

International Maritime Organization. (2007). *Resolution MSC.252(83): Adoption of the revised performance standards for integrated navigation systems (INS)*. London, UK: IMO.

International Maritime Organization. (2010). *Resolution MSC.302(87): Adoption of performance standards for bridge alert management*. London, UK: IMO.

International Maritime Organization. (2020). *International convention for the safety of life at sea (SOLAS), 1974, as amended*. London, UK: IMO.

International Organization for Standardization. (1999). *ISO 13407: Human-centred design processes for interactive systems*. Geneva, Switzerland: ISO.

International Organization for Standardization. (2005). *ISO 17894:2005: Ships and marine technology — Computer applications — General principles for the development and use of programmable electronic systems in marine applications*. Geneva, Switzerland: ISO.

International Organization for Standardization. (2007). *ISO 8468:2007: Ship's bridge layout and associated equipment — Requirements and guidelines*. Geneva, Switzerland: ISO.

International Organization for Standardization. (2013). *ISO 11064-4:2004: Ergonomic design of control centres — Part 4: Layout and dimensions of workstations (confirmed 2013)*. Geneva, Switzerland: ISO.

International Organization for Standardization. (2019). *Ergonomics of human-system interaction — Part 210: Human-centred design for interactive systems (ISO Standard No. 9241-210:2019)*. Geneva, Switzerland: ISO. https://www.iso.org/standard/77520.html

Johnsen, S.O., Thieme, C., Aminoff, H., Park, J., Marius Fernander, M., Boren, M., ··· Kirwan, B. (2025). *CRIOP A scenario method for crisis intervention and operability analysis*. Norway: SINTEF Technology and Society.

Kirwan, B. (2025). Human factors requirements for human-AI teaming in aviation. *Future Transportation*, 5(2), 42. https://doi.org/10.3390/futuretransp5020042

Lloyd's Register. (2013). *Guidance notes on the application of ergonomics to marine systems*. London, UK: Lloyd's Register.

Lützhöft, M. (2004). *"The technology is great when it works": Maritime technology and human integration on the ship's bridge*. Graduate School for Human-Machine Interaction, Division of Quality and Human-Systems Engineering, Department of Mechanical Engineering, University of Linköping.

Mallam, S. C., Nordby, K., Johnsen, S. O., & Bjørneseth, F. B. (2020). The digitalization of navigation: Examining the accident and aftermath of US Navy Destroyer John S. McCain. In Proceedings of the Royal Institution of Naval Architects Damaged Ships V, March 11, London, UK.

Marine Accident Investigation Branch. (2017). Collision between pure car carrier City of Rotterdam and ro-ro freight ferry Primula Seaways (MAIB Investigation Report 3/2017). UK Department for Transport. https://www.gov.uk/maib-reports/collision-between-pure-car-carrier-city-of-rotterdam-and-ro-ro-freight-ferry-primula-seaways

Marine Insight. (2025). The relation between human error and marine industry. *Marine Insight*. Retrieved June 5, 2025, from https://www.marineinsight.com/marine-safety/the-relation-between-human-error-and-marine-industry/

National Transportation Safety Board. (2019). Collision between US Navy destroyer John S McCain and tanker Alnic MC: Singapore Strait, 5 miles northeast of Horsburgh Lighthouse, August 21, 2017 (Marine Accident Report NTSB/MAR-19/01, PB2019-100970). NTSB. https://www.ntsb.gov/investigations/AccidentReports/Reports/MAR1901.pdf

National Transportation Safety Board. (2020). Collision between US Navy destroyer Fitzgerald and Philippine-flag container ship ACX Crystal, Sagami Nada Bay off Izu Peninsula, Honshu Island, Japan, July 17, 2017. NTSB.

Nocera, F. D., Camilli, M., & Terenzi, M. (2007). A fuzzy logic approach to the assessment of mental workload. *Ergonomics*, 50(4), 548–561.

Nordby, K., Mallam, S. C., & Lützhöft, M. (2019). Open user interface architecture for digital multivendor ship bridge systems. *WMU Journal of Maritime Affairs*, 18(3), 297–318. https://doi.org/10.1007/s13437-019-00168-w

Norwegian Maritime Authority. (2025, May 28). *Inspection report completed after Port State Control*. https://www.sdir.no/nyheter/inspeksjonsrapport-klar-etter-havnestatskontroll/

Norwegian Safety Investigation Authority. (2019). Part one report on the collision on 8 November 2018 between the frigate HNoMS Helge Ingstad and the oil tanker Sola TS outside the Sture Terminal in the Hjeltefjord in Hordaland county (Marine Report 2019/08). Norwegian Safety Investigation Authority. https://nsia.no/Marine/Published-reports/2019-08-eng

Norwegian Safety Investigation Authority. (2021). Interim report 2 on the collision between the frigate KNM Helge Ingstad and the tanker Sola TS outside the Sture terminal in Hjeltefjorden, Hordaland, 8 November 2018 (Report 2021/05).

Norwegian Safety Investigation Authority. (2024a). Loss of propulsion and near grounding of Viking Sky, Hustadvika, Norway, 23 March 2019 (Marine Report 2024/05). https://havarikommisjonen.no/Sjofart/Avgitte-rapporter/2024-05

Norwegian Safety Investigation Authority. (2024b). Report on collision between the high-speed passenger craft Tyrhaug and the service vessel Frøy Loke in the Edøyfjord on 28 June 2023 (Marine Report 2024/03). https://nsia.no/Marine/Published-reports/2024-03

Nygård, B. (2019). *Learning after platform supply vessel collision with Statfjord A*. SINTEF. https://www.sintef.no/globalassets/project/hfc/documents/learning-after-platform-supply-vessel-collision-with-statfjord-a-bjorn-nygard-hfc-2019.pdf

OpenBridge. (2025). *OpenBridge design system* [Open-source design system and guidelines]. https://www.openbridge.no/

Park, J., Johnsen, S. O., Saghafian, M., & Alsos, O. A. (2025, June). Challenges and opportunities in remote operations of automated passenger ferries identified using the CRIOP method. In Proceedings of the 35th European Safety and Reliability Conference (ESREL'25) and the 33rd Society for Risk Analysis Europe Conference (SRA-E'25). Research Publishing.

Pernice, K., Gibbons, S., Moran, K., & Whitenton, K. (2021, June 13). *The 6 levels of UX maturity*. Nielsen Norman Group. Updated January 24, 2024. https://www.nngroup.com/articles/ux-maturity-model/

Petroleum Safety Authority Norway. (2019). *Investigation of collision between Sjøborg and Statfjord A*, 7 June 2019. https://www.havtil.no/en/supervision/investigation-reports/2019/equinor-statfjord-a-investigation-of-collision-by-vessel/

Porathe, S. T. (2023). Alarm and hand-over concepts for human remote operators of autonomous ships. In Proceedings of the 33rd European Safety and Reliability Conference (ESREL'23). Research Publishing.

Razmjooei, D., Alimohammadlou, M., Ranaei Kordshouli, H.-A., & Askarifar, K. (2023). Industry 4.0 research in the maritime industry: A bibliometric analysis. *WMU Journal of Maritime Affairs*, 22(1), 1–28. https://doi.org/10.1007/s13437-022-00298-8

Sitompul, T. A., Park, J., & Alsos, O. A. (2023a). Designing user interface elements for remotely operated ship-to-shore cranes. In Proceedings of the 36th International BCS Human-Computer Interaction Conference (pp. 169–178). BCS.

Sitompul, T. A., Park, J., & Alsos, O. A. (2023b). Designing user interface elements for remotely operated rubber-tired gantry cranes. In Proceedings of the IEEE International Conference on Systems, Man, and Cybernetics (SMC) (pp. 4832–4839). IEEE.

Society of Automotive Engineers. (1996). *SAE ARP4761: Guidelines and methods for conducting the safety assessment process on civil airborne systems and equipment*. Warrendale, PA: SAE International.

Stanton, N. A. (2006). Hierarchical task analysis: Developments, applications, and extensions. *Applied Ergonomics*, 37(1), 55–79. https://doi.org/10.1016/j.apergo.2005.06.003

Stratmann, T. C., & Boll, S. (2016). Demon hunt: The role of Endsley's demons of situation awareness in maritime accidents. In *Human-centered and error-resilient systems development* (pp. 24–42). Springer International Publishing.

Vehus, M. (2024). Design system in practice: Exploring implications of using a design system to implement user interface components for remotely operated cranes (Master's thesis), NTNU.

Weick, K. E. (1993). The collapse of sensemaking in organizations: The Mann Gulch disaster. *Administrative Science Quarterly*, 38(4), 628–652. https://doi.org/10.2307/2393339

Wróbel, K. (2021). Searching for the origins of the myth: 80% human error impact on maritime safety. *Reliability Engineering & System Safety*, 216, Article 107942. https://doi.org/10.1016/j.ress.2021.107942

2 How to Engineer Meaningful Human Control in Digitalisation, Automation/AI, and Remote

Stig Ole Johnsen and Jooyoung Park

2.1 INTRODUCTION

We employ the concept of Meaningful Human Control (MHC) to underscore that humans – not machines – should retain ultimate authority and responsibility for decisions made by digital, automated, or AI-driven systems, particularly when those decisions have significant consequences. MHC ensures that automated systems remain understandable, monitorable, and, when necessary, overridable by human operators – particularly in safety-critical situations. In the following, we describe how to engineer MHC in the new digital environment.

The rapid adoption of digitalisation, automation, and AI across safety-critical industries has transformed workflows, enabled remote operations, and increased system complexity. Experience from maritime operations shows that advanced technology does not automatically translate into safer work. Field observations and accident investigations reveal that poor alarm management, fragmented interfaces, and inconsistent automation behaviour can degrade situational awareness (SA) and remove operators from the control loop when they are most needed. These issues illustrate why MHC must be engineered – not assumed – as part of the wider transformation of safety-critical systems (Alsos & Bjørneseth, 2026).

Automation decisions driven primarily by cost reduction rather than human-centred design can create serious gaps in safety. An example is the Boeing 737 MAX development based on the need for cost-cutting, ignoring HF design practices in aviation. The Maneuvering Characteristics Augmentation System (MCAS), which was installed, forced the aircraft into a dive and ultimately led to a crash due to faulty sensor input (Hopkins, 2025). The pilots, unaware of the system's behaviour and lacking proper training, were unable to intervene in time.

However, an example of best practice in MHC was the design and operation of Flight 1549 (Airbus A320), which struck a flock of birds shortly after take-off, losing all engine power. The pilots managed to glide the plane to ditch on the Hudson River (NTSB, 2010) without casualties. The aeroplane was designed to be able to land on water. The pilots were supported by an excellent HMI/cockpit design, as usual in the aviation industry, and had extensive experience, and made their decision in just two minutes.

MHC is not simply the ability to press an override button. It means that humans remain genuinely 'in the loop': able to understand, monitor, anticipate, and shape system behaviour effectively. It requires that systems be defined and designed so that humans can take timely and informed action when automation fails or behaves unexpectedly. In this sense, MHC is central to achieving safety by design – addressing the limitations of automation, AI, and remote operations by ensuring human involvement throughout system definition, design, and operation.

DOI: 10.1201/9781003741824-3

Research consistently shows that autonomous systems do not remove the need for human oversight (Bradshaw et al., 2013). Indeed, the paradox of automation (Bainbridge, 1983) reveals that as automation becomes more advanced, operators need greater expertise for rare but critical interventions. Supervising highly automated systems can lead to complacency and reduced vigilance, while emergencies still demand rapid, skilled human responses (Hancock, 2021). Designing for MHC, therefore, requires realistic expectations of what human operators can perceive, decide, and manage under stress and uncertainty.

As automation and AI continue to advance, there is a growing need for a systematic approach to digitalisation that explicitly integrates safety, efficiency, and usability. Achieving this requires attention to MHC from concept development through design, operation, and learning from use. This aligns with the National Institute for Occupational Safety and Health's (NIOSH) Prevention through Design (PtD) principle, which recognises that eliminating or minimising hazards early in the design process is one of the most effective ways to prevent accidents, injuries, and fatalities (Behm et al., 2014). Extending this logic, ensuring MHC from the earliest stages of system design, is fundamental to preventing failures in the age of automation and AI.

However, accident investigations still provide limited systematic evidence on the contribution of design to failures in safety-critical systems. Often, accident investigations highlight 'Human Error' as a root cause, but ideally, 'Human Error' should be the starting point of investigations. Investigations must then identify underlying system contributors such as poor SA support, excessive or poorly prioritised alarms, cognitive overload, fatigue, inadequate training, or other error traps that undermine robust work practices. Recently, some systematic analyses of accident investigations estimate that a large proportion of adverse events (approximately 40–50%) are associated with poor design of systems, routines, or technology (Kinnersley and Roelen, 2007; Moura et al., 2016).

MHC has been discussed since 2010, particularly in relation to autonomous weapon systems, and later in the context of autonomous car systems (Mecacci et al., 2024). Initially, MHC was applied to weapon systems, emphasising that humans – not computers or algorithms – should retain control over decisions affecting Health, Safety, Security, and Environment (HSSE) (Calvert et al., 2024). Operational issues reveal the degree of MHC, where human limitations must be addressed to achieve meaningful control. Humans must be able to understand and effectively control situations.

The knowledge of human limitations and possibilities has been based on the science of Human Factors (HF), defined as '*the scientific discipline concerned with the understanding of interactions among humans and other elements of a system, and the profession that applies theory, principles, data, and methods to design in order to optimize human well-being and overall system performance*' (IEA, 2000).

In this chapter, we have addressed MHC as a goal and operational principle that can be achieved by focusing on three key areas: design, operations, and learning. The goal of MHC should be a widely supported principle. However, reaching it requires a system perspective that integrates man/humans, technology, and organisational factors (MTO).

We suggest that MHC should be established and maintained through an ongoing learning cycle, covering design, operations, and learning, that is:

- Design issues, including governing factors such as laws, regulations, and audits that impact the system; system scoping and problem definition; and the use of methods and system design activities, including task analysis (TA) and safety-critical task analysis (SCTA).
- System operational conditions, that is, workload, training, procedures, organisation, reporting culture and interface qualities that support situational awareness and provide a clear overview (situation at a glance) through the Human–Machine Interface (HMI).
- Learning and improvement from accidents based on an investigation model, avoiding shortcuts such as 'Human Error'. Prioritise the exploration of design issues, trying to avoid blame and examine relevant governing factors.

Our definition of MHC is *the ability of a system (involving humans, technology, and organisation) to be managed by humans in order to prevent accidents affecting HSSE, considering human capabilities and limitations (within the context of the science of HF)*.

Evidence from maritime ship-bridge operations demonstrates that MHC is lost when systems overload operators with alarms, distribute critical information across incompatible displays, or force operators to compensate for design flaws through workarounds. In several high-profile accidents, crews misinterpreted external cues, navigated cluttered interfaces, or faced automation that behaved in ways that were opaque or context-insensitive. These failures underline a core requirement of MHC: systems must be designed so that humans can understand, monitor, and intervene in automated behaviour, even under time pressure and degraded conditions.

AI is being considered in many situations. However, in MiT (2025), State of AI in Business 2025, only 5% of the AI projects documented benefits. One of the issues was poor adaptation to the workflow (process-specific customisation) and poor learning (poor evaluation based on business outcomes). These issues are key in safety-critical situations – thus, proper analysis of MHC is essential. This is in line with the oversight concept, under which the EU/AI Act (2024) mandates that high-risk AI systems be designed to allow people to monitor and intervene in their operation to mitigate risks to health, safety, and fundamental rights.

MHC is a fundamental operational principle that ensures humans remain actively engaged and capable of intervening in automated systems, thereby maintaining human control over safety-critical operations. MHC depends on design factors (such as scoping and HF knowledge), operational practices, and the capacity to learn and improve through investigations and continuous learning. Based on this context, the research questions we have selected are as follows:

- RQ1: What design issues enable MHC?
- RQ2: What operational issues sustain/ensure MHC?
- RQ3: How can we learn and improve MHC from incidents?

2.2 APPROACH AND METHODS

2.2.1 Research Approach

This research adopts an Action Research (AR) approach (Greenwood & Levin, 2006), emphasising collaboration between researchers and stakeholders to address real-world problems and develop practical, context-sensitive solutions. Through prolonged engagement, interviews, and participatory workshops, key lessons were identified to define and operationalise MHC.

AR was chosen because it allows researchers to be actively involved in both understanding and improving safety-critical systems. The approach has proven effective in enhancing safety outcomes in complex socio-technical environments (Antonsen et al., 2007). By integrating research and action, the study ensures that findings are not only theoretically sound but also directly applicable to operational settings.

2.2.2 Theoretical Foundation

The concept of MHC is grounded in the domains of HF and Safety Science. HF provides the natural foundation for MHC, focusing on the safety of interactions between humans and systems to optimise human efficiency, usability, and overall system performance. In this study, safety is viewed as the result of effective MHC, and accidents are treated as indicators of its absence or degradation.

Accident investigations incorporating HF perspectives can thus reveal how MHC can be strengthened through improved design, operational practices, and learning processes. The research draws on the science of HF to analyse both design issues (e.g., system interfaces, workload, and task allocation) and operational conditions (e.g., training, communication, and procedures).

To guide prioritisation and structure findings, the study applies the Hierarchy of Controls (Manuele, 2005), a well-established framework for risk reduction that organises safety interventions from most to least effective.

2.2.3 Research Design and Data Collection

To address the research questions, a multi-method qualitative design was employed, combining literature review, expert interviews, and case studies. This section, based on Johnsen and Park (2025), has been improved with analyses of new accidents and papers, alignment with Alsos and Bjørneseth (2026), and a description of a curriculum for Safety in Design.

2.2.4 Literature Review

The literature review examined HF and safety research related to automation, remote operations, and accident learning. Particular attention was given to safety-critical environments such as aviation, maritime operations, road transport, and process industries, which offer established best practices in human control and automation (Kirwan, 2025; Johnsen & Winge, 2023; Park et al., 2025). A snowballing strategy was used to identify additional relevant sources, building upon prior results from Johnsen and Winge (2023).

2.2.5 Expert Interviews and Workshops

Semi-structured interviews and workshops were conducted with domain experts and operators from relevant industries to capture real-world insights into maintaining MHC in digitalised and automated systems. These sessions supported both data triangulation and the co-development of practical recommendations.

2.2.6 Case Studies

Case analyses examined specific automation and remote operation scenarios, assessing how MHC was established, lost, or recovered in practice. These included analyses of empirical studies in AI/automation (Park et al., 2025) and aviation system design (Johnsen et al., 2024).

2.2.7 Analytical Framework

From Design to Learning – The Hierarchy of Controls was applied as both a validation structure and a prioritisation tool for the analysis. This safety framework (ANSI, 2012; Dyreborg et al., 2022; NIOSH, 2024) systematically organises interventions by effectiveness, ranging from the most effective – elimination of hazards – to the least effective – personal protective equipment (PPE).

The five levels of the hierarchy were applied as follows:

1. Elimination: Removing hazards and risks through scope definition, design reformulation, or system redesign.
2. Substitution: Replacing hazardous methods or components with safer alternatives, such as different automation or redundant systems.
3. Engineering Controls: Implementing technical barriers, alarms, workload assessments, and HMI enhancements.
4. Administrative Controls: Using procedures, training, and organisational measures.
5. PPE: Establishing individual-level barriers as a last line of defence.

This structure was used to prioritise design and operational improvements aimed at strengthening MHC.

2.2.8 Accident Analysis and Learning

The understanding of accidents was based on systemic accident investigation models used by the Norwegian Safety Investigation Authority (NSIA, 2022) and the Australian Transport Safety Bureau (ATSB, 2008). These approaches draw on Reason's (1997) theory of organisational accidents and SA models (Endsley & Jones, 2012) to examine how individuals and organisations perceive and respond to complex situations.

The ATSB model was adapted to explore how MHC can be analysed across system levels, moving from event occurrence to organisational and regulatory factors:

1. Occurrence of events – describing the incident sequence.
2. Local control factors – conditions influencing the immediate response.
3. Risk controls – barriers and defences intended to prevent escalation.
4. Organisational influences – responsibilities, management of change, and resources.
5. Regulatory oversight – laws, audits, and supervisory structures.

To align with the Safety Forum approach to learning from accidents, this investigation framework was adapted to focus on three dimensions:

1. Event sequence and SA of the actors involved (Safety Forum, 2019).
2. Local and risk control factors that affect MHC during operations.
3. Design and governance deficiencies arising from poor problem definition, inadequate methods, or insufficient regulatory and organisational controls.

This analytical structure directly links accident learning to design and operational improvements, creating a feedback loop that enhances MHC over time.

2.3 RESULTS AND DISCUSSION

The findings are organised around the three main research questions: design issues, operational issues, and learning and improvement. They are based on the literature review, analyses of accident reports, and insights gathered through expert interviews and workshops.

2.3.1 Key Design Issues

The most powerful way to achieve safety in complex socio-technical systems is to embed it directly into design. Regulations, HF principles, and early hazard identification provide the foundation for this approach. Effective design enables **MHC** by ensuring that humans can understand, monitor, and manage automation even under abnormal conditions. The analysis identified several interrelated design themes that together constitute best practice for strengthening MHC.

Learning from best practice in Aviation: Aviation remains the benchmark for safety in high-risk operations. With an exceptionally low accident rate, it is often described as an ultra-safe industry (Amalberti, 2017). The International Air Transport Association (IATA), which manages the majority of global air traffic, recorded no hull losses in either 2012 or 2017. This performance reflects a long-standing commitment to HF integration, dating back to World War II (Kirwan, 2025). Aviation's safety record is a product not only of technical reliability but also of its human-centred approach. A defining feature is the concept of *Just Culture*, now codified in European regulation (EU 376/2014), which promotes learning rather than punishment by protecting operators from disciplinary action for errors made within the bounds of their training and experience. By embedding HF principles and a culture of open reporting into regulation and daily practice, aviation demonstrates how MHC can be systematically supported through design, training, and organisational learning. In contrast, the oil and gas sector and other process industries often treat HF as a secondary consideration, particularly in early design phases (Bergh et al., 2024). In addition, some companies lack an open reporting culture, a key element for learning and adapting from 'work as done' and for supporting collaboration among regulators, industry, and the workforce (Equinor, 2025; VG, 2025).

This gap can be addressed through the use of *Technology Readiness Levels (TRL)* and *Human Readiness Levels (HRL)* – two complementary frameworks for assessing the maturity of technology and its human integration (ANSI/HFES, 2021; Yasseri and Bahai, 2018). These frameworks provide a structured way to evaluate whether both humans and technologies are adequately prepared for real-world applications. Iterative assessment of readiness levels helps identify potential hazards early, align interfaces and procedures with human capabilities, and ensure operators are appropriately trained before deployment (Johnsen & Aminoff, 2024). This approach is particularly valuable when introducing artificial intelligence (AI) in safety-critical systems, where challenges such as brittleness, perceptual limitations, and hidden biases can undermine human understanding and control (NAS, 2021; Cummings, 2024). Challenges with use of AI (NAS, 2021) are *brittleness* (AI will only be capable of performing well in situations that are covered by its programming or training data); *perceptual limitations* (AI algorithms continue to struggle with reliable and accurate object recognition in 'noisy' environments); *hidden biases* (AI software may incorporate many hidden biases from being created using a limited set of training data); and *no model of causation* (ML-based AI is based on simple pattern recognition; the underlying system has no causal model. Because AI cannot use reason to understand cause and effect, it cannot predict future events). We have experienced challenges with poor sensor fusion, negligent neural network training, and poor testing (Cummings, 2024). Use of HRL and TRL can mitigate this and support MHC through human-centred design. Applying TRL and HRL frameworks supports MHC by ensuring that both technological and human readiness evolve in parallel before systems are placed into operation. HRL optimises interfaces to align with human capabilities, reducing the probability of accidents. Furthermore, HRL ensures workers are adequately trained, minimising mistakes, and HRL readiness prepares workers to handle crises effectively. TRL ensures technologies are thoroughly tested and safe before use (especially when combined with HRL) and confirms that technologies are reliable, ensuring safe operations.

Regulatory support of HF plays a decisive role in embedding HF and MHC principles into design practice. The Health and Safety Executive (HSE, 2015) demonstrated that regulatory requirements mandating HF methods have a lasting positive effect on industry adoption. The Norwegian petroleum industry's management regulations, for instance, explicitly reference ISO 11064 for control-room ergonomics. ISO 9241-210 outlines broader principles for human-centred design. Although organisations sometimes cite the cost of HF activities during development, accidents such as Deepwater Horizon and Boeing MAX have shown that neglecting HF considerations leads to far higher costs in the long term.

Regulations that explicitly require HF integration help establish a consistent baseline across industries and ensure that safety, usability, and human control are considered from the outset.

Equally important is the process of *system scoping and project definition*. Research has shown that projects with a broad system perspective, which includes human, technological, and organisational factors (MTO), achieve better safety and usability outcomes (Begnum, 2021; Helgar, 2022; Bjørneseth, 2021). Clear scoping ensures that the right problem is defined and the right solution developed. Effective scoping involves determining how functions should be distributed between humans and automation – decisions traditionally informed by the 'Fitts List' and its modern adaptations (De Winter and Dodou, 2014). The development of the unified bridge concept in maritime operations (Bjørneseth, 2021) demonstrates the benefits of combining broad system definition, user involvement, and iterative prototyping. The project achieved high user satisfaction and received recognition for both safety and design quality. These results show that systematic scoping and task allocation are critical enablers of MHC.

Human-centred design, as formalised in ISO 9241-210, provides an overarching framework for aligning systems with real user needs. When properly implemented, it reduces cognitive workload, improves usability, and supports safer and more efficient performance. *Safety* is enhanced through iterative testing, which reduces human errors and improves regulatory compliance. *Efficiency* improves through optimised workflows, reduced training time, and fewer errors. *Usability* increases with intuitive design, user satisfaction, and accessibility. Studies by Helgar (2022) and Bjørneseth (2021) demonstrate that iterative usability testing throughout the design process reduces human error and improves operational reliability. However, while ISO 9241-210 (2019) and ISO 11064 (2013) define high-level principles, they often lack explicit methods for applying Human Factors Engineering (HFE) in practice (Leva et al., 2015). Practical techniques such as *Safety Critical Task Analysis (SCTA)*, *alarm design standards* (EEMUA 191, 2024; IEC 62682, 2022), and *HMI guidelines* (IEC 63303, 2024, based on ISA 101.01) translate these principles into actionable design activities. Together, these standards help designers implement MHC by ensuring that interfaces, alarms, and procedures match human cognitive and physical capacities.

TA plays a central role in achieving meaningful control. TA and SCTA help identify how operators interact with systems, what information they need, and where potential errors could occur (Energy Institute, 2020). It supports the design of HMIs, procedures, and team coordination structures that enhance decision-making and reduce cognitive overload. Studies have shown the value of supplementing TA with eye-tracking and simulation to validate design decisions and better understand cognitive workload (Bjørneseth, 2021; IOGP, 2024).

Bjørneseth (2021) employed TA to understand 'work as done' through interviews, observations of work practices, and an exploration of personal adaptations. In addition, they used eye-tracking to understand user needs and used simulators and prototyping to develop usable concepts. Eye-tracking has also been studied in the oil and gas industry. It seems an effective tool to assess design and cognitive workload, and could lead to safety improvements, IOGP (2024). TA helps determine what information is critical for operators to perform their tasks effectively and to design an interface that supports them. By analysing tasks, designers can identify decision points and the cognitive processes involved, ensuring the interface provides the right information at the right time. This is a starting point for Ecological Interface Design (EID). EID is a human-centred design approach that enhances decision-making in complex systems. This is done by providing visual representations of system constraints, relationships, and affordances. It aims to support both routine and unexpected situations by reducing cognitive load and improving SA. The need for EID was highlighted in Meshkati (2006). An important issue was matching task demand analysis with the operator's ability to handle difficult deviations (such as alarms). This helps ensure that automation supports rather than replaces human reasoning.

Workload management is another essential element in the design for MHC. Many accidents occur under conditions of excessive workload or alarm flooding, such as the BP Texas City Refinery explosion in 2005. Conversely, periods of low workload may lead to loss of vigilance and poor preparedness for unexpected events. Workload should therefore be evaluated as part of the TA, using methods such as timeline analysis or NASA-TLX (Stanton et al., 2017). Designing systems that maintain an optimal workload balance helps sustain SA and operator engagement under both high- and low-demand conditions.

Alarm system design and management remain persistent challenges across various industries, particularly in high-risk environments where operators are involved, such as chemical plants, refineries, oil and gas production facilities, and transportation. Poorly designed or mismanaged alarm systems are a significant factor in many accidents, as they often fail to provide timely and clear warning signals when hazardous conditions arise.

Disasters due to poor alarm management are, among others, the King's Cross Fire in 1987 with 31 fatalities, the Piper Alpha Disaster in 1988 with 167 fatalities, the BP Texas City Refinery Explosion in 2005 with 15 fatalities and the BP Deepwater Horizon in 2010 with 11 fatalities. These incidents, HF issues, and best practices of standards are discussed in Briwa et al. (2022).

They highlighted EEMUA 191 as one key standard with usable/achievable guidelines. In the North Sea, persistent alarm issues have been reported. The Norwegian regulator has published guidelines and conducted audits, as reported by Bjerkebaek and Eskedal (2004). They found poor alarm handling and that the safety-critical function of the alarms during a crisis may be impaired. Walker et al. (2014) conducted a review of 30% of North Sea control rooms and found persistent issues with alarms and poor support for operators during non-routine and emergencies. An alarm audit was conducted in 2021–2022 by the Petroleum Safety Authority in Norway (PSA, 2022). Only one installation got no remarks. One reason might be that they had used good practice from CRIOP (Johnsen et al., 2025).

Similar alarm challenges are also documented in maritime operations, where poor alarm philosophy and inconsistent follow-up actions contributed to escalating crises. Poor alarms and missing follow-up actions were identified as key issues in the accidents of *Sjøborg* (PSA, 2019) and *Viking Sky* (Porathe, 2023). In both cases, the systems issued several alarms that were not understood or resolved. The alarm avalanches overwhelmed operators and removed MHC at critical moments (Alsos & Bjørneseth, 2026). Automatic follow-on actions caused the power/propulsion to be turned off. In the case of *Viking Sky*, they lost propulsion in a storm close to shore. In the Sjøborg case, the vessel collided with a nearby platform after losing propulsion.

Effective alarm systems should be designed to quickly and accurately alert operators to potential dangers, enabling them to understand the situation and take appropriate action before the problem escalates. Best-practice standards, EEMUA 191 and IEC 62682, are recommended. EEMUA 191 has practical HFE guidelines, a valid questionnaire, and relevant limits on the number of alarms.

Verification and validation through user testing under realistic conditions remain essential to ensure that alarm systems truly support human control.

Finally, *user testing* is the most direct and practical technique for validating design assumptions and ensuring that systems support MHC in real operations. As noted by Norman (2013), early and iterative testing identifies usability issues before they become costly or embedded in system architecture. User testing enhances clarity, efficiency, and user confidence, and is strongly aligned with standards such as ISO 9241-210 and ISO 11064.

Recent maritime accident analyses (Alsos & Bjørneseth, 2026) show that design shortcomings, such as inconsistent vendor interfaces, non-intuitive navigation flows, and excessive interaction steps, directly undermine human control during time-critical events. For example, operators on several vessels experienced cognitive overload when attempting simple tasks, such as display dimming, which required navigating dozens of menu layers at night. These seemingly minor design choices consumed scarce cognitive resources and delayed hazard detection. Ensuring MHC requires that interfaces be coherent, predictable, and designed around human cognitive limits.

Testing with representative users under realistic conditions strengthens trust in automation, reduces training time, and ensures that system behaviour remains transparent and predictable.

2.3.1.1 CRIOP as an Engineering Tool for Validating Meaningful Human Control

CRIOP is a practical engineering method for assessing whether MHC is supported in digitalised, automated, or remote operations. Through structured checklists, scenario analyses, and team evaluations, CRIOP reveals whether operators have sufficient information, time, and cognitive capacity to understand system behaviour and intervene when needed.

CRIOP directly supports MHC by the following:

- **Supporting Management of Change**, ensuring MHC is maintained as automation, AI models, or procedures evolve
- **Checking that the layout and HMI design enable a 'situation at a glance'**, a core prerequisite for human control
- **Validating task allocation, communication flow, and teamwork** under normal, degraded, and emergency conditions

- **Revealing gaps between work as imagined and work as done**, including workload peaks, error traps, and poorly integrated interfaces
- **Evaluating alarm behaviour**, ensuring signals support SA rather than create overload

Accident analyses from maritime operations show that many failures of control – such as alarm cascades, mode confusion, poor display integration, and contextual blind spots – are precisely the issues CRIOP is designed to capture early (Alsos & Bjørneseth, 2026). Embedding CRIOP in design and operations, therefore, provides a powerful mechanism to ensure that humans remain able to monitor, understand, and intervene – fulfilling the core requirements of MHC.

In summary, effective design for MHC depends on integrating regulatory and HF principles from the earliest stages of system development, ensuring both technological and human readiness, defining system scope across the MTO dimensions, applying structured HF methods such as SCTA and alarm management, and verifying outcomes through iterative, user-centred testing. These practices collectively ensure that automated systems remain under human control, that humans retain the ability to intervene when needed, and that safety, usability, and efficiency reinforce rather than undermine one another.

2.3.2 Key Operational Issues

Operational issues play a decisive role in determining whether MHC is achieved in daily work. They influence how well design intentions are translated into safe and effective operations. Even the most advanced systems can fail if operational responsibilities, workload, or communication across interfaces are poorly managed. This section examines key operational challenges that affect MHC, including maintaining safety responsibility in supply chains, managing night work and fatigue, enhancing SA, and preventing error traps.

2.3.2.1 Maintaining Safety Responsibility in Supply Chains

Modern industrial operations rely heavily on outsourcing and contractor arrangements, especially in the oil, gas, and maritime sectors. Such supply chains can obscure responsibility for safety and risk management. The International Association of Oil and Gas Producers (IOGP, 2017) identified unclear ownership of safety barriers between operators and contractors as a common cause of accidents and near misses. To address this, IOGP recommends that both operators and contractors actively participate in site visits, audits, and verification activities to ensure that risk controls and barriers work as intended.

This principle reflects what Norwegian regulation calls 'see-to responsibility' (påse-ansvar), which requires organisations to ensure that safety measures are effectively implemented across the entire value chain. In Norway, this responsibility is established in several laws, including the Petroleum Act, the Framework and Management Regulations, the Working Environment Act, and the Ship Safety Act (2007). It is enforced by the Norwegian Ocean Industry Authority (Havtil, 2025). The see-to responsibility is proactive: operators must monitor, verify, and, when needed, intervene to maintain safety standards.

From an MHC perspective, this principle ensures that human authority, accountability, and control are maintained even when operations are distributed across organisations and technologies. A case presented by Equinor (2024) at the Drilling Conference in Kristiansand demonstrated this challenge. In outsourced drilling operations, inadequate cabin design, distractions from core tasks, and poor ergonomics created a demanding environment that weakened SA and decision-making. These issues highlight inadequate focus on human-centred design and Management-of-Change (MoC) processes.

Maintaining MHC in such settings requires continuous oversight and the integration of HF principles into contract management, training, and performance evaluation. An open reporting culture and psychological safety are also essential, ensuring that employees can report concerns without fear of blame. Operators must verify that subcontractors apply proper risk management, that workstations and control rooms are ergonomically designed, and that training evolves in line with new technology. These measures sustain human responsibility and intervention capability throughout the supply chain.

2.3.2.2 Night Work and Fatigue Management

Night work presents well-documented safety challenges because human alertness follows a circadian rhythm, with the lowest levels of vigilance typically occurring between midnight and 5:00 a.m. (Smith et al., 1994). During this period, slow reaction times and attention lapses become more frequent. Some of the most serious industrial disasters – Chernobyl (1986), Three Mile Island (1979), and Bhopal (1984) – occurred during night shifts when operator fatigue contributed to loss of SA and delayed corrective action (Smith et al., 1994).

Maritime accident records confirm this vulnerability. In several cases, including *Tyrhaug* and *NCL Salten*, operators lost SA during late-night watchkeeping, sometimes due to monotony, sometimes due to distraction, and sometimes because critical alerting systems were disabled to avoid nuisance alarms. These examples demonstrate that MHC cannot be sustained unless systems, staffing strategies, and procedures explicitly account for human fatigue, night-time limitations and the time required for recovery of SA.

Empirical studies confirm that night work increases the risk of accidents (Lie et al., 2014). Fatigue reduces memory, coordination, and communication, weakening operators' ability to maintain MHC over automated or complex systems. In highly automated or remote operations, reduced alertness can result in deviations being detected too late for effective intervention.

Maintaining MHC during night operations requires organisational foresight. Whenever possible, safety-critical activities should be performed during daytime hours when cognitive performance is highest. Where continuous operation is unavoidable, managers should strengthen supervision, use shorter shifts, and avoid transferring responsibility between crews during the early morning.

From a *safety by design* perspective, systems must be configured to accommodate human fatigue rather than ignore it. Control rooms should include lighting and environmental conditions that support alertness, and alarm systems should be calibrated to human performance limits. Technologies such as fatigue monitoring and adaptive automation can help detect reduced alertness and dynamically adjust workload. Effective fatigue management safeguards human oversight, accountability, and the capacity for timely intervention – the core attributes of MHC.

2.3.2.3 Situational Awareness

SA – the ability to perceive, understand, and anticipate one's environment – is essential for maintaining MHC. Many accidents can be traced to degraded SA. In an analysis of offshore supply-ship incidents, Sandhåland et al. (2015) found that poor Level 1 SA (failure to perceive critical cues) contributed to 13 of 23 accidents, often due to deficient interface design or inadequate training.

Time is a prerequisite for SA. Alarm standards, such as EEMUA 191, recognise this, estimating that operators can manage approximately six high-priority alarms per hour, or roughly ten minutes per event, to assess and respond. The US Airways Flight 1549 incident in 2009 illustrates the importance of time, training, and interface design. When the aircraft lost all engine power, the pilots – supported by an intuitive cockpit layout – diagnosed the situation and decided to ditch in the Hudson River within two minutes (NTSB, 2010). This case demonstrates how effective design and experience enable rapid, informed decisions under pressure. Expecting faster decisions in less structured environments, such as maritime or process control, is unrealistic.

Poor SA should not be viewed as an operator failure but as a design and organisational failure. As Miranda (2019) notes, attributing an accident to 'loss of SA' often conceals deeper flaws in system design, training, or workload management. To reformulate van Winsen and Dekker (2017) we have an ethical and moral responsibility to defend the human operator who inherits a poorly designed system, in other words, SA should not be used to blame the operator. Designers and regulators share the ethical obligation to provide humans with adequate time, tools, and information to establish SA. High-performance HMIs must offer a 'situation at a glance' (Hollifield et al., 2008), enabling operators to quickly and accurately grasp complex system states

2.3.2.4 Error Traps

A further operational concern is the presence of error traps – latent system conditions that predispose humans to error. These may arise from poor equipment layout, ambiguous roles, inadequate staffing, outdated procedures, or communication breakdowns (Norsk Industri, 2023). Emerging technologies can introduce new error traps, such as AI model drift, where unmonitored algorithmic changes lead to unexpected or unsafe behaviour (Cummings, 2024).

Maintaining MHC, therefore, requires systematic reality checks that compare *work as done* with *work as imagined* and the rigorous use of MoC to identify and mitigate emerging risks. Regular reporting and follow-up of error traps should be a core activity in high-risk operations. A key enabler is psychological safety, which ensures open reporting of issues and fosters a supportive culture that encourages reporting to management. This is especially vital during outsourcing (Equinor, 2025).

2.3.2.5 Summary

Maintaining MHC in operations depends on ensuring clear responsibility across supply chains, recognising human limitations such as fatigue, designing processes that allow enough time for SA, and systematically identifying and removing error traps. When accidents occur, investigators should first analyse system design, organisational factors, and management decisions, rather than blaming individual operators. Operational safety is maintained not through perfect human performance, but through well-designed systems that support and respect human capabilities. Thus, operational safety depends on designing systems that support human perception, cognition, and intervention – even under degraded conditions. Evidence from ship-bridge operations shows that when HF principles are neglected, humans lose meaningful control long before an emergency becomes visible.

2.3.3 Key Issues to Support Learning from Incidents

This section explains how learning from incidents and accidents is needed to strengthen MHC. In many organisations, accident analyses still focus on blaming individuals, identifying 'human error' or causes grounded in bias (such as hindsight bias and confirmation bias). Such approaches overlook the deeper systemic insights that reveal how design, local control, and organisational conditions impact human performance. To enhance learning and system resilience, investigations should systematically record events, assess SA among actors, analyse local and risk control factors, and identify design flaws caused by poor problem definition or a lack of HF methods.

2.3.3.1 Documenting Events and Exploring the Situational Awareness of Actors

A fundamental condition for learning is to avoid blame and instead seek understanding. Incidents should be documented systematically to capture how the system of humans, technology, and organisation (MTO) supported or constrained MHC at the time of the event. A practical and transparent tool for documenting steps and timelines is the Sequential Timed Events Plotting (STEP) method, developed by Hendrick and Benner (1986). The STEP diagram organises events chronologically, showing the actions of each actor and their interactions over time. It creates a clear, visual sequence of what happened, allowing investigators to see how decisions and system responses evolved. By mapping events in this way, investigators can integrate multiple perspectives – technical, human, and organisational – into a shared understanding of the incident. To complement the event chronology, the SA of key actors should also be analysed. The framework by Endsley and Jones (2012) provides a structured way to do this, distinguishing between three levels of SA:

1. **Perception** of relevant cues in the environment (e.g., alarms, system status, visual indicators).
2. **Comprehension** of the meaning of this information in context.
3. **Projection** of future states to anticipate and act proactively.

Understanding where and why SA broke down helps determine whether operators had sufficient information, time, and tools to maintain control. When SA collapses, the cause is often poor interface design, inadequate training, or insufficient time, not operator incompetence. Documenting SA also enables targeted improvements to design and procedures that support MHC under stress.

2.3.3.2 MHC from Local Control and Risk Control

The second focus area concerns how local conditions and risk controls affect MHC. Local control refers to the immediate context influencing human and technical actions – factors such as equipment layout, environmental conditions, staffing, fatigue, and workload. These issues are thoroughly checked and documented in methods such as CRIOP (Aas et al., 2009). Local conditions can also include technical influences, such as component failures, high temperatures, or software malfunctions.

Risk controls, also known as barriers or defences, are measures implemented to prevent or mitigate hazardous events. They may be preventive controls (e.g., design improvements, procedures, alarms, or training) or reactive controls (e.g., warning systems, emergency shutdowns, and evacuation procedures). Effective systems use defences in depth, combining multiple, independent layers of protection and try to ensure safety by design. The interaction between local conditions and weakened barriers often explains how small deviations align to produce major accidents, as illustrated by Reason's Swiss Cheese Model (1997).

Investigations should also account for normal performance variability, the gap between 'work as imagined' and 'work as done' (Hollnagel, 2017). Variability is not necessarily negative – it is often how humans adapt to complex environments – but it can reveal how system design and management assumptions fail to match reality.

In emerging domains such as AI and automation, new failure modes must also be recognised. Cummings (2024) proposes extending Reason's taxonomies to capture 'AI-related human error', such as model drift, biased training data, or opaque decision logic. These issues should be investigated using the same structured principles applied to human performance, avoiding simplistic labels like 'AI error' that obscure underlying system flaws.

A useful tool for visualising preventive and reactive risk controls is bow-tie analysis, which clearly links causes, barriers, and consequences. When applied systematically, it supports both proactive safety management and post-incident learning.

From an MHC perspective, investigations should always ask whether operators had the means to remain in control. This also involves examining whether MoC processes were conducted, whether the SCTA informed HMI design and procedures, and whether alarm systems, workload, and environmental conditions were designed in accordance with best-practice standards such as EEMUA 191 or IEC 62682. Investigators should also verify that user testing was conducted, that staff were competent in the system's operational design domain, and that HF methods, such as workload or fatigue analysis, were applied in design and operation. Each of these checks reveals whether MHC was structurally supported – or unintentionally undermined – before the incident occurred.

2.3.3.3 Key Issues in Local Control and Risk Control

- Is the management of change performed, and work as done checked versus work as imagined in safety-critical areas?
- Has a SCTA been used in design as a basis for flow of SA, HMI, Bowtie, Organisation, Procedures, and Layout?
- Has the alarm been designed to a best-practice standard, such as EEMUA 191?

- Can situations be assessed 'at a glance' using High Performance HMI design?
- Has a workload analysis been done to handle stress, boredom, fatigue, and mental workload?
- Is noise, lighting, temperature, and work environment (ergonomics) designed in line with best practices?
- Has systematic user testing been performed on relevant data, including deviations and edge conditions?
- Has user selection and training been performed based on operational design domain and need for competence as a result of the SCTA?

2.3.3.4 Exploring Design and Organisational Issues

Accident research consistently shows that poor design is a leading cause of accidents. Studies estimate that 40–50% of industrial incidents stem from design deficiencies (Kinnersley and Roelen, 2007; Moura et al., 2016). Norman (2013) summarised this insight succinctly: 'Human error? No – poor design'. Despite this, design-related causes are often underreported because the standards for investigation vary widely in quality and depth. As a reversal of the unsupported claim that '80% of accidents are due to human error' (Wróbel, 2021), we asked in a discussion with Don Norman: why not begin by assuming that 80% may be due to poor design? This is not intended as a statistical claim, but as a way to shift the investigation from operator blame towards asking whether the accident could have been prevented through elimination, substitution, or better engineering design. Rather, human error should be treated as a starting point for inquiry, not the endpoint (Dekker, 2017). In many cases, the so-called human error is the visible symptom of systemic weaknesses – poor project definition, poor MoC, poorly designed interfaces, unrealistic procedures, or inadequate training. As Don Norman has argued, most failures attributed to humans originate in design decisions that did not account for human limitations: 'human error usually is a result of poor design: it should be called system error' (Norman, 2013).

Investigators should therefore explore whether accidents could have been prevented through better system definition, a more robust design process, or the application of engineering controls. Three questions are particularly useful:

1. Was there a clear project definition and operational domain design that considered human readiness and technological readiness levels (HRL/TRL)? (De Winter and Dodou, 2014; Johnsen & Aminoff, 2024).
2. Did the design process follow recognised human-centred design standards such as ISO 9241-210 or ISO 11064?
3. Were design decisions guided by the Hierarchy of Controls – elimination, substitution, engineering, administrative, and personal protective measures – in that order of priority?

Addressing these questions ensures that investigations look beyond operator behaviour to the deeper structural causes of failure. When poor design is recognised as a primary contributor, organisations can feed lessons back into design, regulation, and training – closing the learning loop that underpins MHC.

2.3.3.5 Using CRIOP as an Accident Investigation Tool to Support Meaningful Human Control

While CRIOP was originally developed as a proactive verification tool, its structure makes it equally valuable in accident and incident investigations, particularly when the goal is to understand whether MHC was supported, weakened, or absent at the time of the event.

Traditional investigations often emphasise operator actions – focusing on what individuals did wrong instead of examining the design, automation behaviour, workload, barriers, and organisational conditions that shaped performance. CRIOP provides a systematic, HF-based framework for shifting the analysis towards *systemic contributors*, consistent with both Safety-II principles and MHC requirements. CRIOP supports accident investigation in four essential ways:

- **CRIOP verifies whether design supported human control at the time of the event**: CRIOP contains structured checklists and scenario elements that directly assess whether the system design followed recognised HF principles – such as task-centred design, workload analysis, alarm philosophy, layout design (ISO 11064), and HMI standards (ISO 9241-210, IEC 63303).

  Using CRIOP retrospectively allows investigators to evaluate whether these design activities were adequately performed or omitted before the system entered service. Poorly designed or inconsistent interfaces, excessive cognitive workload, alarm avalanches, and insufficient SA support – become visible when CRIOP's design-oriented items are applied post-incident. This enables distinguishing operator-induced errors from design-induced error traps, a fundamental requirement in MHC-oriented learning.
- **CRIOP reconstructs *work as done* and highlights MHC breakdowns:** CRIOP includes scenario walk-throughs, communication analysis, and teamwork evaluations that help investigators reconstruct how operators actually experienced the situation. This is particularly important when automation, remote supervision, or complex digital systems are involved, as the cognitive workload and decision loops may differ markedly from *work as imagined* in manuals or procedures. By replaying the event scenario using CRIOP's structured prompts, investigators can map where SA collapsed (Levels 1–3); identify when alarms became overwhelming or ambiguous; examine whether automation behaved transparently or unexpectedly; and assess whether operators had the time, information, and mental bandwidth to intervene meaningfully. These insights directly relate to whether human control was possible, not merely whether it was exercised.
- **CRIOP reveals organisational or governance weaknesses affecting MHC**: CRIOP addresses not only technical aspects but also organisational structures, procedures, staffing, shift patterns, and Management-of-Change processes. Applied in an investigation, these checks help reveal whether training was aligned with operational design domain (ODD); staffing levels supported safe SA recovery (2–10 minutes); alarm maintenance processes were sound; MoC was applied before new automation or interfaces were introduced; HF design requirements were embedded in procurement, verification, and testing; and cultural issues such as psychological safety and ability to maintain an open reporting culture could be assessed. These organisational aspects are frequently invisible in traditional investigations but central to whether operators retained meaningful control under abnormal conditions.
- **CRIOP produces a structured, HF-based output suitable for learning loops and redesign**: Accident investigation outcomes are often limited to narrative findings, making it difficult to translate learning into redesign. CRIOP, however, produces explicit lists of missing barriers, documented HF requirement gaps, scenario-based breakdown points, recommendations tied to engineering controls, and a traceable logic from observations → conditions → design implications. This aligns directly with safety by design and the MHC philosophy, enabling organisations to feed incident learning back into design, standards, regulations, and training. Applying CRIOP post-incident thus supports a robust, evidence-based learning loop, moving accident learning beyond individual blame towards redesign, governance improvement, and system-level safety.

2.3.3.6 Towards Better Learning and Investigation Standards

Current investigation practices often stop once a 'human error' has been identified, leaving the underlying design, organisational, or regulatory issues unexplored. This limits learning and prevents systemic improvement. There is a clear need for international investigation standards that integrate HF science and explicitly consider MHC. Such standards would shift investigations from blame to understanding, ensuring that lessons are captured not just to correct human performance but to improve the design of systems, organisations, and governance structures.

These lessons mirror findings from maritime accident analyses, where design weaknesses, poor alarm handling, and fragmented interfaces consistently eroded MHC long before an emergency became visible (Alsos & Bjørneseth, 2026).

To summarise, meaningful learning from incidents begins with accurate documentation, understanding of human SA, and analysis of local and organisational conditions that shape control. It requires investigators to move beyond 'who failed' towards 'how the system enabled failure'. Only when we systematically connect events, context, and design can we transform accidents into knowledge that strengthens MHC and builds safer, more resilient systems.

2.4 VALIDITY/CREDIBILITY

The research presented in this chapter builds on both empirical evidence and theoretical grounding in HF and Safety Science. The proposed framework for establishing MHC was developed through extensive interviews with HF experts, documented best practices, and triangulation with existing literature. Member checks were performed through workshops and collaborative discussions with industry practitioners to validate key findings and ensure practical relevance. The study also draws on prolonged engagement with safety-critical industries, providing contextual depth and ensuring that the results are grounded in real operational experience. The combination of theoretical alignment, expert validation, and long-term field experience strengthens both the credibility and transferability of the findings.

2.5 CONCLUSION AND FURTHER WORK

2.5.1 Establishing Meaningful Human Control

MHC in digitalisation, automation, and remote operations is not an inherent property of a system – it is an ability that must be consciously designed, operated, and maintained within the broader man/human–technology–organisation (MTO) system. Achieving MHC depends on continuous learning across three interdependent domains:

1. **Design** – creating systems that align with human capabilities and limitations.
2. **Operation** – ensuring that human authority, awareness, and accountability are preserved during real-time work.
3. **Learning** – systematically improving design and practice through analysis of incidents and near misses.

Learning from incidents should go beyond pinpointing 'human error' to include understanding missing or flawed design features, weak HF integration, and organisational conditions that fail to support MHC. Structured investigation methods, such as those used by the Norwegian Safety Investigation Authority (NSIA, 2022) and the Australian Transport Safety Bureau (ATSB, 2008), provide a solid foundation for this analysis. However, existing methods need to be expanded with HF-based criteria that assess how effectively HF methods and HFE practices are applied during design, operation, and maintenance. Organisational and structural issues must also be addressed by incorporating HF experts in accident investigations (Safety Forum, 2019) and by improving regulatory inspections. The Norwegian National Audit Office (2023) concluded that the Norwegian Maritime Authority's inspection system is not adequately designed to identify poor working and living conditions. The result is a learning system that observes symptoms but does not address their root causes. Creating an international standard for HF-informed accident investigation could greatly improve learning across industries and encourage consistent best practices.

2.5.2 From Design to Operations and Learning

The creation of MHC should begin early in the design process, with HF methods integrated into engineering and management activities. Projects need to explicitly evaluate TRL and HRL to ensure that both technology and human skills develop concurrently. During the operational phase, MHC should be monitored in critical areas to confirm that new technology, procedures, or organisational structures continue to support safe human performance. This could be achieved through MoC processes. These issues are part of the CRIOP framework (Aas et al., 2009), which offers a structured approach for evaluating control-room operations, teamwork, and system design from an HF perspective. Explicitly utilising CRIOP to address MHC and digitalisation challenges would enhance safety, usability, and efficiency.

2.5.3 Key Design Priorities

Design sets the foundation for MHC. Regulation should promote human-centred design, open sharing of information, transparent communication, and a Just Culture that emphasises learning over blame. By learning from aviation – where open reporting and HF integration are required – other industries can similarly incorporate MHC into both their regulations and practices.

Effective design begins with a clear MTO-based project definition, supported by MoC and readiness-level assessments (TRL/HRL). The design process should follow recognised human-centred standards like ISO 9241-210 and ISO 11064, ensuring an understanding of user needs, context, and constraints. Critical systems should be designed according to key principles of ecological system design. Applying TA or SCTA, HFE techniques, and alarm design standards such as EEMUA 191 helps ensure that workload, fatigue, and cognitive demands are balanced with system automation. Systems should be designed to give operators a 'situation at a glance' (Hollifield et al., 2008), supporting quick comprehension and prompt action. Ongoing user testing in realistic scenarios confirms design choices before implementation.

2.5.4 Key Operational Priorities

Operational environments must be designed around human cognitive and physical limitations – especially the time required to regain SA during unexpected events. Operators usually require between two and ten minutes to diagnose and respond effectively to unexpected events, depending on their complexity.

The ability to manage a deviation or alarm should be based on established standards; for example, an operator might only handle six critical alarms per hour. Systems and procedures should be designed accordingly, and operators should not be blamed for errors that happen under unrealistic time constraints or inadequate system support. Maintaining MHC in daily operations requires proactive management of the supply chain through the 'see-to' responsibility, ensuring that safety standards and risk controls stay effective across all contractors.

Clear communication, appropriate workload management, and strong MoC processes are essential.

Organisations must systematically identify and manage key error traps – latent conditions that increase the likelihood of human error – such as unclear roles, poor ergonomics, or fatigue. Several examples are listed by Norsk Industri (2023). Error traps should typically be avoided in design. These issues should be systematically reported and mitigated through collaboration with the workforce across different companies and through training, rather than being blamed.

2.5.5 Key Learning Priorities

Authentic learning begins when organisations look beyond individual actions to explore why people acted as they did, given their tools, information, and environment. Investigations should examine whether good-practice HF techniques were applied in design and operation, and whether known human-centred methods were used, as specified in CRIOP (Aas et al., 2009).

A learning culture requires regulation and management that support Just Culture principles, encouraging transparent reporting and open discussion of errors without punishment, as defined by EU in aviation regulation. The analysis should always explore the systemic reasons for behaviour – how design, workload, and organisation influenced actions at the sharp end. Systematic methods such as NSIA (2022) and ATSB (2008), supported by tools like CRIOP and the Hierarchy of Controls, provide a structured framework for identifying and prioritising corrective actions. Exploring poor design should be part of the analysis.

Adopting a common international standard for human-centred accident investigation would ensure consistency, improve data quality, and accelerate cross-industry learning. Such a standard should explicitly address how MHC, HF integration, and design maturity (HRL/TRL) are assessed before and after incidents.

2.5.6 Moving Forward

Academic and professional training must equip engineers, managers, and investigators to apply HF and MHC principles in practice, ensuring that technology serves human decision-making, not the other way around. Ultimately, MHC is the bridge between human responsibility and growing system complexity, ensuring that automation remains understandable, predictable, and controllable – even under degraded operational conditions. It can be engineered, supported, and continuously improved – but only through conscious design, responsible operation, and committed learning. By embedding these principles into regulation, design practice, and education, we can ensure that the next generation of automated and AI-driven systems remains safe, ethical, and under human control.

Future research should focus on improving and developing metrics and validation criteria for assessing MHC in complex systems. Building on the integration of HF methods, design readiness, and learning frameworks, these metrics could guide both certification and operational audits.

2.6 TRAINING AGENDA – BUILDING COMPETENCE FOR SAFETY BY DESIGN

Modern safety-critical industries are undergoing rapid digital transformation, introducing automation, AI, and remote operation across their value chains. In this chapter, we have stated that MHC depends not only on technical design but also on operational competence and continuous learning. To support these activities, targeted training is needed in three domains: design integration, operational performance, and learning from experience. Across these domains, training should aim to build cross-disciplinary knowledge. Engineers, operators, regulators, and AI specialists must share a common understanding of:

- The principles of MHC – human authority, SA, time for cognition, and ethical accountability.
- The systemic nature of safety, linking design, operation, and learning.
- The need to apply Human Factors methods within engineering and organisational decision processes.

Safety by design is supported by training consisting of:

1. Human-Centred Design and Engineering process.
2. Key Human Factors Techniques for Modern Systems.
3. Human Performance and workload.
4. Learning from Practice and incidents.

These areas are described in the following.

2.6.1 Human-Centred Design and Engineering Process

Training must enable engineers and designers to embed HF and MHC principles from concept to operation.

This is elaborated by the following sections in safety by design:

- **Design for automation and new contexts of work**: Increased use of technology, maritime bridge automation, remote operation of automated ships, remote supervision of many facilities (Park; Alsos & Bjørneseth; Eitrem et al.).
- **Examples of integrating HF into engineering processes** (Biede et al.): Understand certification, regulation, and HF traceability in aviation and other safety-critical domains. Balance autonomy and oversight through mature project scoping, HRL, and TRL.
- **Apply user-centred design methods** (Helgar et al.): Plan early user involvement, TA, and iterative prototyping when 'every second counts'.
- **Translate research to world-leading safety** using road-traffic safety, democratic processes, and remote-management examples (Elvik et al.).

The chapters support case workshops where participants map the flow of MHC through design stages, apply standards (ISO 9241-210; ISO 11064; EEMUA 191), and evaluate interfaces using SCTA and EID.

2.6.2 Key Human Factors Techniques for Modern Systems

Professionals must master established and emerging analytical tools for supporting and building human performance in digitalised environments.

- **Design and operation of remote-control centres** (Høstmark et al., also Park): layout design, communication load, and distributed teamwork.
- **Contextual factors in Arctic conditions or extreme environments** (Sætren et al.).
- **SCTA** for automation and remote operations (Fernander/DNV).
- **Ecological Situation Awareness design** (Mulder/TU Delft) to ensure operators perceive and act within dynamic, complex systems.
- **Human-centred alarm management** (Briwa/TU Dublin) applying EEMUA 191 & IEC 62682 for alarm prioritisation and user validation.

These chapters support a curriculum that includes simulation exercises on workload and alarm management, scenario-based application of SCTA, and cross-domain examples (aviation, maritime, process industry).

2.6.3 Human Performance and Workload

Operational reliability depends on supporting human performance and managing human workload, stress, and fatigue in environments with automation, AI, and remote operations. The chapters build competence in the following areas:

- **Exploring and discussing new workload issues in automation, AI and remote**. Using methods such as *NASA-TLX* and evaluating physiological measures (Matthews; Rasmussen).
- **Simulator-based training and performance assessment** (Nazir): applying metrics of SA, teamwork, and decision-making under automation.

2.6.4 Learning from Practice and Incidents

The last parts highlight the necessity of systematic, blame-free learning:

- **Learning from normal work** (*Cichowicz & Nazaruk*) to identify adaptive successes and early warning signs.
- **Structured HF analysis in investigations** (Winge et al.): Discussing best practices in accident investigations, documenting events, SA, and risk control conditions.
- **Understanding and adapting to evolving HF standards** (Earthy/Lloyd's) for robotics and AI.
- **Concluding remarks: How to build Safety and Efficiency** in Automation, AI and Remote Operations, through HFE – Verification, Validation and Systematic Safety Management (Johnsen & Bjørneseth)

Investigation training should teach how to move beyond 'human error', apply HF taxonomies, and feed results back into design and training. Courses should include exercises in reconstructing incidents, mapping MTO factors, and identifying design-related causes of degraded MHC.

ACKNOWLEDGEMENTS

The HFC forum, the MAS project, NFR:326676 and the MIDAS project have funded this research.

REFERENCES

Aas, A.L., Johnsen, S.O., & Skramstad, T. (2009). CRIOP: A human factors verification and validation methodology that works in an industrial setting. In Proceedings of Computer Safety, Reliability, and Security (SAFECOMP'09), Hamburg, Germany, September 15–18 (Vol. 28, pp. 243–256). Springer, Berlin, Heidelberg.

Alsos, O.A., & Bjørneseth, F.B. (2026). Designing for the night watch: Human factors challenges on modern ship bridges. In Safety by Design (pp. 3–22). CRC Press.

Amalberti, R. (2017). The paradoxes of almost totally safe transportation systems. In *Human error in aviation* (pp. 101–118). Routledge.

ANSI - American National Standards Institute (2012). *Occupational health and safety management system* (ANSI/AIHA Z10, 2012). American Industrial Hygiene Association, Falls Church, VA.

ANSI/HFES (2021). *Human readiness level scale in the system development process* (ANSI/HFES 400-2021).

Antonsen, S., Ramstad, L., & Kongsvik, T. (2007) Unlocking the organization: Action research as a means of improving organizational safety. *Safety Science Monitor*, 11(1).

ATSB (2008) *Analysis, causality and proof in safety investigations*. Aviation Research and Analysis Report AR-2007-053. Canberra City: Australian Transport Safety Bureau.

Bainbridge, L. (1983). Ironies of automation. In *Analysis, design and evaluation of man–machine systems* (pp. 129–135). Pergamon.

Begnum, M.E.N. (2021). 11 user-centred agile. In *Sensemaking in safety critical and complex situations human factors and design* (p. 173). CRC Press.

Behm, M., Culvenor, J., & Dixon, G. (2014). Development of safe design thinking among engineering students. *Safety Science* 63.

Bergh, L.I.V., Teigen, K.S., & Dørum, F.. (2024). Human performance and automated operations: a regulatory perspective. *Ergonomics*, 67(6), 744–758. https://doi.org/10.1080/00140139.2024.2321457

Bjerkebaek, E., & Eskedal, T.S. (2004). Safety assessment of alarm systems on offshore oil and gas production installations in Norway. In *SPE International Conference and Exhibition on Health, Safety, Environment, and Sustainability?* (p. SPE-86597). SPE.

Bjørneseth, F.B. (2021). Unified bridge–design concepts and results. In *Sensemaking in safety critical and complex situations* (pp. 135–153). CRC Press.

Bradshaw, J.M., Hoffman, R.R., Woods, D.D., & Johnson, M. (2013). The seven deadly myths of "autonomous systems". *IEEE Intelligent Systems*, 28(3), 54–61.

Briwa, H., Leva, M.C., & Turner, R. (2022). Alarm management for human performance. Are we getting better?. In Proceedings of European Safety and Reliability Conference (ESREL'22).

Calvert, S., Johnsen, S., & George, A. (2024). Designing automated vehicles and traffic systems towards meaningful human control. In *Research handbook on meaningful human control of artificial intelligence systems* (pp. 162–187). Edward Elgar.

Cummings, M.L. (2024). A taxonomy for AI hazard analysis. *Journal of Cognitive Engineering and Decision Making*, 18(4), 327–332.

De Winter, J.C., & Dodou, D. (2014). *Why the Fitts list has persisted throughout the history of function allocation*. Cognition, Technology & Work.

Dekker, S. (2017). *The field guide to understanding 'human error'*. CRC Press.

Dyreborg, J., Lipscomb, H.J., Nielsen, K., Törner, M., Rasmussen, K., Frydendall, K.B., ... & Kines, P. (2022). Safety interventions for the prevention of accidents at work: A systematic review. *Campbell Systematic Reviews*, 18(2), e1234.

EEMUA - The Engineering Equipment and Materials Users Association (EEMUA) (2024). Alarm systems: A guide to design, management and procurement (Standard No. 191).

Endsley, M.R. & Jones, D. (2012) *Designing for situation awareness* (2nd ed.). CRC Press, Taylor & Francis, Boca Raton, FL.

Energy Institute (2020). *Guidance on human factors safety critical task analysis*. Energy Institute.

Equinor (2024). Presentation at the drilling conference in Kristiansand. https://hfc-forum.com/wp-content/uploads/2025/09/09-god-praksis-rundt-borekabiner-rapport-og-erfaringer.pdf

Equinor (2025) Granskingsrapport Fallulykke ved montering av tårnreis på HLNG, 24.04.2025 Rapport: A 2025-5 PDP L2.

EU Regulation (376/2014), reporting, analysis and follow-up of occurrences in civil aviation.

EU/AI-The Artificial Intelligence Act - Regulation (EU) (2024/1689).

Greenwood, D.J., & Levin, M. (2006). *Introduction to action research: Social research for social change*. SAGE Publications

Hancock, P.A. (2021). Months of monotony–moments of mayhem: Planning for the human role in a transitioning world of work. *Theoretical Issues in Ergonomics Science*, 22(1), 63–82.

Havtil (2025). https://www.havtil.no/en/regulations/acts/about-the-regulations

Helgar, S. (2022). *User centered design*. Retrieved from https://www.sintef.no/globalassets/project/hfc/documents/03-helgar-stein.pdf

Hendrick, K., & Benner, L. (1986). *Investigating accidents with STEP* (Vol. 13). CRC Press.

Hollifield, B., Oliver, D., Nimmo, I., & Habibi, E. (2008). *The high performance HMI handbook*. Pas, Houston.

Hollnagel, E. (2017). *Safety-II in practice: Developing the resilience potentials*. Routledge.

Hopkins, A. (2025). *Boeing, the 737 MAX crisis and aviation safety: The perils of profit-driven engineering*. CRC Press.

HSE (2015). *Literature review: Barriers to the application of human factors/ergonomics in engineering design*. Retrieved from https://www.sintef.no/globalassets/project/hfc/documents/barriers-to-use-hf.pdf

IEA - International Ergonomics Association (2000). Retrieved on January 20, 2025, from https://iea.cc/about/what-is-ergonomics

IEC - International Electrotechnical Commission 62682 (2022). Management of alarm systems for process industries.

IEC - International Electrotechnical Commission 63303 (2024). Human machine interface for process automation systems.

IOGP (2017). Report 423 – HSE management – guidelines for working together in a contract environment.

IOGP (2024). Report 656 – Assessment of eye tracking technology in well control operations - onshore.

ISO - International Organization for Standardization 11064 (1999–2013). Ergonomic design of control centers.

ISO - International Organization for Standardization 9241-210 (2019). Human-centred design for interactive systems.

Johnsen, S.O., & Aminoff, H. (2024). Use of ANSI/HFES human readiness level to ensure safety in automation. *Human Factors in Design, Engineering, and Computing*, 159, 159.

Johnsen, S. O., & Park, J. (2025). Meaningful human control in digitalisation, automation/AI, and remote oversight. In Proceedings of European Safety and Reliability Conference (ESREL'25).

Johnsen, S. O., Thieme, C., Aminoff, H., Park, J., Fernander, M., Boren, M., ... & Kirwan, B. (2025). *CRIOP: A scenario method for crisis intervention and operability analysis (Version 2025)*. SINTEF Digital.

Johnsen, S.O., Thieme C.A., & Myklebust, T. (2024). Experiences of safety and reliability in remote control of safety critical operations. In Proceedings of European Safety and Reliability Conference (ESREL'24), Contributions Part 5.

Johnsen, S.O., & Winge, S. (2023). Human Factors and safety in automated and remote operations in oil and gas: A review. In Proceedings of European Safety and Reliability Conference (ESREL'23).

Kinnersley, S., & Roelen, A. (2007). The contribution of design to accidents. *Safety Science*, 45(1–2), 31–60.

Kirwan, B. (2025). Human factors requirements for human-AI teaming in aviation. *Future Transportation*, 5(2), 42.

Leva, M.C., Naghdali, F., & Alunni, C.C. (2015). Human factors engineering in system design: A roadmap for improvement. *Procedia Cirp*, 38, 94–99.

Lie, J.A.S., Arneberg, L., Goffeng, L.O., Gravseth, H.M.U., Lie, A., Ljoså, C.H., & Matre, D. (2014). *Working hours and health. Update of a systematic literature review* (STAMI Report No. 1). Arbeidstid og helse. Oppdatering av en systematisk litteraturstudie.

Manuele, F.A. (2005). Risk assessment & hierarchies of control. *Professional Safety* 50(5), 33.

Mecacci, G., Amoroso, D., Siebert, L.C., Abbink, D., van den Hoven, J., & de Sio, F.S., eds. (2024) *Research handbook on meaningful human control of artificial intelligence systems*. Edward Elgar Publishing.

Meshkati, N. (2006). Safety and human factors considerations in control rooms of oil and gas pipeline systems: conceptual issues and practical observations. *International Journal of Occupational Safety and Ergonomics*, 12(1), 79–93.

Miranda, A.T. (2019). Misconceptions of human factors concepts. *Theoretical Issues in Ergonomics Science*, 20(1), 73–83,

MiT (2025). State of AI in Business 2025, MiT report.

Moura, R., Beer, M., Patelli, E., Lewis, J., & Knoll, F. (2016). Learning from major accidents to improve system design. *Safety Science*, 84, 37–45.

NAS - National Academies of Science (2021). Human-AI teaming: State of the art and research needs.

NIOSH - National Institute for Occupational Safety and Health (2024). Retrieved on January 20, 2025 from https://www.cdc.gov/niosh/hierarchy-of-controls/about/index.html

Norman, D. (2013). *The design of everyday things* (revised and expanded edition). Basic Books.

Norsk Industri (2023). *A practical guide to HOP*. Retrieved from www.norskindustri.no/siteassets/dokumenter/hms/hop/hop-veileder-engelsk-2024-3005.pdf

Norwegian National Audit (2023). Sjøfartsdirektoratets arbeid med å fremme gode arbeids- og levevilkår til sjøs Dokument 3:9 (2022–2023).

NSIA - Norwegian Safety Investigation Authority (2022). Human Factors in the NSIA's safety investigations.

NTSB - National Transportation Safety Board (2010). *Aircraft accident report NTSB/AAR-10/03*. Washington, DC.

Park, J., Johnsen, S.O., Saghafian, M., & Alsos, O.A. (2025). Challenges and opportunities in remote operations of automated passenger ferries identified using the CRIOP method. In Proceedings of European Safety and Reliability Conference (ESREL'25).

Porathe, S.T. (2023). Alarm and hand-over concepts for human remote operators of autonomous ships. In Proceedings of European Safety and Reliability Conference (ESREL'23).

PSA (2019). Report "Investigation of collision between Sjoborg supply ship and Statfjord A on 7 June 2019".

PSA (2022). #992923 Responsibility, competence, and maintenance of alarm systems in control rooms.

Reason, J. (1997). *Managing the risks of organizational accidents*. Ashgate, Aldershot, England.

Safety Forum (2019). Learning from incidents/report from safety forum.

Sandhåland, H., Oltedal, H., & Eid, J. (2015). Situation awareness in bridge operations study of collisions between attendant vessels and offshore facilities in the North Sea. *Safety Science*, 79, 277–285.

Ship Safety Act (2007). /lovdata.no/dokument/NL/lov/2007-02-16-9.

Smith, L., Folkard, S., & Poole, C.J.M. (1994). Increased injuries on night shift. *The Lancet*, 344(8930), 1137–1139

Stanton, N.A., Salmon, P.M., Rafferty, L.A., Walker, G.H., Baber, C., & Jenkins, D.P. (2017). *Human factors methods: a practical guide for engineering and design*. CRC Press.

van Winsen, R., & Dekker, S.W. (2017). Human factors and the ethics of explaining failure. In *Human factors and ergonomics in practice: improving system performance and human well-being in the real world* (pp. 65–76). CRC Press.

VG (2025). Retrieved on October 2025 from https://www.vg.no/nyheter/i/QMG5AV/fallulykken-paa-melkoeya-her-er-11-ting-som-gikk-galt

Walker, G.H., Waterfield, S., & Thompson, P. (2014). All at sea: An ergonomic analysis of oil production platform control rooms. *International Journal of Industrial Ergonomics*, 44(5), 723–731.
Wróbel, K. (2021). Searching for the origins of the myth: 80% human error impact on maritime safety. *Reliability Engineering & System Safety*, 216, 107942.
Yasseri, S., & Bahai, H. (2018). System readiness level estimation of oil and gas production systems. *International Journal Of Coastal, Offshore And Environmental Engineering (IJCOE)*, 3(2), 31–44.

Part 2

Design Process to Address Challenges of Automation, AI, and Remote

3 Integration of Human Factors into Engineering Processes for Flight Deck Design and Certification of Large Aeroplanes

Opportunities and Challenges

Sonja Biede-Straussberger and Gernot Konrad

3.1 INTRODUCTION

Aviation can only deliver high safety and performance if all involved operators complete their tasks correctly and efficiently. Pilots, cabin crews, air traffic controllers, maintainers, and ground handlers are among the many end users contributing to this objective. To successfully perform their tasks, the required equipment must support their needs and allow operation at an appropriate level of workload while minimising the risk of human errors under normal and non-normal conditions.

Different terms, such as 'system' and 'equipment', are used interchangeably in civil aviation regulatory material when referring to flight deck features and functions. However, these notions often only refer to machines. Others, like Mindell (2002), define a 'system' as human(s) plus machine(s).

Further, it has to be taken into account that human behaviour does not function in the same way as machine behaviour. Human behaviour is context-driven, depending on the environment in which a machine is operated, the operator's training, experience, and instructions provided to the operators, including procedures and checklists. This is recognised, for example, by the SHELL model, described by the Civil Aviation Authority of the United Kingdom (CAA) (2002).

In this chapter, the term 'system' is used in a broader sense, encompassing not only human and machine elements but also environmental and operational elements (i.e., system = human(s) + machine(s) + environment + operation) (Konrad, 2025).

Risks often propagate across the interfaces between these system elements, and efforts should be directed at improving their coupling (Lee, 2014). In consequence, the most efficient solution is achieved when humans and machines collaborate well within the context of the environment and the kind of operation. But human behaviour is subject to variability at physical and physiological, as well as psychological and cognitive levels. Humans have many strengths, but they also have weaknesses. Machines should be used to augment human strengths and compensate for their weaknesses.

As a scientific discipline, Human Factors (HF), enables the design of usable systems and machines.

The European Union Aviation Safety Agency (EASA) sees HF focusing on why humans function the way they do, incorporating both physical and mental processes and their interdependency.

DOI: 10.1201/9781003741824-5

Human Performance (HP), on the other hand, is seen as the output of HF, focusing on understanding how humans perform their tasks. HF and HP both examine the capabilities and limitations of operators, but with different emphases (EASA, 2018).

Human strengths and weaknesses, however, do not apply only to operators, maintainers, etc., but also to engineers who design the machines and to managers who supervise the designers. Consequently, over recent years, the organisational part has gained importance, enlarging the notion of HF to Human and Organisational Factors (HOF) (Reuzeau, 2024).

Addressing the complexity of human behaviour takes time. An efficient solution can only be found when human capabilities and limitations are continuously considered throughout the entire engineering process. Naturally, during the early design phases, engineers have to make many assumptions about user characteristics and their expected behaviour. It is essential to document these assumptions and validate or modify them as needed.

Processes must be applied to minimise the risk that wrong assumptions lead to bad designs, inducing an unacceptable workload and/or human errors. For that purpose, EASA has introduced regulation 25.1302 with Amendment 3 in 2007 into the Certification Specifications (CS) for large aeroplanes (CS-25), with its latest version being Amendment 28 (EASA, 2021). However, 25.1302 is not the only HF regulation in CS-25; it rather complements other regulations with HF aspects.

This chapter aims to facilitate students' understanding of how HF can be properly addressed in flight deck design and certification and integrated throughout typical engineering processes. After an initial insight into how HF emerged in the past and providing the current regulatory context, the relationship between HF and engineering is represented. Then, a framework explains how to associate the characteristics of human behaviour with flight deck design and certification rules, and illustrates the HF approach with two project examples. Finally, lessons learned to avoid common pitfalls for HF in engineering are highlighted.

3.2 HF EVOLUTION FROM THE SAFETY PERSPECTIVE

According to a study by the Massachusetts Institute of Technology (MIT), Commercial Air Transport (CAT) has become safer each decade. MIT researchers found that the fatality risk was 1 per 350,000 passengers globally between 1968 and 1977, decreasing to 1 per 7.9 million passengers between 2008 and 2017, and further decreasing to 1 per 13.7 million passengers between 2018 and 2022. However, this trend is not guaranteed to continue; CAT safety remains an ongoing challenge (Dizikes, 2024).

Boeing's statistical summary of worldwide CAT operations from 2014 to 2023 (Boeing, 2025) indicates that out of 1,078 onboard fatalities, only 158 were directly related to machine failures. All others included HF aspects.

The two Boeing 737 Max8 crashes in October 2018 (Lion Air 610) and March 2019 (Ethiopian Airlines 302) highlighted again the importance of HF in flight deck design. The crashes were attributed to the malfunction of the Manoeuvring Characteristics Augmentation System (MCAS), which pushed the aircraft's nose down, with the pilots unable to regain control. The newly introduced MCAS was not properly described in the Aircraft Flight Manual (AFM), pilots were not adequately trained on it, and its design did not sufficiently account for HF, including flight crew situation awareness, workload, and response time. Further, organisational factors were found as well, for example, with respect to sufficiently including the HF team in design decisions (Endsley, 2019).

The Federal Aviation Administration (FAA) and other US and international organisations commissioned several studies to evaluate the 737 Max's initial certification and the certification process itself, and multiple reports surfaced as a result, for example, the report from the Joint Authorities Technical Review (JATR) (2019) and the accident investigation report from the Komite Nasional Keselamatan Transportasi (2019). Many of those reports include HF-related recommendations (GAMA, 2023).

The General Aviation Manufacturers Association's (GAMA) Flight Deck HF Working Group developed practical HF recommendations on whether and how the HF issues related to the design and certification process identified in the various reports could be incorporated into regulations, regulatory guidelines, or industry standards. This resulted in GAMA Publication #21, which lists 15 recommendations for regulators and the aerospace industry. GAMA Recommendation #4 specifically identifies the need for more HF training: 'Industry should create a formal training course for HF practitioners on how to incorporate HF into the overall system development and aircraft certification process' (GAMA, 2023).

This motivated a team of flight deck HF practitioners to develop a competency-based syllabus for a training course optimised for the industry's needs, focusing on HF Engineering in Civil Flight Deck Design and Certification (Kemp et al., 2025). The University of Kansas (KU) (2025) now offers an Aerospace Short Course based on the proposed syllabus. The same GAMA recommendation also served as the driving force behind this chapter.

3.3 EVOLUTION OF THE FLIGHT DECK

Flying in an aircraft results from the simultaneous confluence of a collection of machine functions that enable the flight crew to manage the operational status of the entire aircraft throughout its flight envelope. Achieving the correct combination of measures for those functions involves controlling and monitoring them using the interfaces provided in a central crew location, known as the flight deck.

Thus, it is essential for HF practitioners involved in flight deck design and certification to be familiar with the expected flight crew tasks and the intended function of the installed equipment. Understanding how aircraft are designed and operated, how the technology behaves, and how pilots utilise it is essential. Graduates entering the aerospace workforce often lack this knowledge (Konrad et al., 2022). This chapter helps to build a bridge between knowledge of technology and knowledge of the human. HF has always played a crucial role in engineering, particularly in the design of flight decks. This section provides students with an overview of important technologies used by flight crews and offers a historical perspective along with some certification context.

Around 1487, Leonardo da Vinci presented the Vitruvian Man, a drawing derived from anthropometric studies. Da Vinci also published a sketch of a flying machine showing a wing mechanism that a horizontally positioned pilot could manipulate (Wittmer & Bieger, 2011). Da Vinci is credited with being the first to consider the physical aspects of integrating a human into an aircraft. Today, anthropometric data are used in the design of flight decks, for example, to define reach and view envelopes for a wide range of pilots. The minimum and maximum heights to be considered for flight crew members, for example, are now regulated by CS 25.777.

In the early 1890s, Otto Lilienthal was the first to build and fly a heavier-than-air machine by hanging in it and supporting himself with his arms. In a US patent filed in 1894, he directed pilots to grip a handlebar to fly his glider.

The Wright brothers further integrated the pilot into the aircraft. In a later version of the Wright Flyer, two pilots were seated side by side, surrounded by the few instruments and controls necessary to operate the plane (Coombs, 2005). The location of the pilots was critical in their design. Although not explicitly required by regulation, side-by-side pilot seating arrangements are now the standard on modern flight decks.

The Sperry Aircraft Company (today Honeywell Aerospace Technologies), founded in 1918 by Lawrence Sperry, developed the first autopilot (Kearney, n.d.). Its derivatives are standard on all modern large aeroplanes and have contributed to a reduction in physical pilot workload. Autopilots are required today for certain kinds of Instrument Landing System (ILS) approaches (EASA, 2025a). Sperry is also credited with developing the artificial horizon, also called the attitude indicator, which Jimmy Doolittle used in his historic 1929 'blind flight' (Dailey Jr., 2004). Today, an attitude indicator is required for large aircraft (CS 25.1303).

In the early 1940s, Chapanis investigated why, after landing, some pilots operated the landing gear handle rather than the flap controls in some aircraft, causing the gear to unexpectedly retract and resulting in an accident. This led to the shape coding of these controls, so that landing gear levers now resemble an aircraft wheel, and flap levers resemble a section of an airfoil (flaps) (Rose, 2015). Today, shape coding is part of the certification regulations for some large aeroplane controls (CS 25.781).

Fitts et al. (1950) used eye movement monitoring to determine the optimal arrangement of primary flight instruments. This early time-motion study led to what is now known as the Basic-T, described in the EASAs Acceptable Means of Compliance (AMC) 25.11 to support CS 25.1303 and CS 25.1321.

The introduction of technologies such as the Traffic Collision Avoidance System (TCAS) and the Weather Radar (Wx) in the 1950s; the Flight Management System (FMS), digital Primary Flight Displays (PFD), and Navigation Displays (ND), as well as Fly-By-Wire (FBW) in the 1980s; and Engine Indicating and Crew Alerting Systems (EICAS) in the mid-1990s have undoubtedly contributed to the improvement of CAT safety (Konrad, 2025).

Don Bateman's Enhanced Ground Proximity Warning System (EGPWS) from 1996, for example, is one of aviation's safety success stories. It is estimated to have reduced the risk of Controlled Flight into Terrain (CFIT) by 50% in Western Europe and North America within only five years of its introduction (Flight Safety Foundation (FSF), 2025).

Terrain Awareness Warning Systems (TAWS), Airborne Collision Avoidance Systems (ACAS), and Airborne Weather Radars (Wx) are all mandated today by Air Operations Rules for CAT (EASA, 2025a).

Vertical Situation Displays (VSD), Synthetic Vision Systems (SVS), Head-Up Displays (HUD), and Enhanced Flight Vision Systems (EFVS) are becoming the standard, at least for high-end business jets. The latest trends in modern flight deck development incorporate touch-enabled displays and even voice-activated flight controls (Konrad, 2025).

However, development does not stop there. While in the past, the major risks to CAT safety were CFIT, as mentioned earlier, and Loss of Control (LOC), as stated in Bateman (2010), the introduction of EGPWS and FBW with Flight Envelope Protection (FEP) has significantly reduced these risks. Today's challenges include runway excursions and incursions (FSF, 2009; FAA, 2025). As a result, EASA mandated Runway Overrun Awareness and Alerting Systems (ROAAS) by introducing CS 25.705, and some have already been certified. Surface Alerting (SURF-A) systems are under development for large aeroplanes but are not yet mandated.

With the introduction of more advanced technologies and increasingly complex operational procedures, it is crucial to ensure that the flight crew's capabilities and limitations are appropriately considered. This will allow for maintaining and improving aviation safety.

3.4 REGULATORY CONTEXT

Addressing HF throughout the development process is crucial to CAT safety and performance. But products can only enter the market and consequently generate revenue if they are certified. Understanding the applicable regulations, such as for Airworthiness certification and Environment compliance (EASA Part 21), and considering them early is essential. This can help avoid costly design changes later in the process.

The degree of governmental involvement in people's lives, that is, restricting personal freedom and choice through regulation, increases with the increasing risk or concern for the public (e.g., fewer regulations for hang-gliding and many regulations for commercial flights). Some activities, such as national defence, are inherently governmental (Abbott, 2025).

Aviation Authorities worldwide, such as EASA in Europe and the FAA in the United States, are responsible for ensuring civil aviation safety.

EASA's legal powers derive from a Basic Regulation (BR) of the European Parliament and Council, which defines the Agency's roles and responsibilities and establishes common rules in the field of civil aviation. EASA undertakes its role by applying a wide range of regulatory provisions that expand upon and interpret the requirements of the BR in sufficient technical detail. For aircraft certification, binding Implementing Rules (IR) adopted by the European Commission (EC) specify a high, uniform level of safety, conformity, and compliance (Skybrary, n.d.).

Certification Specifications (CS), non-binding technical standards, are provided by EASA to meet the essential requirements of the BR and IRs. CSs are used to establish the certification basis of an aircraft (EASA, 2025c).

Acceptable Means of Compliance (AMC), non-binding, serve as means by which the requirements in the BR and IRs can be met. However, applicants may choose to use alternative means (EASA, 2025c).

This chapter focuses on HF aspects of CS-25 aircraft certification. CS-25 is the home of over 120 regulations (Reg.) that include flight deck HF aspects (Konrad, 2025). These regulations specify the minimum required to demonstrate compliance. Some have complex interdependencies, such as the following:

- CS 25.1301: Function and installation
- CS 25.1302: Installed systems and equipment for use by the flight crew
- CS 25.1309: Equipment, systems, and installations
- CS 25.1322: Flight crew alerting
- CS 25.1329: Flight guidance system
- CS 25.1523: Minimum flight crew

In those examples, certain HF aspects are addressed by multiple regulations and/or their associated AMCs, but with a different focus. Intended function, for example, is addressed by 25.1301 and 25.1309; human error by 25.1302, 25.1309, and 25.1329; and flight crew workload by 25.1523, 25.1302 and 25.1309.

There are also many HF-related regulations that are not in the primary focus of HF because they are under the responsibility of other disciplines, such as flight test, for example, handling qualities, aircraft stability, and controllability. But the regulations, CS 25.143 through CS 25.255, are rife with ambiguous requirements on human performance, such as 'must be possible ... without exceptional piloting skills, alertness, or strength'. For that purpose, the Cooper-Harper rating scale (Cooper & Harper, 1969) was developed and later modified to also assess workload, becoming known as the Bedford Workload Rating Scale (Gawron, 2019).

EASA Rules for Air Operations and for All Weather Operations (AWO) also include important HF requirements (Req.).

Additional publications from regulators and the industry related to HF aspects of flight deck design exist, and include the following:

- DOT/FAA/TC-16/56: HF Considerations in the Design and Evaluation of Flight Deck Displays and Controls
- DOT/FAA/HF-STD-001B, an HF Design Standard
- RTCA DO-372 for addressing HF/pilot interface issues for avionics

3.5 HF IN ENGINEERING PROCESSES

Today, a great deal of scientific knowledge about human functioning exists. But its large variability represents a challenge for integrating HF into engineering processes.

Different frameworks exist to describe the stages of systems engineering, including integration with project management processes, and are formalised by the International Organisation for Standardisation (ISO) in ISO 15288 (ISO, 2023).

The engineering community typically follows a V-model (e.g., Schallert, 2016), characterised by a descending line from the definition to the detailed system levels using iterative evaluations, and an ascending line for integration and tests.

In a more linear representation, NASA (2016) identifies various reviews that should be completed at (or before) four process phases. These are (a) concept development, (b) preliminary design and design completion, (c) final design, and (d) integration and test. The reviews include Mission Concept Review (MCR), Mission/System Requirements/Design Review (SRR/MDR/SDR), Preliminary Design Review (PDR), Critical Design Review (CDR), and Test/Flight Readiness Review (TRR/FRR). Aligned with these reviews, and complemented with activities for the certification phase, which is required to receive an aircraft Type/Amended Type/Supplemental Type Certificate (TC/ATC/STC), Konrad et al. (2022) identify major activities that flight deck HF practitioners typically perform throughout a flight deck design process. Those are shown in Figure 3.1, synchronised with six proposed process stages.

For the remainder of this chapter, the terms 'verification' (Ver.) and 'validation' (Val.) are used as defined in AMC 25.1709; that is, validation is the 'determination that requirements for a product are sufficiently correct and complete', and verification is the 'evaluation to determine that requirements have been met'.

In order to know to what extent HF activities need to be performed, a so-called Level of Scrutiny (LOS) is identified depending on the degree of novelty, complexity, and integration of a design item (i.e., 'A defined and bounded set of either (one or more) hardware elements or (one or more) software elements that are treated as a single entity for analytical purposes. . . .' – detailed in AMC 25.1302.

Once the HF LOS is identified, all HF activities have to be planned. It is important to document this in the so-called HF Certification Plan (HFCP), which is used as a formal way to communicate with the Authorities. Guidelines on how to write an HFCP can be found in PS-ANM111-1999-99-2 (FAA, 1999). Equally important is having a company internal flight deck philosophy defined, which

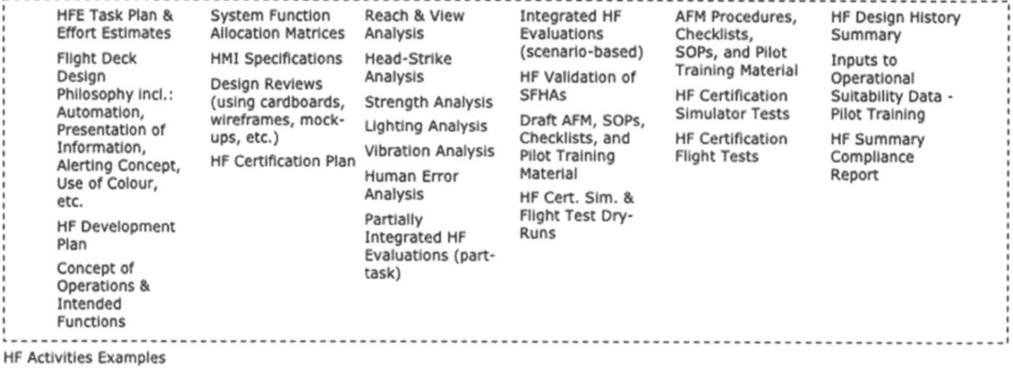

FIGURE 3.1 Typical major activities for flight deck HF practitioners.

Source: Konrad (2025).

all the various engineering disciplines adhere to (for more details and an example, cf. Palmer et al., 1995).

Similarly, the European Union's (EU) Single European Sky Air Traffic Management Research (SESAR) community has implemented the Human Performance Assessment Process (HPAP), which integrates the human element throughout the engineering process (Pelchen-Medwed et al., 2021). This initiative is based on the systematic integration of a user-centred design approach into engineering processes that has been developed over the past few decades. It aims to make machines usable and useful by focusing on the users and their needs, applying HF principles and usability knowledge. It is intended to enhance effectiveness, efficiency, and user satisfaction while also mitigating potential adverse effects on human well-being, safety, and performance. Such a process was standardised by the International Organization for Standardization (ISO) in ISO 9241 (ISO, 2019), which delineates the importance of understanding and specifying the context of use and creating user requirements, followed by the definition of specific design requirements, which must then be validated. The stream of design thinking philosophy added the focus on empathy, ideation, and user research to this process.

A user-centred approach must be established right from the beginning when developing new or modified equipment or functions. It begins by understanding the context of a new technical or operational evolution and identifying the associated user needs. Technical solutions are most effective when they meet users' needs and requirements. Hence, it is essential to understand who the target users are and the tasks associated with a new or modified design. Operators' needs refer to what is necessary or desired by the user and may not always be declared (European Committee for Standardisation, 2014). Often, needs may be associated with challenges users have already encountered while performing their tasks, being identified as 'pain points' or associated with a projection of new ideas. Other sources of user needs can be related to the human (biological or social needs), the machine (interface needs), or the environment and operations (Bach & Biede, 2016). A certain design may not be able to cover all the needs users express; prioritisation and hierarchisation of needs are required before they can be translated into requirements.

In systems engineering, requirements define the functions and features a machine must perform. They can cover both technical (e.g., machine performance) and non-technical (e.g., operational, safety, or use) aspects, with the latter being translated into technical requirements. This includes defining the machine's intended function, a top-level requirement that describes what a machine is supposed to do, not how.

CS 25.1301 requires demonstrating that 'Each item of installed equipment must - Be of a kind and design appropriate for its intended function; . . .'. Therefore, the main attributes of an intended function statement must be clear, unambiguous, and verifiable (Konrad, 2025).

During the definition of a technical solution through requirements development, potential HF issues may emerge regarding the previously introduced human capabilities and limitations. These relate to the effects of design solutions to be resolved to avoid negative impact on human performance (SESAR Joint Undertaking, 2020). Or, as similarly stated by EASA (2021), '. . .a deficiency that results from the interaction between the flight crew and the system' including human errors, '. . . but also other kinds of shortcomings, such as hesitations, doubts, difficulties in finding information, suboptimal strategies, inappropriate levels of workload. . .'. It is only through a deeper investigation of these potential issues that it is possible to confirm whether they are negatively impacting human performance or whether the machine requires adaptation to better meet users' needs (e.g., by providing clearer information, easier tasks, or better alignment with the human body).

Often, engineering solutions innovate technology, leading to technocentric designs or improved operations (e.g., lower aircraft separation requirements or more fuel efficiency), but the benefits for users may not be apparent. However, acceptance of a solution can decrease considerably if users' needs are not considered. For example, an interface should be designed to facilitate an intuitive user interaction without requiring specific technical knowledge. For this reason, an approach that connects the investigation of potential HF issues with an explanation of the benefits the user can expect from the new solution is appropriate.

A direct way to address potential HF issues is to utilise existing guidelines for user-centred design (e.g., ISO 9241) and for interface or specific flight deck-related items (e.g., US DOT FAA HF-STD-001B, Ahlstrom, 2016).

Various methods can be applied to investigate potential HF issues. Early in the process, it is important to gather user data. This can be data from field operations, incidents/accidents, and user experience. One challenge is to identify the appropriate HF activities and plan the appropriate involvement for each design process stage (Figure 3.1). Exploration may be needed through dedicated paper studies, user workshops, or experiments to derisk specific design questions as early as practicable. The use of prototypes of increasing levels of fidelity allows capturing direct user feedback to mature solutions to an adequate level. As this approach is iterative, it may not be possible to identify the right requirements immediately. For this reason, HF practitioners may need to propose design recommendations to be incorporated into the improvement cycle.

As the design advances, the goal is no longer to validate requirements but to verify that the requirements have been implemented correctly and to confirm that the equipment is acceptable for the user and certifiable. There is added value in having the Authority involved early in the development process, for example, through familiarisation sessions and witnessing of evaluations. This allows for early mitigation of potential certification risks. Further, the applicant may be granted partial certification credit (Certification Authorities for Large Transport Aircraft (CATA), 2018).

Typical validation methods for flight deck HF requirements include reviews, analyses, and assessments. CATA (2018) uses the term 'assessment' for both evaluations and tests. 'Evaluations are intended to be conducted using partially representative assessment means, whereas tests make use of conformed assessment means'.

The purpose of evaluations is often to derive and validate design requirements, whereas tests are intended to verify those requirements and demonstrate compliance with applicable regulations. Tests are usually final and require an approved test plan. The certification authority personnel or their delegate will typically witness or participate in testing. However, an HF certification strategy based solely on one test is generally not acceptable (CATA, 2018).

Therefore, evaluations must be planned throughout the entire development cycle and should start early in the design process, before production hardware and/or software are available. During HF evaluations, a wide variety of part-task to full-installation representations of the equipment or the flight deck may be used. Evaluations in low-fidelity mock-ups and part-task simulations help to explore how users react to a solution. They often address scenarios that are difficult or impossible to capture during tests and thus can provide broader, more structured, and better-controlled conditions than testing.

To demonstrate that a product is suitable for deployment, its maturity must be shown. In the mid-1970s, NASA introduced Technology Readiness Levels (TRL) to provide the aerospace industry with a standardised scale to measure the maturity of technology. The nine levels comprise technology conception (TRL 1 to TRL 3), design and demonstration (TRL 4 to TRL 6), and the engineering development cycle that culminates in successful mission demonstration (TRL 7 to TRL 9). However, this scale does not adequately assess the technology's maturity in human-related aspects, nor does it assess certification readiness. To close this gap, Human Readiness Levels (HRL) were introduced for the European aviation HF landscape, which focus on the readiness of the technology for the operator (Schmitz & Rodriguez, 2024), and were based on the Technical Standard developed by Human Factors and Ergonomics Society (2021). EASA (2025b) provides Certification Readiness Levels (CRL) to address the remaining certification gap. Table 3.1 illustrates the association between these maturity frameworks.

TABLE 3.1
Comparison of Readiness Levels for Technology, Human, and Certification

	Technology Readiness Level Modified from NASA (2016)	Human Readiness Level Modified from Schmitz & Rodriguez (2024)	Certification Readiness Level Modified from EASA (2025b)
1)	Basic principles observed and reported	Basic principles for human characteristics, performance, and behaviour observed and reported	Safety regulator familiarisation with technology and the Concept of Operation (CONOPS)
2)	Technology concept and/or application formulated	Human-centred concepts, applications, and guidelines defined	Confirmation of the CONOPS assumptions and safety objectives
3)	Analytical and experimental critical function and/or characteristic proof of concept	Human-centred requirements to support HP and human-technology interactions are established	Identification of key regulations that would be mainly affected, and need to be amended/developed in priority
4)	Component and/or breadboard validation in a laboratory environment	Modelling, part-task testing, and trade studies of the design concepts and applications were completed	Identification of the main technical standards and roadmap for the overall necessary rulemaking and/or recommended industry standards
5)	Component and/or breadboard validation in a relevant environment	Human-centred evaluation of prototypes in mission-relevant part-task simulations was completed to inform the design	Identification of the roadmap/action plan for a comprehensive aviation framework readiness
6)	Design/subdesign model/ prototype demonstration in the relevant environment	Design is fully mature and influenced by HP analysis, metrics, prototyping, and high-fidelity simulations	Confirmation of the necessary technical knowledge and availability of the first draft of the rulemaking material
7)	Design prototype demonstration in a target/ flight environment	Design is fully tested and verified in an operational environment with system hardware and software, and representative users	When applicable: mature draft generic elements of a certification basis in the form of Special Conditions (SC) or AMCs
8)	Actual design completed and flight qualified through test (ground or flight)	Overall performance is fully validated and approved in mission operations using completed system hardware and software, and a representative user	The product (integrating the technology) is certified. Design Organisation Approval (DOA) granted
9)	The actual design is flight-proven through successful mission operations	The machine was successfully used in operations across the operational envelope with systematic monitoring of human-system performance	The actual system (design, operation, maintenance) is proven in the operational environment

3.6 HUMAN PROPERTIES IN FLIGHT DECK DESIGN

When examining how humans interact with machines, it is essential to consider various human properties to determine if the impact on the operator is acceptable. Ultimately, we seek a high level of safety, performance, utility, and usability. This section explains how to characterise human properties and details six critical areas associated with flight deck features: task integration, information processing, workload, situation awareness, automation, and human error.

At the time when the Polish professor Wojciech Jastrzębowski first used the word 'ergonomics', which he derived from the Greek words 'ergon', for 'work', and 'nomos', for 'law', in 1857, it meant the 'science of work' (Koradecka, 2010). Back then, ergonomic concerns focused primarily on the physical and physiological factors of human operators; psychological and cognitive factors were not much addressed. Today's understanding of HF also considers the mental aspects of humans. Other areas, such as social and cultural factors, are equally important; however, they are not further detailed in this chapter.

In a simplified way, human properties can be grouped into two main areas (Konrad, 2025):

- Physical and physiological factors refer to elements related to the human body itself (e.g., body dimensions) and the body's functions (e.g., hearing and seeing), which influence human thinking. For example, when our body's chemistry is off due to hypoxia, the neurotransmitters that control our thinking processes can be affected (FAA, 2021).
- Psychological and cognitive factors focus on the elements of humans that limit or enhance thinking and feeling. They include personality traits (e.g., extraversion and introversion types, and attitudes such as technology affinity) and various functions that represent our awareness, encompassing the things humans experience (e.g., beliefs and mental constructs), the formation of memories (e.g., short- and long-term), and our ability to analyse and make decisions (e.g., heuristic or analytical).

Figure 3.2 visualises the translation of the physical and physiological factors (green, web version), as well as the psychological and cognitive factors (amber, WEB version), to the integration of the pilot into a flight deck, which includes the display and control factors (cyan, WEB version) as well as the functional factors (magenta). This figure is intended to provide a visualisation of the link between human and machine properties. Any changes to the system (i.e., to the human, the machine, the environment, or the operation) can have a flight deck effect, impacting the information available to the human and how the human reacts to it.

Human nature has specific characteristics and ways of functioning that vary from one person to another. Multiple human sciences, such as psychology or neuroscience, study these in depth. Human behaviour can be described by its processes (e.g., perception) and the states and conditions

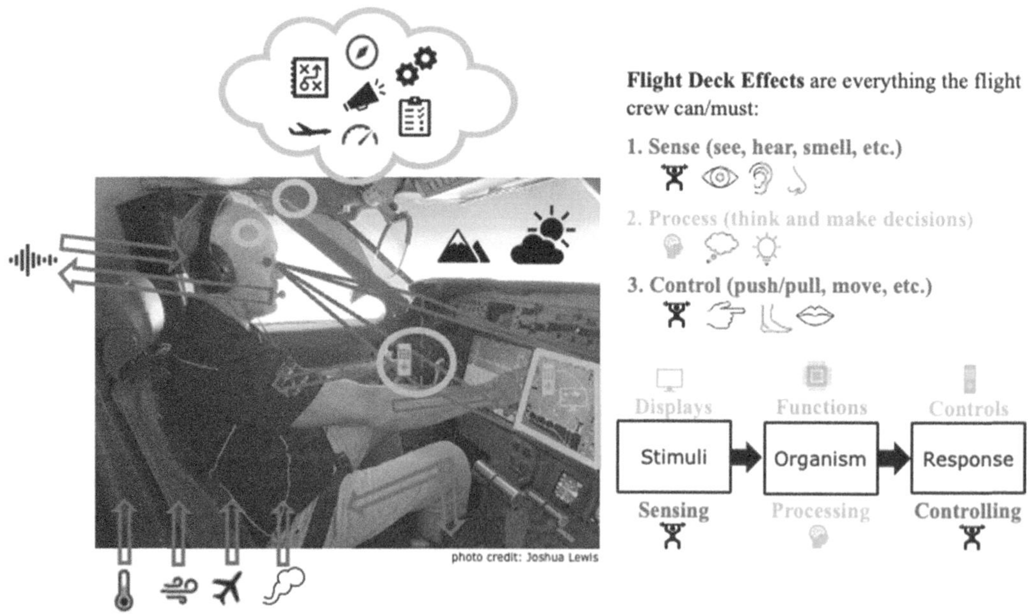

FIGURE 3.2 Factors associated with flight deck effects and human senses (colours on WEB version).

Source: Konrad (2025).

it creates in humans (e.g., fatigue). The wide range of variability among operators creates numerous expectations from individual users towards designs that engineers may not necessarily be aware of. Humans possess impressive capabilities (e.g., assigning meaning in uncertain situations), which differ from those of machines. Humans are also affected by limitations that differ from those of machines (e.g., memory limitations and cognitive biases).

On the flight deck, pilots are surrounded by a multitude of information sources in various formats to help them accomplish their mission. The objective of humans is to perform the expected tasks in line with their goals. To process and transform the cues (i.e., a stimulus) around them (both externally and internally within the human body), humans utilise energetic resources tied to their physiological and cognitive structures. As Feldman-Barrett (2021) emphasises, humans strive to minimise energy use when interacting with the environment.

The brain is the central tool for managing the complexity of the human nervous system. It perceives and integrates information from the external (e.g., a changing environment) and internal (e.g., memory or body signals) worlds, and translates it into outputs that can act on both worlds. Norman (2013) refers to this as 'knowledge in the head' versus 'knowledge in the world'.

On the flight deck, information is typically provided visually through displays, acoustically through loudspeakers, or tactually through haptic interfaces such as moving or vibrating controls. The pilots sense the provided information with their eyes, ears, and/or hands (or other body parts). The cues are then processed by the pilot and given meaning. This processing is context- and task-driven and requires an understanding of the machine functions (i.e., the machine behaviour).

Beyond psychological and cognitive specificity, human behaviour is also constrained by physical and physiological functions, which drive how the human body reacts in space, such as acceleration that acts upon the human, distances impacting reachability, or environmental conditions such as vibration and lighting conditions impacting user states. Resulting limitations affect the design of the flight deck. For example, the risk of hypoxia due to insufficient oxygen supply resulting from cabin pressurisation malfunction requires aircraft operating at high altitudes to be equipped with dedicated oxygen masks (CS 25.1147). Or, somatogravic illusions caused by weight force during climb, which is perceived similarly to the resulting force during acceleration (Fernandez et al., 2017). Considerations for that are made in CS25.143 for go-around manoeuvres.

While human properties are associated with human capabilities and limitations, machine properties are the inherent capabilities and limitations of the machine (Konrad, 2025).

Measurable system performance is achieved by combining specific human, machine, environmental, and operational properties as necessary for a particular design (Konrad, 2025).

For example, to achieve good readability of a text (i.e., a measurable system property), vision acuity (i.e., a physiological factor), font size, font style, colour and contrast (i.e., a display factor), and flight deck vibrations, for instance, caused by turbulent air (i.e., an environmental factor) must be considered together. In another example, for the understandability and predictability of autopilot behaviour (i.e., a measurable system property), information processing (i.e., a cognitive factor), autopilot mode changes (i.e., a machine behaviour), and pilot training (i.e., an operational factor) are interconnected.

Table 3.2 provides an example overview of how different measurable 'system' properties could be associated with different human, machine, and environmental/operational properties. All these factors, however, are interdependent and cannot be addressed individually. Humans should not be considered 'factors' but rather as groups of variables interconnected through states and processes that affect each other.

The remainder of this section details some examples of how these properties can be addressed through HF criteria. The goal is not to provide an exhaustive overview of all topics; specific sources are available (e.g., Endsley & Jones, 2025). The goal, rather, is to help students understand how topics commonly cited in HF literature can be connected in daily practice and why this is important.

TABLE 3.2
Examples for Human, Machine, Environmental/Operational, and System Properties

Human Properties	Machine Properties	Environmental and Operational Properties	System Properties (Measurable)
Physical/physiological factors	**Controls/display factors**	**Environmental factors**	**System properties by physical/physiological and controls/displays**
Anthropometry, *e.g., stature, arm length, angular limitations* Strength, *e.g., force, torque* Motor skills Sensation, *e.g., visual, aural, tactile*	Information content, type, & modality, *e.g., Symbol/text size/ style, tone volume* Interaction means, *e.g., control size, activation force* Layout, *e.g., function allocation, clutter* *Display, e.g., size, brightness, colour*	Temperature Lighting Noise Vibration Acceleration	Reach/viewability Operability Read/hearability Physical well-being
Psychological/cognitive factors	**Functional factors**	**Operational factors**	**System properties by psychological/cognitive and machine-behaviour**
Memory, *e.g., working, long-term* Comprehension Prediction, diagnosis Decision-making & action selection Mindset, mental model Attention & vigilance Motivation	Functions & allocation between humans & machines Complexity Automation State/mode transitions Degradations	Task structure, *e.g., complexity* Time and priorities Standard Operating Procedures (SOP) Normal/abnormal/ emergency checklists Technical documentation Training manual Personnel selection criteria	Understandability, *e.g., unambiguous, predictable* Usability, *e.g., effectiveness, efficiency, satisfaction* Situation awareness Cognitive workload Response time Mental fatigue & stress Startle & surprise Task difficulty

Source: Konrad (2025).

3.6.1 Integration in the Task Context

Operators' tasks represent a key concept for design. To achieve their missions, operators must perform their tasks safely and efficiently. Pilots' tasks include aviating, navigating, communicating, and managing systems. Tasks are defined as 'activities required to achieve a goal' (ISO 9241-11), and AMC 25.1302 (EASA, 2021) considers that 'design items have to be evaluated in terms of tasks' to understand '…how tasks might be changed or modified as a result of introducing new design items'. Any controls and information introduced have to be '… accessible and usable by the flight crew in a manner appropriate to the urgency, frequency and duration of their tasks'.

Different task analysis methods are referenced in textbooks, such as Stanton et al. (2013), to investigate the impacts of tasks. Tasks can be represented hierarchically or sequentially, encompassing both visible and invisible activities, inputs, and outputs (e.g., actions, decisions, and errors), along with their corresponding contextual settings. On the flight deck, numerous tasks must be completed in parallel or concurrently (such as one crew member 'reading' a checklist while the other 'doing' it). It is important to identify the tasks impacted by a design change and the level at which a task analysis is relevant. For example, a new navigation interface might affect how pilots communicate with Air Traffic Control (ATC) and, hence, require investigation of both navigation and communication tasks. Using task analyses, tasks can be assessed for potential difficulties and complexities, deviations from expected sequences, sources of human errors, effect on crew workload, and the impact of interruptions. The goal is to identify the most adequate task flow and task allocation between humans and machines suitable for the operational mission. For this reason, it might be required to decompose tasks not only at a procedural observable level – the number of steps required to perform a task, and the complexity of the task flows – but also at a cognitive level, such as describing information-processing steps when interacting with a system (e.g., how a user associates different information types or performs mental simulations to imagine possible outputs on task subgoals before they occur, such as flight path management (Abbott et al., 2013). Furthermore, when effectively exposed to situational challenges and constraints during their operations, operators adapt their tasks to achieve their goals most efficiently, for example, to manage their workload. Hence, an analysis must cover both prescribed tasks (e.g., as outlined in the pilot's Standard Operating Procedures (SOP)) and expected tasks (e.g., as specified by mission requirements), as well as effectively performed tasks, to identify potential variability that can be addressed through design. For example, to analyse the impact of a new communication technology on pilot tasks, it is necessary to study its impact on standard pilot-ATC exchanges (e.g., frequency changes) and on the occurrence of unusual workload-increasing events, such as machine failures and weather events.

3.6.2 Information Processing – From Prediction to Sensation and Decision-Making

Some technologies present major advantages for providing information that would otherwise be difficult to access and integrate. For users to benefit from them, they need to be well aligned with the specific functioning of the brain, whose primary goal is to maintain energetic balance within its environment (Feldman-Barrett, 2021).

The human brain continuously develops a prediction of the situation surrounding it and connects it with information already present in its mind (Bubic et al., 2010). This process is largely unconscious and becomes conscious only when the prediction no longer meets the brain's expectations and adaptation is required (e.g., in cases of information conflict or physiological needs).

To manage operational situations and achieve their tasks, operators must filter and integrate relevant cues from both their internal (physiological) and external (environmental) environments, transforming them into meaningful information. Different human senses may process large amounts of information in parallel (e.g., acoustic and visual), requiring appropriate levels of attention and vigilance to filter relevant information. As the human sensory system has certain physiological limitations (e.g., capability to process light or the capacity limits to process data), information must be salient enough (e.g., regarding sound or light) to be attended to.

Attention is either focused on where people expect information or directed towards a source (e.g., by attention-getters that direct the gaze to a primary field of view), and it may need to be divided between different sources (Vidulich et al., 2010). This process is a control loop with many stages and occurs on different behavioural levels.

For the brain to determine the relevance of sensory information, the sensed inputs are compared with previous learning and experiences. One way to characterise related behaviours was established by Rasmussen (1983) when developing the skill, rule, and knowledge framework, also known as the SRK model. Whereas skill-based behaviours are nearly automatic and do not require many resources, standardised situations allow operators to follow rules and hence limit the required effort. However, when situations are uncertain or unfamiliar, operators must rely on learned knowledge or experience and exert effort to find solutions through unplanned decision-making.

Humans use heuristics and are subject to biases when handling the complexity and uncertainty of situations. Heuristics are 'mental shortcuts' used when time or cognitive resources are limited (Wickens & Hollands, 2000). They are not guaranteed to yield the best answer, but they are efficient, such as rules of thumb. A cognitive bias is '... a systematic error in thinking that occurs when people process and interpret information in their surroundings, influencing their decisions and judgments' (Kahneman, 2012). Cognitive biases differ from machine algorithms (except if based on Artificial Intelligence (AI) technologies), designed to yield correct answers.

It is further a human characteristic to consider only a subset of potentially relevant information at a sufficient level, rather than seeking optimal solutions. For that purpose, for example, humans tend to omit available information when making decisions, such as using mostly the first- or last-arriving cues. Humans also tend to avoid resource-demanding processing by prioritising automated processing based on previous learning, which can, however, result in wrong outcomes (Kahneman, 2012).

Finally, to develop experience and later reuse it, humans store information about a situation or state in their minds. This can lead to mental images or models associated with specific situations. They can be reactivated in similar situations, when projecting the current situation to the future (Billman et al., 2021).

Many of these processes are performed by the brain at a cognitive level; they are not directly visible to external observers. To access them, specific methods are required. Scientists have developed cognitive models to describe, in simplified representations, how humans process information. They support the design process by investigating detailed cognitive processes related to tasks and help drive the specification of display information properties (Holder et al., 2024).

For example, Parasuraman et al. (2000) illustrate cognitive processes in a linear format, differentiating between stages of information acquisition, information analysis, decision-making, action selection, and action implementation. A more extensive model was developed by Endsley (e.g., described in 2017), which connects meta-concepts related to human behaviour (e.g., vigilance and attention) with task and system factors in addition to individual factors along the processing stages. The Decision Ladder model by Rasmussen et al. (1990) highlights that cognitive processes act as a connected network. This model also allows a better explanation of experts' behaviours, who are known to apply specific rules to find meaningful patterns with highly organised knowledge structures (Vidulich et al., 2010).

These different models highlight different behaviours. One challenge is identifying the most suitable model to solve the design problem at hand. The goal is to identify the characteristics of the information to be displayed and to ensure adequate interaction mechanisms that enable humans to be most efficient in collecting internal or external information and translating it into appropriate output or action within the system.

To connect these models with an understanding of human processing, designs must present information at the right time and in the right format for the tasks at hand. First, operators shall only be presented with the information needed for the task within the scope of their mission. The operator must be able to assign meaning to information in relation to the task requirements. The most pertinent information relevant to a particular situation must be communicated at a rate that does not overwhelm the person receiving it. Information representation must also accommodate both the order of occurrence and the order of priority of each reported event. Information from different sources shall not contradict each other. For example, contradictory information between TCAS and ATC traffic instructions led to a mid-air collision at Überlingen in 2002, highlighting the importance of clear priorities among data sources (Aviation Safety Network (ASN), 2025). Furthermore, pilots receive information through various sensory channels: visual (text or symbols), auditory (sounds and verbal messages), tactile/haptic (vibrations, movements), and even olfactory (smells). This requires the operator to scan the information using previously learned skills. Among all possible surrounding information, relevant information can be filtered by humans when it is detectable and perceptible, for example, if displayed for a sufficient amount of time and with a certain level of saliency. EASA expectations on information presentation can be found in AMC 25.1302.

Alerts are among the most critical kinds of information pilots must manage on the flight deck. They indicate a potential hazard in or around the aircraft that requires flight crew awareness or response. CS 25.1322 and its AMC provide specific requirements for alerts. To communicate the severity and urgency of machine malfunctions (e.g., engine failure) or dangerous situations (e.g., TCAS) to pilots, three escalating alert levels are defined. They depend on how quickly the flight crew must be aware of and respond to a certain situation. Warnings (presented in red) are necessary when immediate flight crew awareness and response are required. Cautions (presented in amber or yellow) are used if immediate flight crew awareness and subsequent flight crew response are required. Warnings and cautions must provide attention-getting through at least two senses by combining aural, visual, or *tactile/haptic* indications. Advisory alerts (presented in any colour except red or green) indicate that flight crew awareness is required, which may result in a subsequent response from the flight crew.

Finally, the alert must be presented at the most suitable location and align with alerting, display, and language philosophies to ensure an adequate operator response. For example, in the case of insufficient runway length under wet conditions, Airbus displays the amber caution 'IF WET: RWY TOO SHORT' on the PFD, requiring the pilot to be immediately aware of the situation and decide whether to adapt the flying task (Jacob et al., 2009). EASA expectations on acceptable alert presentation can be found in AMC 25.1322 and AMC 25.11.

3.6.3 Workload

As the previous section emphasised, many information-processing components are involved in performing complex tasks. Processing information requires energy and, therefore, incurs a cost for the human body. The concept of workload helps highlight the links between tasks – both physical and mental – and their impact on the operator. Parasuraman, Sheridan, and Wickens (2008) define mental workload as 'the relationship between the mental resources demanded by a task and those resources supplied by the human operator'. If resources do not match, the risk of overload or underload emerges. The basic assumption is that if the task demand is too high, the operator may not have sufficient spare capacity to handle unexpected situations or become overly fatigued over time. For example, the amount of information to be processed is constrained by working memory capacities and impacts efficient processing. On the contrary, if task demand is too low, the necessary level of body activation cannot be maintained, resulting in inadequate human conditions (e.g., boredom or reduced vigilance). A similar concept applies to physical workload.

These aspects are addressed by several regulations. As per CATA Worklist Item (CWI) EASA-003–25.1302, the primary intent of CS 25.1523 is to evaluate the workload with the objective of demonstrating compliance with the minimum flight crew requirement. The intent of CS 25.1302, however, is to identify design-related pilot performance issues. Under CS 25.1302, the acceptability of workload levels is but one parameter among many to be investigated to highlight potential issues. To provide evidence of appropriate workload, AMC 25.1302 requires demonstrating that '... the flight crew can be expected to complete their allocated tasks successfully in both normal and non-normal operational conditions, within the bounds of acceptable workload and without requiring undue concentration or causing undue fatigue'. AMC 25.1309 specifically addresses the acceptable workload associated with recovery from and management of failure conditions.

This raises the question of how to effectively assess workload. Hilburn and Jorna (2001), also cited by the National Test Pilot School (NTPS) (2013), distinguish between the actual task demand (i.e., taskload) and the individually experienced task demands (i.e., workload). Individuals may not react the same way to the same tasks, as different factors influence individual reactions (e.g., training and experience). Often, standardised rating scales such as NASA-TLX or Bedford are used to assess workload (Gawron, 2019). They do, however, have the limitation that they could be intentionally influenced and manipulated by humans; therefore, multi-dimensional approaches that combine subjective scales with objective measures, such as task performance and physiological measures, are preferred (NTPS, 2013).

One of the difficulties in assessing workload is that people use different strategies to adapt their resources to situations, such as controlling by anticipation or supervisory control, where one monitors that a machine's behaviour remains within a certain range and corrects it when necessary. This allows people to maintain an equilibrium between the level of resources, the effort they invest, and the level of performance they seek. Still, it is another factor that needs to be integrated into the analysis of workload effects (Durso & Alexander, 2010).

AMC 25.1302 states that 'the techniques used to collect data in the context of the CS 25.1302 evaluations could make use of workload rating scales, but in that case no direct conclusion should be made from the results about the compliance with CS 25.1302' and 'workload ratings should complement other data from observations of flight crew behavior or other types of data'.

3.6.4 Situation Awareness

Workload is associated with an operator's ability to assess and understand the situation around them, as it is related to the energetic cost of this assessment. 'Situation assessment refers to the processes in which operators engage to make sense of the situation, to understand it', and the conscious part therein is about situation awareness (Durso & Alexander, 2010). To conduct a situational assessment, operators rely on mental models of the situation and how it is unfolding. However, this mental model must be updated with information from the actual situation to align top-down expectations with bottom-up inputs from reality. Furthermore, to anticipate the possible outcomes of decisions, it may be necessary to run a mental simulation (Holder et al., 2024). All these concepts of information processing are merged into a single concept known as situation awareness, which connects the components of perception, understanding, and predicting future states of the situation in the information-processing cycle (Endsley & Jones, 2025).

While general situation awareness scales and methods exist (e.g., Situation Awareness Rating Technique (SART) and the Crew Awareness Rating Scale (CARS), Gawron, 2019), flight deck designers must establish clear connections between an individual's understanding of the situation and specific design items.

Unfortunately, over the past few decades, this concept has also evolved into a buzzword within the engineering community, where statements about situation awareness are often used to define the intended function of design items. However, simply stating that a design item's intended function is to provide situation awareness is unacceptable because it is not testable. Instead, the statement must be specific; for example, the intended function of the weather radar is to provide the pilot with awareness of potentially hazardous weather conditions along the projected flight path (Konrad, 2025).

3.6.5 Automation

Automation has enabled aviation to achieve the operational performance it exhibits today. Automation can perform operations that humans cannot, such as processing large amounts of data quickly. On the other hand, humans have the advantage of being able to give meaning to situations, even when data is lacking, or to find adapted solutions to unexpected situations.

One of the key benefits of automation is improved information integration, which facilitates the pilot's tasks. For example, TCAS helps the pilot avoid a potential in-air collision with another aircraft by identifying the intruder's relative position and expected trajectory. To better support the pilot during an avoidance manoeuvre on some aircraft, an automation called AP-TCAS (autopilot coupled with TCAS) has been implemented. AP-TCAS performs an automated, timely avoidance manoeuvre in a context that can often be difficult for pilots to handle (Botargues, 2009).

Achieving the right level of performance requires considering both the automated functions and the user interface simultaneously. If the human-machine interface (HMI) is poorly designed, it introduces inherent paradoxes known as the ironies of automation (Bainbridge, 1983), in which features intended to help operators can inadvertently create new challenges. It may shift the workload from a visible physical one to an invisible cognitive one. This may lead to over-reliance (Parasuraman & Manzey, 2010) and dependency (Roberts & Abbott, 2025) on automation, or to an increased need for information (Degani et al., 2013).

A well-known problem is mode confusion, where pilots lose track of which automation mode is active or misunderstand how mode transitions work, leading them to react with surprise. Due to the complexity of data inputs, humans might develop an inadequate understanding of functions and form incorrect expectations about their next actions, changes in mode, and how transitions work (Abbott, 2022).

The types of HF issues encountered with automation vary depending on the type of automation involved. For example, decision-support automation raises questions about differences in decision-making between expert pilots (e.g., test pilots) and pilots with usual airline training in accordance with EASA Part-FCL (i.e., Flight Crew Licensing), potentially leading to different design requirements. Assumptions about received pilot training need to be documented as Flight Crew Data (FCD) and communicated to operators (CS-FCD).

To avoid negative consequences of unsuitable automation, CS 25.1302(c) introduced requirements for 'operationally-relevant behaviour' to be '... (1) predictable and unambiguous and (2) designed to enable the flight crew to intervene in a manner appropriate to the task ...)' so that '... qualified flight-crew members trained in the use (of the equipment) can safely perform their tasks associated with its intended function ...'.

Machine behaviour has to be correctly understood and match the operator's expectations to produce the designer's expected behaviour. The operator must predict what the machine is doing and whether it is enabled or disabled to avoid confusion. For this, operators have mental representations, or inner pictures, of how machines behave, which must match the actual function and align with the situation outside the machine and the operator's expectations. Such representations may, however, be difficult to elicit when asking operators, as they are often subject to highly automated processes and can only be inferred indirectly through methods that focus on cognitive assessments.

Abbott (2022) further highlights that aircraft employ not only one type of automation but many different types for various tasks. When combined, they bring together different benefits and vulnerabilities. Automation can be found in specific low-level tasks (e.g., direct inputs to an autopilot) or in higher-level tasks, such as navigation management by the use of a Flight Management System (FMS).

There are various ways to classify types and levels of automation, often referred to as taxonomies, and different authors focus on different aspects. Billings (1996) distinguished control, information, and management. Save and Feuerberg (2012), in the context of the SESAR automation framework, highlighted differentiated levels combining Sheridan's and Verplank's (1978) taxonomy regarding how much of tasks are done by humans versus the machines, and the previously mentioned information-processing steps of Parasuraman et al. (2000). Endsley (2017) largely detailed a taxonomy to emphasise collaboration dynamics.

Each framework supports different aspects of the design process, from early task allocation decisions to final product validations. But whatever the taxonomy, its choice should match the design context: some situations require focus on automation levels, others on task types, and still others on collaboration patterns.

3.6.6 Human Error

AMC 25.1302 defines human error as

> a deviation from what is considered correct in a given context, especially with the hindsight of the analysis of accidents, incidents or other events of interest. Types of human error include an inappropriate action, a difference from what is expected in a procedure, an incorrect decision, an incorrect keystroke or an omission.

Ideally, any design shall guide the operator to easily conduct the required tasks by interacting with the machine and procedures without leading to errors. Human error on the flight deck can have critical consequences, which makes error-minimising design essential, especially in situations with potentially severe outcomes.

For machine malfunctions, AMC 25.1309 requires a logical and acceptable inverse relationship between the severity of the effect of a failure condition and the allowed probability. For example, the allowed probability for a minor failure condition is 1e-3 per flight hour, whereas the allowed probability for a catastrophic failure condition is 1e-9 per flight hour. The severity of such failure conditions is typically determined in a Functional Hazard Assessment (FHA), which AMC 25-11 Appendix 3 defines as 'a systematic, comprehensive examination of aeroplane and system function to identify potential Minor, Major, Hazardous, and Catastrophic failure conditions that may arise as a result of malfunctions or failures to function'. Specific engineering methods, for example, Markov analysis or Fault Tree Analysis, are used to determine the quantitative probability of failure conditions.

While machine failures can be predicted by using quantitative probabilities, this is not feasible for human errors. As Thomas and Abbott (2024) argue, probability-based techniques do not capture the range and real-world context of human error, nor do they explain why a decision or action made sense to a specific person at a specific point in time.

The flight deck has the potential to be the source of a range of errors, varying in severity. For example, improper interpretation of information may subsequently lead to improper actions or wrong control inputs. Another kind of error may be engendered by a distraction that interrupts a task, leaving it incomplete if the crew forgets to resume it. Errors can be characterised in different ways, for example, by the cognitive process involved (internal or external to the human), by their causes, or by their consequences.

CS 25.1302 (d) declares that the crew has to be enabled '. . . to manage the errors resulting from the kinds of flight crew interactions with the equipment that can be reasonably expected in service, assuming the flight crew acts in good faith'. AMC 25.1302 emphasises that pilots must be able to detect and recover from errors, which is facilitated by alerts and established procedures. So any observed or flight crew-reported human error during evaluations has to be analysed in order to avoid errors by design as encouraged by CS 25.1302 (a)(b) and (c). First, the source of an error must be understood before identifying the most appropriate mitigation. When humans make errors in their tasks, it can be for many different reasons. For example, at the perception level, information may not have been detected due to attention being directed to the wrong source, or it may not have been well understood or integrated with other information. Or, decision-related errors may occur when an inadequate solution is selected due to a wrong mental representation of the problem or a lack of knowledge.

For that purpose, as with addressing workload, it is necessary to collect both objective performance data and observations, as well as subjective data (e.g., comments provided during interviews and debriefings), to assess human error. AMC 25.1302 highlights that '. . . observables should not be limited to human errors, but should also include pilot verbalisations in addition to behavioural indicators such as hesitation, suboptimal or unexpected strategies and inappropriate use of controls.

In addition to error observation during evaluations/tests and data collected through debriefs, errors can also be anticipated when creating new designs through various predictive Human Error Analysis (HEA) techniques.

In either case, their objective is to provide robust qualitative insights into plausible flight crew errors that may occur, given the phase of flight, environment, and operational context, and to allow for the assessment of the potential consequences and outcomes of errors, as well as mitigation strategies (Thomas & Abbott, 2024).

GAMA Publication #22 provides a methodology for conducting Flight Crew HEA. It is suggested that an HEA can be structured similarly to an FHA. However, a flight crew error should not be treated like a machine failure. A HEA typically starts with a task analysis and then identifies reasonable error modes. It then assesses the flight crew's ability to manage the flight crew error and determines its severity (with and without mitigations). Each HEA should include the identification and documentation of HF assumptions regarding flight crew perception, decision-making, and response to the error (GAMA, 2024).

Although determining the quantitative probabilities of machine failure, for the FHA seemed comparatively straightforward, determining the appropriate allowable probabilities depends on correctly assessing the severity of the failure conditions' effects. This is not as straightforward as many assumptions about flight crew behaviour have to be made, including the flight crew recognising a failure condition, understanding the failure condition, making the correct decisions, and finally applying the correct response within the assumed time. Predicting pilot response time can be very challenging, as human responses to a stimulus involve several phases, each of which may be affected by different factors. Total response time is the combination of the times required to complete each of these phases (detection, recognition, response formation, and response execution), as described in Fercho et al. (2024).

Many assumptions can go wrong during this process. And the level of confidence for each assumption may vary greatly. For that reason, it is essential to document all assumptions with sufficient detail and to validate or correct them as needed. The EASA Certification Memorandum CM-SA-002 (EASA, 2023) offers a structured methodology for an HF validation of assumptions about flight crew behaviour in an FHA. Additional methods exist to validate the classification of failure conditions through HF simulator and/or flight evaluations/tests, for example, Taylor et al. (2022).

3.7 EXAMPLES OF HF INTEGRATION INTO A SYSTEMS ENGINEERING PROCESS

The following section presents two examples that illustrate how HF activities are integrated during the development of a new flight deck function, from the problem statement, through the definition of the intended function and the design concept, to proof of concept. It also highlights the required skills for flight deck HF practitioners involved in such projects.

The first example provides an overview of the development of a ROAAS by Honeywell and Gulfstream, as described in Konrad et al. (2018).

Problem statement: Runway overruns during landing have been widely recognised as one of the highest operational risks in CAT and are a major contributor to aeroplane accidents. Data from FSF (2009) shows that, between 1995 and 2008, an average of approximately 30% of worldwide CAT accidents (i.e., 431 accidents) were runway-related. Of those 431 accidents, approximately 97% were runway excursions. 712 people died in runway excursion accidents over the 14-year period. Thus, any solution that helps pilots to avoid this type of accident has immense safety value.

Intended function: The intended function of ROAAS was defined to first provide flight crew awareness of aircraft stopping-point(s) relative to the approaching runway, based on real-time aircraft energy state and, second, to aid flight crew decision-making for go-around and for timely use of all available stopping devices during a pending runway overrun situation.

Design concept: Based on the intended function, a preliminary design was developed, including graphical indications of predicted aircraft stopping points on the runway and visual and aural alerts indicating a possible overrun.

Proof of concept: Once the initial requirements were developed, they had to be validated. An essential part of such validations, in addition to various types of analyses (e.g., task analysis, HEA, and function allocation analysis), are part-task and scenario-based HF evaluations using low-, medium-, and high-fidelity simulators and/or aircraft. The HF practitioners had to ensure early on that the final ROAAS would be certifiable. This included identifying applicable regulations, for example, for workload in normal and non-normal situations (CS 25.1523, CS 25.1309, CS 25.1302), for human error (CS 25.1302, CS 25.1309), and for alerting (CS 25.1322). Note that regulation CS 25.705 did not exist when ROAAS was developed. Defining clear evaluation objectives was the key to the success of the validation effort. For ROAAS, these included assessing aspects such as the ability of visual and aural alerts to gain the flight crew's attention, the impact of symbology on cognitive workload, and the overall impact on go-around decision-making. Customised operational scenarios needed to be designed to exercise all HF evaluation objectives. Choosing the right methods was essential to collect the data needed to support the design decisions. The selected methods included the Bedford scale for workload, the Haworth-Newman scale for readability and understandability of text, the System Usability Scale (SUS) for overall system usability, and objective measures, such as missed alerts, time to react, and execution of go-around procedures. For post-evaluation data analysis and due to the small sample sizes, typical of HF evaluations in the aerospace industry, nonparametric descriptive statistics were applied to assess objective data and determine pilot consensus for rating scales and questionnaires. Equally important was identifying assumed pilot characteristics, such as qualifications, training, and experience, as well as age and anthropometric range.

Skills: The project showed that creating a solid problem statement required developers' knowledge and experience beyond HF. Understanding runway overruns required knowledge of aircraft performance. Operational knowledge was required to understand how approaches are flown and how pilot workload is affected. Once this was understood, the HF practitioners could define the intended function of ROAAS (required for compliance with CS 25.1301 and CS 25.1309). When the actual design began, the challenge for the HF team was not only to fit the indications into a very limited available space on a previously certified PFD, but also to make them easy to interpret and aesthetically pleasing, while adhering to a predefined flight deck philosophy. To validate the design, a strong foundation in flight operations was required to help craft the evaluation scenarios.

The second example, taken from Biede et al. (2018), describes a Flight Formation (FF) function for commercial aircraft.

Problem Statement: Placing an aircraft in the upwash of wake vortices from a preceding aircraft results in thrust reduction for the following one, enabling reduced fuel burn and emissions. It, however, poses the challenge of safely placing the aircraft in this upwash. Although Airbus has already demonstrated the in-flight applicability of this concept, it remains in the research phase, as many questions need to be addressed due to its wide-reaching impact on the aviation community. Again, HF plays a key role in the design and assessment of this evolution.

Intended function: The preliminary objective was to optimise flight and air traffic efficiency through the introduction of commercial FF based on the natural principle of energy optimisation from birds.

Design concept: The first key step was defining and refining the concept of operation (CONOPS). To perform FF operations, aircraft operators first analyse and plan an operation to arrive at an aircraft pairing scheme. This affects the tasks, roles, and responsibilities of many operators throughout the aviation system (e.g., pilots, ATC, and airline operators), as well as additional impacted organisations. On the flight deck, this requires specific procedures for pilots to plan, join, and maintain aircraft trajectories in formation flight, as well as close collaboration with ATC to provide separation from surrounding aircraft and to regulate the flow of aircraft. This operation leads to new tasks and changes to existing ones, all of which affect the pilot's interaction with the machine. Consequently, a large number of potential HF issues and benefits emerge for further study, including those related to task integration. Introducing new functions to automatically position and maintain the trailing aircraft in the optimum position near the vortex generated by a leading aircraft raised questions about the ability to predict automation unambiguously (CS 25.1302), and to intervene in tasks accordingly. The trailing aircraft has to be placed in the optimum position near the vortex generated by a leading aircraft to benefit from the vortex's energy. At the same time, sufficient distance must be maintained to ensure protection against potential wake-vortex encounters and mid-air collision risks. ATC procedures related to separation assurance must be smoothly integrated into tasks, not only for pilots but also for ATC, without leading to an unacceptable increase in the flight crew's workload (CS 25.1523).

Proof of concept: For such a highly novel, complex, and integrated design, validation occurs in several steps, including a series of design workshops with user representatives, studies across various platforms (simulators and aircraft), and collaboration with partners representing the affected operator and stakeholder groups. The first low-fidelity study was conducted with eight flight test pilots in a simulator environment implementing the new operations. Its objective was to investigate the feasibility of the proposed CONOPS and identify additional communication and information needs, based on interviews, questionnaires, and observations. An explorative analysis was performed based on observed behaviours, subjective feedback, and comparison of descriptive statistics. Perceived workload and usability assessments showed that pilot workload remained within median ranges and at acceptable levels during formation flights. However, improvements were required to optimise procedures. Furthermore, it is worth noting that this study took place in the early design phase, with detailed designs for both normal and non-normal operations not yet available. Achieving operational maturity will require further comprehensive assessments of all HF areas discussed in this chapter, including detailed evaluations of task integration, automation design, situation awareness, and error management, to be conducted in subsequent phases.

Skills: This project requires a broad skill set covering both wide operational knowledge, not only of the flight deck but also of global aircraft and ATM operations. Especially to address the diversity of HF problems, a combination of specific competencies in the human sciences is required to address cognitive, physical, physiological, and linguistic questions.

Both examples highlight the importance of early assessment of potential HF issues to identify the right design requirements and avoid later rework. HF studies scheduled at the right time contribute to cost management. For highly innovative settings with new operational concepts and technical advancements, this may even involve complex stakeholder management and collaboration within a multidisciplinary team with diverse competencies. All required HF activities are needed to align with the company's overall engineering process.

3.8 PITFALLS AND HOW TO AVOID THEM

Even if the need and approach for addressing HF are now more widely established in the aerospace industry, in daily practice, numerous pitfalls still exist.

The goal of this section is to highlight common misunderstandings and provide recommendations to address them. It highlights the risk that engineers may oversimplify HF concepts, fail to adequately address human complexity and variability throughout the engineering process, and thus disconnect system engineering from HF.

3.8.1 Oversimplification

People not trained in the particularities of human behaviour may often think of HF as common sense (i.e., folk psychology). Humans tend to project their way of thinking and behaving onto others. Consequently, significant risk lies in underestimating the complexity and variability of human behaviour, which may result in the pursuit of a design based on inadequate decisions. It is a human bias to believe that the expressions of one individual can be generalised to others. In so doing, they overlook the fact that even if operators are well-trained in their tasks, humans may have many reasons to deviate from expected tasks or to interpret their environment in ways other than those assumed. Therefore, properly understanding and interpreting how humans function in interaction with a machine within the environmental and operational context requires scientific knowledge of HF.

The following disciplines constitute the essential knowledge required for the HF's focus in flight deck design (Biede et al., 2023):

- Physical ergonomics, as defined by the International Ergonomics Association (IEA) (2025), concerns human anatomical, anthropometric, physiological, and biomechanical characteristics related to physical activity;
- Psychology focuses on the study of the mind and behaviour and covers cognitive (e.g., perception, attention, language, and memory in the interaction with a machine), emotional, physical, and social aspects (APA, 2025);

- Neuroscience specifically studies the structure and functions of the human brain and the nervous system;
- Anthropology focuses on physical and cultural human evolution, while sociology focuses on social groups, relations, and organisation; and
- Language sciences (e.g., linguistics) study how humans communicate using language, including verbal aspects (e.g., phonetics), para-verbal aspects (e.g., intonation), graphical aspects (e.g., typography), as well as other symbols used to convey information (e.g., colour codes, pictograms).

However, as shown in Konrad et al. (2022), HF knowledge alone is not sufficient for the successful flight deck HF practitioner. A solid understanding of the following engineering disciplines is as important:

- Aircraft design and operation, including knowledge about aircraft performance, aircraft systems, flight deck integration, flight operations, and pilot training;
- Systems engineering, including knowledge about the systems development process, system safety, requirements definition, validation and verification; and
- Aircraft certification, including knowledge about Certification Specifications (CS), Acceptable Means of Compliance (AMC), and compliance demonstration.

HF engineering provides concrete methodologies for applying the knowledge of the above human science disciplines to the design of products, environments, and operator work.

3.8.2 Addressing Human Variability with Inadequate Competencies

Leonardo da Vinci is credited with the quote: 'He who loves practice without theory is like the sailor who boards a ship without a rudder and compass and never knows where he may be cast' (Leonardo Da Vinci's Life, 2025). Similarly, practising HF is not possible without a solid theoretical foundation.

Adequately addressing the complexity of the human element throughout the process is a real challenge, as it requires translating knowledge from the aforementioned human sciences into practical design guidance for engineers.

Experience has shown that a robust skill set for HF practitioners who design, assess, and certify flight decks is essential. It is a critical factor in ensuring safety, regulatory compliance, usability, and overall system performance, as well as in providing a superior pilot experience. However, the industry faces a shortage of competent flight deck HF practitioners. This is exacerbated by the ageing workforce, leading to a loss of mentors and an increased global demand for HF practitioners. Most universities do not provide education that enables graduates to move directly into flight deck HF, which exacerbates the situation (Konrad et al., 2022).

Recent European (Biede et al., 2023) and US initiatives (Konrad et al., 2022) have developed frameworks to clarify the competencies required to achieve this transition. HF practitioners require a combination of knowledge to identify the right human aspects relevant to specific systems and operational evolutions.

For flight deck HF, the challenge is identifying the relevant scientific knowledge to define the right systems and tuning academic knowledge for efficient application in industry. Organisations use different models to integrate HF into their teams. Either HF practitioners are organised in dedicated teams, or they are combined with other competencies (e.g., operations or safety).

Some organisations may require a minimum of a master's degree in human sciences for psychologists, physiologists, neuroscientists, and linguists who work in multifunctional teams with engineering competencies. Other aerospace companies may choose different staffing approaches, often with a more diverse team, focusing on achieving a balance between practitioners with advanced degrees in the human sciences who have a strong understanding of aerospace applications and practitioners with advanced degrees in aerospace engineering who have a strong understanding of the human sciences.

Another challenge concerns competencies with similar or overlapping skill sets. For example, over the past few decades, the design-focused competence of User Interface/User Experience (UI/UX) has emerged from the fields of computer science and industrial design. Similarities in some methods, misunderstandings about competency profiles, and the language used have led to considerable confusion within the field.

Many different competencies are represented in aviation, but each has limitations that can only be overcome when combined in multidisciplinary teams, which are necessary to solve complex problems (Konrad et al., 2022). Whereas the human sciences encompass the physical and physiological, as well as the psychological and cognitive aspects of HF, the operational side of HF is driven by the operators (i.e., pilots). Additionally, trained engineers bring the necessary technical expertise in systems engineering and certification.

3.8.3 Late, Disconnected, and Inadequate Ways of Working

As highlighted in Section 3.4, knowledge of HF and methods must be systematically integrated into the engineering process. In systems engineering, the tendency is to overlook the human element, as biases and false beliefs may lead to an underestimation of the impact of human variability. In addition, large industrial organisations, such as large aircraft manufacturers or large avionics system suppliers, are often under pressure and spread across national frontiers. If organisations plan HF studies too late, if at all, it can lead to the late discovery of human factors risks requiring significant effort to mitigate them.

Due to the complexity of the topics to be addressed, individual experts can rarely cover the full range of required knowledge. In a complex environment, people from various disciplines often work together, frequently across international boundaries. Collaboration challenges are exacerbated by different academic cultures (e.g., French, German, and US-American), which have not evolved the same way and are not necessarily grounded in the same theories, potentially leading to additional misunderstandings. One such misunderstanding concerns where to start the HF process. The ergonomics' approach starts with the understanding of the user needs in the task context, whereas an immediate focus on investigation of task impacts on criteria such as workload and human error might be prioritised by HF practitioners with psychology backgrounds.

Additionally, HF is a relatively small discipline in industrial applications compared to other technical fields, and it is often centralised around specialised consulting agencies or academic institutions. For this reason, HF practitioners often require connections outside their organisations to exchange information about professional experience and evolving activities. For example, both Europe and the United States have developed dedicated associations for aviation psychology and provide certificates to aviation psychologists and aviation HF practitioners (e.g., the European Association of Aviation Psychology (EAAP) and the Board of Certification of Professional Ergonomics (BCPE).

3.8.4 Handling the Assessment Challenge

As with any technical field, when dealing with humans in design, it is crucial to follow a scientific approach to collect data and make informed decisions. For many of the HF topics introduced in Section 3.6, specific measures have been developed in academic research following well-defined scientific approaches. Additionally, generations of HF practitioners from the human sciences have undergone scientific and academic training in developing, applying, and interpreting measurements.

However, the challenge is to effectively apply academic research measures in an industrial environment, as producing scientifically valid data is crucial for informed decisions on design maturity. Furthermore, the use of a measure shall not be limited to determining compliance with regulations (e.g., whether the workload is acceptable) but shall require a more in-depth analysis of the underlying mechanisms. In industrial settings, it is not feasible to use, for instance, five different rating scales, each taking five minutes to complete, in the context of simulator-based evaluations. Quantitative measures and balanced experimental protocols may not always be feasible.

It could require finding different approaches to measurement, such as decomposing indicators to a level that clearly links a design change to the measurable impact on humans. Alternatively, a combination of quantitative, experimentally controlled studies and qualitative, scenario-specific simulator exposures can support robust design decisions.

3.8.5 Recommendations for Prevention Strategies

These above observations lead to the following recommendations for preventing potential pitfalls.

Oversimplification:

- Development of awareness for engineers on the large number of properties characterising humans in the system;
- Development of awareness among specialists in the human sciences to systematically explain specific human properties and variabilities, and their relationships to a design item throughout the engineering process.

Addressing human capabilities, limitations, and variability with inadequate competencies:

- Specific competencies must be available, developed, and maintained;
- Be aware of the competencies of others with whom one is collaborating; and
- Ensure to address the right topic with the right competence.

Late, disconnected, and inadequate ways of working:

- Ensure the diversity of factors impacting human behaviour in specific design projects is well understood and integrated into the project plan right from the beginning, at the right time, and with the right means. Priorities may need to be decided depending on their impact. An example is sound on the flight deck, which requires knowledge of physiological limitations, information processing, and the characterisation of sound properties.
- Involve representative end users in participatory design throughout the project. It is also recommended to involve certification authorities early in the process.
- Support distributed but connected teams in developing a mindset of open communication, adaptiveness, and, most of all, the ability to listen and deal with conflicts.
- Connect with other specialists inside and outside the company.

Handling the assessment challenge:

- Systematically identify the most suitable approach from all possible methodological sources, not only from the ones with which an HF practitioner has experience.
- Optimise the integration of measures throughout the development cycle to use the right measure at the right time, not just for a single assessment.
- Have the assessments checked against expenses and expected outputs.

3.9 CONCLUSION AND CHALLENGES TO COME

The challenges students might encounter in the HF discipline within the aerospace industry are not exhaustive, nor do they aim to be. A large body of specialised literature exists to detail various topics, and many references have been provided in this chapter. However, it is crucial to understand how to integrate these concretely into design, link them to regulations, and avoid potential pitfalls that humans can easily fall into due to their specific way of functioning. As such, this chapter can help to better prepare students for entry into their professional lives and facilitate their 'entry into service'. It may also help to prepare for the complexity of the context that newcomers can encounter in daily practice. However, it requires passion, curiosity, and an openness to learn, to connect with and improve other disciplines, and to find one's best way of working to get the system right.

Since the first powered, heavier-than-air, manned flight in 1903, HF has made significant progress in society and industry. Today, many companies have come to understand the value of building products adapted to users' needs. However, meeting these needs remains challenging amid the ever-increasing complexity of aircraft and operations. A tendency to over-specialise in specific areas, even in HF, leads to a risk of losing the big picture of what is needed. Despite increased automation and autonomy, human operators will remain in charge. Adding new technologies will make the system even more complex, and it may become increasingly difficult to identify flaws before they result in operational issues. HF issues may be more difficult to detect, requiring a deeper understanding of the system to identify the connections between all its elements.

Techno-centred design will persist, despite the human-centred focus set forward, for example, by the European Commission (2021). Engineers will naturally continue to innovate, creating new tools, devices, and machines, possibly without considering whether a real user need is associated with them. This practice is important because it generates new ideas and insights. However, as history shows, a purely techno-driven approach to automation has not proven to be successful without the integration of a human-centred view.

Industrial aircraft development takes years, and the commercial nature of the process demands that early development work protect proprietary information. A design may undergo many changes years before it is finally presented to the certification Authority to find compliance. Also, it is rare for industrial commercial aircraft to be 'all-new'; therefore, HF practitioners may find themselves constrained by the burden of commonality with previous designs. It may have the benefit of reduced cost and ease of certification, but it constrains the optimisation of HF.

Finally, a rapidly expanding set of new challenges is associated with the widespread introduction of AI-based technologies, which bring new opportunities to improve the tools and interfaces provided to operators, but also potentially pose new challenges for operators to adopt these evolutions. AI enables processing information that current technology cannot handle, and it is expected to profoundly alter how human operators collaborate with systems, shifting many current paradigms of operator-machine interaction.

EASA (2024) provides the first guidelines for addressing AI's specific capabilities from an HF perspective, and Kirwan (2025) reflects on a first set of requirements for Human-AI-Teaming (HAT). However, new questions arise for HF practitioners: What should they consider or do differently in design analysis and evaluation with AI on the flight deck? How do CSs and AMCs need to change – or do they need to change at all? For example, CS 25.1302 (c)(1) requires that the operationally relevant behaviour of the installed equipment must be predictable and unambiguous. Can this be claimed for AI?

Some aspects, such as explainability, may already be addressed in AMC 25.1302 but may not be explicitly stated. How can specific technology characteristics be integrated into the engineering process?

To be able to meet these challenges in addition to the already existing ones today, it requires concrete initiatives for:

- Avoiding the common pitfalls outlined in Section 3.8 by reinforcing competency development in both human sciences and aviation technology, and building relationships across engineering disciplines early in engineering careers;
- Including regulatory requirements (CS 25.1302) in foundational training; and
- Practising the integration approaches described in Section 3.7.

Despite the challenges of working in such a complex field as aviation, humans will remain central to its future. HF practitioners make key contributions to the 'system' by complementing the machine, environment, and operation, by specifically addressing the human element. Since its inception, aviation has sparked a profound passion in humans to make aircraft fly and explore the world. In HF, this has found one of its focuses, as proper design for humans is the fundamental basis for achieving this passion for flying and satisfying societal as well as individual needs.

REFERENCES

Abbott, K., McKenney, D., & Railsback, P. (2013). Operational use of flight path management systems–Final report of the Performance-based operations Aviation Rulemaking Committee. Commercial Aviation Safety Team Flight Deck Automation Working Group. https://www.faa.gov/sites/faa.gov/files/aircraft/air_cert/design_approvals/human_factors/OUFPMS_Report.pdf

Abbott, K. H. (2022). Automated systems in civil transport airplanes: Human factors considerations. In J. R. Keebler et al. (Eds.), *Human Factors and Aerospace* (3rd ed., pp. 373–397). Elsevier.

Abbott, K. H. (2025). *A Primer on the US Aviation Regulatory System*. Washington, DC: FAA.

Ahlstrom, V. (2016). *Human Factors Design Standard* (DOT/FAA/HF-STD001B). Egg Harbor Township, NJ: Federal Aviation Administration William J. Hughes Technical Center. https://hf.tc.faa.gov/publications/2016-12-human-factors-design-standard

American Psychological Association (2025). *In APA Dictionary of Psychology*. https://dictionary.apa.org

Aviation Safety Network (ASN) (2025). https://aviation-safety.net/database/record.php?id=20020701-0&lang=es

Bach, C. & Biede, S. (2016). Assessing Needs from a Human Factors perspective when designing complex systems. In Proceedings of the 32nd Meeting of the European Association for Aviation Psychology.

Bainbridge, L. (1983). Ironies of automation. *Automatica*, 19(6), 775–779.

Bateman, D. (2010). *Some Thoughts on Reducing the Risk of Aircraft Loss of Control*. AIAA.

Biede, S., Detaille, F., Narota, T., Petrovic, K., Rathje, H., Rea, A., Schwarz, M., & Vereker, A. (2023). A competency framework for aviation psychologists and human factors specialists in aviation. *Aviation Psychology and Applied Human Factors*, 13(1), 58–68.

Biede, S., Lebas, J., & Petit, E. (2018). Challenges of early identification of human factors impacts for new concepts of operations: An example for the airborne view on commercial formation flight operations. In Proceedings of the 33rd Conference of the European Association for Aviation Psychology (EAAP33), Dubrovnik, Croatia, September 2018.

Billings, C. E. (1996). *Human-Centered Aviation Automation: Principles and Guidelines*. NASA Technical Memorandum 110381. Santa Clara County, CA: Ames Research Center.

Billman, D., Mumaw, R., & Feary, M. (2021). A model of monitoring as sensemaking: Application to flight path management and pilot training. *Proceedings of the Human Factors and Ergonomics Society Annual Meeting*, *64*(1), 244–248 (original work published 2020). https://doi.org/10.1177/1071181320641058

Boeing (2025). *Statistical Summary of Commercial Jet Airplane Accidents - Worldwide Operations 1959-2024*. https://www.boeing.com/content/dam/boeing/boeingdotcom/company/about_bca/pdf/statsum.pdf

Botargues, P. (2009). Airbus AP/FD TCAS mode: a new step towards safety Improvement. *Airbus Safety First*, 7. https://skybrary.aero/sites/default/files/bookshelf/90.pdf

Bubic, A., Yves von Cramon, D. & Schubotz, R. I. (2010). Prediction, cognition and the brain. *Frontiers in Human Neuroscience, Section Cognitive Neuroscience*, 4. https://doi.org/10.3389/fnhum.2010.00025

Certification Authorities for Large Transport Aircraft (CATA) (2018). *CATA Worklist Item (CWI)* EASA-003 – 25.1302. https://www.faa.gov/sites/faa.gov/files/2022-08/CWI_EASA_003.pdf

Civil Aviation Authority (CAA) (2002). *CAP 719 Fundamental Human Factors Concepts*. https://www.caa.co.uk/publication/download/12250

Coombs, L. V. E. (2005). *Control in the Sky - The Evolution and History of the Aircraft Cockpit*. Pen & Sword Books Ltd.

Cooper, G. E., & Harper, R. P. (1969). *The Use of Pilot Rating in the Evaluation of Aircraft Handling Qualities*. Washington, DC: NASA Technical Note. https://ntrs.nasa.gov/api/citations/19690013177/downloads/19690013177.pdf

Dailey Jr., F. E. (2004). *The Triumph of Instrument Flight*. Dailey International Publishers.

Degani, A., Barshi, I., & Shafto, M. (2013). Information organization in the airline cockpit: Lessons from flight 236. *Journal of Cognitive Engineering and Decision*, 7(4), 330–352. https://doi.org/10.1177/1555343413492983

Dizikes, P. (2024). Study: Flying keeps getting safer. *MIT News*. https://news.mit.edu/2024/study-flying-keeps-getting-safer-0807

Durso, F. T., & Alexander, A. L. (2010). Managing workload, performance, and situation awareness in aviation systems. In E. Salas & D. Maurino (Eds.), *Human Factors in Aviation* (2nd ed., pp. 217–247). Academic Press.

Endsley, M. R. (2017). From here to autonomy: Lessons learned from human–automation research. *Human Factors*, 59(1), 5–27. https://doi.org/10.1177/0018720816681350

Endsley, M. R. (2019). *Human Factors & Aviation Safety*. Testimony to the United States House of Representatives Hearing on Boeing 737-Max8 Crashes — December 11, https://docs.house.gov/meetings/PW/PW00/20191211/110296/HHRG-116-PW00-Wstate-EndsleyM-20191211.pdf

Endsley, M. R. & Jones, G. D. (2025). *Designing for Situation Awareness: An Approach to User-Centered Design* (3rd ed.). CRC Press.

European Commission Directorate-General for Research and Innovation (2021). *Industry 5.0 - Towards a Sustainable, Human-Centric and Resilient European Industry*. https://op.europa.eu/en/publication-detail/-/publication/468a892a-5097-11eb-b59f-01aa75ed71a1

European Committee for Standardisation (2014). *Value Management - Vocabulary - Terms and Definitions* (EN 1325:2014).

European Union Aviation Safety Agency (EASA) (2018). *European Plan for Aviation Safety (EPAS) 2019-2023*. https://www.easa.europa.eu/sites/default/files/dfu/EPAS_2019-2023%20final.pdf

European Union Aviation Safety Agency (EASA) (2021). *Certification Specifications and Acceptable Means of Compliance for Large Aeroplanes (CS-25, Amendment 28)*. https://www.easa.europa.eu/en/document-library/certification-specifications/cs-25-amendment-28

European Union Aviation Safety Agency (EASA) (2023). *Final Certification Memorandum ref. CM-SA-002 Issue 02 on "Flight Crew Human Factors Assumptions in Aircraft and System Safety Assessments"*. Applicable to Large Aeroplanes. https://www.easa.europa.eu/en/document-library/product-certification-consultations/final-certification-memorandum-ref-cm-sa-002

European Union Aviation Safety Agency (EASA) (2024). *EASA Concept Paper: First Usable Guidance for Level 1 and 2 Machine Learning Applications*. A deliverable of the EASA AI roadmap, Issue 2. https://www.easa.europa.eu/en/document-library/general-publications/easa-artificial-intelligence-concept-paper-issue-2

European Union Aviation Safety Agency (EASA) (2025a). *Easy Access Rules for Air Operations*. https://www.easa.europa.eu/en/document-library/easy-access-rules/easy-access-rules-air-operations-regulation-eu-no-9652012

European Union Aviation Safety Agency (EASA) (2025b). *Innovation Services*. https://www.easa.europa.eu/en/document-library/application-services/innovation-services#goodtoknow

European Union Aviation Safety Agency (EASA) (2025c). *FAQ n.19026*. https://www.easa.europa.eu/en/faq/19026

Federal Aviation Administration (FAA) (1999). *Guidance for Reviewing Certification Plans to Address Human Factors for Certification of Transport Airplane Flight Decks*. AIR Policy Statements, Memos and Deviations, PS-ANM111-1999-99-2. https://drs.faa.gov/browse/excelExternalWindow/4684B8382B57F18E8625774500705CB7.0001?modalOpened=true

Federal Aviation Administration (FAA) (2021). *Introduction to Aviation Physiology*. https://www.faa.gov/sites/faa.gov/files/2022-11/Intro_Aviation_Physiology.pdf

Federal Aviation Administration (FAA) (2025). *Ending Serious Close Calls*. https://www.faa.gov/closecalls

Feldman-Barrett, L. (2021). *Seven and a Half Lessons About the Brain*. Picador.

Fercho, K. A., Beringer, D. B., & Donovan, C. (2024). Factors that affect pilot response times to alerts: Findings from a literature review and aviation safety reporting system (ASRS) reports. *Proceedings of the Human Factors and Ergonomics Society Annual Meeting*, *68*(1), 50–55. https://doi.org/10.1177/10711813241260286

Fernandez, B., Lesceu, X, & Marconnet, D. (2017). Introduction to the Soft Go-Around Function. *Airbus Safety First* 23. https://safetyfirst.airbus.com/introduction-to-the-soft-go-around-function/

Fitts, P. M., Jones, R. E., & Milton, J. L. (1950). Eye fixations of aircraft pilots, IV. frequency, duration, and sequence of fixations during routine instrument flight. USAF Technical Report No. 5975. https://apps.dtic.mil/sti/pdfs/ADA329371.pdf

Flight Safety Foundation (2025). *80 Years of Aviation Safety*. https://flightsafety.org/asw-article/80-years-of-aviation-safety-2

Flight Safety Foundation (FSF) (2009). *Reducing the Risk of Runway Excursions.* https://skybrary.aero/sites/default/files/bookshelf/900.pdf
Gawron V. (2019). *Human Performance, Workload, and Situational Awareness Measures Handbook* (3rd ed.). CRC Press.
General Aviation Manufacturers Association (GAMA) (2023). *Boeing 737 MAX Related Reports & Recommendations and their Impact on Human Factors.* GAMA Publication #21. https://gama.aero/wp-content/uploads/GAMA-737MAX-HF-Report-05242023.pdf
General Aviation Manufacturers Association (GAMA) (2024). *Predictive Flightcrew Human Error Analysis - A Report from the GAMA Flight Deck Human Factors Working Group*, GAMA Publication # 22.
Hilburn, B. & Jorna, P. (2001). Workload and air traffic control. In P. A. Hancock & P. A. Desmond (Eds.), *Stress, Workload, and Fatigue* (pp. 384–394). Lawrence Erlbaum Associates.
Holder, B., Finseth, T., & Lubold, N. (2024). Cognitive skills for flight path management. *Journal of Cognitive Engineering and Decision Making, 18*(3), 209–240. https://doi.org/10.1177/15553434241248298
Human Factors and Ergonomics Society (HFES) (2021). *Human Readiness Level Scale in the System Development Process.* ANSI/HFES 400-2021.
International Ergonomics Association (IEA) (2025). *What is ergonomics (HFE)?* https://iea.cc/about/what-is-ergonomics
International Organisation for Standardisation (ISO) (2019). *Ergonomics of Human-System Interaction Part 210: Human-Centred Design for Interactive Systems* (ISO Standard No. 9241: 2019). https://www.iso.org/standard/77520.html
International Organisation for Standardisation (ISO) (2023). *Systems and Software Engineering — System Life Cycle Processes* (ISO/IEC Standard 15288). https://www.iso.org/standard/81702.html
Jacob, A., Lignée, R., & Villaumé, F. (2009). The runway overrun prevention system. *Airbus Safety First.*
Joint Authorities Technical Review (JATR) (2019). *Boeing 737 MAX Flight Control System – FAA Certification Observations, Findings, and Recommendations.* https://www.faa.gov/sites/faa.gov/files/2021-08/Final_JATR_Submittal_to_FAA_Oct_2019.pdf
Kahneman, D. (2012). *Thinking, Fast and Slow.* Penguin.
Kemp, A., Konrad, G., Majumdar, N., & Feyereisen, T. (2025). Development of a competency-based short course syllabus for flight deck human factors practitioners. *Transportation Research Procedia*, 88, 176–184. 10.1016/j.trpro.2025.05.022
Kearney, K. (n.d.). *Five Top Flight Deck Innovations.* Honeywell Aerospace Technologies. https://aerospace.honeywell.com/us/en/about-us/blogs/five-top-flight-deck-innovations#:~:text=In%201912%20Lawrence%20Sperry%2C%20son,an%20airplane%20stable%20during%20flight
Kirwan, B. (2025). Human factors requirements for human-AI teaming in aviation. *Future Transp.*, 5(2), 42. https://doi.org/10.3390/futuretransp5020042
Komite Nasional Keselamatan Transportasi (2019). *Aircraft Accident Investigation Report, Lion Mentari Airlines, Boeing 737-8 (MAX).* PKLQP, FINAL KNKT.18.10.35.04, PT, https://studylib.net/doc/26109698/2018—035—pk-lqp-final-report
Konrad, G. (2025). *Human Factors Engineering in Civil Flight Deck Design.* Aerospace Short Courses, Jayhawk Global, The University of Kansas.
Konrad, G., Khatwa, R., Feyereisen, T., & Kemp, A. (2022). Development of a flight deck human factors design and certification competency taxonomy: preliminary findings. *Transportation Research Procedia*, 66, 279–291. https://doi.org/10.1016/j.trpro.2022.12.028
Konrad, G., Landers, T., Martin, A., Taylor, S., Feyereisen, T., Johnson, S., He, G., & Khatwa, R. (2018). Development of a predictive runway overrun awareness and alerting system. In Aviation Electronics Europe Conference Proceedings.
Koradecka, D. (2010). Introduction. Occupational Safety and Health. From the past, through the present, into the future (xiii–xxi). In D. Koradecka (Ed.), *Handbook of Occupational Safety and Health.* CRC Press.
Lee, R. (2014). *Human Factors in Flight Safety; (Training Course).* Barcelona, Spain: European Association for Aviation Psychology.
Leonardo da Vinci's Life (2025, Oct. 10). *Da Vinci Quotes.* www.davincilife.com/davinci-quotes.html
Mindell, D. (2002). *Between Human and Machine.* Johns Hopkins Studies in the History of Technology. The Johns Hopkins University Press.
NASA (2016). *NASA Systems Engineering Handbook* (NASA SP-2016-6105, Rev2).
National Test Pilot School (NTPS) (2013). *Cockpit Development & Assessment, a Test Pilot School Perspective.* Empire Test Pilot School.
Norman, D. A. (2013). *The Design of Everyday Things.* MIT Press.

Palmer, M. T., Rogers, W. H., Press, H. N., Latorella, K. A., & Abbott, T. S. (1995). *A Crew-Centered Flight Deck Design Philosophy for High-Speed Civil Transport (HSCT) Aircraft*. NASA Technical Memorandum 10917.

Parasuraman, R., & Manzey, D. H. (2010). Complacency and bias in human use of automation: an attentional integration. *Human Factors*, *52*(3), 381–410. https://doi.org/10.1177/0018720810376055

Parasuraman, R., Sheridan, T. B., & Wickens, C. D. (2000). A model for types and levels of human interaction with automation. *IEEE Transactions on Systems, Man, and Cybernetics-Part A: Systems and Humans*, 30 (3), 286–297. https://doi.org/10.1109/3468.844354

Parasuraman, R., Sheridan, T. B., & Wickens, C. D. (2008). Situation awareness, mental workload, and trust in automation: Viable, empirically supported cognitive engineering constructs. *Journal of Cognitive Engineering and Decision Making*, 2(2), 140–160. https://doi.org/10.1518/155534308X284417

Pelchen-Medwed, R., Save, L., Heintz, A., Reuzeau, R., & Biede-Straussberger, S. (2021). The evolution toward a common Air/ground framework for Human Performance Assessments in Europe. In I. V. Koglbauer & S. Biede (Eds.), *Aviation Psychology - Applied Methods and Techniques* (pp. 1–16). Hogrefe Publishing.

Rasmussen, J. (1983). Skills, rules, and knowledge; signals, signs, and symbols, and other distinctions in human performance models. *IEEE Transactions on Systems, Man, & Cybernetics, SMC*-13(3), 257–266. https://doi.org/10.1109/TSMC.1983.6313160

Rasmussen, J., Pejtersen, A. M., & Schmidt, K. (1990). *Taxonomy for Cognitive Work Analysis*, Riso, Denmark: Risø National Laboratory.

Reuzeau, F. (2024). Human and organisational factors integration in the aviation industry: maturity analysis and methodological recommendations to answer society trends. In I. V. Koglbauer & S. Biede-Straussberger (Eds.), *Aerospace Psychology and Human Factors* (pp. 259–275). Hogrefe Publishing.

Roberts, E. & Abbott, K. (2025). Investigating the under-reporting of automation dependency in air transport accidents worldwide. *Transportation Research Procedia*, 88, 217–224.

Rose, T. (2015). *The End of Average: How We Succeed in a World that Values Sameness*. HarperOne/ HarperCollins.

Save, L., & Feuerberg, B. (2012). Designing human-automation interaction: A new level of automation taxonomy. In D. de Waard, K. Brookhuis, F. Dehais, C. Weikert, S. Röttger, D. Manzey, S. Biede, F. Reuzeau, & P. Terrier (Eds.) *Human Factors: A View from an Integrative Perspective*. Proceedings HFES Europe Chapter Conference Toulouse. ISBN: 978-0-945289-44-9. http://hfes-europe.org

Schallert, C. (2016). Integrated safety and reliability analysis methods for aircraft system development using multi-domain object-oriented models. Doctoral thesis, Technische Universität Berlin. http://doi.org/ 10.14279/depositonce-4930

Schmitz, A. D., & Rodriguez, R. (2024). Single European sky air traffic management research case study: Where have we been and where are we going. In I. Kogelbauer & S. Biede-Straussberger (Eds.), *Aerospace Psychology and Human Factors – Applied Methods and Techniques* (pp. 245–258), Hogrefe.

SESAR Joint Undertaking (JU) (2020). *Human Performance Assessment Process V1 to V3 Including VLD*. Brussels: Belgium. https://www.sesarju.eu

Sheridan, T. B., & Verplank, W. (1978). *Human and Computer Control of Undersea Teleoperators*. Cambridge, MA: Man-Machine Systems Laboratory, Department of Mechanical Engineering, MIT.

Skybrary (n.d.). *EASA Regulatory Instruments*. https://skybrary.aero/articles/easa-regulatory-instruments

Stanton, N. A., Salmon, P. M., Rafferty, L. A., Walker, G. H., Baber, C., & Jenkins, D. P. (2013). *Human Factors Methods: A Practical Guide for Engineering and Design* (2nd ed.). Routledge.

Taylor, S. A., Rummel, A., Nilson, C., Palmsteen, S. & Schroeder, J.A. (2022). A modified rating scale for validating system failures in complex integrated flight decks. In AIAA SCITECH 2022 Forum, San Diego.

The University of Kansas (KU) (2025). *Aerospace Short Courses - (AERO0690)*, https://www.enrole.com/ kupce/jsp/session.jsp?sessionId=PA27125F&courseId=AERO0690&categoryId=10001

Thomas, L. J., & Abbott, K. H. (2024). *The Devil Between the Details: Limitations of Probability-Based Approaches to Human Error*. HCI International Late Breaking Papers.

Vidulich, M. A., Wickens, C. D., Tsang, P. S., & Flach, J. M. (2010). Information Processing in Aviation. In E. Salas & D. Maurino (Eds.), *Human Factors in Aviation* (2nd ed., pp. 175–215). Academic Press.

Wickens, C. D., & Hollands, J. G. (2000). *Engineering Psychology and Human Performance* (3rd ed.). Prentice-Hall.

Wittmer, A., & Bieger, T. (2011). Fundamentals and structure of aviation systems. In: A. Wittmer, T. Bieger, & R. Müller (Eds.), *Aviation Systems. Springer Texts in Business and Economics*. Berlin, Heidelberg: Springer.

4 User-Centred Design in Critical Operations
Every Second Counts

Stein Helgar, Nick Grigg, and Andreas Myskja Lien

INTRODUCTION

Missiles are fired. Drones are scrambled. You have the next minute to locate, assess, target and destroy dozens of incoming projectiles. The difference between an 80% and 100% success rate is the loss of key infrastructure, equipment, supplies, or even loss of life.

Every second counts.

Safety-critical systems are defined as systems whose failure could result in the loss of life, significant property damage, or damage to the environment (Knight 2002). Human-operated air defence systems exemplify such environments: extreme time pressure, complex information demands, and high cognitive workload place exceptional requirements on system design. In these contexts, human-centred design is not a matter of preference but a determinant of operational success and survival.

This chapter outlines the Halogen Design Methods for Critical Systems (HDMC) and how it was applied in the project *Next Generation Air Defence Console* (ADX), where Halogen, together with Hareide Design, assisted Kongsberg Defence & Aerospace in facilitating the development of the new NASAMs Fire Distribution Centre (FDC). The system is widely spread throughout NATO and is currently in operation in Ukraine.

As demonstrated in simulator evaluations conducted during ADX development, several operators achieved decision times of 3–5 seconds after adapting to the redesigned workflow, compared to 12–15 seconds in earlier baseline tests.

Halogen's approach enhances and builds upon existing 'human-centric' standards, such as ISO 9241-210 (2019), by integrating practical insights and the latest research findings. HDMC integrates the best methods from human-centred design, human-factors theory, and situation awareness (SA) theory to enhance decision-making in high-stakes scenarios. By focusing on understanding real user needs, acknowledging human capabilities and limitations, and prioritising critical information, HDMC aims to reduce decision-making time and operator error. The application of HDMC to the ADX process has demonstrated improvements in SA, workflow, precision, concentration, and training efficiency.

HDMC is designed to bridge a critical gap in existing human-centred design standards. While frameworks such as ISO 9241-210 offer high-level principles, they do not account for the cognitive, temporal, and organisational constraints that define safety-critical work. In air defence operations, milliseconds shape outcomes; therefore, design must explicitly address perceptual thresholds, cognitive load, decision latency, and shared SA. HDMC integrates these theoretical foundations into a coherent methodology tailored for high-risk, time-constrained operations.

This chapter will present the HDMC framework, methods, and tools, along with a practical example of its application.

The core phases of the HDMC design process are *1.insight, 2.concept and 3.design and development*. The process is in accordance with the design principle *What for Whom before How*

DOI: 10.1201/9781003741824-6

and builds on the ISO 9241-210 human-centred design process and IDEO's (2015) design processes.

HDMC emphasises defining the user's goals, perceptual constraints, cognitive limits, and coordination demands before proposing any technological solution. This principle is supported by Human Factors (HF) research showing that mis-specified user needs are among the most frequent root causes of system-related errors in safety-critical domains.

Although human-centred design provides the overarching framework , such as ISO 9241-210, they do not address the domain-specific cognitive, temporal, and organisational constraints that define safety-critical work. As noted in the earlier HDMC work, existing guidelines often fail to build a sufficiently deep understanding of the user's real decision environment, particularly under extreme time pressure, high data density, distributed teamwork, and risk of catastrophic failure. In such contexts, a more specialised framework is required – one that integrates human-factors science, SA theory, and observational insight methods to reveal the latent conditions that shape real operator performance.

The approach builds on empathy with the (human user) community, enables the designer to understand their needs, and subsequently builds designs that maintain integrity and substance for both the client and end user.

In the *insight phase*, we seek to learn the users' needs and desires, single out actual problems from perceived problems and create a common platform for understanding Figure 4.1.

In the *concept phase*, we explore and generate possible solutions, identify and verify design opportunities through testing and refinement.

In the *design and development phase*, we detail and validate design solutions before implementation.

The three phases are separated in theory, but in reality, it is not a strictly linear process. The designers gather insights throughout all phases, and ideas might appear outside the concept phase.

HDMC is therefore deliberately modular and iterative. Rather than prescribing a fixed sequence of activities, it provides a flexible 'toolbox' of methods that can be recombined depending on context, user group, and time-criticality. This reflects Schön's account of design as reflective practice, where insight emerges through cycles of action and reflection, and aligns with contemporary design research describing complex sociotechnical problems as 'wicked problems' (Buchanan, 1992; Coyne, 2005). In safety-critical domains, this iterative modularity is not a stylistic choice – it is a scientific necessity. Operator cognition, information demands, and teamwork structures shift throughout an operation, and the design process must accommodate this complexity by repeatedly validating assumptions with real users and real workflows.

This chapter is structured around how these three phases played out in the ADX project and what methodologies were applied throughout the process.

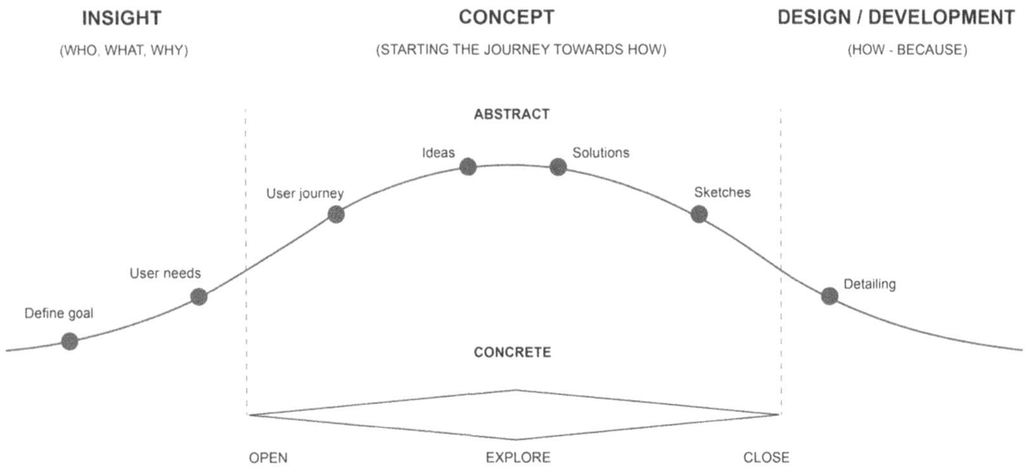

FIGURE 4.1 Insight, concept, design/development.

Illustration: Halogen.

HDMC PHASE 1: INSIGHT PHASE

As designers, our focus is to help users or operators achieve their objectives. We want to provide them with systems that truly meet their needs and enable them to complete their tasks in the most effective way possible. Before we think about how to solve their problem(s), we need to ask: What is the real problem? Are we focusing on a symptom or a cause? Who are we designing for?

Building on the 'What for Whom before How' principle, the insight phase focuses on the what for whom, uncovering real operator goals, constraints, and needs.

The purpose of the insight phase is to establish a thorough understanding of the user needs and the system as a whole. The designers use a number of techniques to find out what you know, what you don't know and possibly also get an idea of what you don't know that you don't know. By using visualisation techniques, designers uncover user context, operational constraints, workflow dependencies, and hidden organisational processes that influence performance.

In the insight phase, the team focuses on elements such as the following:

- **User**: Who are the operators? What are their training, experience, their roles, what is important to them, what challenges do they face, and how do they interact?
- **Workflow**: What tasks does each operator perform, what is their workflow, and where are the inefficiencies?
- **Information**: What information does each operator need, and at what point in time during their workflow?
- **Decision-making**: How does each operator make decisions, how does that impact the other operators and workflow, and what influences their decisions?
- **Situation Awareness**: How do operators perceive and process threats, and project the effect of their actions? What adds to their cognitive workload, and how are they aware of each other's task status?
- **Context**: What distracts them, gets in the way of their tasks, or impedes their movement, vision, and comfort?
- **Scope**: What is outside and inside the system – and what are the interfaces?

In his doctoral thesis, Jan Fredrik Schønheyder lists 12 different design methods used in the insight phase (Schønheyder, 2019). To a greater or smaller degree, the insight phase of the ADX project included activities from all these methods. In the coming section, we'll describe how the methods in italic were used in the project.

- *Field observations*
- *Semi-structured interview*
- *SA and team SA field observations*
- *Frequency and criticality mapping*
- *MTO analysis*
- Relations mapping
- Project insight mapping
- Political and operator stakeholder mapping
- Visual interview mapping
- Roles and context mapping
- Specific and complete timeline mapping
- ZIP analysis

From earlier insight activities, the designers knew that two operators were working side by side in a time-critical operation. But they had a lot of questions that they needed to dig deeper into: Where is each operator's focus? What is their workflow? How does the task flow from one operator to another? How does the room layout support or impede them?

In the ADX project, the initial activity was a kick-off mapping workshop. The workshop aimed to establish a common understanding of the project's operational needs and purpose. The workshop included visual mapping of the ecosystem around the NASAMs FDC (Figure 4.2), the user context, and the sales process. This helped the team to clarify:

- project scope
- overall needs
- stakeholders
- system anatomy
- vocabulary

Field Observations and Interviews

Moving further into the insight phase, the designers dug deeper into the needs and requirements of the users.

People are generally good at talking about what they do and what they think. Yet they are often not conscious or aware of their own behaviour or actions. Observations take us beyond what people

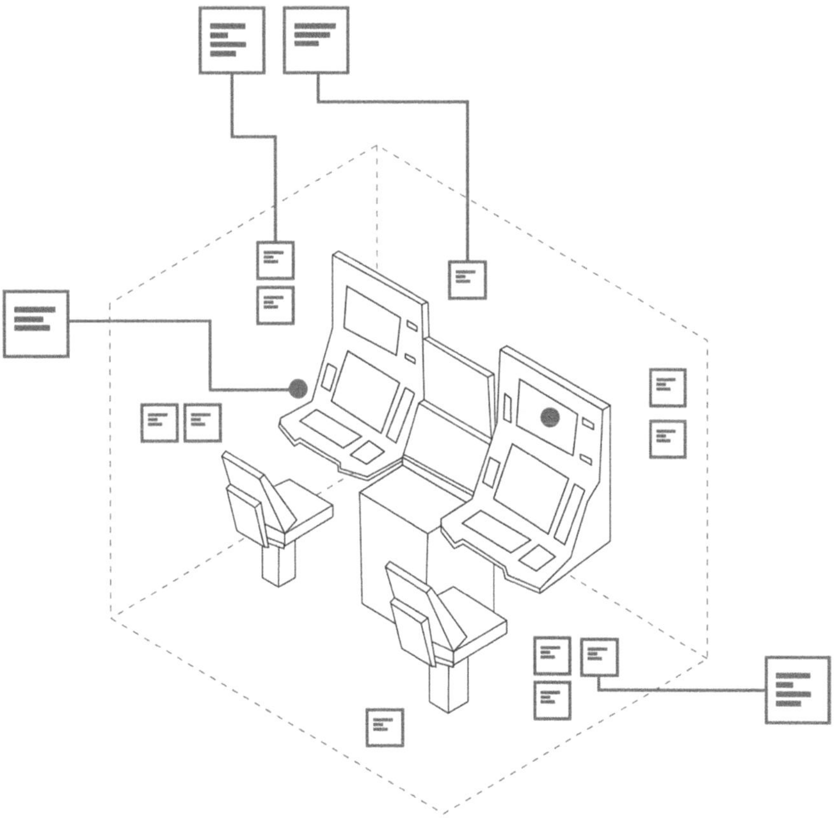

FIGURE 4.2 Mapping to establish a common understanding and clarify the purpose.

Illustration: Halogen.

say they do and give us an insight into *what they actually do* in a given work process. It helps us to gain a deep understanding of context and latent problems.

Observations are documented through video, photos, or notes, focusing on movements, distances, tasks, communication, decision-making, mood, stress levels, environmental constraints, user quotes, and critical findings. This method provides specific input for the design phase and should be combined with semi-structured interviews.

In the ADX project, the designers conducted a series of interviews and observations. It started with observing and interviewing users in context to gauge the level of stress operators were under. We looked for ways to challenge the system, testing all functionality and interfaces (Figure 4.3), probing into everything we see. In this case, the designers had access to a simulator only because it is logistically difficult to get users out into the field. Nevertheless, it provided excellent insight into the operator's tasks, workflow, SA, physical environment, and context.

During the interview, the designers used paper mock-ups and cardboard drawings to place functions and information (Figure 4.4). This level of user involvement provides designers with highly valuable insights and fosters engagement and buy-in from core users, which is precious during implementation.

The interviews and observations presented a clear case and scope for improvement. Given the context where seconds make the difference in intercepting a rocket before it strikes its target, there was a very evident need to speed up decision-making and create more efficient workflows.

Situation Awareness Analysis

SA (Endsley, 1995) considers a person's ability to make decisions in critical scenarios. To have an effective decision-making process, a person needs to perceive their environment, comprehend the

FIGURE 4.3 Testing the existing air defence simulator.

Illustration: Halogen.

FIGURE 4.4 Paper mock-ups can be a more precise way of gathering insights, as they are more similar to the users' actual working situation than a written description can ever be.

Illustration: Halogen.

situation, and project what will happen when they intervene (Figure 4.5). This corresponds to three levels of SA (Endsley, 1995):

Level 1: perception and processing of individual units
Level 2: comprehension of the current situation
Level 3: projection into the future.

An SA analysis can help us pinpoint the factors contributing to excessive workload, stress, and errors. In a design context, it is used to create systems that focus on the user's needs by identifying what (supporting) information is necessary at what stage (Schønheyder, 2019).

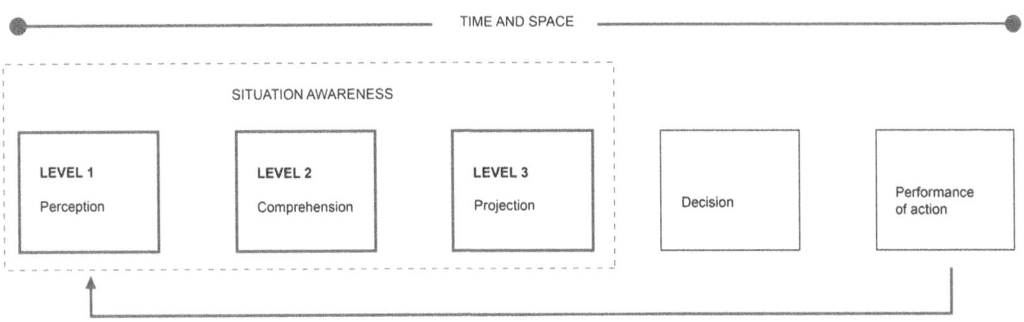

FIGURE 4.5 Endsley's three levels constituting situation awareness, redesigned by Halogen.

Illustration: Halogen.

In our observations, we apply the SA theory to both individual operators and teams. For example, we identify users' levels of attention when facing heavy data overload. It involves observing when and how users detect problems, identify problems, and how their anticipatory thinking forms explanations, sees relationships, and projects the future.

When observing a team's SA, we assess the degree to which each team member possesses the SA required for their specific duties, ensuring the team's overall success.

The method identifies information that is common across team members, the status of other team members' tasks as they impact one's own, the status of one's own tasks as they affect others, the reciprocal impact of actions between team members, and projections of other team members' future actions.

The critical goal is to identify shared SA devices and information. This includes all forms of communication (both verbal and non-verbal), shared visual and auditory displays, and information directly perceived by users within a shared environment. All these elements are crucial for effective team coordination, as a breakdown in shared SA can lead to critical errors, even if individual team members possess high levels of SA.

Frequency and Criticality Analysis

The aim of conducting a frequency and criticality analysis is to prioritise information elements to Frequency and criticality analysis helps determine which information and functions operators need, when they need them, and where they should be placed. It considers both how often information or functionality is needed and how critical it is for safe and effective operation. Using a two-axis matrix, the designers distinguish between elements that must always be visible, quickly available, usually visible, or easily available (Figure 4.6). In the ADX project, the matrix became a central tool for structuring interface elements and preparing later concept work on interface layout and interaction philosophy. The matrix forces prioritisation. By comparing all information elements relative to one another, designers can distinguish truly critical cues from merely useful ones. In the ADX

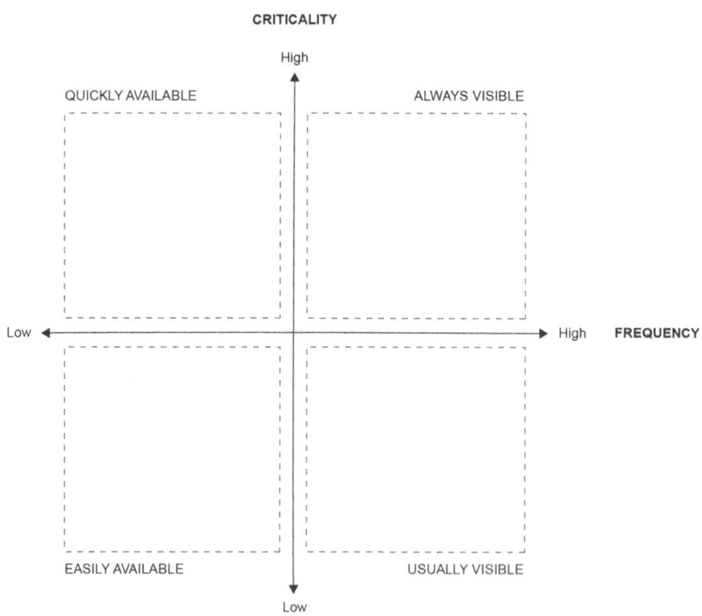

FIGURE 4.6 Halogen's frequency and criticality matrix.

Illustration: Halogen.

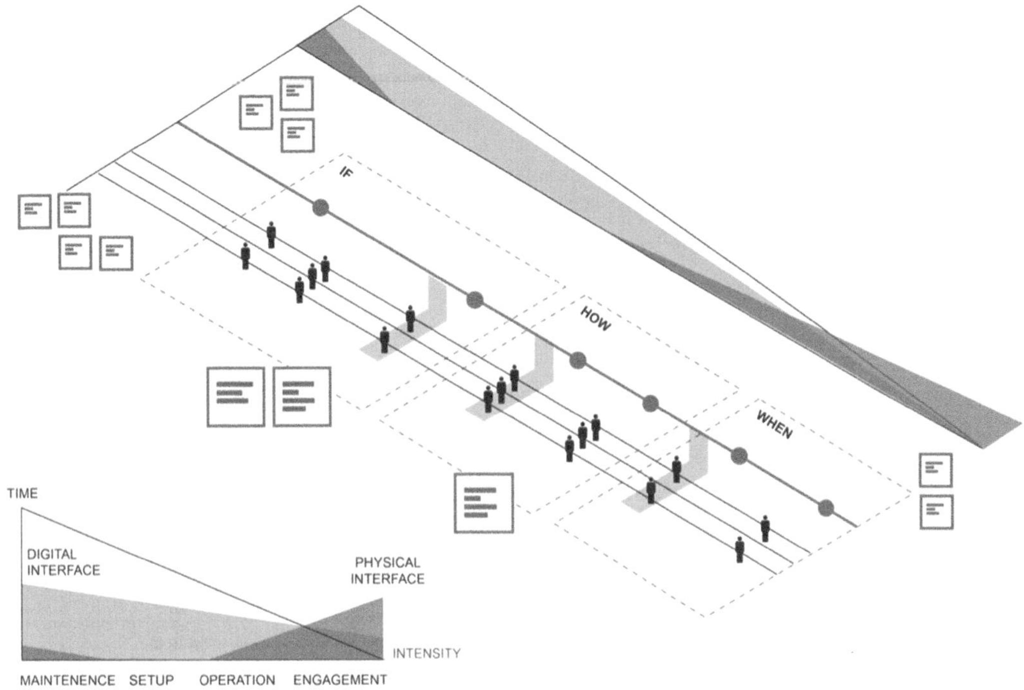

FIGURE 4.7 Chronological mapping of the system.

Illustration: Halogen.

project, this resulted in a 60–70% reduction in information density on the primary interface – a key factor in reducing decision time.

In addition to creating a frequency and criticality matrix, the designers mapped operator tasks, task duration, assumed pain points, information elements, and critical findings chronologically (Figure 4.7). The mapping helped the project group to prioritise information elements, functionality, and placement for an optimal decision-making process.

MTO Analysis

These methods of gathering insights yield vast amounts of data. To make sense of it all, the designers must organise the information. A Man-Technology-Organisation (MTO) analysis (Drøivoldsmo, Lunde-Hansen, Heimdal, 2012) provides a structured overview of the tasks, their relevance, timespan, criteria, and the information required for users to perform their duties (Figure 4.8). This detailed understanding is crucial for reallocating tasks between operators to match new, more efficient functions.

MTO analyses are typically performed early in the project. It involves end users and other key actors, product owners, relevant decision-makers, and designers as facilitators. A detailed specification of each operator's tasks is created, including the flow of information within the users' work processes. Human, technological, and organisational structures are systematically documented. All problem areas from the insight are classified in the MTO triangle based on the problem's root cause. In safety-critical operations, the majority of root cause findings are typically related to technology and organisation.

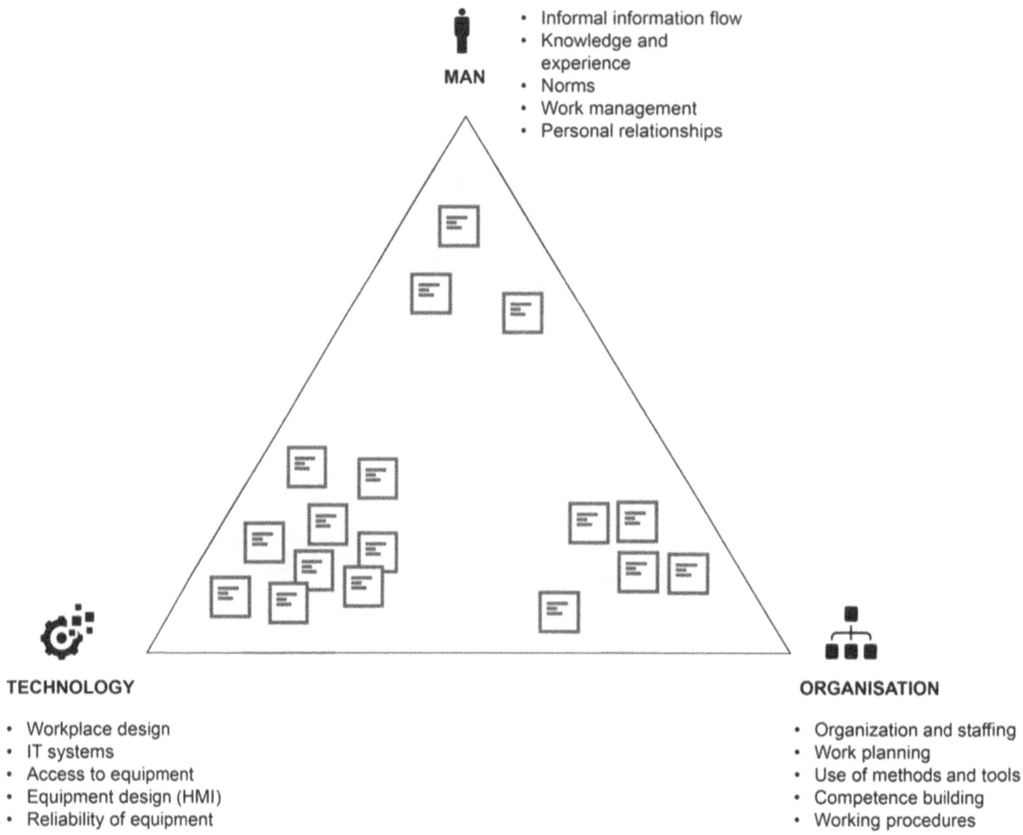

FIGURE 4.8 The IO MTO method by Drøivoldsmo, Lunde-Hansen and Heimdal, redesigned by Halogen.

Illustration: Halogen.

The MTO method reveals causal chains linking human actions to technological and organisational conditions. This makes it an effective diagnostic tool for reallocating roles, redesigning workflows, developing new procedures, making technology and tool choices, and addressing systemic weaknesses that degrade SA or increase cognitive load.

Relations Mapping

In the ADX project, the initial findings from the insight phase were then analysed and visualised in a map, revealing the information infrastructure of all systems and pairing them with actual user requirements. The visualisation of such complex information in system maps is explorative in its nature and enhances understanding of complex situations and systems (Sevaldson, 2013).

The mapping correlates operator roles with specific contexts or locations, so we can present the right information to the right user. The designers analysed existing insight data to map roles to information and functionality in context. This enabled the evaluation of existing and the redrawing of potential new correlations. Stakeholders and end users validated these maps.

Based on the insight activities, the designers identified the need for an interaction philosophy that could distinguish between operational modes according to available time and level of criticality. When monitoring a system, users typically have more time and a lower degree of criticality, making a broader range of interaction tools appropriate. In engagement mode, however, limited time and high criticality require concise tools and focused information displays that support fast decisions and actions. To illustrate the time-critical nature of this design challenge, during five seconds of

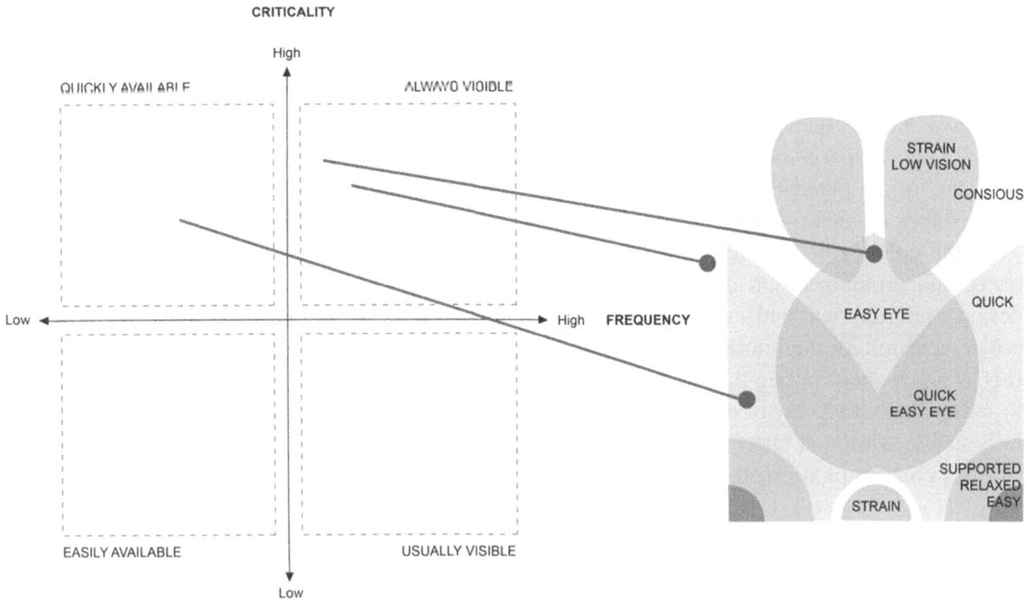

FIGURE 4.9 Frequency and criticality analysis helps prioritise where information and functionality should be placed in a user interface. The principle provides a bridge from insight findings to concept-phase activities, such as Interface Layout Mapping and Interaction Philosophy.

Illustration: Halogen.

hesitation, a fighter jet may have moved more than 2.5 kilometres. The air defence system therefore had to be designed to support users under extreme time pressure. Figure 4.9 shows how frequency and criticality analysis can be used to prioritise the placement of functionality and information in a user interface, taking into account reach, sight lines, and heat mapping. This helps address fatigue, build spatial recognition, and support muscle memory. The principle is developed further in Phase 2 activities such as Interface Layout Mapping and Interaction Philosophy, thereby providing an important bridge between the insight and concept phases.

A SCIENTIFICALLY GROUNDED PICTURE OF OPERATOR CAPABILITIES

Across all insight activities, the underlying aim is to understand how existing workflows impose cognitive demands, where SA breaks down, and how information flow, communication, and ergonomics constrain operator performance. Each method contributes complementary perspectives: observations reveal latent conditions, interviews elicit mental models, timeline and frequency-criticality analysis uncover temporal characteristics of workload, and MTO analysis identifies organisational and technological constraints. Together, these form a scientifically grounded picture of operator capabilities and system demands.

Although each HDMC insight method serves a different practical purpose, together they map directly onto core constructs in human-factors science. MTO analysis exposes latent organisational and technological conditions that shape human error potential (Reason, 1997). ZIP and timeline analyses reveal temporal workload peaks and SA degradation points aligned with Endsley's Levels 1–3. Frequency-criticality mapping links directly to perceptual load theory by reducing competing stimuli during time-critical decision windows. Relations mapping clarifies team SA requirements and boundary objects necessary for coordinated action. By explicitly grounding each method in HF theory, HDMC ensures that design decisions address root causes of degraded performance rather than surface-level symptoms.

HDMC PHASE 2: CONCEPT PHASE

Designing for safety-critical systems requires navigating inherently 'wicked problems' (Rittel & Webber, 1973; Buchanan, 1992), where the problem evolves as the solution is explored. HDMC addresses this by iterating rapidly between divergent exploration and convergent synthesis, using sketches, mock-ups, probes, and SA-based evaluations to reflect-in-action (Schön, 1983) with real users. After the insight phase establishes 'What for Whom', the concept phase translates this understanding into testable, concrete solution directions without losing sight of the user's operational constraints.

In the concept phase, we are interested in how to translate our insight into concrete, testable ideas, and prototypes. On a technical level, we generate possible solutions, identify and verify design opportunities, and create and test prototypes. On a human and organisational level, we project a vision for the future, explore and break boundaries, and provoke and challenge the 'as-is'. It is as much about hitting the reset button to start from scratch, helping stakeholders and operators to see an alternative way of doing things or opening up to new possibilities, as it is about organising information within an interface.

Each concept-phase method serves a distinct cognitive purpose:

- Explorative sketching externalises mental models
- Workflow mock-ups reduce abstraction and support understanding of shared SA
- Interaction philosophy aligns interface demands with operator temporal constraints
- Interface layout mapping reduces visual search and cognitive load
- Scaled mock-ups support team coordination and spatial awareness

These links between design methods and HF theory ensure that concept development is not only creative but cognitively valid.

When we know from our insight work that operators can be better supported, workflows can be optimised, or the system can be more efficient, it leads us to question: What are possible and impossible product directions? Where *could* it go? Where do we *want* it to go? The idea is to be open, generative, and explorative.

We typically raise these questions at the start of the concept phase with the internal project design team, aiming to generate new and enhanced possibilities that effectively address the insight findings. It involves exploring the number of information elements needed, types of functionalities, and mapping these to the users' workflow, context, and needs. Findings from the SA analysis are explored through sketches in this phase. This use of sketching aligns with Buxton's view that sketching is both a way of exploring alternatives and a way of thinking through design problems before committing to a solution (Buxton, 2007).

Schønheyder lists 12 design methods for the concept phase (Schønheyder, 2019). In the coming section, we'll describe how the methods emphasised in italic were used in the ADX project.

- *Moodboard workshop*
- *Explorative workflow sketching*
- *Interface layout mapping*
- *Interaction philosophy mapping*
- *Scaled workflow mock-up*
- *Contextual concept design*
- Workflow mock-up
- Biased concept prototyping
- Concept ideation workshop
- Probe prototyping in context
- Holistic concept and operation philosophy testing
- *SA concept design evaluation (described in the Design & Development section)*

FIGURE 4.10 The user representatives starting from scratch to design the new air defence console, closely facilitated by experienced designers.

Illustration: Halogen.

In a design project where the aim is to redesign something that already exists, the client should ideally be willing to *hit the reset button*. In the ADX project, both clients and users were open to letting the existing solution go for now and start the design process from scratch (Figure 4.10).

Moodboard Workshop

After a rigorous insight phase, the mind can be overwhelmed and perhaps constrained by restrictions and assumptions at this point, it is essential to take a step back and reset the brain. A way to do this is to kick-start the concept phase by drawing inspiration and motivation from neighbouring or alternative fields of concept development, such as the movie, gaming, or automotive industries. These are fields with long traditions of creating a structured information repository that can be drawn on to motivate and direct the project. The bank can consist of movie clips, images from flowers to oil gauges, music, sounds, or even smells. Use the bank through the project to create associations for design solutions and unlock creative blockage. The bank is an internal resource. However, selected imagery can work as input to the moodboard workshop.

In the ADX project, the designers facilitated a moodboard workshop with the relevant stakeholders (Figure 4.11). The purpose is to anchor a common platform for communicating visual intent and design guidelines. It might sound like a fun, light, and easy exercise, and it *is* fun, but it is also of great importance; the moodboard lays the foundation for articulating the product's identity.

A moodboard workshop can help in 'rebooting' the ingrained mindset and create a common direction for visual identity, communication of values, and experience of the final product. The output serves as a collective platform for communicating the design – input to design guidelines and form language.

FIGURE 4.11 Moodboard workshops should include relevant stakeholders such as operators, engineers, managers, and R&D personnel.

Illustration: Halogen.

Explorative Workflow Sketching

To design the best product for operators, we need to step away from technology and what already exists, and focus solely on workflow. In this activity, we involve end users and stakeholders to start from scratch once again: we prototype alternative layouts with them, constantly gathering input and feedback to co-create a solution that fits their workflow and mental model. Through ad hoc mock-ups of the working environment and the placement of key information and functionality elements, we explore spatial recognition, sight lines, reach, and muscle memory.

This form of early prototyping is intentionally 'basic' and low fidelity. By being open to change and input, users and stakeholders are motivated to reflect on their own needs, challenge current workflows and technologies, and gain a sense of ownership and expectation for the project. The benefit is that we can explore and test user tasks, operation philosophy, and alternative concepts, independent of technological constraints.

In the ADX project, the early hand-drawn sketches addressed findings from analysis and explored possibilities (Figure 4.12). They visualised types of information layout and display, and explored which elements should be interactive, and which were only informative.

Interface Layout Mapping

This exercise is used to prioritise what information goes where on an interface. It follows from the criticality-frequency matrix in the insight phase to place the most important pieces of information in the most-used areas of the interface (Figure 4.9). It involves exploring possible layout solutions using scenario mapping and analysis findings, with delivery screen formats providing input for size,

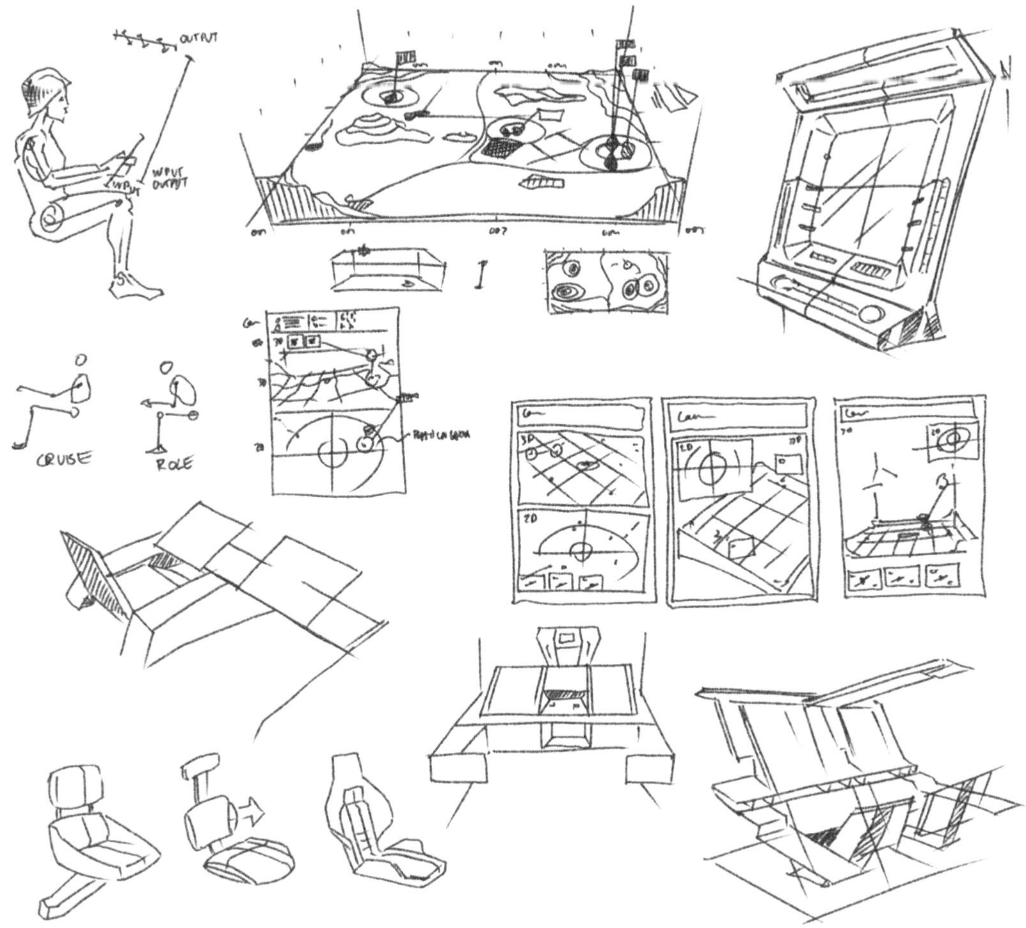

FIGURE 4.12 Early-stage sketching based on information gathered in the insight phase.

Illustration: Halogen/Hareide.

orientation, and resolution. The goal is to tie all key findings from the analysis and workflow mock-up to exploring solutions that support users' task workflow and goals. It helps in addressing fatigue and building spatial recognition and muscle memory by considering sightlines, reach, and criticality/frequency.

The ADX project involved a lot of early prototyping and testing. In workshop settings, we explored workflow concepts using large blank sheets of paper and post-it notes to keep concepts low-fidelity and open to change. This involved mapping out tasks and information needs for the two operators sitting side by side, with operators, stakeholders, and designers participating in generative and iterative sessions (Figure 4.13).

The designers also ran highly explorative concept sessions to test the insight. By taking them to something so far off in the future and so far removed from existing systems, they were primed to leave 'as-is' behind and think openly. It was far easier to take them back from the future to a concept that is relatively realistic and possible than to take them away from what they already know today.

Interaction Philosophy

By this stage in a project, the designers have learnt about the system and user context, analysed information, and generated ideas in earlier concept work. Now it is important to define the type and

FIGURE 4.13 Creating the new workflow.

Illustration: Halogen.

level of interaction between the user and system. This involves categorising information and functionality according to available time and level of criticality, which then informs the design of input devices and the abstraction of information. For example, situations requiring 'split-second' decisions with high criticality and minimum time available (e.g., 'putting out a fire') require physical controls for muscle memory and haptic response, with a high degree of spatial recognition in shape and size (Figure 4.14).

Defining an interaction philosophy safeguards appropriate levels of interaction and provides input for contextual concept design and operation philosophy (operation and interaction philosophies are different in that operation philosophy is about the overall operation, while interaction philosophy is specifically about HMI and interfaces).

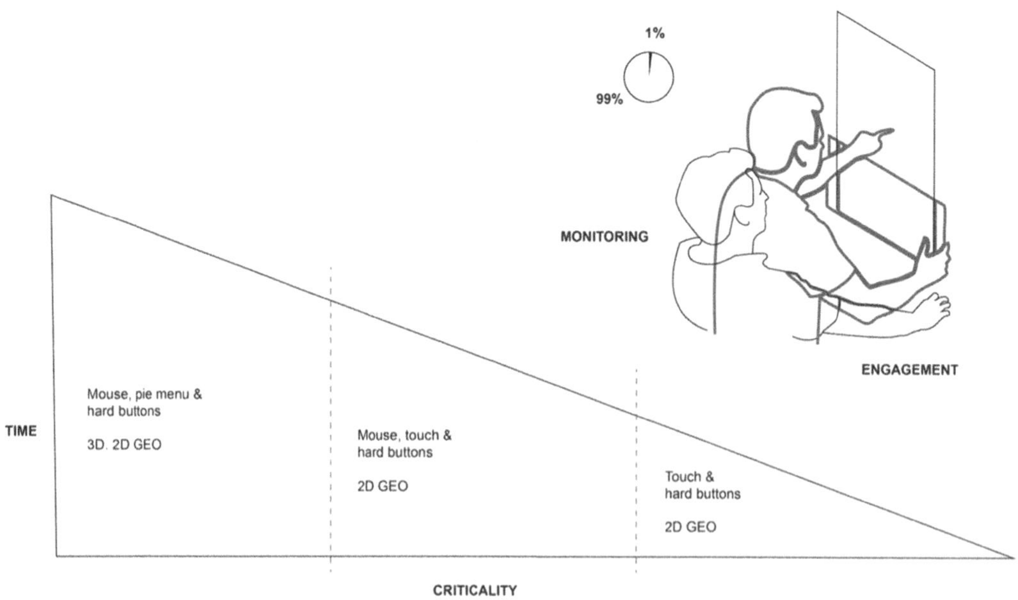

FIGURE 4.14 The interaction philosophy distinguishes between functions and tasks based on criticality and available time.

Illustration: Halogen.

Scaled Workflow Mock-Up

Much of the concept work until now has involved generating ideas, sketching, and creating very early prototypes. But how will the system 'feel' in the space and function in situ? By building scaled mock-ups and layouts of the working environments, we establish a sense of scale, dimension, and operational space for exploring and evaluating early concept ideas (Figure 4.15). Figurines can be used to enact user workflow, communication, and movement within the environment. This enactment, combined with team SA theory and the Criticality and Frequency matrix, helps the designers to generate and reflect on placement, shape, and form for components and the working environment. It's a valuable tool for workshops with users and stakeholders to evaluate concept proposals, as working with physical objects stimulates reflection and strengthens ownership.

Contextual Concept Design

This stage is about the 'nitty gritty' of a specific scenario and operator role. The designers critically explore a specific context, focusing on a full-screen resolution exploration of placing information elements and functionality needed. The outcome is visualisations of types of information layout and display, distinguishing between interactive and informative elements, and mapping these to SA, user workflow, context, and needs. This entails working deliberately with shape, form, and contrast on information and functionality elements to build spatial and form recognition and generating layouts that can build muscle memory.

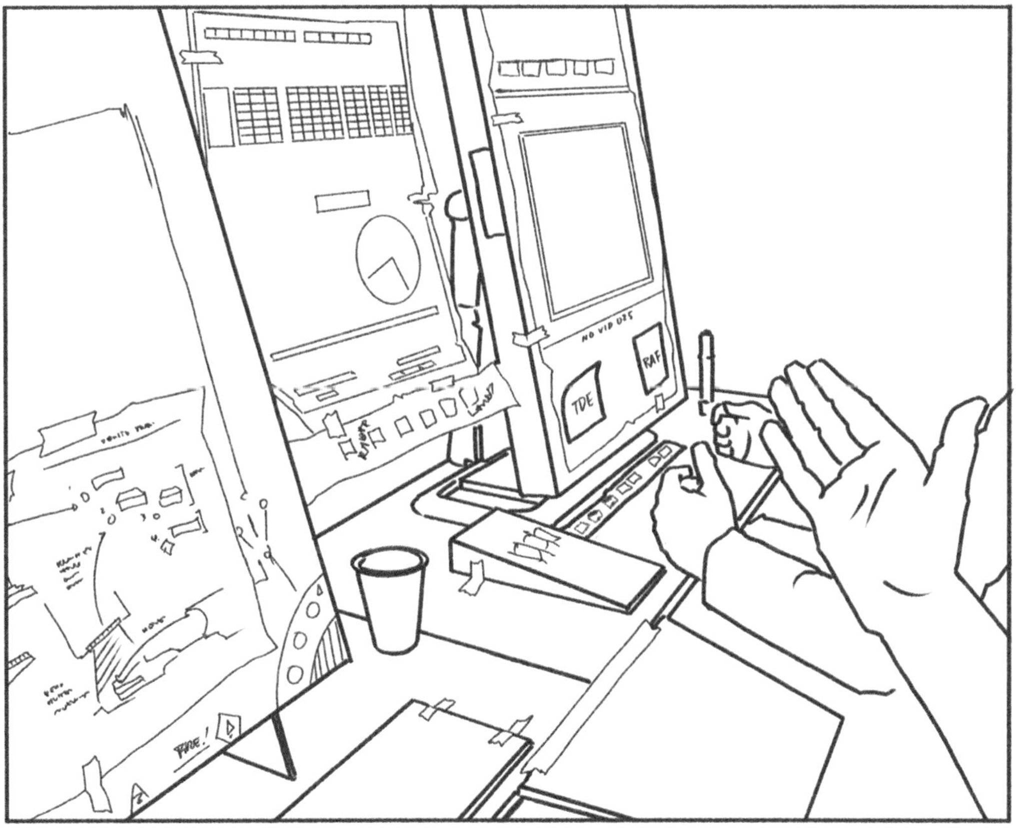

FIGURE 4.15 Low-cost prototyping with live-size mock-ups.

Illustration: Halogen.

HDMC PHASE 3: DETAILED DESIGN AND DEVELOPMENT PHASE

The Halogen design process continuously collects and verifies insights to enhance the final design through the entire process. As time progresses, the input level of insight decreases and the output level of concept and design increases through the design process. Although the insight process never really ends, the amount of new insight generated is lower in the final phases. The specifics and detail level, however, is high.

Validation is secured through the phases and activities, and insight becomes infused with the design work and iterated with users and stakeholders. This feedback loop motivates reflection and prompts project members to articulate tacit knowledge, thereby strengthening the quality of the end product.

Jan Fredrik Schønheyder lists three design methods for the design and development phase (Schønheyder, 2019). All three were used in the ADX project.

- Iterative detailed workflow prototyping
- Pilot prototyping (in Context)
- SA design evaluation (SA Concept Design Evaluation)

As the phases are not strictly sequential, especially in prototyping and detailed design, this chapter does not rigidly follow the model's structure. In this section, we'll describe operation philosophy and full-scale prototyping, iterative prototyping in context and SA Concept Design Evaluation (in the model placed in the concept phase).

Operation Philosophy and Full-Scale Prototyping

Full-scale mock-ups or screen-based prototypes are used to mimic and test tasks and scenarios, addressing key findings from the analysis. It can be performed with both internal and external end users, and the mock-up can be constructed from foam boards and printouts of interface concepts, with third-party objects like headsets and coffee cups added for realism. Design iterations are performed consecutively as the project team evaluates the concept through enactment, laying the foundation for formalisation, detailed interaction, and industrial design.

In the ADX project, the operation philosophy describes how everything the user needs should be closer, and that the workflow has a logical direction and flow. It was formulated based on all of the insight and concept work to date and was a way of prioritising what was most important to the user and the system, whilst serving as a reminder to the project team for the rest of the project.

The operation philosophy also maps the type of operator interaction to the type of task. For example, a task that is highly critical and must be performed very quickly requires tactile interaction with touch and hard buttons.

The prototyping phase was centred around how information would be presented to each operator on their individual interfaces, whilst taking into account the overall workflow. The team worked with both paper prototypes and tabletop testing with operators. Rather than a back-and-forth workflow, the operation was now designed for an intuitive, natural flow. Information and functionality were placed based on findings from the Criticality and Frequency analysis, knowledge of how the operators sit and work in context, maintaining focus on their task, and the areas of the screen which are best suited to high-priority information.

Iterative Prototyping in Context

It is important to continuously validate concept directions and operational content with users through quick iteration loops. At Halogen, the designers do this using conceptual ideation sketches as 'probes' to prototype and iterate ideas directly with users in their working environment. As users evaluate and comment on the probes, iterations to the concept are quickly sketched out on-site for new discussions. This 'reflection in action' stimulates users to reflect on their actual needs without the constraints of current solutions, creating valuable directions for the concept and opportunities to collect new, unknown insights.

SA Concept Design Evaluation

Before a final concept is approved, it is important to carry out a final sense check. Endsley defines Situation Awareness (SA) as the perception of environmental elements and events with respect to time or space, the comprehension of their meaning, and the projection of their status after some variable has changed, such as time or an event. SA involves being aware of what is happening in the vicinity to understand how information, events, and one's own actions will impact goals and objectives, both immediately and in the near future (Endsley 2011). With this in mind, the designers should conduct a comprehensive assessment of the design's effectiveness in critical operations. Jan Fredrik Schønheyder and Halogen have developed a set of 12 criteria that evaluate the concept design, ensuring that it supports user goals, presents information effectively, and enables efficient system interaction. The overall goal is an intuitive design where interaction with the system follows the workflow, and the layout and order feel seamless and correct (Figure 4.16). The criteria are divided into three main categories:

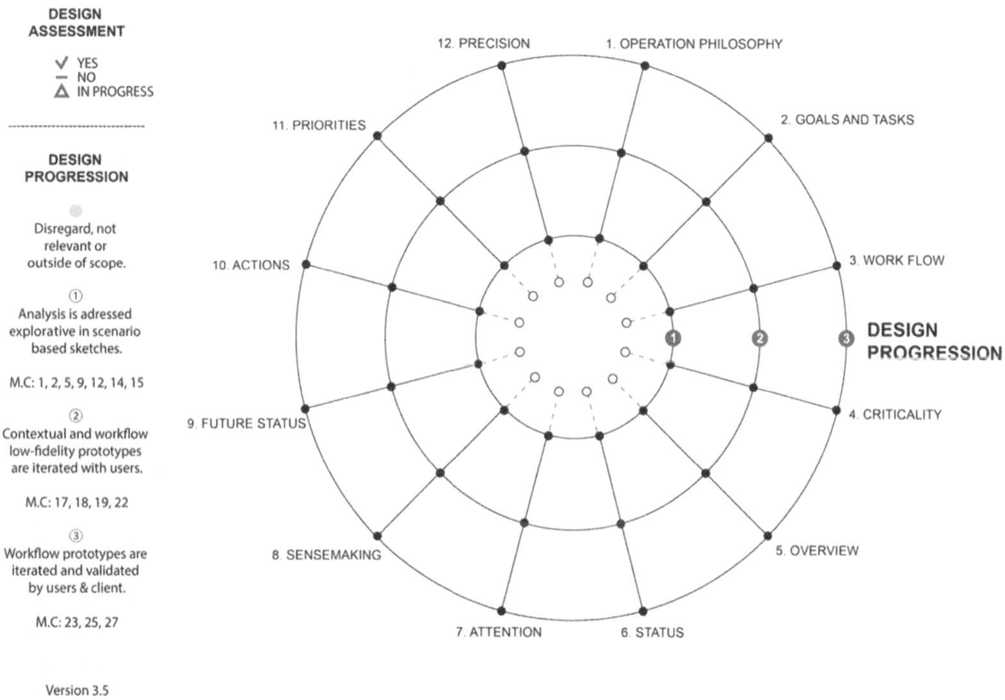

FIGURE 4.16 This spider diagram is a tool for designers to assess and evaluate their design using 12 questions based on SA theory.

Illustration: Halogen.

Perception

1. Is there a clear and consistent *operation philosophy*?
2. Are user *goals and tasks* supported and mapped to the design of the system?
3. Does the design provide an efficient and natural *workflow*?
4. Does the design provide optimal support in the most *critical situations*?

Comprehension

5. Does the design provide an *overview* of the situation at a glance?
6. Are *status* and information designed in a manner that provides visual and spatial recognition?
7. Does the design reduce mental workload and guide the user's *attention* in use and alarm situations?
8. Does the design utilise the human ability to recognise *patterns* and reduce mental workload?

Projection

9. Does the design enable easy identification of possible *actions* and the ability to execute the right actions at the right time?
10. Is the highest visual real estate *prioritised* to critical and frequent content?
11. Does the design provide *feedback* in a timely manner with visual or haptic means?
12. Does the design provide the right *input devices*, given the operational context?

HDMC PHASE 4: IMPLEMENTATION AND OPERATION

The application of the HDMC framework to the Next Generation Air Defence Console project led to significant improvements. Through a comprehensive process of understanding the system and the user, conducting fieldwork and discovery, applying rigorous analysis frameworks, and prioritising for effectiveness, the project brought new levels of user-centricity and consideration of HF to the domain. The iterative approach enabled the project team to 'hit the reset button', leaving behind the restrictions of what already existed and fostering open and explorative concepts.

Where possible, HDMC outcomes were validated through field observations, simulator testing, and structured operator feedback. Reductions in decision latency were linked to decreased visual search time, improved grouping of high-criticality functions, and the relocation of decision-relevant information into the operators' foveal vision zones. Improvements in workflow fluidity were observed during tabletop simulations and later confirmed in high-tempo operational scenarios. These observations demonstrate not only that the system 'felt better' to operators but that measurable components of their perceptual and cognitive workload were reduced through targeted design interventions.

Operational feedback from users of the NASAMS FDC provided additional empirical support for HDMC. Operators reported that the redesigned workflows enabled faster comprehension of threat patterns, more reliable spatial and muscle-memory-based actions under pressure, and reduced mental load during sustained high-tempo operations. These outcomes align with well-documented HF effects: reducing the number of visual search targets lowers cognitive workload, grouping high-criticality information increases Level 1 SA, and consistent interaction patterns improve response speed and reduce slips. Several operators emphasised that the new design was 'easier to learn, simpler to use' despite increased system complexity, confirming that well-designed interaction philosophy can lower the threshold for operational mastery and reduce training time.

The operators are trained in Norway before they are sent to the front. 'The training is proving to be good enough. They go directly into battle with the foundation we provide them with', says

Colonel Per Steinar Trøite, Head of Division – Support to Ukraine (Aftenposten, 2025). In the finalisation of the project, a training officer expressed that the ADX project, resulting in the new generation NASAMS FDC, had made their job much easier. The new, upgraded system has also proven efficient in the field. Since the system was donated to Ukraine, Norwegian media have reported that user-friendliness for operators has been an important factor in the system's success. An operator quoted in Forsvarets forum compares NASAMS to older air defence systems: 'Even if NASAMS is more complex and advanced, it is easier to learn and simpler to use' (Forsvarets forum, 2024). In 2025, Aftenposten reported a 94% hit rate for the system in Ukraine.

HDMC contributes to the scientific discourse on safety-critical design by demonstrating how modular human-centred methods can be systematically integrated with HF theory. By explicitly linking information design to perceptual thresholds, workflow to cognitive load, and team interaction to shared SA, the method addresses well-documented limitations in standard design processes for high-risk domains.

The HDMC framework therefore offers not only a practical methodology but also a transferable model, based on theoretically grounded and empirically informed methods for designing decision-critical systems across other high-consequence domains.

REFERENCES

Aftenposten (February 24, 2025). *Lovpriser norsk luftvern i Ukraina: – Det er et svært, svært høyt tall.* https://www.aftenposten.no/verden/i/QMWGJP/lovpriser-norsk-luftvern-i-ukraina-det-er-et-svaert-svaert-hoeyt-tall

Buchanan, R. (1992). Wicked problems in design thinking. *Design Issues*, 8(2), 5–21.

Buxton, B. (2007). *Sketching User Experiences: Getting the Design Right and the Right Design.* San Francisco, CA: Morgan Kaufmann.

Coyne, R. (2005). Wicked problems revisited. *Design Studies*, 26(1), 5–17.

Drøivoldsmo, A., Lunde-Hansen, L., & Heimdal, J. (2012). *IO MTO Handbook: Integrated Operations – Man, Technology, and Organisation.* IFE - Institute for Energy Technology, Norway. https://hdl.handle.net/11250/5514655.

Endsley, M. R. (1995). Toward a theory of situation awareness in dynamic systems. *Human Factors: The Journal of the Human Factors and Ergonomics Society*, 37(1), 32–64.

Endsley, M. R. (2011). *Designing for Situation Awareness: An Approach to User-Centered Design* (2nd ed.). CRC Press.

Forsvarets forum (November 16, 2024). *NASAMS-operatør: Utenkelig å forsvare Ukraina uten.* https://www.forsvaretsforum.no/nasams-operator-utenkelig-a-forsvare-ukraina-uten/406048

IDEO.org (2015). *The Field Guide to Human-Centered Design* (1st ed.). San Francisco, CA: IDEO.org/Design Kit.

ISO 9241-210 (2019). Ergonomics of human-system interaction Part 210: Human-centred design for interactive systems.

Knight, J. C. (2002). Safety critical systems: challenges and directions. In Proceedings of the 24rd International Conference on Software Engineering (ICSE'02) (pp. 547–550). New York: ACM.

Reason, J. (1997). *Managing the Risks of Organizational Accidents.* Aldershot, UK: Ashgate.

Rittel, H. W., & Webber, M. M. (1973). Dilemmas in a general theory of planning. *Policy Sciences*, 4(2), 155–169.

Schön, D. A. (1983). *The Reflective Practitioner: How Professionals Think in Action.* Basic Books.

Schønheyder, J. F. (2019). Method development for the design of safety-critical systems. Doctoral Thesis, The Oslo School of Architecture and Design.

Sevaldson, B. (2013). Systems oriented design: The emergence and development of a designerly approach to address complexity. In Reitan, J.B., Lloyd, P., Bohemia, E., Nielsen, L.M., Digranes, I., & Lutnæs, E. (eds.), *Cumulus: Design Learning for Tomorrow*, Oslo, Norway: DRS. https://doi.org/10.21606/learnxdesign.2013.133

5 Human-Centred Design for Autonomy in Remote Ferry Operation*

Jooyoung Park

5.1 INTRODUCTION

Autonomous urban passenger ferries represent a major shift in city transit, using waterways to provide dependable, efficient, and eco-friendly public transportation. The implementation of Remote Operation Centres (ROCs) represents a significant shift in ferry operations, aiming to reduce onboard crew requirements while enabling multiple ferries to be monitored and managed from a single location (Man et al., 2015). Advances in maritime autonomy and digitalisation enable this development (Goerlandt & Pulsifer, 2022; Namireddy et al., 2019).

The launch of *MF Estelle* in 2023, the world's first commercial autonomous passenger ferry in Stockholm, has played a significant role in advancing this development. Zeabuz, a company based on research at the Norwegian University of Science and Technology (NTNU), has been a key contributor to this progress, helping turn autonomous ferries into a reality.

However, transitioning from onboard to remote operations, where human operators remain in the loop, introduces significant safety and Human Factors (HF) challenges. Although long-term ambitions aim for fully autonomous, crewless ferries, current EU regulations (European Parliament, 2024) still require human oversight within ROCs to ensure timely intervention during unexpected events and to mitigate risks. In practice, human supervision is expected to continue for the foreseeable future, as both regulatory frameworks and operational safety demands require that a remote operator remains accountable and able to intervene when necessary. This requirement emphasises the vital role of Human–Machine Interfaces (HMIs) as the primary channel through which remote operators maintain Situational Awareness (SA) and control.

Addressing these challenges demands early and continuous application of HF and Human-Centred Design (HCD) principles to ensure safety, efficiency, and usability throughout development. The cost of design changes rises significantly as a project moves from early concepts through design, build, and operation phases (Johnsen et al., 2025).

This chapter illustrates Safety by Design as a process of learning through iterative design and evaluation. By integrating HCD methods (ISO 9241-210, 2019; ISO 11064, 2000) with scenario-based safety validation using CRIOP (Crisis Intervention and Operability Analysis), the development of *MF Estelle* demonstrates how safety can be systematically embedded in autonomy design well before operational deployment.

---

* This chapter synthesises and extends findings from two previously published conference papers: Park et al. (1) Challenges and Opportunities in Remote Operations of Automated Passenger Ferries Identified Using the CRIOP Method, and (2) Designing HMI for Remote Operation of Urban Autonomous Ferries with CRIOP. In proceedings of the 35th European Safety and Reliability Conference (ESREL 2025) and the 33rd Society for Risk Analysis Europe Conference (SRA-E 2025). Research Publishing.

DOI: 10.1201/9781003741824-7

The chapter synthesises insights from three iterative phases: (1) operational findings from the milliAmpere2 trials, which provided an early understanding of user needs for HMI development; (2) the operational context and HMI design process for *MF Estelle*; and (3) the application of the CRIOP method to evaluate a prototype ROC concept for autonomous ships.

Together, these phases demonstrate how CRIOP translated Safety by Design principles into practice by aligning ROC development with industry best practices and international ergonomic standards. Drawing on insights from all three phases, the chapter identifies key HF challenges and opportunities and offers recommendations to enhance the safety and reliability of autonomous ferry operations.

This chapter is organised as follows: Section 5.2 reviews related works. Section 5.3 introduces the three phases of the HMI and ROC design journey. Section 5.4 presents the methods and results of each phase, with particular focus on HMI design. Section 5.5 discusses the findings from Phase 3, focusing on broader ROC development. Section 5.6 provides an overall discussion and key insights, and Section 5.7 concludes the chapter.

5.2 BACKGROUND

5.2.1 The Transition to Remote Operations

The transition from traditional onboard operations to shore-based remote management of autonomous ferries marks a significant shift in maritime practice. Remote operators must evolve from supervising a single vessel with direct sensory feedback to managing multiple vessels through mediated interfaces, potentially overseeing an entire fleet from a central location (Porathe, 2022). This shift aims to reduce human involvement in operations by enabling task sharing between human operators and autonomous systems.

This transformation introduces several interconnected challenges that must be addressed in the design and development of ROCs. Firstly, ensuring safety becomes a top priority when removing human operators from vessels. Remote systems must compensate for the loss of direct sensory input and the capability for immediate intervention, while supporting an evolving onshore operator role and maintaining equivalent or preferably superior safety levels. Stable network connectivity is another critical requirement for maintaining uninterrupted and reliable remote control. Designing ergonomic, optimised control centres that help operators maintain awareness across multiple vessels – and manage diverse operational states and potential emergencies – is essential for safe fleet supervision. Operator roles and responsibilities will undergo significant redefinition, expanding beyond traditional navigation and system monitoring to include safety, efficiency, and timely intervention across multiple vessels. Finally, developing robust, intuitive HMIs that support 'situation-at-a-glance' awareness is crucial for quick, accurate decision-making.

This transition also transforms human roles and stakeholder interactions. Autonomous passenger ferry systems involve a wide range of actors, especially those directly responsible for operating and supervising the vessels. These stakeholders can be grouped into four main categories: developers, operators (including emergency responders), passengers, and other vessels navigating in shared waterways (Veitch & Alsos, 2021).

In traditional manned operations, onboard safety hosts or operators facilitated communication between the ferry and these groups. However, as operations shift to land-based ROCs, the absence of a physical crew onboard requires a redesign of interaction methods. This transition calls for the development of multiple HMIs, each customised to meet the specific information and communication needs of different user groups.

- **Developers** are responsible for monitoring the system, identifying and resolving technical issues, and deploying software updates to maintain and enhance system performance (Veitch & Alsos, 2021).
- **Operators** oversee navigation, vessel control, and passenger safety. These responsibilities remain vital in remote operations, requiring practical tools for oversight and intervention (Veitch et al., 2021).

- **Passengers** interact with the ferry through systems for ticketing, boarding, in-transit communication, and disembarkation (Petermann et al., 2024).
- **Other vessels** operating nearby must be able to understand the autonomous ferry's intentions, especially in complex navigation scenarios. External interfaces that visually communicate navigational intent have been proposed to enhance maritime safety (Simic & Alsos, 2023; Porathe, 2021).

Given this transition to remote operations, a suite of HMI solutions is required to ensure effective communication and interaction between all stakeholders and the autonomous ferry system. This chapter focuses on the operator's HMI, examining how its design can support remote SA, operational safety, and Meaningful Human Control (MHC) (Endsley, 1988).

5.2.2 Human–AI Teaming and Meaningful Human Control (MHC)

The adoption of autonomous systems and artificial intelligence (AI) in operations introduces both opportunities and risks (NAS, 2022). Opportunities include lowering workload through automation of repetitive tasks, enabling remote operations, and increasing efficiency across multiple vessels. However, both AI and human operators have their limitations. AI performs well when properly trained but faces challenges in unfamiliar or unpredictable environments (Woods, 2016), and it lacks causal understanding, depending instead on pattern recognition (Pearl & Mackenzie, 2018). Additionally, biased training data can result in biased AI decisions (Ferrer et al., 2021).

Human performance can also be affected by AI. High perceived reliability of AI decisions over time can lead to operator over-reliance, potentially reducing vigilance and increasing complacency (Parasuraman et al., 1993). This can create a false impression that the AI system is infallible (Bradshaw et al., 2013). In critical situations that require timely human intervention, operators may be out of the loop, delaying their ability to regain SA and respond effectively (Moray, 1986). Even short-term reliance on AI systems can lead to a rapid decline in manual skills among operators (Casner et al., 2014). Additionally, if users do not fully understand how the AI functions, they may interact with it inappropriately, potentially compromising system safety and effectiveness (Endsley, 2019). This underscores the importance of designing systems that align with human capabilities and providing thorough training to build a clear understanding of AI's operation and limitations (European Union, 2026).

Autonomous systems with self-learning capabilities can operate independently, but their complexity, unpredictability, and lack of transparency often hinder human operators from fully understanding their behaviour or intervening effectively (NAS, 2022; Mittelstadt et al., 2016). This lack of transparency can delay or obstruct appropriate actions during critical situations.

To address these challenges, ensuring MHC is essential for the design and operation of autonomous systems, especially as human–machine interaction becomes increasingly central in operations. MHC requires that automated systems behave in ways that match human expectations and mental models (Calvert et al., 2019). In remote operations, MHC refers to humans and systems working together reliably and safely. This includes giving operators enough SA and the ability to influence the system's future state, within human and technological limits. Achieving MHC involves the systematic application of HF-based design practices that focus on usability, SA, realistic training, manageable workloads, and adaptive system behaviour. The aim is to ensure that systems, organisations, and individuals can learn from experience and respond effectively to the root causes of incidents (Johnsen & Park, 2025).

Alongside these efforts, the International Maritime Organization (IMO) is developing a regulatory framework for Maritime Autonomous Surface Ships (MASS), with a focus on safety, operational procedures, and HF (MSC, 2024). Central to this framework is the principle of MHC, ensuring that remote operators remain responsible for vessel safety, even when autonomous systems are active. This aligns with existing concerns about communication robustness, SA, and emergency preparedness in ROCs. Regardless of the level of autonomy, a designated individual must stay accountable and able to intervene if needed (MSC, 2023).

5.2.3 HMI Design for Meaningful Human Control

Recent international research on HMI design for autonomous systems has revealed several notable trends. In the automotive domain, studies have emphasised the importance of fostering trust and ensuring transparency in human–AI interaction (Calvert et al., 2019). Within aviation, research has focused on reducing cognitive workload through intuitive interfaces and reliable alerting systems (Endsley, 2019). In the maritime sector, Porathe (2021) and Rødseth (2017) have examined the challenges of remote monitoring and control for autonomous vessels, highlighting the need for clear communication of system intentions and robust SA tools. Collectively, these insights provide valuable guidance for developing HMIs that strengthen MHC in autonomous ferry operations.

Achieving MHC relies heavily on well-designed HMIs. Designing for MHC requires intuitive and effective interfaces that facilitate seamless communication between human operators and autonomous systems. A Safety-Critical Task Analysis provides the foundation for such design by identifying the essential tasks required for safe and efficient operations. Furthermore, task analysis plays a vital role in assessing cognitive workload, ensuring that the interface supports informed operator decisions without causing mental overload (Energy Institute, 2011).

The following key HMI design principles are central to supporting safe and efficient remote operation:

- User-Centred Design: Engaging operators and stakeholders throughout the design process to ensure the HMI reflects their needs and operational realities, thereby reinforcing MHC (ISO 9241-210, 2019).
- Support for Situational Awareness (SA): Enabling operators to perceive, comprehend, and project the ferry's operational state and its environment to improve decision-making (Endsley, 1988).
- Automation Transparency: Providing operators with clear insight into the system's decision-making logic and upcoming actions to ensure predictability and trust (van de Merwe et al., 2023).
- Cognitive Workload Management: Designing information displays and interaction flows that balance mental effort and operational performance. Ecological Interface Design (EID) principles support this by aligning interface information with human cognitive capabilities, promoting intuitive reasoning and reducing unnecessary workload (Vicente & Rasmussen, 1992).
- Alarm Systems: Implementing prioritised and reliable alarm mechanisms that distinguish critical alerts from minor ones, supporting timely and accurate operator responses (EEMUA 191, 2024).

These design principles are essential for maintaining safety, SA, and operational efficiency in remote autonomous operations, where operators rely entirely on the HMI to monitor, interpret, and control vessel behaviour.

Building on these insights, the following section describes the specific study context and three-phase journey that operationalised Safety by Design for *MF Estelle*.

5.3 STUDY OVERVIEW AND CONTEXT

Driven by rapid technological advancements and rising expectations for autonomous systems' capabilities, the development of autonomous ferries has advanced considerably over a relatively short period. Correspondingly, the HMI design has undergone continuous refinement. The development of HMIs for autonomous ferries and their ROCs has been an iterative, evolving process, shaped by real-world trials, prototyping, usability testing, and HF analysis. This ongoing journey consolidates a series of studies conducted since 2022, as summarised in Table 5.1.

TABLE 5.1
Timeline of Key Events, Activities, and Stakeholders of This Study

	2022 September (3 weeks)	2023 Jan–May (5 months)	2024 August (2 days)
Key event	Trial operation of milliAmpere2	Design of *MF Estelle*'s HMI	Conducting a CRIOP workshop
Stakeholders	NTNU researchers, safety host, Zeabuz developers	NTNU Designers, Zeabuz developers, safety host	Torghatten operators & managers, Zeabuz developers, NTNU & SINTEF HF experts

The work has progressed through three distinct yet interconnected phases, each addressing specific design challenges while contributing to a broader understanding of effective HMI development for autonomous ferry operations. The research follows an iterative, human-centred methodology, moving systematically from operational insights to design refinement and ultimately to safety validation, with each phase informing and strengthening the next.

5.3.1 milliAmpere Ferry Research

milliAmpere2 is a full-scale autonomous ferry prototype designed and developed by NTNU, serving as a valuable research and testing platform (Eide et al., 2025; Brekke et al., 2022).

The vessel (Figure 5.1) was the world's first autonomous ferry to conduct public trial operations (Alsos et al., 2024). A qualified safety operator, a professional navigator on standby for manual manoeuvring, was onboard to manage emergencies and ensure compliance with maritime regulations, providing a safeguard during the early testing of autonomy features. Insights gained during the 2022 trial were crucial in identifying the need for an HMI that clearly communicates system status and decisions – marking the starting point of the HMI development process described in this chapter.

5.3.2 *MF Estelle*

The development of *MF Estelle* was built on research and experience from NTNU's milliAmpere ferries. Launched on 8 June 2023, *MF Estelle* (Figure 5.2) represents the world's first commercial autonomous urban passenger ferry (Kosmajac, 2024). By August 2024, *MF Estelle* had carried more than 30,000 passengers, with daily ridership sometimes reaching 400. The ferry operates an 800-metre route across Riddarfjärden in central Stockholm, connecting Kungsholmen and

FIGURE 5.1 milliAmpere2.

Source: NTNU.

FIGURE 5.2 *MF Estelle.*

Source: Brødrene Aa.

Södermalm, with a crossing time of about 6 minutes at a cruising speed of 4.5 knots. It is designed to hold 24 passengers and one onboard operator. The operator, Torghatten, aims to expand services to new routes and manage multiple ferries efficiently and safely via an ROC.

MF Estelle is equipped with both a dynamic positioning (DP) system for manual manoeuvring and an autonomy system that follows a predefined path when activated by the onboard captain pressing a 'start crossing' button. While in autonomy mode, the captain can override the vessel's speed or switch to full manual control using the DP system.

A significant milestone was the demonstration of *MF Estelle*'s remote operation in Stockholm in August 2024 (Inspenet, 2024). This demonstration involved the remote control of the vessel from an ROC mock-up located 600 km away in Trondheim. Although conducted without passengers, an onboard operator was present on standby.

The ROC mock-up was integrated with the ferry's control systems and equipped with key monitoring and control functions, as illustrated in Figure 5.3. In addition to CCTV (Closed Circuit Television) cameras from the DP supplier, used to support SA during docking, the setup included

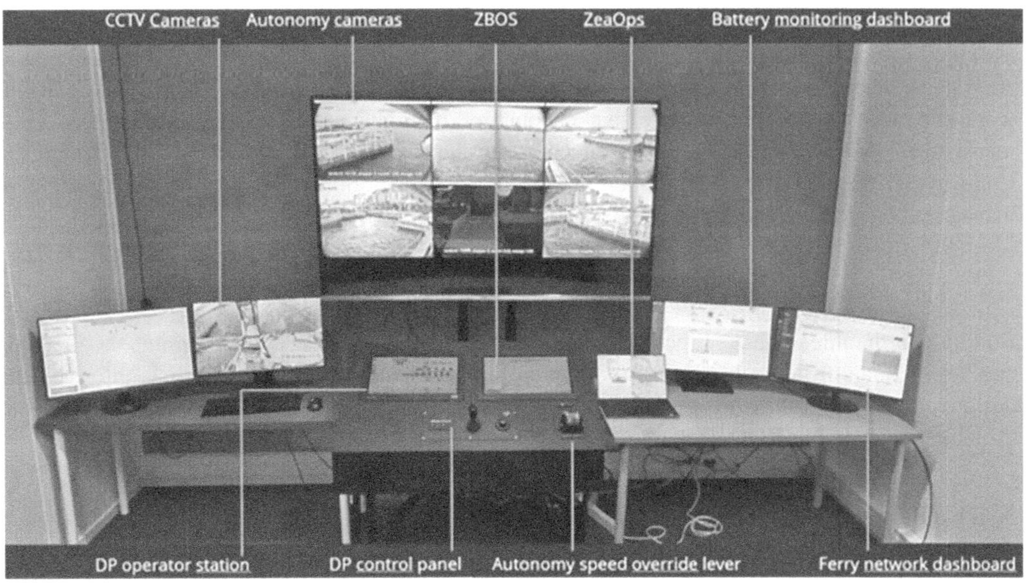

FIGURE 5.3 Early-stage setup of the ROC.

Source: Zeabuz.

autonomy cameras developed by Zeabuz that serve as optical sensors for obstacle detection and tracking. The Zeabuz Autonomy Operator Station (ZBOS) displayed autonomy status and allowed core commands such as starting or stopping the system and enabling speed override. A DP operator station and control panel were included, as the DP system ultimately executes all thrust commands, with autonomy enabled when DP is set to autonomy mode. The ZeaOps system provided an overview of vessel position, tracked objects, and operational status, while a network dashboard displayed real-time 4G/5G connectivity. Using this setup, the ferry completed a remotely supervised crossing with real-time obstacle detection, and the manual speed override was successfully tested.

While the demonstration validated the technical feasibility of autonomy, it also showed that fully remote operations are not yet viable. Challenges still exist, especially in maintaining continuous and secure network connectivity.

Ensuring safe passenger operations without an onboard operator, as well as designing ergonomically and cognitively optimised control centres for managing multiple vessels, are key priorities for future development.

5.3.3 The CRIOP Method

The CRIOP method (Johnsen et al., 2025) was used to evaluate the current HMI design and identify key challenges and design considerations in the transition of *MF Estelle* to unmanned operation and the development of its ROC. The adaptation of CRIOP for autonomous ferries focused on early design evaluation rather than mature system validation. Conducted over two days (15–16 August 2024) with hybrid participation, the workshop reflected the developmental stage of the ROC and emphasised an HCD approach. As the first application of CRIOP in the context of an autonomous vessel's ROC, it demonstrated the method's flexibility as a proactive Safety by Design tool for identifying human–system interaction risks before commissioning.

CRIOP builds shared understanding among designers and operators through structured dialogue – a practical example of Safety by Design's participatory learning principle. Originally created for the oil and gas industry, CRIOP is a structured method for assessing control room design and performance, particularly during crises, with a strong emphasis on HF. Over time, it has expanded into a broader toolset for verifying if control centres can operate safely and effectively across different modes.

The CRIOP methodology combines three main components to ensure a strong control centre design. Checklists are used to identify and assess key elements, reflecting intended workflows and assumptions about how tasks are meant to be performed, known as 'work as imagined'. Scenario analysis examines actual operational situations to understand 'work as done'. Finally, the CRIOP workshop acts as a learning space that encourages dialogue among operators, designers, and management, supporting collaborative reflection and improvement based on real-world experience. Collectively, these components support the design and operation of control centres capable of managing both routine and emergency situations effectively.

5.4 THE HUMAN-CENTRED DESIGN PROCESS AND FINDINGS

Building on the three-phase framework described in Section 5.3, this section presents the methods and key findings from each phase. It highlights how insights from real-world trials informed the design and testing of the onboard HMI, and how the CRIOP workshop extended these findings to evaluate the emerging ROC concept.

5.4.1 Phase 1: Operational Insights from MilliAmpere2 Trial

The 2022 public trial of milliAmpere2 (Eide et al., 2025) demonstrated the capabilities of the autonomy system, including effective target tracking and collision avoidance, even in a challenging

TABLE 5.2
Logbook Summary

Category	Qty	Description
Positive log entries	–	Operational logs without issues
Boat interactions	53	Encounters with other vessels
System problems	15	Technical issues affecting system performance
Tracking problems	40	Issues related to object detection and tracking
-Strange track	*(7)*	*Correct detection but display error*
-Lost track	*(2)*	*Object detection lost*
-Late track	*(4)*	*Delayed object detection*
-False track	*(27)*	*Misidentification of detected object*
--Stops	*((25))*	*Stops due to False track*
--Speed	*((2))*	*Speed changes due to False track*
Unknown error	1	Error with unidentified cause

environment with leisure boats and strong currents (Alsos et al., 2024). The first phase examined data collected during the milliAmpere2 trial operation (Alsos et al., 2024; Veitch et al., 2024). Operators and technicians documented system errors, unexpected behaviours, and examples of successful performance. These logs, along with autonomy system records, were analysed to identify operational challenges and patterns in operator–system interaction.

However, the operator logs revealed several shortcomings in HMI design. Operator logbooks documented approximately 25 manual interventions during the three-week period (see Table 5.2). Many of these were due to behaviours that were not immediately understandable, such as the vessel accelerating or stopping unexpectedly when misinterpreting floating debris as obstacles (Eide et al., 2025).

These incidents underscored the importance of HMI transparency, clarity, and usability in enabling operators to understand the system's decision-making process and make timely, informed interventions.

5.4.2 Phase 2: HMI Design and Integration for *MF Estelle*

Building on the insights from Phase 1, Zeabuz developed *MF Estelle* for commercial operation. The HMI design was intended to support SA, enable autonomous crossings, and increase transparency of the autonomy system's behaviour. The design process followed an HCD approach in accordance with ISO 9241-210, including the following: (1) a co-design workshop and interviews were conducted with participants from the milliAmpere2 trial, enabling direct input from operators and other stakeholders on needs and challenges; (2) these insights informed the development of user scenarios and operational requirements; (3) HMI concepts were then iteratively designed and tested, with feedback from researchers, developers, and operators guiding successive refinements; and (4) the finalised design was implemented in close collaboration with front-end developers to ensure proper integration with the autonomy system. A major limitation of this phase was the absence of the actual *MF Estelle* operator during the initial design activities.

5.4.2.1 Key Design Principles

Key takeaways and refinements from this phase focused on improving usability and reducing cognitive load. While a formal task and cognitive work analysis was not conducted during this phase, several design activities addressed cognitive aspects of operator work. User interviews, iterative prototyping, and usability testing paid particular attention to operators' mental workload, information needs, and decision-making processes. These efforts aimed to ensure that the ROC interface supported intuitive interaction and effective SA. In line with the principles of IEC 63303, information was kept simple, relevant, and directly linked to operational tasks, helping to balance transparency with cognitive manageability.

The HMI had to remain intuitive while avoiding information overload. For instance, acknowledging every detected target was found to be unnecessary and could overwhelm the operator. Usability testing also revealed a common dilemma: while users wanted the system to indicate and predict the ferry's upcoming actions, excessive explanatory content increased cognitive burden. Although alerts for approaching objects supported SA, a fully autonomous system should ideally handle such events without requiring user notification. Based on these insights, refinements focused on simplified visualisations, consistent use of colour coding and symbols, and the effectiveness of bird's-eye-view displays.

5.4.2.2 Implementation and Tools

The finalised HMI (Figure 5.4) was designed to enhance usability and reduced cognitive load, providing (i) an overview of system health, (ii) control menus for autonomy and crossing initiation, (iii) detailed crossing adjustment menus, and (iv) facilitation of manual handovers via HMI or a DP console. A speed override lever, located beneath the interface console (Figure 5.5), provided additional control for operators in specific scenarios requiring manual intervention.

To ensure consistency and compliance with maritime interface standards, a slightly modified ECDIS (Electronic Chart Display and Information System) interface was integrated using the OpenBridge design system. This provided standardised buttons, layouts, and design conventions. Notably, icons representing (v) AIS and (vi) radar-detected targets, developed by OpenBridge based on interpretations of IEC 62288 (2021), were adopted. The widespread use of OpenBridge in the maritime industry enhances familiarity with interfaces and supports user understanding (OICL, 2024).

To further ease cognitive load, visual alerts for nearby obstacles were removed, and (vii) a colour-coded trajectory was added to show speed changes; green for normal, dark green for fast, and red for slow, helping operators anticipate vessel movements.

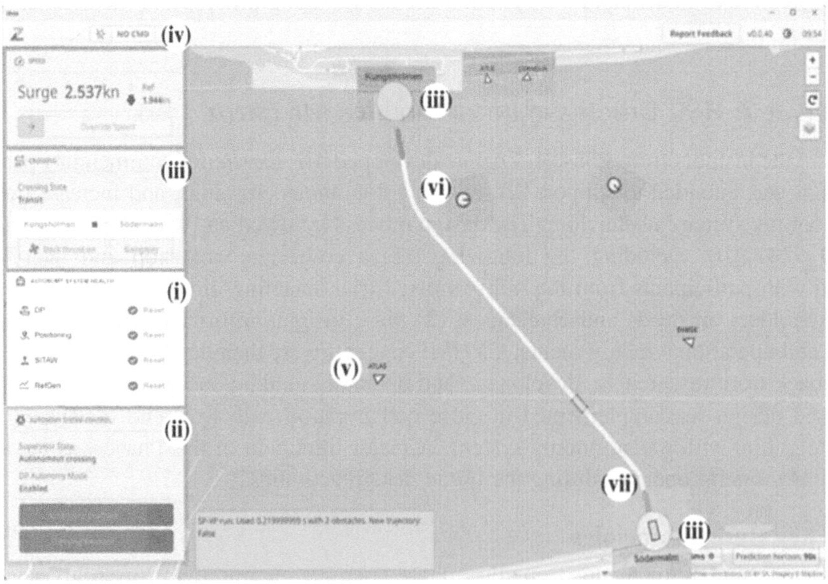

FIGURE 5.4 Operator interface of *MF Estelle*.

Source: Zeabuz.

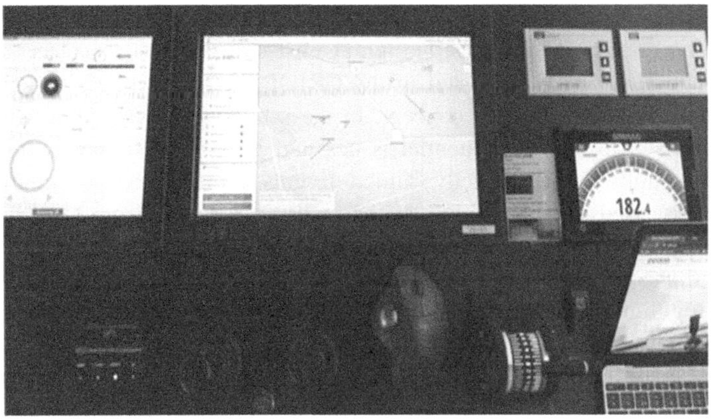

FIGURE 5.5 Console on the bridge of *MF Estelle* – lever beneath the interface.

Source: Zeabuz.

5.4.3 Phase 3: CRIOP Workshop Insights for ROC

In this study, the CRIOP methodology was adapted to address the emerging challenges of digitalisation, autonomy, and remote operations, making it particularly relevant for the context of autonomous ferry operations. The CRIOP analysis offered valuable insights for shaping the HMI design within the ROC. While broader findings, including organisational and operational concerns, are discussed in Section 5.5, this section focuses explicitly on HMI-related considerations.

Preparation for the CRIOP workshop followed a systematic four-step process: stakeholder recruitment, scheduling, documentation gathering, and checklist adaptation.

1. Stakeholder recruitment brought together diverse expertise, including management representatives from Torghatten, system developers from Zeabuz, facilitators from NTNU and SINTEF, risk analysts from Zeabuz, and *MF Estelle* operators. A hybrid format allowed participants from different locations to join.
2. The scheduling phase resulted in a condensed two-day workshop format, reflecting a shorter-than-typical five-day schedule in the oil and gas industry, and considering the project's early design stage and relatively simple operations concept.
3. Documentation and information gathering involved collecting background materials such as hazard analyses, suggestions for critical scenarios based on operational experience, equipment descriptions, a mock-up layout, and operator input. However, complete documentation was limited at this stage.
4. Finally, the checklist adaptation involved selecting and customising relevant elements from CRIOP's evaluation framework. The modified checklist addressed eight key areas: initial issues, remote operation considerations, control and safety systems, job organisation, procedures, working environment, layout design, and training requirements. This customisation ensured the evaluation stayed focused on the specific challenges of autonomous ferry ROCs.

The workshop promoted shared understanding and alignment with industry best practices and standards by examining how work was actually performed compared with how it was expected to be performed. It began with an introductory presentation, where the *MF Estelle* operator shared routines, team dynamics, and challenges related to the transition to unmanned operation and an ROC, helping to identify relevant scenarios. Following this, an HF expert introduced the CRIOP methodology and its risk-based approach. During the general analysis phase, checklists were used to review the current operation and ROC design against industry best practices and standards such as OpenBridge and IEC 63303, as listed in the CRIOP method, while also helping participants become more familiar with the ROC concept. In the scenario analysis, three key scenarios were selected and discussed to identify potential issues and mitigation strategies through design and development. A simplified Sequentially Timed Events Plotting (STEP) method (Hendrick & Benner, 1987) was used during the workshop, with a formal diagram subsequently created. Finally, in the wrap-up session, participants shared perspectives and future goals. A results report was prepared and shared after the workshop.

The analysis highlighted the need for a more systematic task and cognitive workload analysis to strengthen the design foundation.

Task analysis maps what operators do – their actions, decisions, and information needs – providing the foundation for assessing cognitive workload. Evaluating each task step in terms of attention, memory, reasoning, and decision-making helps reveal where workload peaks occur or where task design exceeds human capacity. This integration supports the design of interfaces and procedures that balance workload and maintain SA across varying operational conditions.

Aligning development with IEC 63303 (2024) and using structured frameworks such as OpenBridge (OICL, 2024) enhances interface usability and consistency. These practices also support SA by ensuring that information is presented clearly and coherently, enabling operators to detect, interpret, and anticipate events. The standardised use of symbols, layouts, and colour coding further improves usability and reduces cognitive load.

Automation transparency is also identified as a crucial design requirement. Operators must be able to interpret system behaviour and anticipate upcoming actions, particularly when manual intervention may be required.

The alarm system is a key component of usability. It should be based on hazard analysis and aligned with best-practice standards such as EEMUA 191 (2024). Alarms must be concise, clear, and actionable to support decision-making effectively. To further improve SA, the integration of additional tools, such as 360-degree bird's-eye-view displays and CCTV systems, is recommended. When aligned with relevant IEC or ISO standards, such as IEC 62676, these tools should provide clear visuals with minimal latency, enabling effective real-time monitoring and navigation.

Finally, the analysis highlighted the importance of systematic user testing. Scenario-based evaluations are essential for identifying usability problems, validating interface performance, and ensuring reliability across different operational contexts.

5.5 CHALLENGES AND CONCERNS IDENTIFIED THROUGH CRIOP

The CRIOP workshop revealed a range of challenges and concerns regarding the design and operation of the ROC from multiple perspectives. These findings emphasise the importance of adopting the HCD approach, particularly in enhancing SA, usability, and operational safety to achieve MHC.

5.5.1 Challenges and Concerns from the Operator's Perspective

Insights shared by the *MF Estelle* operator during the workshop were grouped into three primary categories:

5.5.1.1 Maintaining SA and Navigational Oversight

Maintaining clear operational awareness from a remote location remains essential. Tools such as top-down map interfaces, panoramic visual displays, and context-filtered chart layers can help operators maintain an accurate mental model of the vessel's environment. However, recognising technical faults and conducting system diagnostics were identified as particularly challenging at a distance. Additionally, autonomous docking systems must be sufficiently advanced to eliminate the current reliance on visual confirmation during mooring, especially in fully unmanned operations.

5.5.1.2 Bridging the Digital Skills Gap

Increasing system complexity requires stronger digital competence. Many experienced seafarers have limited familiarity with modern software interfaces and control layouts. This gap underscores the need for highly intuitive interfaces that reduce cognitive load, as well as comprehensive training strategies. The ROC's design must lower the learning curve and ensure operators can manage core functions with confidence. Important activities to addressing this include user-centred design based on task analysis and systematic user testing of the core functions (Endsley & Kaber, 1999; Norman, 2013).

5.5.1.3 Operational Risk, Passenger Interaction, and Workload Management

Operational safety is a key concern in remote ferry operation. Effective alarm management is therefore essential. The system must distinguish between critical and non-critical alerts to ensure that only necessary information reaches the operator, preventing distraction or overload. Passenger handling poses a distinct challenge because it is inherently unpredictable. Maintaining a sense of operator presence, especially during emergency situations, is crucial. Systems such as onboard speaker announcements, screen-based alerts, and live video streams between ROC operators and passengers can help preserve trust and reduce uncertainty. Visible indications that the ROC is actively monitoring may further enhance reassurance. A temporary solution, such as placing a trained safety manager on board, may be necessary during system testing in a transitional period.

Managing operator workload is equally vital for sustaining performance. The ROC must be designed to minimise mental strain and environmental stressors. This includes regulating operator shifts, providing adequate recovery periods, and controlling workspace conditions such as lighting, noise, and air quality.

5.5.2 Takeaways from CRIOP Checklist Analysis

The checklist analysis aimed to guide future ROC development by integrating HF and HCD principles early. Results were grouped into three categories:

5.5.2.1 ROC Design and Usability

The design should reflect task requirements and user input, with clear alignment with industry standards such as ISO 11064 and ISO 9241-210. Spatial design must support ergonomic posture and task efficiency. Environmental variables like temperature, illumination, and ventilation should be optimised. Particular focus is needed on ensuring reliability and error tolerance in systems supporting mission-critical actions, such as docking procedures.

5.5.2.2 Situational Awareness and Interface Reliability

The ROC interface must support rapid interpretation and seamless command transitions. Interfaces should be guided by standards such as IEC 63303, integrate task analysis results, and support consistent iconography (e.g., from OICL, 2024). Error recovery processes and cognitive workload assessment tools must be embedded. Transparency from AI systems is critical to support operator confidence and decision-making. Additional features, such as directional sound cues, haptic or reactive seats, or overhead visualisation, can improve overall perception and control. Alarms should conform to EEMUA 191 guidance, with tested prioritisation and messaging logic. Monitoring through CCTV should meet the quality and coverage standards of IEC 62676. To ensure robustness, all systems must undergo iterative testing under realistic, high-pressure conditions, with user feedback informing design refinements.

5.5.2.3 Communication and Operational Safety

Effective communication between the ferry, shore-based teams, and emergency responders is critical. Clearly assigned duties and procedures are needed, and a formal task analysis will be essential in future development to structure these interaction protocols. Communication tools (e.g., PA, live video, alert systems) must help passengers feel connected to human oversight. Risks associated with system complexity, human limitations, or organisational gaps should be proactively addressed. Standardised procedures must support safe fallback actions. Training programmes should include simulation-based learning and walkthroughs of abnormal scenarios to boost preparedness. Moreover, systematic live testing should be conducted to ensure that ROC operators and emergency teams are trained and have practised responding to undesirable incidents, such as fires. Support mechanisms for ROC operators during emergencies must be integrated into both organisational structure and interface design.

5.5.3 Emergency Scenario Analysis

Three representative emergency situations were selected for detailed analysis, reflecting *MF Estelle*'s short route, constrained environment, and compact design. Each scenario revealed critical design requirements for the ROC.

The onboard fire scenario (Figure 5.6) exposed the complexity of coordinating water-based evacuation procedures through remote systems. In the absence of onboard crew, the ROC must guide passengers through emergency procedures using only audio and visual communication systems. This scenario highlighted the need for intuitive alarm systems, clear evacuation instructions, and real-time monitoring of passenger movements through CCTV systems. In addition, the scenario underscored the critical importance of timely external emergency assistance, such as fire brigade support, as an unmanned passenger vessel must be reachable within approximately 3–5 minutes to ensure safe and effective emergency response.

The ROC fire or communication failure scenario addressed catastrophic failures in the control centre itself. Analysis revealed the necessity for distributed backup systems, automatic vessel fail-safe behaviours, and clear protocols for transferring control to alternative control locations. The scenario emphasised that passenger safety must be maintained even during a complete loss of remote control capability.

The collision or grounding scenario focused on navigation hazards specific to shallow, congested urban waterways. The analysis identified the need for enhanced sensor systems providing real-time depth monitoring, dynamic route adjustments, and AI-assisted navigation support. These findings informed recommendations for solutions that can operate independently when communication latency prevents timely remote intervention.

Scenario exploration was based on task analysis, systematic STEP analysis, and bow-tie analysis, all conducted in a collaborative setting with users and experienced safety/HF experts to ensure the study included human, organisational, and technical factors. These scenarios highlighted the need for robust emergency protocols, real-time hazard detection, effective communication, and coordinated response plans. The selected scenarios were analysed using the STEP method to identify proactive barriers that reduce the likelihood of such events and mitigate their potential impact. This process supports the early integration of safety measures into system design, as illustrated in Figure 5.6.

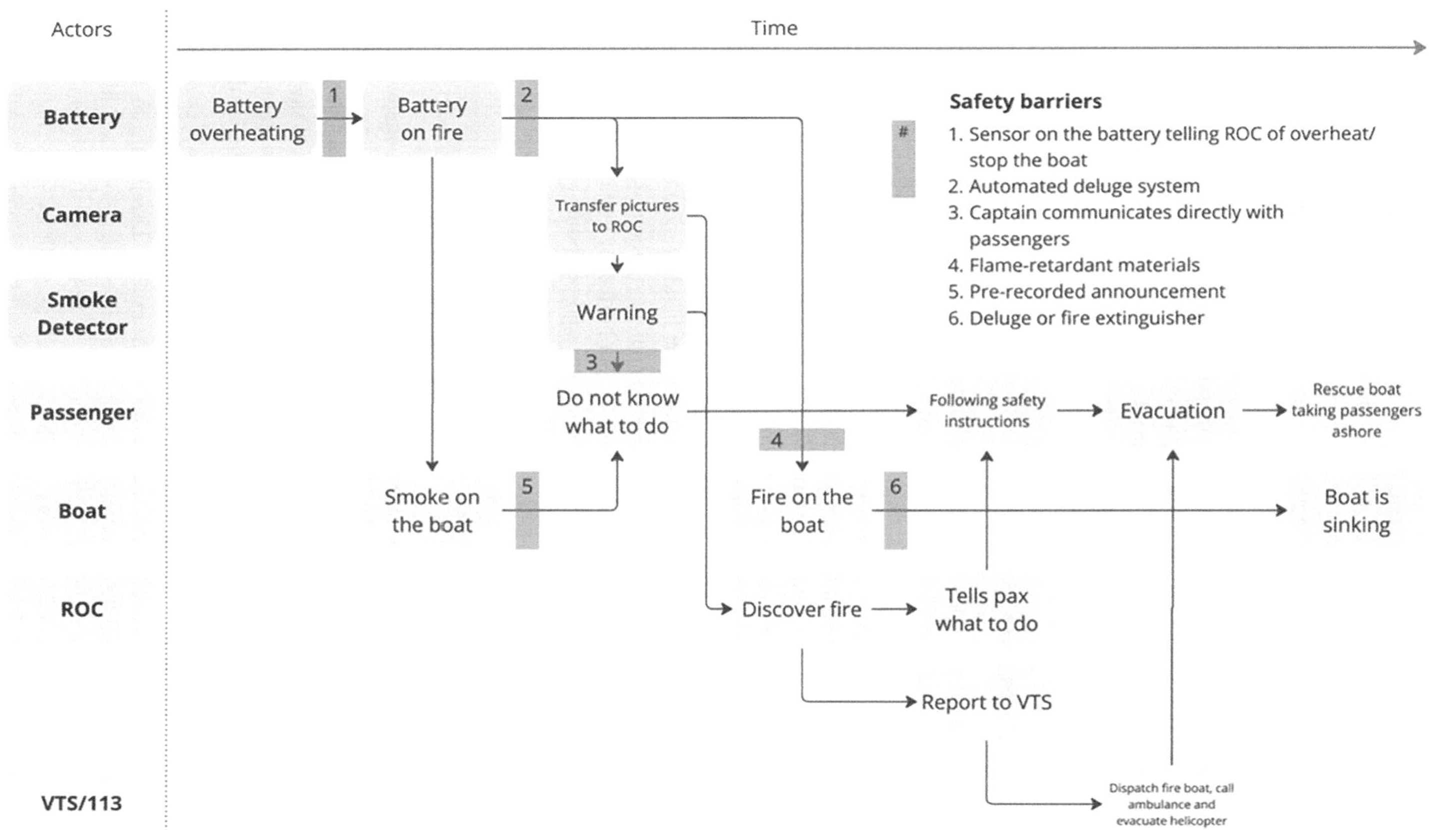

FIGURE 5.6 STEP analysis combined with safety barriers for Scenario 1: Fire onboard the ferry.

5.6 DISCUSSION AND KEY INSIGHTS

As autonomous ferry operations advance towards remote supervision, several key themes have emerged from this study that have broader implications for design, safety, and organisational readiness. The following sections synthesise lessons learned from trial operations, HMI development, and CRIOP evaluation, focusing on guiding future implementation and research.

CRIOP scenario analysis also strengthens organisational readiness. By bringing managers, operators, designers, and developers together in realistic scenarios, the method helps identify MTO (man/human–technology–organisation) issues, such as unclear roles, communication gaps, and weak procedures. These workshops allow participants to evaluate how well technology and procedures support human performance during demanding situations and help build shared understanding across stakeholder groups. This participatory process not only identifies system vulnerabilities but also improves the organisation's ability to adapt and respond effectively as operations become more autonomous.

5.6.1 Operator Competence and Training

While the transition to remote operation aims to reduce direct human involvement in ferry navigation, human oversight remains essential. This is required not only to comply with the EU AI Act and IMO requirements but also to ensure that operators can safely intervene during unexpected situations. Remote operators must be able to understand and operate complex digital systems to respond effectively and activate fallback modes when required. This requires a combination of strong digital proficiency and foundational knowledge of conventional navigation, particularly in managing interactions with other vessels and interpreting environmental conditions.

Maintaining this dual competence is especially important in a supervisory role, where direct control is limited and rapid decision-making depends on digital SA. Without regular training and skill maintenance, there is a risk of deskilling. If automation handles most operational tasks and the human role becomes purely supervisory, essential manual navigation skills may quickly erode. Therefore, the transition to remote operations must be supported by organisational strategies that include tailored training programmes, realistic simulations, and periodic skill refreshment to ensure readiness and resilience.

5.6.2 ROC Design Implications

The shift from onboard operation to shore-based supervision represents a fundamental change in how vessel control is conceived and executed (Man et al., 2018). In this new paradigm, operators function more as supervisors than as direct navigators. This shift reinforces longstanding automation challenges: autonomous systems continue to rely on human oversight (Bradshaw et al., 2013), rare interventions require high levels of expertise (Bainbridge, 1983), and prolonged supervisory monitoring can reduce vigilance, even though crises demand rapid and skilled responses (Hancock, 2021).

SA must therefore be achieved through mediated digital interfaces, since operators no longer have access to direct sensory cues from the vessel environment. Designing an effective ROC requires more than arranging multiple screens; it requires an integrated socio-technical design that supports perception, communication, and decision-making. A holistic design should combine well-structured visual displays with multisensory feedback, such as auditory cues, haptic signals, and tactile controls, to compensate for the absence of physical presence onboard. All relevant information, including vessel status, environmental conditions, passenger movements, docking activity, and ship integrity, must be digitised and presented in a way that supports timely, accurate decision-making.

Alarm management also plays a critical role. Centralised alarm control, alert prioritisation, and intuitive messaging are essential to minimise distractions and maintain focus during critical situations. Furthermore, communication systems must be designed to support seamless coordination within the ROC team and with external stakeholders, including passengers and emergency responders. These tools must operate within clear protocols and dependable infrastructure to preserve both safety and trust.

5.6.3 Balancing Transparency and Workload

One of the primary challenges in HMI design for autonomous systems is striking a balance between transparency and cognitive load. As autonomy increases, it becomes essential for operators to understand how the system perceives its environment and makes decisions. However, displaying too much information at once can overwhelm users and reduce their ability to act effectively.

This trade-off was observed during the development of *MF Estelle*'s HMI. As shown in Figure 5.7, early design features, such as colour-coded icons and frequent pop-up alerts, increased transparency but also contributed to cognitive overload. These experiences emphasised the need to present the right amount of information, at the right time, using a consistent and familiar interface. Design systems such as OpenBridge provide useful conventions for ensuring clarity and usability.

To maintain this balance, input from a diverse group of users is essential during the design process. Operators with varying levels of experience should be involved in testing prototypes and evaluating interface behaviour. Scenario-based testing can help identify the cognitive demands placed on users and guide interface adjustments. Ultimately, the goal is to ensure that automation transparency supports SA and informed decision-making without introducing new risks.

5.6.4 Future Research

This study was conducted during the early design phase of the ROC, and therefore, its findings and recommendations require further validation through continued development and testing. Scenario

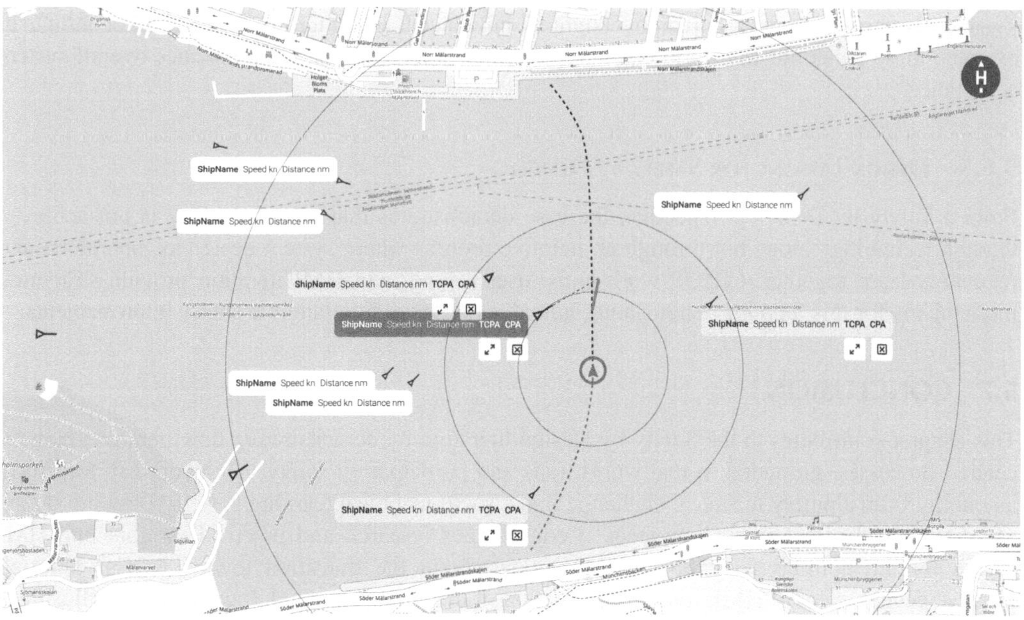

FIGURE 5.7 Initial design of the operator interface.

analysis, while useful, was somewhat limited in scope and should be expanded to include a broader range of emergency and operational conditions.

A key motivation behind ROC development is cost reduction. However, the findings suggest that realising this objective, especially in multi-vessel oversight, demands competent autonomous systems and potentially extensive redesign of ferry and port infrastructure. In some cases, delays in transferring control between vessels may necessitate maintaining dedicated stations, which may challenge anticipated efficiency gains. Achieving safe and cost-effective remote operations requires sustained investment in the early design phase, where requirements, user needs, and operational contexts can be validated before large-scale implementation. Early attention to HCD prevents costly redesigns later in operation and supports user acceptance. Establishing pilot projects and small-scale prototypes allows organisations to experiment, identify unforeseen challenges, and co-create solutions with end users, building both competence and trust in remote operations. Future work should focus on several interrelated steps to ensure robust and scalable implementation:

1. **Iterative testing of ROC prototypes** under realistic trial conditions to evaluate usability, reliability, and workload.
2. **Extending scenario-based evaluations** to include systematic safety validation and human performance assessments.
3. **Addressing industry-wide challenges** such as standardisation of digital components, interface consistency, and structured operator training.
4. **Integrating advanced technologies** – for example, AI-based decision support, adaptive automation, and real-time hazard detection – to enhance operational resilience.
5. **Developing Human Readiness Levels (HRL)** in parallel with Technical Readiness Levels (TRL) to ensure that human, organisational, and technical maturity progress in balance, supporting safety, usability, and efficiency (NASA, n.d.; ANSI/HFES, 2021).

Together, these steps establish a systematic pathway for transitioning from experimental pilots to fully integrated, safe, and trusted remote operations.

Finally, passenger acceptance remains a critical factor. Perceived safety, technological familiarity, and reassurance through human oversight all influence trust. Future research should continue to examine these dynamics to inform effective communication strategies and enhance overall system acceptance.

5.6.5 Design Lessons for Safety by Design

From a Safety by Design perspective, the case demonstrates that safe autonomy is not achieved through technology alone but through an iterative process where system designers, operators, and regulators learn together. CRIOP workshops, user testing, and HMI iteration provide structured mechanisms for this learning – translating human experience into tangible safety improvements.

5.7 CONCLUSION

This chapter contributes to the Safety by Design literature by demonstrating how iterative, human-centred processes, grounded in real-world trials and participatory safety validation, can transform autonomy from a purely technical challenge into a socio-technical learning system. Drawing on the *MF Estelle* and milliAmpere2 projects, it consolidates research and practical insights into HMI design for autonomous ferry operations, particularly in the transition toward ROCs. Through iterative development of HMIs and the application of the CRIOP methodology, this work illustrates how early integration of HF, standardised design frameworks, and collaborative evaluation fosters safer and more resilient autonomous operations. It further highlights the importance of HCD, automation transparency, and cognitive workload management in developing systems that are reliable, user-friendly, and maintain MHC in complex automated environments.

The milliAmpere2 trial highlighted the need for transparent, intuitive interfaces, while the *MF Estelle* development demonstrated practical solutions to support SA by reducing cognitive load and improving alarm systems. The CRIOP workshop emphasised the value of early HF integration and the use of structured design approaches, such as task and cognitive work analysis, as well as adherence to standards like IEC 63303 and the OpenBridge design system.

The transition from onboard to remote operations fundamentally transforms the operator's role and environment. Remote operators must rely entirely on mediated information, requiring careful design attention to maintain SA and operational effectiveness.

This shift demands robust training programmes and iterative system evaluation to ensure operators can manage both routine operations and unexpected scenarios. For passenger management, communication systems, such as public announcements, visual cues, and two-way video calls, are critical for maintaining a sense of human presence and reassurance during unmanned operations.

The research provides practical answers to three critical HF questions facing the industry:

(i) **Who will the remote operators be, and what will their responsibilities entail?**
Remote operators will need to combine maritime expertise with technical competence. Their primary responsibilities include monitoring vessel status, managing alarms, handling abnormal situations, and ensuring passenger safety. These roles require intuitive interfaces and comprehensive training programmes that address the digital skills gap identified in this study.

(ii) **How will passengers be managed when the vessel is unmanned?**
Effective passenger management relies on robust communication technologies (PA systems, visual cues, two-way video) to ensure passengers feel connected to a human presence and can follow instructions during both normal operations and emergencies. This may require infrastructure modifications to vessels and terminals, following the hierarchy of controls to prioritise design-level safety interventions (NIOSH, 2024).

(iii) **How will ROC operations handle emergencies or dangerous situations?**
Emergency response capabilities depend on predefined protocols, alarm prioritisation, real-time SA tools, scenario-based training, continuous system evaluation, and close coordination with authorities.

Although this study focuses on small urban ferries, the principles and recommendations are scalable to larger vessels and more complex contexts. Broader implementation will require advanced navigation systems, multi-vessel control capabilities, and reliable communication protocols. Future research should explore how these elements can be adapted, tested, and validated in larger, more demanding maritime environments.

While offering practical design guidance, the study also acknowledges key limitations. The research was conducted during the early phases of ROC development, and some scenarios were exploratory. Further testing in operational settings is necessary to validate and refine the recommendations. Using an agile development approach is particularly beneficial for safety-critical systems like autonomous ferries because it enables early detection of design flaws, continuous integration of HF input, and incremental safety validation. Short iterative cycles allow teams to adapt HMI and ROC design as operational experience, regulatory requirements, and user needs evolve. This iterative learning process aligns closely with Safety by Design, ensuring that critical issues related to usability, SA, and human–automation coordination are systematically addressed well before full-scale deployment.

Future development of autonomous ferry operations must continue to prioritise HF throughout the design process. Key areas for further research include validating ROC prototypes in operational settings, expanding scenario-based safety evaluations, and investigating human readiness levels alongside technical development. As the maritime industry advances toward greater autonomy, maintaining MHC through thoughtful HMI design remains essential for safe, efficient, and publicly accepted autonomous ferry services.

ACKNOWLEDGEMENT

This chapter has been funded and supported by the Norwegian Research Council through the MAS project (No. 326676), the SCAPE project (No. 327714), and the MIDAS project (No. 331921).

REFERENCES

Alsos, O. A., Saghafian, M., Veitch, E., Petermann, F.-M., Sitompul, T. A., & Park, J. (2024). Lessons learned from the trial operation of an autonomous urban passenger ferry. *Transportation Research Interdisciplinary Perspectives*. https://doi.org/10.1016/j.trip.2024.101142

ANSI/HFES. (2021). *Human Readiness Level Scale in the System Development Process* (ANSI/HFES 400-2021). Human Factors and Ergonomics Society.

Bainbridge, L. (1983). Ironies of automation. In *Analysis, Design and Evaluation of Man–Machine Systems* (pp. 129–135). Pergamon.

Bradshaw, J. M., Hoffman, R. R., Woods, D. D., & Johnson, M. (2013). The seven deadly myths of autonomous systems. *IEEE Intelligent Systems*, *28*(3), 54–61.

Brekke, E. F., Eide, E., Eriksen, B.-O. H., Wilthil, E. F., Breivik, M., Skjellaug, E., Helgesen, Ø. K., Lekkas, A. M., Martinsen, A. B., Thyri, E. H., Torben, T., Veitch, E., Alsos, O. A., & Johansen, T. A. (2022). milliAmpere: An autonomous ferry prototype. *Journal of Physics: Conference Series*, *2311*, 012029. https://doi.org/10.1088/1742-6596/2311/1/012029

Calvert, S. C., Heikoop, D. D., Mecacci, G., & van Arem, B. (2019). A human-centric framework for the analysis of automated driving systems based on meaningful human control. *Theoretical Issues in Ergonomics Science*, *21*, 1–29.

Casner, S. M., Geven, R. W., Recker, M. P., & Schooler, J. W. (2014). The retention of manual flying skills in the automated cockpit. *Human Factors*, *56*(8), 1506–1516.

EEMUA (Engineering Equipment and Materials Users Association). (2024). *Alarm systems – A Guide to Design, Management and Procurement*. EEMUA 191:2024.

Eide, E., Breivik, M., Brekke, E. F., Eriksen, B.-O. H., Wilthil, E., Helgesen, Ø. K., Thyri, E. H., Veitch, E. A., Alsos, O. A., & Johansen, T. A. (2025). The autonomous urban passenger ferry milliAmpere2: Design and testing. *Journal of Offshore` Mechanics and Arctic Engineering*, *147*(3), 031409. https://doi.org/10.1115/1.4067370

Endsley, M. R. (1988). Design and evaluation for situation awareness enhancement. Proceedings of the Human Factors Society 32nd Annual Meeting (pp. 97–101). Santa Monica, CA: Human Factors Society.

Endsley, M. R. (2019). Human factors and aviation safety: Testimony to the United States House of Representatives Hearing on Boeing 737-Max8 crashes.

Endsley, M. R., & Kaber, D. B. (1999). Level of automation effects on performance, situation awareness and workload in a dynamic control task. *Ergonomics*, *42*(3), 462–492.

Energy Institute. (2011). *Guidance on Human Factors Safety Critical Task Analysis (SCTA)*. Energy Institute.

European Parliament. (2024). *Artificial Intelligence Act (AIA)*. Retrieved from https://www.europarl.europa.eu/doceo/document/TA-9-2024-0138_EN.html

European Union. (2026). *Article 14: Human oversight. In Artificial Intelligence Act* (Entry into force: August 2, 2026, as per Article 113). Retrieved from https://artificialintelligenceact.eu/article/14

Ferrer, X., van Nuenen, T., Such, J. M., Coté, M., & Criado, N. (2021). Bias and discrimination in AI: A cross-disciplinary perspective. *IEEE Technology and Society Magazine*, *40*(2), 72–80.

Goerlandt, F., & Pulsifer, K. (2022). An exploratory investigation of public perceptions towards autonomous urban ferries. *Safety Science*, *145*, 105496. https://doi.org/10.1016/j.ssci.2021.105496

Hancock, P.A. (2021). Months of monotony–moments of mayhem: Planning for the human role in a transitioning world of work. *Theoretical Issues in Ergonomics Science*, 22(1), 63–82.

Hendrick, K. M., & Benner, L. (1987). *Investigating Accidents with STEP*. Marcel Dekker.

Inspenet. (2024). *Swedish Autonomous Ferry Triumphs in Remote Operation Test*. Retrieved from https://inspenet.com/en/noticias/autonomous-ferry-passes-test-in-sweden

International Electrotechnical Commission. (2013). IEC 62676 *Video surveillance systems for use in security applications*.

International Electrotechnical Commission. (2021). *IEC 62288:2021 Maritime Navigation and Radiocommunication Equipment and Systems – Presentation of Navigation-Related Information on Shipborne Navigational Displays – General Requirements, Methods of Testing and Required Test Results*.

International Electrotechnical Commission. (2024). *IEC 63303:2024 Human Machine Interfaces for Process Automation Systems*.

International Organization for Standardization. (2000). *ISO 11064: Ergonomic Design of Control Centres*.

International Organization for Standardization. (2019). *ISO 9241-210:2019 Ergonomics of Human-System Interaction – Part 210: Human-Centred Design for Interactive Systems*.

Johnsen, S. O., & Park, J. (2025). Meaningful human control in digitalization, automation/AI, and remote oversight. In Proceedings of the European Safety and Reliability Conference (ESREL'25).

Johnsen, S. O., Thieme, C., Aminoff, H., Park, J., Fernander, M., Boren, M., Kjøl, Å. A., & Kirwan, B. (2025). *CRIOP: A Scenario Method for Crisis Intervention and Operability Analysis* (Version 2025). SINTEF Digital.

Kosmajac, S. (2024, October 7). 'World's first' remotely operated electric ferry wraps up trial in Sweden. *Offshore Energy*. Retrieved from https://www.offshore-energy.biz/mf-estelle-world-first-autonomous-electric-ferry-wraps-up-trial-in-sweden

Man, Y., Lundh, M., Porathe, T., & MacKinnon, S. (2015). Human factor issues in remote monitoring and controlling of autonomous unmanned vessels. *Procedia Manufacturing*, *3*, 2674–2681.

Man, Y., Weber, R., Cimbritz, J., Lundh, M., & MacKinnon, S. N. (2018). Human factor issues during remote ship monitoring tasks: An ecological lesson for system design in a distributed context. *International Journal of Industrial Ergonomics*, *68*, 231–244. https://doi.org/10.1016/j.ergon.2018.07.004

Maritime Safety Committee (MSC). (2023). *Development of a Goal-Based Instrument for Maritime Autonomous Surface Ships (MASS): Report of the MSC-LEG-FAL Joint Working Group on Maritime Autonomous Surface Ships (MASS) on Its Second Session* (MSC 107/5/1). International Maritime Organization.

Maritime Safety Committee (MSC). (2024). 109th session (MSC 109). International Maritime Organization. https://www.imo.org/en/MediaCentre/MeetingSummaries/Pages/MSC-109th-session.aspx

Mittelstadt, B. D., Allo, P., Taddeo, M., Wachter, S., & Floridi, L. (2016). The ethics of algorithms: Mapping the debate. *Big Data & Society*. https://doi.org/10.1177/2053951716679679

Moray, N. (1986). Monitoring behavior and supervisory control. In K. R. Boff, L. Kaufman, & J. P. Thomas (Eds.), *Handbook of Perception and Human Performance* (Vol. II, pp. 40/41–40/51). John Wiley & Sons.

Namireddy, P. R., Reddy, M. K., Zadeh, M. K., Thieme, C. A., Skjetne, R., Sorensen, A. J., Aanondsen, S. A., Breivik, M., & Eide, E. (2019). Zero-emission autonomous ferries for urban water transport: Cheaper, cleaner alternative to bridges and manned vessels. *IEEE Electrification Magazine*, *7*(4), 32–45. https://doi.org/10.1109/MELE.2019.2943954

NAS (National Academies of Sciences, Engineering, and Medicine) (2022). *Human-AI Teaming: State of the Art and Research Needs*. The National Academies Press.

NASA (n.d.). *Technology Readiness Levels*. Retrieved from https://www.nasa.gov/directorates/somd/space-communications-navigation-program/technology-readiness-levels

NIOSH (National Institute for Occupational Safety and Health) (2024, January 20). *Hierarchy of Controls*. Retrieved from https://www.cdc.gov/niosh/hierarchy-of-controls/about/index.html

Norman, D. A. (2013). *The Design of Everyday Things* (Revised and expanded ed.). Basic Books.

Ocean Industries Concept Lab (OICL) (2024). *OpenBridge Design System*. Retrieved from https://www.oicl.no/content/openremote-has-started

Parasuraman, R., Molloy, R., & Singh, I. L. (1993). Performance consequences of automation-induced complacency. *International Journal of Aviation Psychology*, *3*(1), 1–23.

Pearl, J., & Mackenzie, D. (2018). *The Book of Why: The New Science of Cause and Effect*. Basic Books.

Petermann, F.-M., Alsos, O. A., Veitch, E. A., Eide, E., & Breivik, M. (2024). Touchpoints and systems for unmanned autonomous urban ferry operation. *Journal of Physics: Conference Series*, 2867(1), 012032. https://doi.org/10.1088/1742-6596/2867/1/012032

Porathe, T. (2021). Human-automation interaction for a small autonomous urban ferry: a concept sketch. In Proceedings of the 31st European Safety and Reliability Conference (ESREL'21). https://doi.org/10.3850/978-981-18-2016-8_543-cd

Porathe, T. (2022). Remote monitoring of autonomous ships: A quickly getting into the loop display (QGILD). *Advances in Transportation*, 60, 691–697. https://doi.org/10.54941/ahfe1002506

Rødseth, Ø. J. (2017). From concept to reality: Unmanned merchant ship research in Norway. In Proceedings of the IEEE Underwater Technology (UT). https://doi.org/10.1109/UT.2017.7890328

Simic, V., & Alsos, O. A. (2023). Automation transparency: Designing an external HMI for autonomous passenger ferries in urban waterways. In Proceedings of the ACM Designing Interactive Systems Conference (pp. 1145–1158). https://doi.org/10.1145/3563657.3596130

van de Merwe, K., Mallam, S., Engelhardtsen, Ø., & Nazir, S. (2023). Operationalising automation transparency for maritime collision avoidance. *TransNav: International Journal on Marine Navigation and Safety of Sea Transportation*, 17(2), 333–339.

Veitch, E., Alsos, O. A., Saghafian, M., Petermann, F.-M., Sitompul, T. A., & Park, J. (2024). Dataset on passenger acceptance during autonomous ferry public trials: Questionnaires and interviews. *Data in Brief, 54*, 110282. https://doi.org/10.1016/j.dib.2024.110282

Veitch, E. A., & Alsos, O. A. (2021). Human-centered explainable artificial intelligence for marine autonomous surface vehicles. *Journal of Marine Science and Engineering*, *9*(11), Article 1227. https://doi.org/10.3390/jmse9111227

Veitch, E. A., Kaland, T. C. G., & Alsos, O. A. (2021). Design for resilient human-system interaction in autonomy: The case of a shore control centre for unmanned ships. *Proceedings of the Design Society*. https://doi.org/10.1017/pds.2021.102

Vicente, K.J., & Rasmussen, J. (1992). Ecological interface design: theoretical foundations. *IEEE Transactions on Systems, Man, and Cybernetics*, 22(4), 589–606. https://doi.org/10.1109/21.156574

Woods, D. D. (2016). The risks of autonomy: Doyle's catch. *Journal of Cognitive Engineering and Decision Making*, *10*(2), 131–133.

6 Designing for Safety in Multi-Unit Operation

Insights from Conceptual Design in Petroleum

Maren Helene Rø Eitrheim, Lars Hurlen, and Vidar Hepsø

6.1 INTRODUCTION

There is a growing trend towards centralised control of industrial operations to enhance efficiency, address workforce shortages, and optimise the use of expertise. At the same time, industries are shifting towards simpler, highly automated designs that are expected to minimise the need for human intervention. However, the implementation of centralised operations varies across industries. Some sectors, such as hydropower and chemical processing, have successfully operated centralised control centres for years, while others are still in the planning, development, or early operational stages.

In the petroleum industry, Equinor established its first onshore control centre for offshore operations in 2017. The business case for centralised operations depends on the number of facilities that can be safely controlled from one location, ensuring both production efficiency and a high level of health, safety, and environmental (HSE) performance (Saeverhagen, Kellas & Bouillouta, 2013). To address these challenges holistically, Equinor funded a three-year design-oriented study conducted by the Institute for Energy Technology (IFE). The study aimed to provide practical input for company operating guidelines by exploring the technologies, work processes, competencies, and governance principles needed to support safe and efficient multi-unit operations from an onshore control centre.

This chapter builds on insights from that study, which used a feasibility study approach, and discusses key operational objectives, human factors challenges, and design considerations for multi-unit control in the petroleum industry. While grounded in the offshore sector, these findings contribute to a broader understanding of centralised, remote, and multi-unit operations across safety-critical industries.

6.2 TERMINOLOGY

6.2.1 Multi-Unit Operation

The definition of a 'unit' varies across industries, reflecting differences in operational structures and control strategies. In hydropower, a unit may correspond to an entire power plant, with some control centres managing more than 60 plants remotely. In aviation, Avinor has established a remote tower centre in Bodø, Norway, which provides air traffic services for multiple airports. In these cases, 'remote' refers to managing operations from a centralised location, often located some distance away from the operational sites. These examples illustrate how centralised control can extend across multiple operational sites.

DOI: 10.1201/9781003741824-8

TABLE 6.1
Glossary of Terms

Term	Definition
Heterogeneous concepts	Multiple different units operate from the same control centre.
Homogeneous concepts	Multiple identical units operate from a common environment.
Multi-field operation	Centralised control of multiple oil and gas fields from a single location.
Multi-unit operation	Multiple plants, installations or facilities operated from a single control room.
Remote operation	Managing operations from a centralised location outside the operational site.
Small Modular Reactors (SMRs)	Nuclear reactors with up to 300 MW(e) capacity, designed for compactness and modular factory-built deployment.
Unit	The physical facility with its associated control system and user interface.

In the nuclear industry, a unit has traditionally referred to a single reactor, each with its own dedicated control room. In some power plants, two or more units may be controlled from shared or adjacent control rooms. However, each unit typically has its own team of operators. While the teams generally work independently, they may provide support to one another during high-demand periods or to cover absences due to sick leave. In contrast, Small Modular Reactors (SMRs) may involve multiple reactor modules controlled from a single control environment by a single team of operators, with the possibility of shared infrastructure and support systems across modules.

Through our research project with Equinor, a unit in offshore petroleum operations was understood as 'the physical facility with its associated control system and user interface'. This definition aligns with centralised control of multiple offshore installations, where onshore centres manage operations across geographically distributed sites. For this chapter, we use the term 'multi-unit operation' to describe cases in which multiple plants, installations, or facilities are controlled from a single control room (see Table 6.1). This working definition provides a common reference point for discussing operational objectives, human factors challenges, and design considerations in centralised control environments.

6.2.2 Heterogeneous and Homogeneous Multi-Unit Operations

The design of the control environment and the organisation of monitoring and control for two or more units from a single control room will be related to the units' characteristics, that is, how similar they are. We can imagine two hypothetical and principled extremes:

1. Homogeneous concepts, where multiple *identical* units are operated from a common environment, or
2. Heterogeneous concepts, where multiple *different* units are operated from the same control centre.

In the petroleum industry, a homogeneous concept may consist of several similar, unmanned production platforms that are remotely operated from a centralised location, either offshore or onshore, with no personnel present during production. In contrast, a heterogeneous concept could encompass multiple distinct facilities, remotely operated from a single location. We have found it helpful to distinguish between similarities in unit *appearance* and *design/behaviour*. A hypothetical development path for remote multi-unit operation can be illustrated through four key scenarios (see Figure 6.1), each shaping the operational complexity and competence requirements for co-located operation.

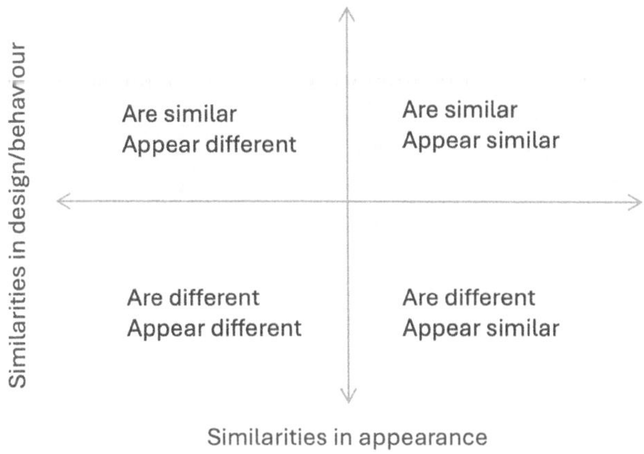

FIGURE 6.1 Similarities and differences between units in heterogeneous concepts.

One heterogeneous example is platforms with comparable technology and functions, but with different control systems (are similar, appear different). Another example is platforms that differ technologically and functionally, with diverse maintenance requirements, control systems, and operational strategies (are different, appear different). We can also imagine combinations of the above, potentially integrated with other types of facilities, such as subsea wells and/or offshore wind farms.

Given the scope of the research project with Equinor, this chapter focuses on heterogeneous multi-unit operations.

6.3 STANDARDS AND FRAMEWORKS

While multi-unit operations are well-established in some industries, onshore control of multiple offshore installations is a more recent development. The introduction of such first-of-a-kind concepts raises important questions about how control environments should be designed and operated. In addition to operational challenges, broader organisational considerations must be addressed, including the economic and technological feasibility, scalability, and sustainability of the operational concept, as well as its alignment with regulatory requirements. Developing future operational concepts can therefore be viewed as a novel and complex endeavour, often shaped by conflicting stakeholder interests. As such, this type of open-ended 'wicked problem' (Rittel & Webber, 1973) may not be fully addressed by traditional scientific approaches or existing guidelines and standards.

In the petroleum industry, ISO 11064 'Ergonomic design of control centres' (ISO, 2000) is the established standard for designing control environments. It promotes a human-centred design approach, emphasising multidisciplinary involvement and iterative user feedback. This helps ensure that proposed solutions align with operational needs, support human performance, and minimise the risk for human error. While the standard does not explicitly address multi-unit operations, it provides a structured framework with design phases applicable to novel control room concepts. The first three phases are particularly relevant for first-of-a-kind developments: Phase A (Clarification), Phase B (Analysis and Definition), and Phase C (Conceptual Design). These stages focus on identifying operational objectives, understanding user needs, and exploring potential design solutions.

The CRIOP (Crisis Intervention and Operability Analysis) methodology (Johnsen et al., 2025) has, since its first publication in 1990, been developed to minimise hazards and risks as early as possible and ensure systematic integration of human factors throughout the design and operational lifecycle. Although not applied in this study, CRIOP offers relevant tools for early-stage analysis. It includes two main components: a general analysis based on a structured checklist to assess work as imagined, and a scenario-based analysis involving end users to explore work as done. While recent updates to the checklist – addressing the complexities of multi-unit operations – were not available at the time of this study, insights from the work contributed to their development. These updates recommend early identification of unit characteristics (e.g., homogeneity or heterogeneity) and systematic mapping of human factors challenges and mitigation strategies. CRIOP may support ISO 11064 Phases A and B by clarifying the operational scope and user demands in future applications.

To support conceptual design in early-stage development, the broader project also drew inspiration from a feasibility study framework (Hurlen & Skraaning, 2023). While not detailed in this chapter, the framework emphasises exploration of novel operational concepts under conditions of uncertainty and ambiguity. It complements ISO 11064 by offering a flexible mindset suited to complex socio-technical systems in transition.

The study of multi-unit operations from an onshore control centre was guided primarily by ISO 11064 Phases A through C. The following sections outline key operational objectives, human factors challenges, and conceptual dilemmas identified during the study.

6.4 OPERATIONAL OBJECTIVES

In Phase A (clarification), a cross-industry experience review, combined with input from operators and stakeholders, led to the identification of a set of overarching operational objectives. These objectives balance core operational needs – such as safety, capacity, and flexibility – with factors necessary for deployment, scalability, and long-term viability, including vendor independence and the ability to attract and retain skilled personnel.

Safety. Ensure safe operation under all conditions. This includes maintaining adequate situation awareness, efficient detection and identification of disturbances and events across all units, and strong problem-solving and collaboration capabilities – both within the control room and with external teams, such as offshore maintenance personnel.

Capacity. Manage multiple units safely and efficiently under expected operating conditions. This includes meeting productivity goals while ensuring effective use of human and technological resources. Key research questions include: How many units, of what complexity and diversity, can be effectively managed by a control team during normal operation and when simultaneous events occur?

Flexibility. Adapt operations dynamically in response to changing system conditions, priorities, or disturbances. This includes redistributing tasks, modifying coordination strategies, and shifting focus as needed – without compromising safety or performance.

Scalability. Accommodate evolving workload and operational scope over time while sustaining effective performance. This includes both short-term adjustments in the volume or complexity of tasks and longer-term changes, such as the integration of new assets or the expansion of control responsibilities.

Realisability. Ensure the concept can be implemented cost-effectively, robustly, and practically. The solution should be manageable in day-to-day operations and have a strong likelihood of success in real-world settings.

Replicability. Enable the concept to be implemented in other locations or at future points in time using the same core design principles and solution strategies. While adjustments may be needed to account for technical, cultural, or organisational differences, the overall approach should be adaptable and repeatable.

Vendor independence. Support the use of products and services from multiple third-party vendors. The centre should not be tied to a single supplier but should remain open and compatible with diverse vendor solutions over time.

Attractive workplace. Create a work environment that attracts and retains competent personnel. To reach its full potential, the centre must foster engagement, provide meaningful work, and support long-term professional satisfaction.

6.5 HUMAN FACTORS CHALLENGES

The human factors challenge of operating multiple units from a single onshore control centre was examined, drawing on Phase A of ISO 11064, which focuses on the analysis of operational contexts and user needs. This analysis informed early considerations for functional requirements, which are addressed in Phase B. Two distinct operational contexts were identified: (1) remotely operated facilities without on-site personnel for extended periods, and (2) minimally staffed facilities where a small team – typically 10 to 12 people – is present at all times. Some remotely operated facilities may be designed to operate normally unmanned, with only one scheduled visit per month and fewer than 15,000 maintenance hours annually (Edwards & Gordon, 2015).

These operational contexts introduce a range of implications for control room design and operator support. For instance, clear criteria and robust procedures are needed for starting up and shutting down individual units. Operators may also need to develop new strategies for handling disturbances, as they cannot rely on field personnel to verify process conditions or assist in diagnosing the situation. Mitigating actions must be prioritised and scheduled in alignment with planned maintenance campaigns.

To support decision-making, enhanced sensor capabilities are required through instrumentation, cameras, acoustic monitoring, and mobile sensor platforms such as drones. Moreover, the operators' familiarity with local conditions may diminish over time. To maintain competence and prevent deskilling, regular personnel rotation and targeted training on unit-specific tasks are essential.

Control room operators must continuously monitor whether field personnel are present at a facility and under what conditions they are permitted to remain or required to leave. This situational awareness is particularly critical in emergency scenarios, where a shared understanding of authority, responsibilities, and evacuation criteria between the control room and field personnel is essential. Reliable communication technologies and standardised verbal protocols can support effective task coordination, enhance safety, and ensure precise alignment between onshore and offshore teams.

A major human-factors challenge in multi-unit operation is maintaining situation awareness across multiple units during periods of high workload (O'Hara et al., 2012; Kirkwood, 2019). Situation awareness (Endsley, Bolté & Jones, 2003) is essential for effective emergency response, particularly when simultaneous disturbances occur at different units. A well-tuned alarm system that presents only relevant alarms can support timely detection and prevent resources from being diverted to non-critical issues. Operators should also be trained in an effective handover of unit responsibility to ensure that critical information and situation awareness are preserved during transitions.

Additional vulnerabilities may arise when operators are responsible for multiple facilities that differ significantly in design or behaviour – especially when those differences are subtle, masked, or inconsistently presented. Errors may occur even when the operators have the right intentions, such as acting on the wrong unit. Misinterpretations can also result from inconsistent human-machine interfaces (HMIs), such as when component status indicators or alarm priorities differ between units.

A further challenge is the inappropriate transfer of experience between units – treating dissimilar units as if they were the same or failing to recognise when comparable behaviours can be expected – leading to misguided expectations or decisions.

In summary, there is a need to balance the clear separation between units, so operators can reliably identify which unit they are working on while also efficiently monitoring and comparing multiple units. Interfaces should support both accurate, unit-specific actions and accessible cross-unit awareness without increasing the risk of confusion or cognitive overload.

6.6 CONCEPTUAL DILEMMAS

The operational objectives and human factors challenges were addressed during the concept design Phase C through a series of design workshops with end users, aimed at exploring potential dilemmas and identifying promising design principles.

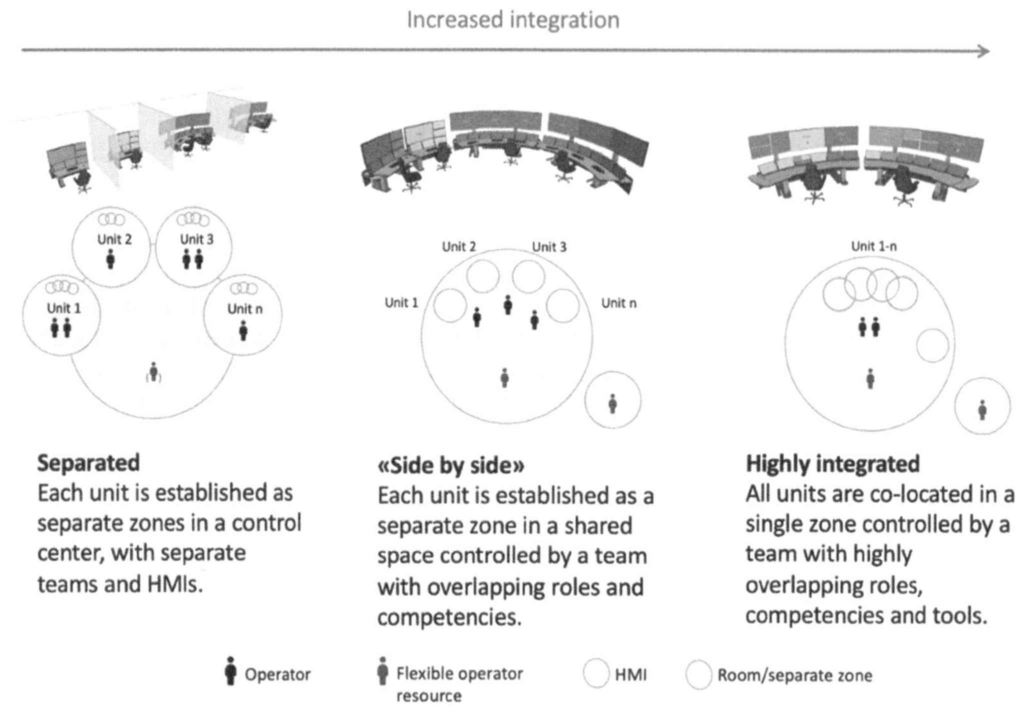

FIGURE 6.2 Degree of unit integration in the control environment.

A key dilemma in HMI and control room layout design concerns the degree of integration between units (see Figure 6.2). At one end of the spectrum, each unit may have a dedicated operator workstation with control systems and displays used exclusively for that unit. At the other end, the units may be fully integrated into a shared HMI, allowing monitoring and control of multiple units through a common interface. Unit separation with dedicated workstations may reduce the likelihood of addressing the wrong unit and make it easier to maintain oversight of who is working on what. It can also be technically more straightforward to implement (e.g., by keeping systems unit-specific) but may require more physical space and thus be less scalable. A fully integrated concept may support cross-unit monitoring, enable flexible personnel use, and scale more easily. However, it may offer less flexibility during high-workload periods and may require additional or redundant workstations to manage the increased task load.

The degree of integration also influences layout decisions and the presentation of overview information to support multi-unit oversight. Overview displays play a key role in directing operator attention during disturbances and supporting decision-making when events affect multiple units. The information may be displayed on a shared large-screen display (LSD) visible from all workstations, or on smaller screens located near each workstation. The content and layout may be fixed, or operators may be given flexibility to customise the display(s) based on their preferences or operational needs. In addition to aiding situational assessment, overview displays can enhance team coordination and communication across units. A strength of fixed overview displays is their ability to support pattern recognition over time, enabling operators to detect deviations from familiar visual layouts quickly. In contrast, customisable overview displays offer greater adaptability to varying operational needs but may be less effective in supporting situation awareness during demanding conditions (O'Hara et al., 2012).

If multiple vendors provide solutions and services for a multi-unit control room, operators may need to switch between different control systems and interact with displays that aggregate information from diverse sources. To support safe and effective operations, the HMI must adhere to shared principles and standards that promote consistency across units. Minor differences in visual design or interaction patterns can help operators distinguish between units. However, significant discrepancies in alarm priorities, layout conventions, or control procedures may lead to confusion, misinterpretation, and increased risk of error.

The degree of integration in the HMI and the organisation of operator roles is closely interrelated – decisions in one area often shape and constrain options in the other. When developing an operational model for a multi-unit environment, the operators may be highly specialised in specific units or tasks or share similar roles and responsibilities. The types of competence needed to operate multiple – and potentially different units – can be described in terms of (1) core competencies, (2) unit-specific knowledge, and (3) expertise in support disciplines.

Core competencies may include knowledge of general process principles; new operational models, such as unmanned facilities; and new ways of working, for example, co-located multi-disciplinary teams that utilise smart diagnostic and decision-making tools. The operators may also need to switch attention between multiple units and tasks, and to prioritise and coordinate efforts during high-workload periods and emergencies. Unit-specific knowledge may include a sufficient understanding of a facility, its design, equipment, and processes, as well as its operating principles. To maintain this knowledge, operators may visit the facility regularly, for example, during maintenance periods. When developing the operational organisation and defining support functions, the allocation of tasks and authorities between the control room and other departments may be altered. Thus, the control room staff may have supplementary competences such as automation, production optimisation, maintenance, and planning.

To achieve flexibility and scalability, operators should possess similar formal qualifications and be capable of assisting one another across units. A high degree of overlapping competence will increase the ability to perform routine and planned tasks effectively while also improving resilience during operational disruptions and unforeseen changes in workload. This allows for flexible personnel use and mutual support among operators. However, a broadly shared competence base may reduce the depth of expertise available for handling complex or unit-specific situations that require specialised knowledge. It may also place greater demands on planning, coordination, and continuous training to maintain sufficient proficiency across roles.

In contrast, a team of highly specialised operators may be better equipped to manage heterogeneous units and complex situations that demand active problem-solving. Operators may be assigned to specific units and have limited ability to support one another across units or tasks. While this model can ensure strong expertise for specialised operations, it may reduce overall flexibility and increase vulnerability to sudden workload shifts. On the other hand, it places less demand on broad-based training and ongoing coordination within the team. Figure 6.3 illustrates the types of competences and their possible overlap among operators.

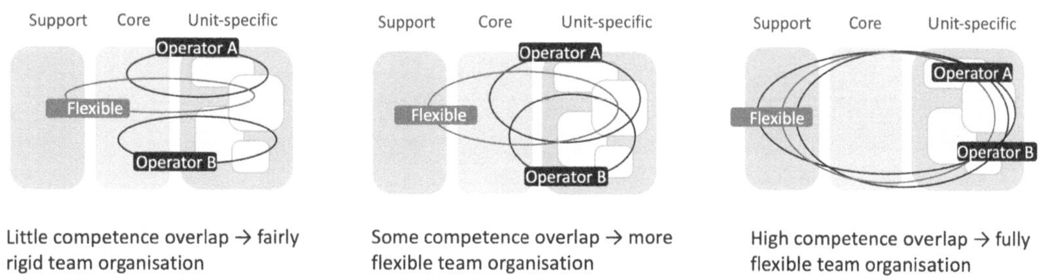

FIGURE 6.3 Types of competences and degree of overlap influence team flexibility.

6.7 CONCEPT OVERVIEW

The project with Equinor aimed to explore an integrated way of working in the control room while remaining grounded in solutions considered feasible to implement within the next five to ten years. Key elements of the proposed concept are illustrated in Figure 6.4. These include flexible allocation of tasks and responsibilities in the control team; modular workstations equipped with unit-specific overview displays to support shared situation awareness; and access to on-demand operator resources and control facilities during periods of high workload.

To maintain operator competence and adaptability across units, team rotation is essential. An extra operator workstation permanently available in the control room can also provide relief during peak demand.

The effectiveness of automated systems depends on their integration into the broader operational context and on how well they support collaboration with the human operators. Key design principles include transparency – clearly communicating goals, status, and intentions – and the ability for operators to guide or adjust system behaviour as needed.

Coordinated, campaign-based maintenance across facilities helps ensure a manageable and predictable workload. The HMI should support intuitive navigation and access across units, enabling efficient multi-unit oversight. Displaying alarms by unit is the most promising grouping principle, although cross-unit alarm overviews may also be beneficial. Finally, a modular approach to team organisation, control room layout, and HMI design further enhances long-term flexibility and scalability.

Figure 6.5 illustrates how the proposed concept supports key safety-critical situations.

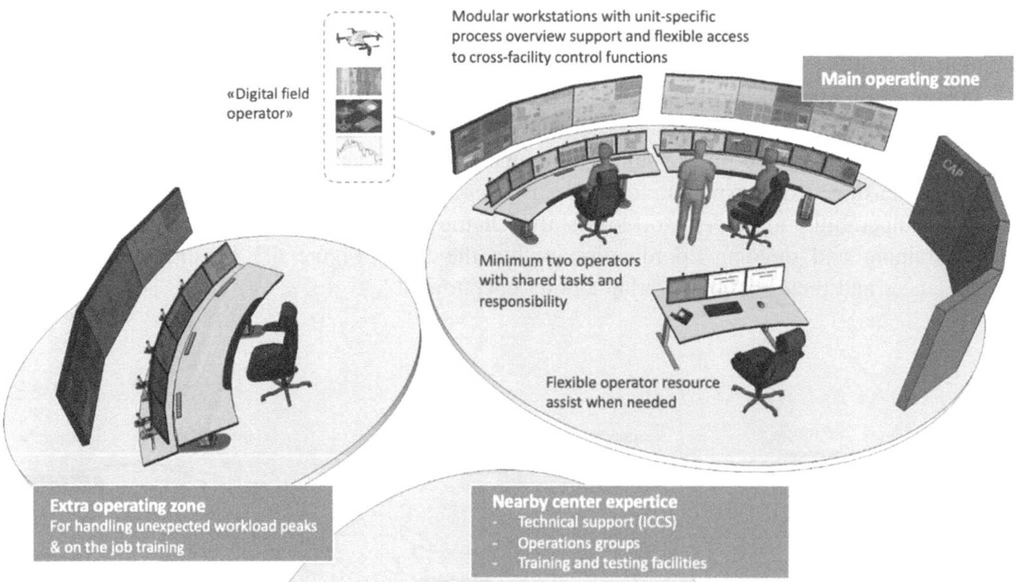

FIGURE 6.4 The proposed concept includes modular workstations, an additional zone for high workloads, and a nearby expertise centre.

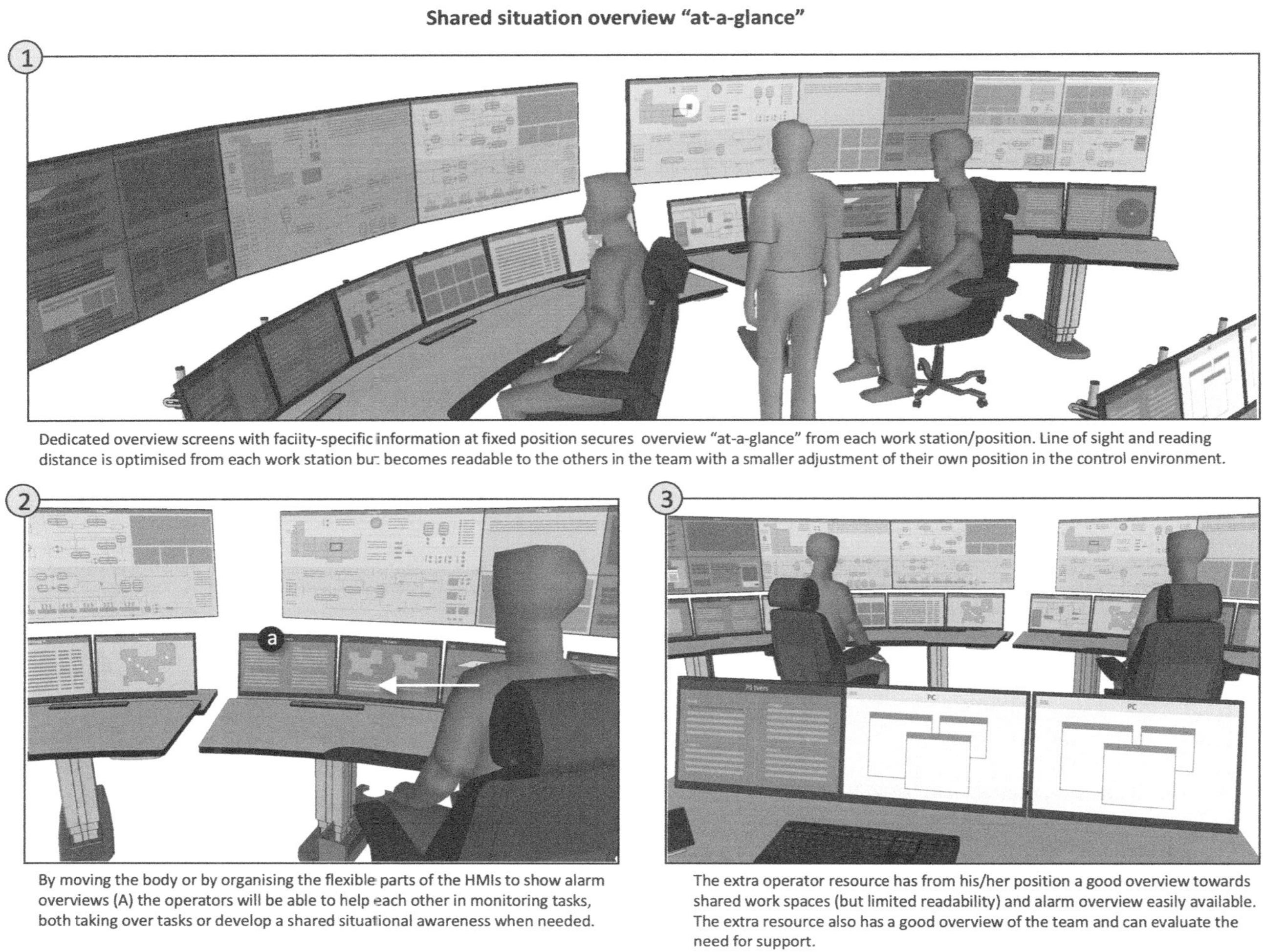

FIGURE 6.5 Illustrations of how the proposed concept may support key safety-critical situations.

Continued

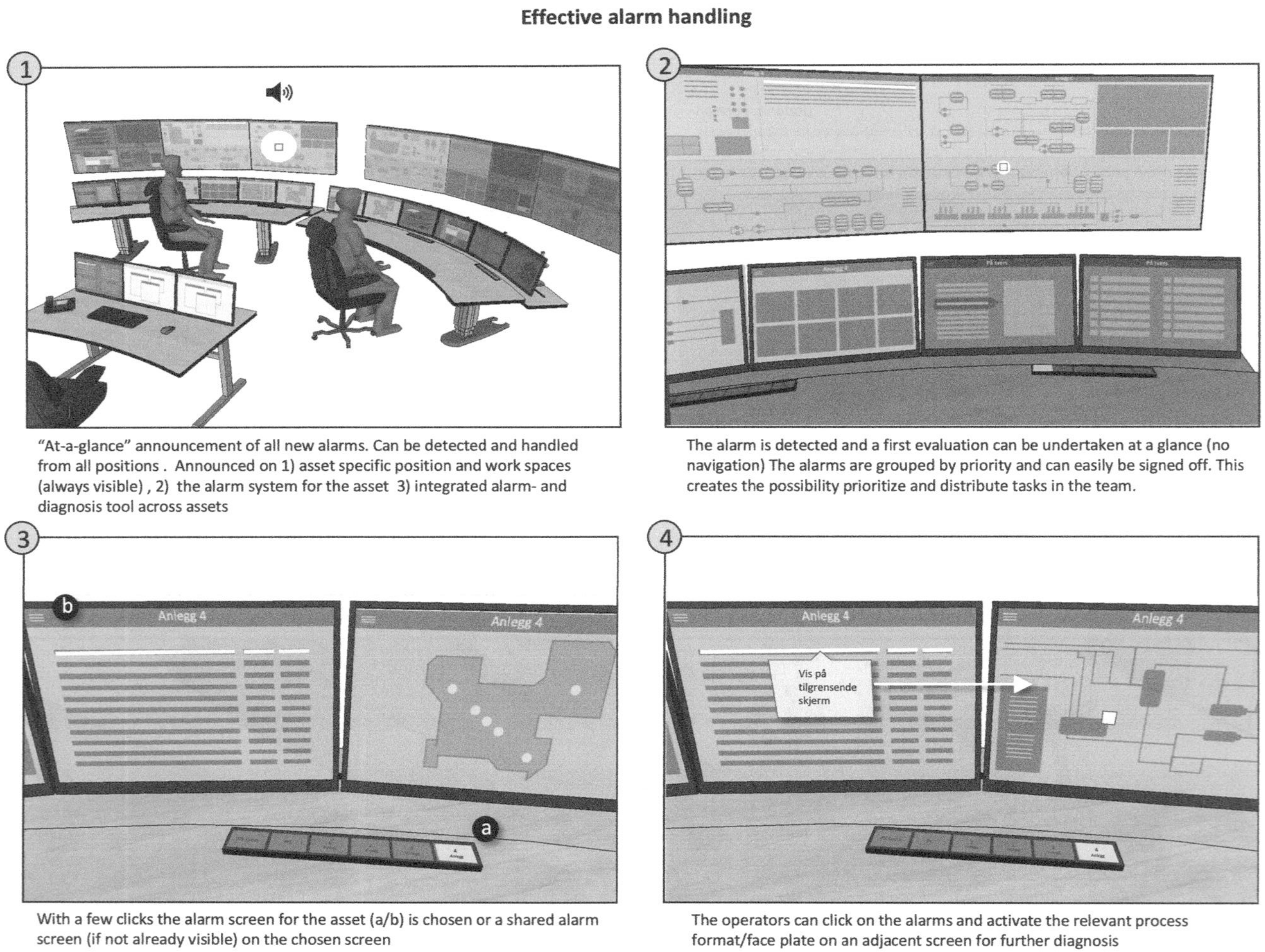

FIGURE 6.5 Continued

FIGURE 6.5 Continued

6.8 OPERATIONAL EXPERIENCE

Since the concept was developed, it has been used in several Equinor projects and control room developments that can be described as the integration of both heterogeneous and homogeneous facilities. The first step was to formalise the design principles in a company guideline for multi-field control room development. Multi-field operation is the centralised control of multiple oil and gas fields from a single location. These principles were then used to redevelop the onshore Valemon control to support multi-field operation with the addition of the Hywind Tampen offshore wind park. This is an example of a heterogeneous multi-field control room with two very different types of facilities. Valemon is a normally unmanned installation in the North Sea. It is unmanned for four out of six weeks and operated from an onshore control room at Equinor facilities at Sandsli outside Bergen. The Hywind Tampen wind park is a dedicated wind park with 11 offshore wind turbines that deliver electric power to the Gullfaks and Snorre facilities.

Insights from the three-year research study described in this chapter were used both to develop the cross-training model of the operators and to develop the refurbished control room. The study also informed the discussion of the optimal integration between the existing Valemon control systems and the new functions of the wind farm. It also addressed how shared situation awareness could be developed across facilities operating under very different process conditions.

The redesign of the control room followed the steps outlined in ISO 11064 with broad participation from existing users/control room operators. Several users had already participated in the earlier design phases. In the human factors process, the existing Valemon set-up was not changed. A new Hywind Tampen control desk was designed, and the existing LSD was expanded to accommodate LSD features from the wind park. The solution is a side-by-side arrangement, meaning there is no technical integration between the Hywind and Valemon control systems. They remain as separate systems on workstations but share a common visual interface on the LSD. Principles derived from the research study were used in the design process, the organisation of cross-trained teams, the development of the HMI, and the alarm management process, that is, to harmonise the alarm priorities of the Valemon and Hywind Tampen systems. Priority 1 alarms on Valemon were to be prioritised, and the control room's operating philosophy was changed to incorporate the new multi-field control room's features. Since Hywind Tampen is integrated into an existing oil and gas control room and also relies on communication with Gullfaks and Snorre, the new control room had to comply with Norwegian Petroleum legislation.

Second, the research study has been used to develop regional control rooms for offshore wind. The regulatory framework for offshore wind differs from that for oil and gas. The Doggerbank control room is currently operating partly two offshore wind parks out of Newcastle-upon-Tyne in the UK. A third will be added soon, and additional wind parks in the area will be integrated over the next few years. In this case, the wind turbines, cable infrastructure and onshore/offshore substations on Doggerbank A, B, and C are to a large degree similar, so it is easier to operate many similar assets and standardise the design and operations of a regional control room. These are fundamentally more homogeneous facilities. The research study was used to develop the concept of a scalable regional control room for wind farm operations in the United Kingdom, focusing on IT systems design, work processes, and team development.

The focus was on standardising the input on level 1 (LSD) and workstations (level 2 and beyond), integrating control system input from both wind parks and the electrical system infrastructure. The work focused to a large degree on how the system could be made scalable, so that similar and new wind farms could be added.

The research study by IFE has provided important input for the realisation of both heterogeneous and homogeneous Equinor multi-field control rooms.

6.9 CONCLUSION

This chapter has examined safety considerations for the design of multi-unit systems, drawing on insights from the conceptual design and development of a multi-field, onshore control room at Equinor. While the insights and recommendations provided are based on this specific context and may not be universally applicable, they offer valuable guidance for similar projects. Central themes include the importance of scalable design to accommodate future expansion, flexibility in allocating operators based on their expertise, and consistency in interfaces to support both efficiency and safety.

Looking ahead, increasing automation and remote control will, in many contexts, transform the operator role from direct process control to supervisory oversight across multiple units. Key design challenges include sustaining situation awareness when interventions are rare, managing alarms and prioritisations across units, and providing interfaces that balance integrated monitoring with access to unit-specific details. These design choices will be decisive for ensuring safe and resilient multi-unit operations.

ACKNOWLEDGEMENT

We would like to thank Gyrd Skraaning Jr. (IFE) for his contributions during the initial phase of the study. We are also grateful to the Equinor control room personnel and shift team leaders who participated in the concept development. Special thanks to Equinor HF and the technical advisors for their valuable input and constructive dialogue throughout the process.

REFERENCES

Edwards, A.R. & Gordon, B. (2015). Using unmanned principles and integrated operations to enable operational efficiency and reduce capex and opex costs. In SPE Middle East Intelligent Oil and Gas Conference and Exhibition (p. D021S008R003). SPE. https://doi.org/10.2118/176813-MS

Endsley, M. R., Bolté, B. & Jones, D. G. (2003). *Designing for Situation Awareness: An Approach to User-Centered Design*. CRC Press.

Hurlen, L. & Skraaning, G. (2023). Designing future control environments: A feasibility study approach developed at IFE FutureLab. In Proceedings of the 33rd European Safety and Reliability Conference (ESREL'23). Southampton, UK. https://doi.org/10.3850/978-981-18-8071-1_P042-cd

ISO (2000). Ergonomic design of control centres — Part 1: Principles for the design of control centres (ISO-11064-1).

Johnsen, S. O., Thieme, C., Aminoff, H., Park, J., Fernander, M., Boren, M., Kjøl, Å. A. & Kirwan, B. (2025). CRIOP: A scenario method for crisis intervention and operability analysis (SINTEF Report No. 2025:00192). SINTEF Digital. https://www.sintef.no/globalassets/project/criop/2025.07.04-criop-new.pdf

Kirkwood, M. (2019, November). Designing for Situation Awareness in the Main Control Room of a Small Modular Reactor. In Proceedings of the Human Factors and Ergonomics Society Annual Meeting (Vol. 63, No. 1, pp. 2185–2189). Los Angeles, CA: SAGE Publications.

O'Hara, J., Higgins, J., & D'Agostino, A. (2012). Multi-unit Operations in Non-nuclear Systems: Lessons Learned for Small Modular Reactors. In Proceedings of the Eighth American Nuclear Society International Topical Meeting on Nuclear Plant Instrumentation, Control and Human-Machine Interface Technologies (NPIC&HMIT'12) (pp. 22–26).

Rittel, H. W., & Webber, M. M. (1973). Dilemmas in a general theory of planning. *Policy Sciences*, 4(2), 155–169.

Saeverhagen, E., Kellas, R., & Bouillouta, F. (2013, October). Remote operations centers and re-engineering work processes: Retaining competent personnel in an extremely competitive marketplace. In SPE/IADC Middle East Drilling Technology Conference & Exhibition (p. SPE-166743). SPE. https://doi.org/10.2118/166743-MS

7 Abstracting Lessons from Remotely Managed Incidents to Inform Road Safety Policy for Automated Traffic

Ross O. Phillips, Rune Elvik, and Tor-Olav Nævestad

Driven by Vision Zero and a systems approach to road safety management, Norway's road safety policy has been considered a success. This success is being challenged by the emergence of highly automated vehicle and road systems. Human factors play an essential role in shaping policy to maintain safety improvements in the face of increasingly complex sociotechnical systems. This chapter explores whether human factors – and in particular principles from cognitive systems engineering (CSE) – can be used to abstract lessons from past remote traffic incidents to inform future system design and safety policy.

We begin by reviewing evidence on the success of road safety policy in Norway (Section 7.1.1), before discussing how emerging automated traffic systems pose new challenges for policy and management (Section 7.1.2). We then describe how the human factors paradigm of *cognitive systems engineering* can help address the challenges (Section 7.1.3), and how opportunities to study human factors in existing automated traffic systems in the EU have been limited (Section 7.1.4). We present the idea that a useful way to learn about future automated traffic systems involving remote operation of vehicles and roads is to use CSE to abstract lessons from past incidents involving traffic control centres (TCCs) (Section 7.1.5). In the rest of the chapter, we describe a preliminary study that aimed to explore this idea (Section 7.2) and present its findings (Section 7.3). The implications of the study for future road safety policy and management are discussed (Section 7.4).

7.1 BACKGROUND

7.1.1 Road Safety Policy Improvements in Norway

Road safety has improved considerably in most highly motorised countries since the all-time high in traffic fatalities around 1970. In Norway, the peak number of traffic fatalities was reached in 1970: 560 people were killed on the roads that year. In 2020, the number was 93; in 2021, 80. Then, a couple of bad years followed, but in 2024, the number of traffic fatalities in Norway was 90, the second lowest since 1944.

The form of government has a significant influence on road safety. In a study of 145 countries, Elvik (2021) found a strong negative relationship between the democracy index score developed by The Economist Intelligence Unit and the number of traffic fatalities per 100,000 inhabitants. The democracy index ranges from 0 to 10. In 2019, Norway scored 9.87 – close to the maximum value. It also had the lowest traffic fatality rate per 100,000 inhabitants among the countries included in the study.

The adoption by Norway in 2001 of the vision that no one is seriously injured or killed in road traffic (Vision Zero) was a major change in road safety policy (Elvik, 2022). Road safety action

DOI: 10.1201/9781003741824-9

plans for four-year periods were developed; the first was for 2002–2005. These plans were developed by a collective policy-making forum, which included an expanding number of stakeholders. Consensus on policy was established, and the responsibility for implementing specific road safety measures was clearly assigned to each stakeholder. The road safety policy became more evidence-based than before. The approach was underpinned by a systems approach to road safety management by the public road authority, involving the three main components: driver, vehicle, and infrastructure.

Although it is impossible to reconstruct the history of road safety policy in detail, Elvik (2024) developed a road safety policy index consisting of ten road safety measures for which annual data on their use was available from 1980 to 2021. These were extracted from the following list of 13 measures shown in Box 7.1.

According to a negative binomial regression model, road safety policy – as indicated by index values – has contributed to improving road safety after 1980, particularly after 2000. This is shown in Figure 7.1.

In Figure 7.1, the black dots represent data points showing the number of killed or seriously injured road users recorded each year. The solid line is the number of killed or seriously injured road users predicted by the negative binomial regression model. As can be seen, the model predictions track the actual number of killed or seriously injured road users quite closely. The dashed line shows what the number of killed or seriously injured road users could have been if the road safety policy index had remained unchanged at its 1980 value throughout the period; that is, it shows how the number of killed or seriously injured road users might have developed if road safety policy had not become more effective. The policy index was 50 in 1980, rising to 185.2 in 2021.

The period from 1980 to 2021 can be divided into three phases with respect to the effects of road safety policy. From 1980 to 2000, the actual number of killed or seriously injured road users was only 1.2% lower than the counterfactual. From 2001 to 2009, it was 20.4% lower than the counterfactual, and from 2010 to 2021 it was 36.4% lower than the counterfactual. This indicates an increasingly effective road safety policy, in which participation of key societal actors has played a part.

BOX 7.1 POTENTIAL MEASURES OF ROAD SAFETY POLICY (ELVIK, 2024)

1. New motorways, indicated by total length in kilometres each year
2. New 2 + 1 roads with a median barrier, length in kilometres each year. A 2 + 1 road is a road with 2 lanes in one direction, 1 in the other, separated by a median safety barrier
3. Number of roundabouts
4. Share (%) of vehicle kilometres performed by cars with frontal airbags
5. Share (%) of vehicle kilometres performed by cars with electronic stability control (a system preventing skidding)
6. Share (%) of vehicle kilometres performed by cars scoring 5 stars in the European New Car Assessment programme (EuroNCAP). EuroNCAP evaluates the crashworthiness of cars, i.e., how well they protect occupants from injury in a crash
7. Car driver seat belt wearing in urban areas (%)
8. Car driver seat belt wearing in rural areas (%)
9. Bicycle helmet wearing (%)
10. Citations for traffic offences per million vehicle kilometres
11. Real monetary value of fixed penalties for speeding
12. Number of speed cameras in operation
13. Kilometres of road with section control (two or more speed cameras measuring speed along a section of road)

FIGURE 7.1 Actual and counterfactual changes in the number of killed or seriously injured road users in Norway from 1980 to 2021.

(Source: Elvik (2024).)

7.1.2 Challenges from Automated Traffic Systems

Road safety management in Norway has involved a systems approach that considers how infrastructure, vehicles, human drivers, and their interactions together determine road safety. To date, the driver has been responsible for following traffic rules, and driver actions have been analysed following serious collisions in which the driver was involved. Now, however, vehicle automation is starting to take over the driving function with technology that is generally more reliable, precise, and reactive than human drivers. At the same time, road infrastructure and traffic systems are increasingly connected through Intelligent Transport Systems (ITS), with elements such as adaptive traffic signals, real-time traveller information, and electronic toll collection (Papadimitratos et al., 2009; EC, 2023). Connected, Cooperative, and Automated Mobility (CCAM), an extension of ITS, is integrating connectivity and vehicle automation, allowing vehicles operating at varying levels of automation (SAE levels 1–5) to communicate with each other (vehicle-to-vehicle, V2V), with infrastructure (V2I), and with other road users and mobility services (V2X) (Alonso Raposo et al., 2020). As CCAM systems expand, operational tasks will be increasingly managed remotely. It is envisaged that remote operators will oversee whole vehicle fleets, assist specific vehicles, or intervene during edge cases (e.g., incidents, malfunctions) in collaboration with remote road operators (RROs).

Even though they promise to deliver safer driving, vehicle automation, ITS, and CCAM systems pose challenges for road safety policymakers seeking further improvements. Some key challenges are as follows:

Increased automation and connectivity can threaten democracy. As we have seen, road safety improvements are larger in democratic societies, where the development and enforcement of effective traffic policies are associated with participation, transparent governance, and accountability. Democracies tend to foster open public debate, regulatory oversight, and responsiveness to safety concerns. As transport systems evolve with increasing automation and connectivity, preserving the democratic processes associated with road safety gains may require embedding 'meaningful human control' into their design and governance (Santoni de Sio & van den Hoven, 2018). Meaningful human control describes the need to maintain clear human oversight and intentional influence over automated systems for ethical, safety, or legal purposes. In the absence of meaningful human control, automated, highly connected systems could create growing accountability gaps in which no person or institution is held responsible for errors or harm. Such gaps could undermine the transparency and public trust that have underpinned safety improvements in democratic contexts. The idea of meaningful human control is that principles must be followed to ensure that decisions made by machines, or through machines by remote operations, remain linked to accountable human agents, allowing society to contest, improve, and regulate system behaviour (Santoni de Sio & van den Hoven, 2018).

Road safety policies, laws, and liability frameworks appear to be effective but have taken time to develop and are designed for human drivers. Road systems are open, complex, sociotechnical systems (Vicente, 1999). They are *open* in the way that they are subject to external influence and the exchange of information, energy, or matter with surrounding systems (e.g., health system, economic system); they are *complex* in that they involve technical and social (human, organisational) components that interact in nonlinear ways such that emergent system behaviour is difficult to predict. They are *sociotechnical* in the sense that effective functioning depends on the joint optimisation of technical and social subsystems.

Returning to the 13 policy measures that have contributed to road policy success in Norway, in Box 7.1, we see that seven apply to human *driver* behaviour (7–13), with three measures each addressing roads (1, 2, 3) and vehicles (4, 5, 6). Accepting that future road systems will always be open, complex sociotechnical systems, human behaviour, and decision-making will remain critical to safety in ways we do not yet understand and for which there is no clear policy. Systems will increase in complexity, meaning that the policies and management approaches required will themselves need to be more complex to ensure their safety (in line with Ashby's Law of Requisite Variety, see Woods and Hollnagel, 2006). Measure sets will need to address not only road users, vehicles, and roads but also onboard operating systems, ITS and CCAM (V2X) technologies, and remote vehicle monitoring or assist operations.

There will be a transition period in which manually driven and automated vehicles coexist. This introduces complex system dynamics and unforeseeable critical situations. In this phase, road safety policy must account for interoperability and consistency across vehicle types and infrastructure readiness (ITF/OECD, 2015).

New hazards and crises will inevitably arise. Even though automation technology will reduce collision risks caused by human behavioural constraints or road traffic violations, new risks will be created (Shladover, 2018). An example of a new risk is ordering a Waymo taxi to the centre of a public demonstration so it could be set on fire as part of a protest, resulting in a complex chain of events that authorities must manage (Béchard, 2025).

7.1.3 Addressing the Challenges: Cognitive Systems Engineering

The aforementioned challenges imply a need for safe human decision-making and meaningful human control in ITS and CCAM systems. This will be particularly relevant for remote operations on these systems, since their governance function is likely to be critical. Without thoughtful design of remote operations, human operators could be overloaded by supervising too many vehicles or be too removed from traffic events to intervene meaningfully. Remote operator decisions must also be supported by situational awareness and timely feedback, which inadequate signal latencies could jeopardise. Without adhering to principles of meaningful human control in the development of remote systems, there will likely be uncertainties about who is responsible for what when incidents are caused by unpredictable system behaviour.

Remote vehicle and road operations are likely to be central in serious traffic incidents such as multi-vehicle collisions, tunnel fires, or floods. How will complex response systems be managed together by dispersed responsible people? How will they tackle unforeseen consequences arising from increasingly complex systems they manage? How will they be able to take the 'right' safety-critical decisions in complex, dynamic scenarios?

Answers to these questions can be provided by a branch of human factors known as CSE, which was developed for the analysis, design, and evaluation of complex systems in which responsible humans interact with technology (Hollnagel & Woods, 1983; Rasmussen, 1986). CSE is particularly appropriate for the study of high-stakes, dynamic environments like transport (Hollnagel & Woods, 2005). A central aspect of CSE is the *abstraction* of the essential purposes and functions of work systems, for example, traffic safety management (Woods & Hollnagel, 2006; Naikar, 2013). Put more simply, CSE can be used to strip away traditional actors and strategies to identify the essential 'job to be done' in past, present, or future work systems. This leaves analysts free to consider how different functions can be organised to achieve system purposes and who can perform them (e.g., social or technological actor).

Real-world transport systems vary widely, across time, mode, regulation, and technological sophistication, but the essential safety-critical goals never change. CSE allows us to abstract enduring safety and other critical functions, such as the need for the vehicle operator to resolve goal conflicts under time pressure or to detect and manage component/system failures. With critical functions identified, the analyst can then consider how work (=execution of functions) should be organised and allocated across social (e.g., road users, operators, teams, organisations) and technological (e.g., onboard systems, HMI, V2X) actors; how control and information should pass between them; and how humans as responsible decision-makers can best be supported. Through abstraction in this way, CSE can be applied to analyse past case studies to identify future-proof design principles.

CSE methods can also guide the design of remote operator interfaces and control systems to account for the complexities of human cognitive capabilities and the need to support human macrocognitive processes – deciding, planning, adapting, etc. – in real, complex situations. Such support will be essential to providing meaningful human control (Endsley, 1995; Militello et al., 2010; Ward et al., 2017). CSE methods can also be used to study organisational structures and how responsibility can be allocated and traced across remote operations teams, service providers, and manufacturers, particularly if used alongside sociotechnical methods rooted in systems and control theory (Rasmussen, 1997; Leveson, 2012).

7.1.4 Human Factors in Relevant EU Projects

Recent and ongoing EU projects on the use of CCAM in logistics and passenger transport are helping ensure that EU transport systems remain competitive and sustainable (e.g., ULTIMO, AWARD, MODI). While these projects are starting to address social requirements in terms of trust, safety, and security, there are, to date, few activities aimed at demonstrating how to account for human behaviour or meaningful human control in the design of CCAM systems. There are two possible explanations for this. First, projects have focused on technological demonstrations, albeit involving the participation of societal actors. While demonstrations of highly automated vehicles have provided useful insights into the likely roles of humans in future systems, their scope and scale often give analysts limited opportunities to learn from edge cases in which the role of remote managers will be most critical. Second, it can be difficult for project sponsors to imagine how CSE methods can be used to understand the conditions for meaningful human decision-making. Merat et al. (2017) show that the successful deployment of shared automated vehicles depends on social-psychological factors such as trust, acceptance, perceived safety, comfort, and users' direct experience with the technology, all of which remain highly relevant to CCAM deployment.

A way to improve the situation is to use CSE methods to abstract lessons from past edge cases involving remote operations and apply them to understand how to support human decisions and generate meaningful human control in remotely operated traffic systems of the future. This would also address calls that human factors should be doing more to address the complexities of future remote operations (Schrank et al., 2024) and broaden human factors' current emphasis on in-vehicle human-machine interfaces (HMI) and takeover operations (Young and Stanton, 2023).

7.1.5 What Can We Learn from Existing Road Operations Involving Remote Management?

Investigations of serious road safety incidents are available in several countries, some of which involve the remote management of road safety by Traffic Control Centres (TCCs). In these incidents, TCCs have managed safety remotely by responding to alerts, coordinating activities, controlling mitigation measures, and even operating infrastructure and vehicles (the latter indirectly through instructions to drivers via road signs or radio). Using CSE, the challenges faced by control centres attempting to manage traffic safety remotely can be understood and applied to future systems by abstracting elements of the work domain involved (see above) using work domain analysis, one of the most established and widely applied methods within CSE for modelling the functional structure of complex sociotechnical systems (e.g., Vicente, 1999). Particularly important elements are those functions needed for safe system performance, many of which will apply to systems of the past, present, and future (Naikar, 2013). By abstracting the safety functions performed by abstracted roles ('vehicle operators', 'remote road operators'), we can learn which functions actors in new systems will need to perform and consider how these might be allocated.

CSE methods can also be used to abstract safety-critical decisions made by actors in past incidents. These are often described directly in investigative accounts or implied by descriptions of events. Abstracted decisions would help us consider which future system actors – social or technological – could inform or make similar decisions in future systems. Such considerations would then have implications for ensuring meaningful human control in future road systems, in which responsible remote human operators will need to understand system activities to take critical decisions, and in which the system of technological and social actors will have to function to carry out interventions in line with human intentions.

7.2 AIM OF AN EXPLORATORY STUDY

To explore these ideas, we reviewed serious road tunnel incidents in Norway (2007–2021) to abstract insights relevant to future remote CCAM systems, including:

- Functions that human remote agents will need to perform in future systems.
- Decisions that responsible humans need to make in critical situations to perform those functions in the face of constraints ('safety-critical decisions').
- Implications for supporting meaningful human control and decision-making in future systems.

The aim was to explore whether CSE principles could be applied to abstract lessons from past remote traffic incidents to inform future system design and safety policy in Norway and elsewhere.

7.3 METHOD

A search on the website of the Norwegian Safety Investigation Authority (NSIA; havarikommisjonen.no) for in-depth reports of serious incidents involving TCCs, using the terms '*ve[g/i] trafikksentral*' or '*VTS*' (TCC in English) for the years 2007 to 2021, resulted in 40 hits. The content of each report was reviewed for the extent of RROs' involvement. In many of these reports, TCC received little mention or had no direct influence on the course of events. Nine reports were selected in which TCC were judged to play a critical role in the management or coordination of a serious road incident. Each report was analysed in two sweeps.

1. System purposes and functions were identified for the system in line with ideas of work domain analysis (Vicente, 1999; Naikar, 2013). This was done by applying the underlying principles of work domain analysis rather than by systematically drawing up abstraction hierarchies. Safety-critical functions were identified from descriptions of actors performing or not performing them or were inferred from descriptions of safety constraints faced by actors in the events. For example, a failure to salt an icy road means that measures are needed to address acute changes in road friction, as expressed in the function 'Detect and respond to hazards'. Actors performing functions in the reports were assigned an abstract role. Thus, drivers were 'vehicle operators' and TCCs, and the road authorities running them were 'remote road operators'.

   After identifying system purposes and safety-critical functions, exploratory analyses of functional allocation and meaningful human control were performed, as reported in Results 7.4.3. These analyses were exploratory rather than structured using, for example, an abstraction hierarchy.
2. Safety-critical decisions needed to execute system functions effectively were identified either directly from event descriptions or implied by them. Each safety-critical decision was then analysed to determine which technological or human actors were likely to take part in *future* traffic systems and which human actor could be responsible for the consequences of that decision. Again, these analyses were exploratory in nature, so relatively unstructured.

Terms: By the term 'remote road operator (RRO)' we mean the whole road authority; by the term 'remote vehicle operator (RVO)', we mean the persons, team, and organisation operating the vehicle, including potentially fleet managers. The remote operator can be the vehicle owner or a contracted company, so we considered the vehicle owner's role separately. For all reports, the system analysed comprised social and technological actors involved in preventing, mitigating, and responding to the road safety event.

For further understanding of how the results were obtained, refer to the Appendix.

7.4 RESULTS

Four of the NSIA reports analysed involved tunnel fires, with the evacuation of car, truck, and bus drivers and coordination by RROs in cooperation with emergency services. Four involved collisions on roads with reduced friction (ice), in which the role of remote operators as coordinators of mitigating measures (e.g., salting) and responses was examined. The remaining report described a serious incident in which a major road was blocked by a semitrailer, examining the response of the RRO and their communication with road users via signs.

7.4.1 System Purposes

Analysis of these reports resulted in three system purposes, which applied across incidents:

- Control developing hazards
- Ensure a safe distance between the vehicle and the hazard
- Ensure a safe distance between the road user and the hazard

Note that 'safe distance' here refers to the physical separation of two objects needed to avoid their collision, as can be defined by, for example, the velocity of a travelling vehicle in relation to the braking distance.

7.4.2 System Functions and Performers

Three main system-level functions were performed to achieve the system's purposes: adjust velocity of vehicle, alert the remote operator, and detect and respond to hazards.

Table 7.1 presents the abstract roles of actors performing these functions. The 'vehicle operator' role was played by a human driver in the investigations analysed, but in future systems the role could be played by humans, onboard systems or RVOs.

The first function, *adjusting vehicle velocity*, was performed by vehicle operators to maintain safe distances from hazards (e.g., fire, blockage, or icy conditions), or in response to instructions from a responsible overseer via radio or message signs. In the incidents studied, these instructions could be relayed over the radio or through variable message signs. In the future, they could be relayed through onboard systems, road signs, or V2I systems. In addition to making velocity adjustments, vehicle operators could also drive the vehicle away from a hazard to achieve a safe distance.

The second function, *alerting remote operators*, involved informing the control centre about hazards and the locations of vehicles and road users. Both vehicle operators and RROs could perform this function. Developing hazards were tunnel fire, reduced road friction, and road blockages. Other hazards can also be envisaged, for example, landslides, flooded roads, freezing fog, or thick smoke. Vehicle and road user locations were needed by the RRO to coordinate safe and efficient evacuations or safe re-routing away from a hazard.

The third function, *detecting and responding to hazards*, involved several sub-functions. These can be considered at local level, tending to involve the vehicle operator, or system level, involving the RRO. Responders described in sub-functions include police, fire or ambulance services, or road contractors applying mitigating measures.

TABLE 7.1
Road Safety System Functions and Abstract Actor Roles Performing Them

System Function	Performer
1. *Adjust vehicle velocity* to account for: • Need to keep safe distance from hazard • Instructions from responsible overseer	Vehicle operator
2. *Alert remote operator* about: • Hazards developing in vicinity of vehicles • Vehicle locations • Road user locations	Vehicle operator or remote road operator
3. *Detect and respond to hazards* by: • Ensuring hazard response plans and measures work in practice with real systems of actors • Monitoring roads and surroundings to capture hazards emerging • Assessing incoming alerts from remote locations about potential hazards • Alerting vehicle operators about developing hazards and need for response • Alerting responders about developing hazards and need for response • Communicating to all actors on developing hazards to promote shared situational awareness, including information on location of vehicles and road users • Mediating communication between road users and rescuers	

7.4.3 Applying Functions to Understand Future Scenarios

The functions in Table 7.1, abstracted from real crisis events, were used to consider how available social and technological resources and components might be combined to carry out functions in the future. As an example, we considered that either the RRO or RVO could 'monitor roads and surroundings to capture hazards emerging'. A hazard could also be detected by a human passenger in a driverless vehicle. In such a situation, assuming they are responsible for their own safety, the passenger would need meaningful human control – either by allowing them to instruct the technological or remote human vehicle operator to evacuate, or by enabling them to take over the vehicle's operation themselves. Either option would allow the system to respond to the passenger's goals and intentions in a crisis.

Using Table 7.1, we also considered the need of remote vehicle and road operators to understand safety-critical actions in the systems for which they are accountable. Returning to the above scenario, both actors would need to understand the human passenger's intentions behind any vehicle functions being performed. This would, in turn, leave remote operators better able to understand traffic hazards emerging and take action to increase system safety, for example, by mediating communication between evacuating road users and emergency rescuers.

Our analysis showed that, in emergencies, RROs often rely on in situ reports of developing hazards from multiple sources. Since they are likely to remain responsible humans, integrally involved in controlling system-level road safety functions, it would be beneficial for safety that they receive reports on emerging and developing hazards from human road users, RVOs, ITS/CCAM systems, onboard vehicle systems, cameras, and sensors.

We also found that the functional analysis could be used to consider feedback on functional performance for relevant actors. For example, distance to the car in front in relation to (dynamic) safe braking distances would usefully inform any technical or social actor needing to adjust vehicle velocity to account for road friction levels and keep a safe distance from the hazard (object in front).

7.4.4 Safety-Critical Decisions Applied to Future Traffic Systems

Safety-critical decisions derived from an analysis of nine in-depth investigations involving RROs are presented in Table 7.2. Table 7.2 also presents an analysis of which actors in *future* traffic systems could take or help take safety-critical decisions, and which actors could be responsible.

The table considers six actor types in future traffic systems who can potentially decide the answer to each safety-critical question. These are onboard vehicle systems (technological), RVOs (humans using technology), the vehicle owner (human), the traffic control centre (TCC; humans using technology) or authority managing traffic and roads remotely (humans using technology), the CCAM physical/digital infrastructure (PDI; technological), and road users (human passengers, cyclists, pedestrians, other drivers, and so on). In a separate analysis of responsibility (see Responsibility Column in Table 7.2), we assumed that i) human actors making safety-critical decisions in future will also be held responsible for them; ii) human actors will also be held accountable for safety-critical decisions made by technological actors. The latter are as follows:

- Original Equipment Manufacturers (OEMs) or developers of vehicle technology responsible for onboard systems
- Vehicle owners responsible for vehicles
- PDI developers or operators

Note that human actors may also be responsible for the influence of technology on safety-critical decisions; for example, vehicle owners may be accountable for blind spots on vehicles, which prevent onboard systems from detecting pedestrians.

TABLE 7.2
Safety-Critical Decisions Abstracted from Analysis of Remotely Managed Traffic Incidents

	Who Decides?						Responsible?					
Safety-Critical Decision	**Onboard Operating System**	**Remote Vehicle Operator**	**Owner Vehicle**	**TCC**	**Road PDI**	**Road User**	**OEM/ Vehicle Tech Developer**	**Remote Vehicle Operator**	**Owner Vehicle**	**TCC**	**Road PDI Operators**	**Road User**
What hazards are detected in vicinity of vehicle?	✓	✓		✓	✓	✓	✓	✓	✓	✓	✓	
What/when/how is the RRO informed about hazards near traffic?	✓	✓		✓	✓	✓	✓	✓	✓	✓	✓	
When is vehicle speed reduced in response to hazards, how much?	✓	✓		✓			✓	✓	✓	✓	✓	
When and which road mitigation measures should be implemented?				✓						✓		
When are safety warnings by onboard/ roadside systems critical?				✓						✓		
How to scope mitigation actions to address hazard warnings?				✓						✓		
Which other actors (e.g., police) to involve in hazard mitigation?				✓		✓				✓		
How VO's safety response is tailored to each vehicle's unique situation?	✓	✓		✓			✓	✓	✓	✓	✓	
How can road users' safety response be optimised to fit their situation?		✓				✓	✓			✓		✓
How not to develop wrong 'mental' model about situation?	✓	✓		✓		✓	✓	✓		✓		
Considering surrounding factors, how are hazard mitigation measures applied to achieve the best safety result?	✓	✓		✓		✓	✓	✓		✓		
How to apply lessons from previous collisions in practice?		✓		✓		✓	✓	✓	✓	✓		✓

(Continued)

TABLE 7.2 (Continued)

	Who Decides?						Responsible?					
Safety-Critical Decision	**Onboard Operating System**	**Remote Vehicle Operator**	**Owner Vehicle**	**TCC**	**Road PDI**	**Road User**	**OEM/ Vehicle Tech Developer**	**Remote Vehicle Operator**	**Owner Vehicle**	**TCC**	**Road PDI Operators**	**Road User**
Who decides content/target of messages to many VOs at once?		✓		✓	✓			✓		✓		
How to prioritise incidents when several occur at once?		✓		✓						✓		
When are sign instructions unsafe to comply with, when are routines unsafe to follow (e.g., no use of VMS >5km from roadblock)?	✓	✓		✓		✓	✓	✓		✓	✓	
How to ensure infrastructure and safety equipment for evacuation of road/tunnel are adequate, given changing usage (e.g., vehicles, density) and external factors (e.g., temperature)?				✓	✓					✓	✓	
How to account for road user reaction patterns when applying road mitigation measures?				✓						✓		
How to balance road conditions needed to evacuate from/block access to hazard, with those needed for rescue access?				✓						✓		
How much speed is reduced to maintain safe braking distance/How much speed is adjusted to account for upcoming hazardous road surface?	✓	✓		✓			✓	✓	✓	✓		
When do surrounding conditions cause hazardous driving and require mitigation?	✓	✓		✓	✓	✓	✓	✓	✓	✓		

The column 'Who decides?' indicates technological or human actors likely to make the corresponding safety-critical decisions, while 'Responsible?' indicates which human actor will ultimately be liable for the consequences of those decisions. See text for details. TCC, traffic control centre run by road authority; PDI, physical/digital infrastructure; VMS, variable message sign; RRO, remote road operator; VO, vehicle operator.

Looking at the first row in Table 7.2, we see that five actor types could decide which hazards are detected near the vehicle. These can be technological, such as by image detection by onboard systems or roadside PDI set-ups; or social, such as by in situ or remote human operators, passengers, or vulnerable road users, possibly supported by technology (e.g., traffic control centre [TCC] monitoring tunnel cameras). Note that for row 1, the OEM or technology developers could be held responsible for the detection of hazards by onboard vehicle or roadside systems.

Actors who determine *which* emerging hazards are detected near vehicles or roads also determine what, when, and how TCC are informed about those hazards (second row, Table 7.2). Turning to the third row, we see that several actors can reduce vehicle speed in response to a hazard in future systems: the onboard vehicle automation technology; the RVO through remote monitoring and assistance or driving; or the TCC by regulating dynamic speed limits, which can be used to control traffic directly or indirectly. Looking at responsibility for reducing vehicle speed in response to hazards, we see that, in addition to these actors, vehicle owners and PDI developers could also be held liable for speed reduction in response to hazards, through the roles that vehicle technology (e.g., braking) and PDI will play. The remaining rows in Table 7.2 set out our findings for the safety-critical questions identified.

Table 7.2 shows how important RROs ('TCC' in Table 7.2) will be for safety-critical decision-making in future systems – at least for the incidents of the type analysed here.

However, decisions and responsibilities could be shared with RVOs, who will replace drivers as responsible human decision-makers. As monitors of onboard operating systems, they will control communications with RROs and ensure the vehicle responds appropriately to developing situations, especially if the system is acting on the 'wrong mental model' for events. RVOs could also decide 'Content and target of messages sent to many vehicle operators at once' if they have a dual role as fleet managers or liaise with fleet managers working for the same company. A key decision for RVOs could be 'When are road signs or instructions [from roadside infrastructure] unsafe to comply with…' or routines unsafe to follow.

Even though their responsibility for road safety will likely be reduced in future road systems, human road users could still have an important role as decision-makers in road safety. They could capture and inform road operators about developing hazards, involve other in situ actors in hazard mitigation, select and apply mitigation measures (e.g., fire extinguishing), and apply lessons learnt from previous incidents and self-evacuate.

7.5 DISCUSSION

Descriptions of remote safety management of incidents have been analysed using CSE methods to abstract i) safety-critical functions and roles that perform them and ii) safety-critical decisions. Though exploratory in nature, the study suggests that functions, roles, and decisions obtained in this way may be helpful for analysts to consider when integrating meaningful human control into system design or when developing road safety policy and management systems for highly automated and connected traffic.

Though limited, our analysis suggests that maintaining the democratic basis for road safety improvements requires accounting for all social and technological actors who may contribute to detecting hazards, thereby enabling remote operators to respond effectively. To perform safety-critical functions and manage systems safely, remote operators may need to understand the intentions of all human actors performing safety-critical actions within those systems, for example, passengers self-evacuating. All human actors who are best placed to perform safety-critical functions should also be given appropriate means of control, for example, passengers or other vulnerable road users needing to control a vehicle to evacuate safely from a smoke-filled tunnel.

The analysis of safety-critical decisions and potential roles of decision-makers in future scenarios shows that, in some cases, the decision-maker and actor responsible are the same and accountability is clear. But some safety-critical decisions could span both technological and social actors. While higher-level system decisions are likely to remain with RROs, the analysis shows where they might depend on other system actors for optimal safety.

For example, an onboard operating system or RVO might decide which hazards are detected near the vehicle, but the OEM, vehicle owner, or technology developer could each play a role in hazard detection as they design and maintain the systems involved. This raises questions of who should be liable. It may therefore be important for vehicle owners to recognise their potential responsibility in several key areas identified by our analysis: on hazard detection and reporting, on in-vehicle and road user hazard response, on appropriate responses to critical situations by onboard systems, and on ensuring lessons from incidents are learnt and applied in practice (e.g., by developing onboard operator algorithms). Moreover, shared responsibility and liability for decision-making and hazard detection is an area that will need careful consideration by policy-makers and lawmakers.

The analysis also raises interesting questions about the level of insight RVOs will need into V2X exchanges for meaningful human control. For instance, in critical situations, RVOs may need to be 'mindful' on behalf of onboard systems, checking not only that the vehicle responds appropriately but that requests to the vehicle are appropriate. This, again, requires that they understand safety-critical actions in the system, emphasising that intentions and actions of V2X communications should be visible and understandable to remote operators.

It is clear from the analysis that to ensure continued and optimal progress in road safety, policy should be developed to ensure democratic participation by an increasing number of actors influencing road safety. These actors will help determine control and communication activities in road safety response systems through their operations, development, and maintenance. Similarly, system approaches to road safety management will need to expand to address onboard operating systems, physical and digital infrastructure, and remote operations, alongside roads, vehicles, and road users.

Any transition phase involving mixed manual and automated traffic will be particularly demanding. Existing policy measures will still be needed, while new policies must address more technically complex vehicle operations, especially those that depend on roadside technology, cloud-based services, and V2X communication. This will require broader coordination among road authorities, vehicle operators, technology providers, infrastructure managers, and regulators.

Study limitations. The study has been exploratory, and more systematic studies involving a larger number of reports are required to verify the usefulness of the approach described here. These studies could account for abundant findings on human factors in remote maritime, air, and rail operations and integrate security alongside road safety considerations. The fact remains that more work is needed to understand the functions that responsible human remote agents will need to perform in future systems, the constraints they will face, and the feedback they will need to maintain awareness and control.

APPENDIX: SUMMARY ANALYSIS OF EACH IN-DEPTH REPORT

	Functions that Keep System Safe	Actor Contribution to Safety Function in Incident Reported		Requirements Implied by Constraints Faced	Safety-Critical Decisions Implied
2021/1 Skid on icy road	Control developing hazards Maintain safe braking distances between vehicles	VO (Vehicle Operator)	Reduce speed to account for reduced road friction	Communicate to vehicle and road operators on dynamic changes to braking distances Communication between V (Vehicle) and VO, between Vs and between road operator and Vs Measures needed to tackle acute changes in road friction	When to reduce vehicle speed and by how much When and how to inform VO about the need to reduce speed, given temperature, traffic density projections When and which road friction measures should be implemented
		RRO (Remote Road Operator)	Communicates to VO the need to reduce speed Control reductions in road friction		
2021/2 Skid on icy road	Control developing hazards Maintain safe braking distances between vehicles	VO	Warn RRO about challenging driving surface	Provide means by which VO can communicate about challenging driving surface Always implement and control mitigating measures to address safety-critical warnings from VO Control to ensure safety-critical warnings given by VO are acted on Control to ensure mitigating measures implemented	What safety warnings are critical. (How many? Content?) (Who decides? RRO, road authority, contractor?) How to delimit the area for mitigation action When mitigation measures should be applied to address warnings about hazardous driving conditions
		RRO	Log warnings from VO and pass on to road contractor		
		Road contractor	Follow-up on safety-critical warnings from VO mediated by RRO		
2019/5 Fire in tunnel – assumption that maintenance truck knocked off fire appliance	Control developing hazards Retreat to safe distance from developing hazard	VO	Meaningful communication to remote road operator of hazardous events in local vicinity	Accurate, reliable information on nature and location of hazard developing near Vs or on road immediately available to all affected actors Remote actors need to understand each V's and road user's relative location and unique situation to be able to tailor advice and coordinate their response	When and how hazard should be reported and to whom Gathering information from multiple sources rapidly to complete picture of developing hazard Communication to VO about developing hazard Advise VO on how to respond and information during response
		RRO or other actor in	surrounding systems		

(Continued)

(Continued)

	Functions that Keep System Safe	Actor Contribution to Safety Function in Incident Reported		Requirements Implied by Constraints Faced	Safety-Critical Decisions Implied
Deciding who can help understand the developing hazard Deciding how to tailor advice on safest response to VO How to guard against developing wrong mental model about situation					
2018/7 fire in tunnel – Fire in foreign semitrailer	Control developing hazards Retreat to safe distance from developing hazardous situation	VO RRO or other actor in	Heed warning signs that road closed surrounding systems	Human road users need human contact and awareness of developing situation Mitigation protocols for different phases of response Prioritised road access for emergency responders Early, clear and open lines of communication established between RRO and first responders Sufficient, experienced resource should be rapidly accessible for RRO and emergency services, in response to hazard situation	Considering surrounding factors, how hazard mitigation measures should be applied to achieve best safety result How road access can be controlled so that road users do not access hazardous areas, but first responders can How to ensure that lessons learnt from previous collisions are learnt in practice
Early detection of developing hazards. Coordinate hazard mitigation measures (tunnel ventilation) to allow for (i) road user evacuation and (ii) firefighting.					

Direct communications with road users awaiting rescue Mediate communications between rescue services and road users Control access to roads surrounding emergency to allow first responders access					
2017/5	Control developing hazard Maintain safe distance between vehicles Increase efficiency for traffic on roads surrounding blockage by controlling traffic flows	VO RRO	Alert to objects blocking road, especially on high-speed roads VO able to take mitigating action when safety-critical perception ability is reduced Timely and effective response to sudden, hazardous road blockages. Communicate to VO about road blockages and need to reduce speed	VO receive messages about re-routing when and where they are needed (not too early or late) Accurate, reliable information on nature and location of hazard developing near Vs or on road immediately available to all affected actors – implies RRO can monitor all roads sufficiently (sensors + pictures)	Data protection issues when monitoring passenger vehicles Approval of content and target of messages sent to many VOs at once. Risk-based prioritisation of incidents needing decisions when several incidents occur at once Decide when routines do not adequately apply in a situation and should not be followed (e.g., no use of VMS >5km from roadblock)
2013/8 fire in tunnel	Control developing hazard Remove road users from developing hazard	VO	Meaningful communication to remote road operator of hazardous events in local vicinity (fire) Early removal from hazard Attempt control of hazard at source	Ensure V driven such as to reduce motor fire risk Vehicle should warn VO about increased motor risk (e.g., increased engine temperature) to enable early mitigating actions	Pro-active monitoring of road/tunnel, usage, and external factors (e.g., temperature) to assess adequateness of infrastructure and safety equipment for evacuation

(*Continued*)

(Continued)

	Functions that Keep System Safe	Actor	Contribution to Safety Function in Incident Reported	Requirements Implied by Constraints Faced	Safety-Critical Decisions Implied
		RRO	Gathering information from multiple sources (people, sensors, cameras) rapidly to complete picture of developing hazard and location of vehicles and humans Communicate immediately to road users of need to remove themselves from hazard Ensure familiarity with VO reaction patterns Apply local knowledge to assist 'self-rescuing' road users Communicate with system actors about developing hazards to ensure shared situational awareness Ensure evacuation protocols, communications, and measures work in practice with real systems of actors and technology	Ensure messages exchanged with RRO are clear and understandable	How to account for VO reaction patterns when applying mitigating measures How to apply mitigation measures to ensure conditions needed for evacuation (smoke-free access to end of tunnel) with those needed for rescue (smoke-free access from other end)
2018/4 tunnel fire	Control developing hazard Remove road users from developing hazard	VO	Meaningful communication to remote road operator of hazardous events in local vicinity (fire)	Educate VO on need to alert RRO or emergency services Regular testing to ensure safety-critical control actions or monitoring channels work (e.g., ventilation system) Ensure all safety-critical areas can be monitored remotely	How to ensure access to hazard site by responders while limiting access of other vehicles
		RRO	Set up and monitor several channels to detect developing hazards (e.g., cameras, picture recognition, road user telephone, smoke sensor) Open lines of communication with emergency services for shared situational awareness Able to use back-up plans for failed emergency measures		

2010/4 Skid on icy road	Control developing hazard Maintain safe distance between vehicles	VO	Ensure speed is adjusted to account for reduced friction		How much speed to reduce to maintain safe braking distance
		RRO	Log warnings from VO and pass them on to road contractor Ensure road users behave safely in response to hazards	Ensure systems in place to monitor and react to environmental conditions that could cause a hazard (sub-zero temperature), moisture Control to ensure safety-critical warnings from VO are acted on	Assess when surrounding conditions could cause hazardous driving and require mitigation Include considerations of projected traffic density and variations in average speed
		Contractor	Inspect roads regularly and respond with mitigating actions	Control quality of response	
2015/01 Skid on icy road		VO	Communicate to RRO about developing hazards in vicinity	Ensure reduced road friction reported to VO before it becomes a hazard	How to adapt speed before encountering hazardous road surface
		RRO		Ensure hazards are detected early on in their development (don't rely on local VO reports as they are generated by Vs encountering the hazard)	How to inform VO about need to reduce speed before they encounter hazardous road surface
		Contractor	Inspect roads regularly and respond with mitigating actions	Control quality of response	

VO, vehicle operator; RRO, remote road operator.

REFERENCES

Alonso Raposo, M., Ciuffo, G., Alves Dias, P. et al. (2020). *The Future of Road Transport: Implications of Automated, Connected, Low-Carbon and Shared Mobility*. JRC Science for Policy Report, European Commission.

Béchard, D. E. (2025, June 9). Thermal runaway explains why Waymo cars burned so completely in the recent Los Angeles protests. *Scientific American*. https://www.scientificamerican.com/article/why-the-waymo-car-fires-in-recent-los-angeles-protests-caused-the-robotaxis

Elvik, R. (2021). Democracy, governance and road safety. *Accident Analysis and Prevention, 154*, 106067.

Elvik, R. (2022). Chapter 9: Vision zero in Norway. In: Bjørnberg, K. E., Hansson, S. O., Belin, M.-Å., & Tingvall, C. (Eds.), *Vision Zero Handbook* (pp. 295–306). Cham: Springer.

Elvik, R. (2024). The development of a road safety policy index and its application in evaluating the effects of road safety policy. *Accident Analysis and Prevention, 202*, 107612.

Endsley, M. R. (1995). Towards a theory of situation awareness in dynamic systems. *Human Factors 37*, 32–64.

European Climate (EC), Infrastructure and Environment Executive Agency (CINEA) (2023). *Towards Cooperative, Connected and Automated Mobility (CCAM)*. https://www.ccam.eu/wp-content/uploads/2023/05/HE-CCAM-2023_brochure-web-FIN.pdf

Hollnagel, E., & Woods, D. (1983). Cognitive systems engineering: New wine in new bottles. *International Journal of Man-Machine Studies, 18*(6), 583–600.

Hollnagel, E., & Woods, D. (2005). *Joint Cognitive Systems: Foundations of Cognitive Systems Engineering*. New York: CRC Press Taylor & Francis Group.

ITF/OECD (2015). *Automated and Autonomous Driving: Regulation under Uncertainty*. International Transport Forum Policy Papers 7, OECD Publishing.

Leveson, N. (2012). *Engineering a Safer World: Systems Engineering Applied to Safety*. Cambridge, MA; London, UK: MIT Press.

Merat, N., Madigan, R., & Nordhoff, S. (2017). Human factors, user requirements, and user acceptance of ridesharing in automated vehicles. International Transport Forum Discussion Paper, No. 2017-10, Organisation for Economic Co-operation and Development (OECD), International Transport Forum, Paris. https://hdl.handle.net/10419/194045

Militello, L. G., Dominguez, C. O., Lintern, G., & Klein, G. (2010). The role of cognitive systems engineering in the systems engineering design process. *Systems Engineering 13* (3): 261–273. https://doi.org/10.1002/sys.20147

Naikar, N. (2013). *Work Domain Analysis: Concepts, Guides and Cases*. CRC Press.

Papadimitratos, P., De La Fortelle, A., Evenssen, K., Brignolo, R., & Cosenza, S. (2009). Vehicular communication systems: Enabling technologies, applications, and future outlook on intelligent transportation. *IEEE Communications Magazine, 47*(11), 84–95.

Rasmussen, J. (1986). *Information Processing and Human-Machine Interaction: An Approach to Cognitive Engineering*. North Holland Series in System Science and Engineering, 12. Elsevier Science Ltd. ISBN: 10.0444009876.

Rasmussen, J. (1997). Risk management in a dynamic society: A modelling problem. *Safety Science, 27* (2/3): 183–213.

Santoni de Sio, F., & van den Hoven, J. (2018). Meaningful human control over autonomous systems: A philosophical account. *Frontiers in Robotics and AI, 5*, 15. https://doi.org/10.3389/frobt.2018.00015

Schrank, A., Merat, N., Oehl, M., & Wu, Y. (2024). Human factors considerations of remote operation supporting level 4 automation. In: Meyer, G., Beiker, S. (Eds.), *Road Vehicle Automation 11*. ART-Symposium 2023. Lecture Notes in Mobility. Cham: Springer.

Shladover, S. E. (2018). Connected and automated vehicle systems: Introduction and overview. *Journal of Intelligent Transportation Systems, 22*(3), 190–200.

Vicente, K. J. (1999). *Cognitive Work Analysis: Toward Safe, Productive, and Healthy Computer-Based Work*. CRC Press.

Ward, P., Hoffman, R. R., Conway, G. E., Schraagen, J. M., Peebles, D., Hutton, R. J. B., & Petsushek, E. J. (2017). Macrocognition: The science and engineering of sociotechnical work systems. *Frontiers in Psychology, 8*, 515.

Woods, D., & Hollnagel, E. (2006). *Joint Cognitive Systems: Patterns in Cognitive Systems Engineering*. CRC Press.

Young, M. S., & Stanton, N. (2023). *Driving Automation: A Human Factors Perspective*. CRC Press.

Part 3

Key Techniques for Designing for Safety and HF in Emerging Environments

8 The Development of a Remote Operation Centre

Addressing Human Factors When Designing for Autonomous Ship Operations – in ROC

Jon Bernhard Høstmark

8.1 INTRODUCTION

This chapter is based on the development of an ROC design solution in a collaboration among industry, academia, research institutions, and ship operators. Examples and text are inspired by research and papers published by the collaborating parties.

Today's autonomy push tends to focus on how novel technologies can be applied to achieve efficiency and cost reduction, thereby addressing the economic perspective. The context of use and adoption of technology is often framed around the regular operation of ships. Unmanned navigation has been demonstrated, for example, SVAN (Safer Vessel with Autonomous Navigation) by Rolls-Royce Marine (SVAN, 2018), where the conditions are given within the design envelope of the technology. A review of recent research on autonomous ships shows that only 7% of papers include Human Factors (HF) considerations, underscoring the need to integrate HF systematically into ROC design (Porathe, 2023).

However, for the solution to be socially acceptable, technology must comply with governing rules and regulations (IMO, 2018) that address handling abnormalities and failures. This aligns with DNVGL (2018), which states that autonomous and remotely operated vessel concepts should be developed through a risk-based approach and should achieve a level of safety equivalent to or better than that of conventional vessel operations. Until technology can fully handle deviations and uncertainty, human intervention is needed to ensure acceptable mitigating actions for all possible scenarios during a voyage. Edge cases that require reasoning and evaluation of possible actions must be carefully analysed and addressed to maintain adequate safety throughout the operation. Today, humans outperform machines for these edge cases. Therefore, human monitoring of safe passage is often the default solution when the system makes an error or encounters an edge case.

With humans no longer on board the vessel, the context has changed. Existing routines and procedures cannot directly be adopted in an ROC. This requires increased knowledge and understanding of how humans perceive information presented through displays in a fully digitised work environment, as compared to being on board a ship.

The main question becomes: What does a human need to be able to navigate an unmanned autonomous ship efficiently and safely from an ROC?

In this chapter, we will look at a systematic approach to this challenge. Critical steps towards a solution require considering HF as a design requirement of a cyber-human ROC tailored for autonomous ship operation.

DOI: 10.1201/9781003741824-11

8.2 DEVELOPMENT PHASES

When venturing into the unknown to solve an unsolved problem, patience is a key element. This requires the use of tools and methods to analyse the challenge from different perspectives, gain knowledge and understanding, and only then synthesise possible solutions.

In the following, you will find an example of a structured approach, based on the development of autonomy and an ROC by Kongsberg Maritime, Rolls-Royce, and Massterly in collaboration with NTNU and IFE in the period 2017–2023.

The ROC design shows how meaningful human control can be maintained even as autonomy increases – through interface clarity, situational awareness (SA) support, and feedback mechanisms.

The following sections are organised into phases, as suggested by ISO 11064, Ergonomic design of Control Centres (ISO 11064, 2020). The five development phases are illustrated in Figure 8.1.

8.2.1 Phase A: Clarification

This phase examines the essential processes involved in maritime navigation, highlighting the roles and responsibilities, the complexities of ship operations, and the advancements in autonomous

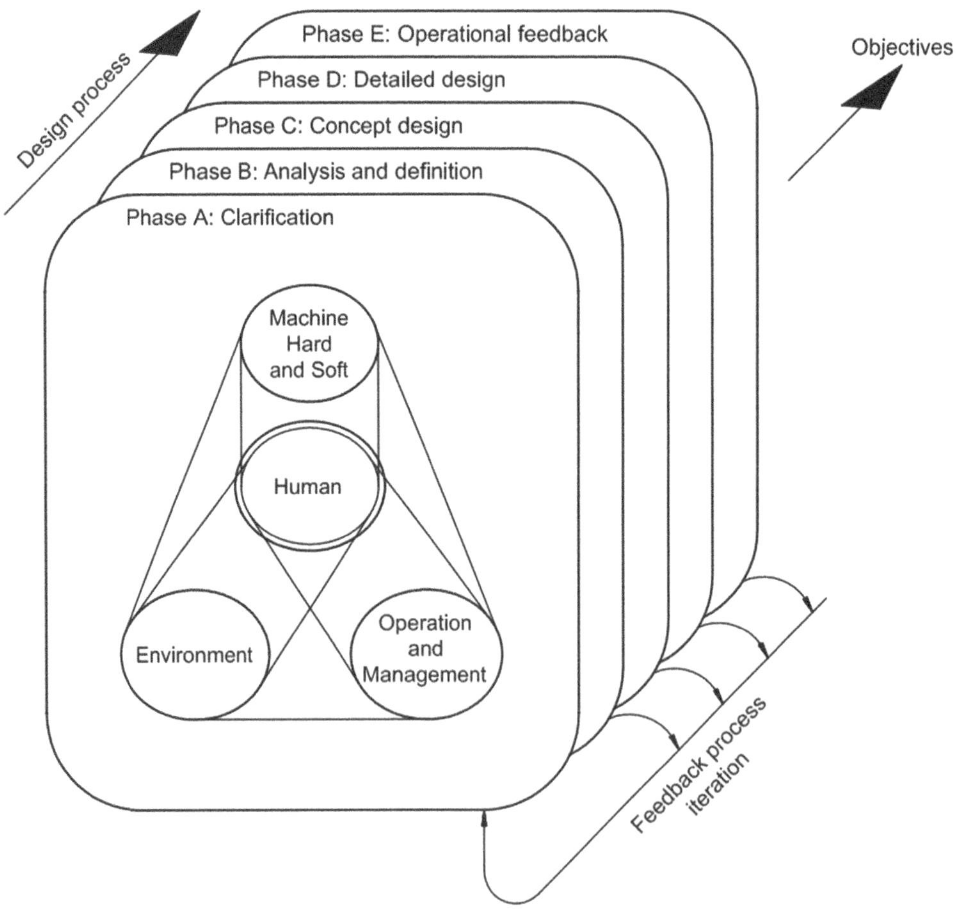

FIGURE 8.1 Ergonomic approach to system designs.

(Source: ISO 11064.)

shipping technologies. It includes understanding the operational context of maritime navigation, key roles, and the intricate nature of operations involving coordination among multiple stakeholders. Thorough background research is needed to gain a proper understanding of the context of use. This lays the foundation for defining the central elements of the problem to be solved, such as constraints and roles, and for determining the appropriate scope.

Traditional ship operations involve a complex, multidisciplinary effort that requires seamless coordination between the ship's crew, shore-based management, port authorities, and regulatory bodies. The responsibilities of the OOW (Officer of the Watch) tie into many of these operational tasks. The OOW plays a vital role in ensuring the smooth and safe operation of the vessel by making sound decisions to protect the ship, crew, cargo, and environment.

In the ROC development case, the operational context was a passenger ferry over 50 m long crossing the Oslofjord between Moss and Horten. The effort was focused on the OOW role, with the task and responsibility for safe navigation and operation of a commercial ship during an assigned watch period while underway. Underway is defined as the part of the voyage where the ship has left the quayside until it is securely tied to the quayside again. While underway, maintaining safety is a dynamic task that requires proper SA to follow the planned voyage, respond to changes in the operational area and access, and update it as needed.

Primary duties for the OOW are to ensure the ship's safety, security, and compliance with maritime regulations. In this example, these duties are limited to the bullet points below.

- Safe Navigation: Monitoring the ship's course, speed, and position using navigational tools and systems, for example, radar, GPS, ECDIS (Electronic Chart Display and Information System).
- Watchkeeping: Maintaining a proper lookout and ensuring compliance with the International Regulations for Preventing Collisions at Sea (COLREG).
- Communication: Handling radio communications with other vessels, ports, and maritime authorities.
- Monitoring Ship Systems: Overseeing the performance of engines, steering gear, and safety systems in coordination with the engine room.
- Log Keeping: Recording navigational and operational information in the ship's logbook.
- Emergency Response: Act promptly in the event of man overboard, fire, or collision.

8.2.1.1 MUNIN

The EU-funded MUNIN: Maritime Unmanned Navigation through Intelligence in Networks (MUNIN 2016) project, active from September 2012 to August 2015, aimed to develop the concept of autonomous merchant vessels.

MUNIN sought to address challenges in maritime transport, including rising transport volumes, environmental requirements, and a shortage of seafarers. The project focused on creating an autonomous ship concept that combined automated decision systems with remote control via shore-based stations. The goal was to enhance efficiency, safety, and sustainability in shipping operations.

Key developments include a structured list of requirements collected from various sources, including the IMO, MARPOL (International Convention for the Prevention of Pollution from Ships), and COLREG. The project proposed core systems needing to comply with associated requirements:

ANS: An Autonomous Navigation System enabled the vessel to follow predefined routes and adjust autonomously in response to obstacles or weather changes.

AEMCS: An Autonomous Engine and Monitoring Control System optimised engine performance and facilitated preventive maintenance.

SCC: A Shore Control Centre monitored and controlled the vessel remotely, ensuring safety and operational efficiency. The acronym ROC will be used instead of SCC in this text.

The project's findings create a good introduction to understanding the complexity of autonomous seaborne transport. MUNIN has influenced the research and development of autonomous shipping technologies, alongside the EU's eMaritime and the IMO's eNavigation initiatives.

8.2.1.2 Bridge Guide Procedure

The Bridge Guide Procedure Handbook (BGPH 2022) for maritime vessels provides comprehensive guidelines and procedures to ensure safe and efficient navigation on a ship's bridge. It is a vital document for ensuring the bridge team operates cohesively, safely, and in compliance with international maritime standards and regulations. It outlines operational responsibilities, safety measures, and best practices for various phases of a voyage, particularly when navigating congested or challenging waters. For the scope of an ROC, the following parts of the handbook are essential.

Bridge Team Organisation: Defines roles and responsibilities for each crew member on the bridge, including the captain, officers, lookout, and helmsman.
Route Planning: Detailed procedures for planning and updating the ship's course.
Use of Navigation Instruments: Guidelines for using radar, GPS, ECDIS and other tools for safe navigation.
Monitoring the Vessel's Position: Regular checks to ensure the ship remains on the correct course; adjustments as necessary.
Communication: Procedures when entering ports, harbours, or restricted waters, along with communication protocols with port authorities and other vessels.
Bridge Operations during Docking/Undocking: Clear steps for safe docking and undocking, including the use of tugboats, communication with port authorities, and crew coordination.
Safety and Emergency Protocols: The handbook outlines emergency procedures, including actions to take in the event of navigation errors, poor visibility, mechanical failure, or collisions. It emphasises crew training for handling these situations efficiently.
Watchkeeping Procedures: Guidelines for maintaining proper lookout, keeping watch, and monitoring vessel systems at all times to ensure safety.
Emergency and Contingency Plans: Steps to follow during emergencies, ensuring the crew is prepared for any unforeseen circumstances.

This phase has examined the essential background research needed to define the OOW's role in autonomous ship operations, focusing on key responsibilities such as navigation, communication, and emergency response. It has also considered the MUNIN project's proposed core systems and the Bridge Guide Procedure Handbook's guidance for safe and efficient ship navigation, including remote monitoring and control.

8.2.2 Phase B: Analysis and Definition

The analysis and definition phase aims to develop an integrated design proposal for an ROC, focusing on defining system functions, allocating functions to humans or machines, defining task requirements, and job design and organisation.

While the development and adaptation of autonomous technology from the automotive and aviation industries had gained increasing popularity within shipping, the operational integration of these technologies into an ROC had not been developed. The ROC was seen as a key enabler for the use of autonomous ships; however, minimal effort had been devoted to the research and development of an operations centre where operators of uncrewed ships perform the duties of conventional ships.

One reason for this may be that technology companies are focusing on advances in machine and automation systems rather than on the operational human tasks required to operate a ship safely.

Thus, the know-how for developing an ROC is closely associated with the competence and daily routines of ship-operating companies, that is, the customers of technology companies.

An ROC serves as an enabler of increased efficiency, better working conditions, and economic growth in the maritime industry. However, the remote operation of commercial ships requires a deeper understanding of the human element to meet shipping regulations, as well as the integration of novel technologies.

The development of an ROC requires a broad approach with a focus on a human-centred design (ISO 9241-110 and ISO 9241-210, 2020). This was initiated by gathering a comprehensive group of industry experts representing a broad range of perspectives and roles in ship operations. All participants involved had a solid background as experts in their respective domains. This included competence in:

- Maritime rules and regulations,
- Ship operations,
- Psychology,
- Interaction design,
- Maritime technology development,
- Navigation system development,
- International research in ship autonomy,
- Vessel management,
- Vessel traffic service,
- Bridge duty,
- HF, and
- Information display design.

Over three days, these experts carefully analysed and detailed tasks necessary for safe passage, quayside to quayside. Necessary tasks were analysed and linked to the responsible role and information needed and mapped to the tasks to be solved in an ROC. Each step of the ship's journey was analysed in group sessions. Details and insights were documented on a large wall canvas for everyone to see. A typical work session is shown in Figure 8.2.

A storyline with deep insight was created, looking at HF, interaction design, ship operations, and technology. Together, a unified set of operational needs was defined for an ROC. The details, guidelines, and principles documented in the workshop represented a holistic approach defining the solution context. This was used as the baseline input for later work and development related to an ROC.

Here, we have examined the definitions required for developing an ROC for autonomous ships, focusing on integrating existing autonomous technologies from the automotive and aviation sectors into shipping operations. Advancements in automation technology need to be balanced with the operational tasks required for safe and efficient remote ship operation. This was addressed through a human-centred design approach, comprising a comprehensive workshop with industry experts from various fields to analyse operational tasks and define the guidelines, principles, and solutions needed for the effective development of an ROC.

8.2.3 Phase C: Conceptual Design

The purpose of phase C is to create a conceptual design for a control centre that meets functional, HF, and organisational requirements, including physical attributes and operator interfaces, while establishing the context and specifications for the detailed design in later steps.

With a clear definition of the context and the problem to be solved, the work to develop an ROC concept can begin. This work was done as part of a team effort in the research project LOAS (Land-based Operation of Autonomous Ships) under the MAROFF-2 (Maritime Activities and

FIGURE 8.2 Workshop sessions.

(With permission from Kongsberg Maritime.)

Offshore Operations) programme, funded by the Norwegian Research Council. The ROC design aligns with Ecological Interface Design (EID) principles (Vicente & Rasmussen, 1992), emphasising perceptual salience and projection to support all three levels of SA (Endsley, 2013).

The project did not regard remote piloting of ships as a commercially viable solution. It was argued that remote piloting of commercial ships would shift crew costs from onboard to onshore and increase build and operational costs due to the additional technology required for sensing, analysis, and communication with a human in an ROC. The initial cost of developing and implementing an ROC should be competitive with the cost of installing a bridge for manned operation on the ships the ROC operates. To become economically viable, unmanned vessel operations needed to become more efficient to offset the additional cost of new technology.

This was formalised by adding a requirement for the simultaneous operation of up to three vessels per OOW in the ROC. This challenged the traditional way of managing, crewing, and operating commercial ships. Maritime law is clear that responsibility of safe navigation, watchkeeping, and communication resides with the human OOW. In addition, Circular (RSV 12-2020) from Sjøfartsdirektoratet states: '5.11 Safety philosophy: The safety philosophy describes how an equal safety level is maintained during unmanned operation'.

It was unclear how this should be addressed, particularly given HF issues such as attention and workload capacity. When supervising autonomous ships from a remote location, it is necessary to ensure that operators maintain an acceptable workload and a high degree of SA of the relevant maritime situation.

The conceptual design should therefore support the three SA levels (Endsley, 2013). The SA levels are rephrased for the project purpose below:

SA Level 1. The necessary data: The first level is related to the available data, providing sufficient data and information in the display to understand the maritime situation.
SA Level 2. Comprehension: The second level concerns how the operator comprehends the data as a meaningful 'picture' of the maritime situation.
SA Level 3. Projection: The third level explains how the operator should be given cues to predict development: how is the situation going to develop in the near future?

It should be noted that all three levels are necessary for supporting a high level of SA.

The ROC concept needed to demonstrate how novel technologies could assist humans in uncrewed ship operations and how humans and machines should team up to maintain safe operations at sea under various conditions.

The following fundamental questions were coined to address this challenge:

- What is the preferred viewpoint for the human in the ROC when supervising one and three ships?
- Which information is needed for the supervision of autonomous ships from an ROC?
- How should information be presented in the ROC displays for efficient and rapid perception of the ship's maritime situation?

8.2.3.1 Phase C: Test Method

These questions were explored through a study using a maritime simulator operated by experienced seafarers. The setup of the ROC station with a test participant and test personnel to the right is shown in Figure 8.3.

8.2.3.2 Phase C: Test Scenarios

Five different setups: A, B, C, D and E – are shown in Figure 8.4.

Two setups for monitoring one ship: A and B, and three setups were used for simultaneously monitoring three ships: C, D and E. The view in each setup is presented in Figures 8.5 and 8.6.

FIGURE 8.3 Test setup conceptual design phase.

(With permission from author J.B Høstmark.)

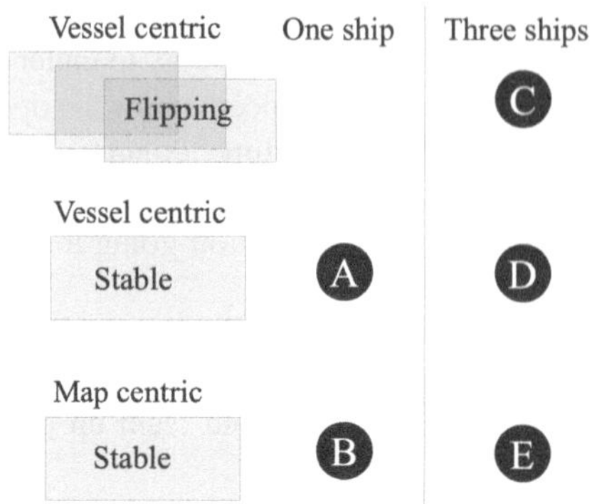

FIGURE 8.4 Setup – one and three ships in command.

(With permission from author M. Kaarstad.)

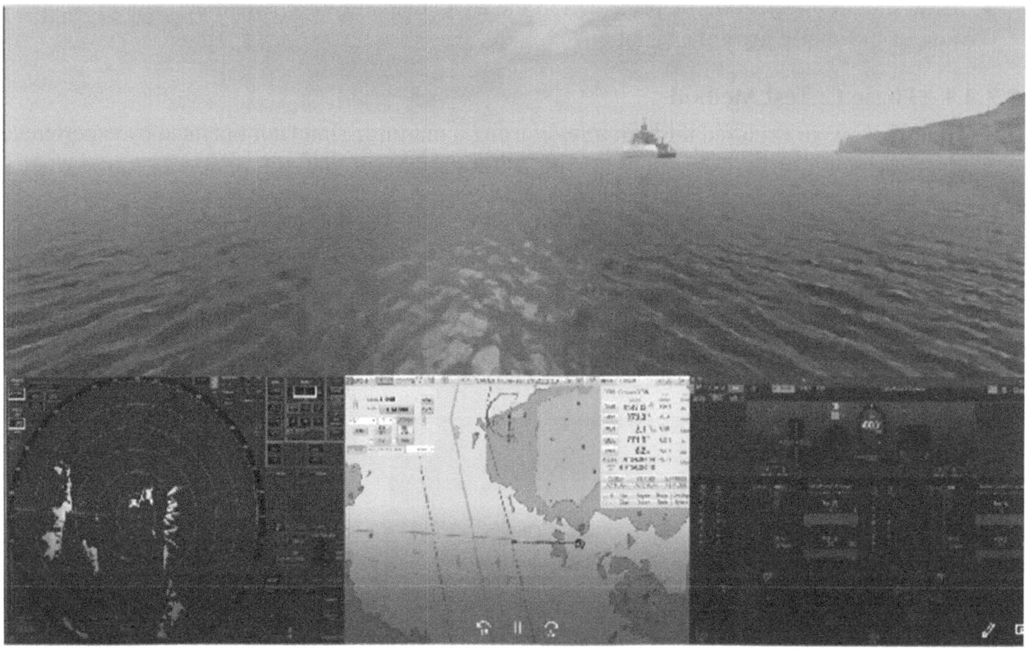

FIGURE 8.5 View in setup A and C (vessel-centric).

(With permission from author J.B Høstmark.)

Setup A presents a traditional vessel-centric layout (bridge view) for one ship.

Setup B presents a map-centric layout (bird's-eye view) for one ship (Figure 8.7).

Setup C presents a traditional vessel-centric layout (bridge view) for three ships. This concept was not stable, as the information and video feeds were flipping and rotating the view automatically, with a timer between the three ships.

Setup D presents a traditional vessel-centric layout (bridge view) for all three ships simultaneously, stacked in three levels.

Setup E is a map-centric setup (bird's-eye view) for all three ships at the same time, with additional video feed and information overlay on the right.

FIGURE 8.6 View in D (vessel-centric three ships).

(With permission from author J.B Høstmark.)

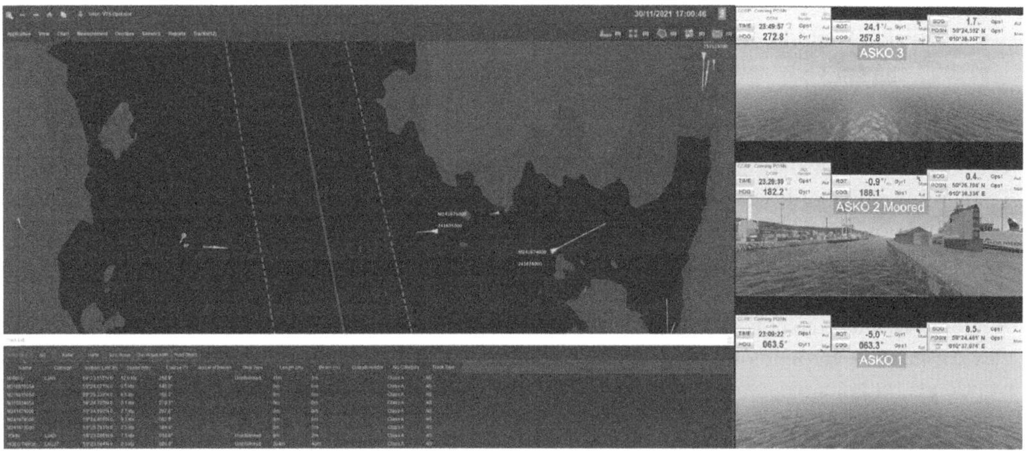

FIGURE 8.7 View in B and E (map-centric).

(With permission from author J.B Høstmark.)

Both one- and three-ship scenarios had similar complexity when crossing the Oslofjord, including additional traffic such as large cargo ships. The weather in the scenarios was hazy to stimulate participants to use instruments and digital tools, adding to the overall complexity. Furthermore, the scenarios were designed around a near-critical situation, in which it was natural for the participants to take manual control of one of the automated ships. Each scenario took 10–15 minutes to complete. All participants experienced the scenarios with one ship first – either A or B – then the scenarios with three ships – randomly C, D, or E. The order of display setups was counterbalanced.

The conclusions from the conceptual design provided direction for the detailed design solution. Evaluating the first question regarding preferred viewing perspective, the results suggest that a stable, larger map-centric layout (a bird's-eye perspective) was the reasonable choice, particularly for operating multiple ships.

The preference scoring of concepts C, D, and E is shown in Figure 8.8, Preference for supervising three ships. Concepts C, D, and E are explained in Section 8.2.3.2. Concept E was chosen as the

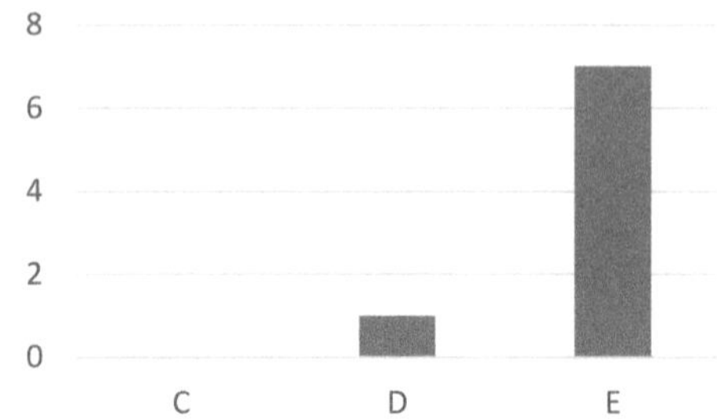

FIGURE 8.8 Preference for supervising three ships.

(With permission from author M. Kaarstad.)

baseline for building an operational environment in which a single operator must supervise a fleet of ships.

The information needed in the displays, based on participant feedback, was as follows:

- Adding a larger overview map by enabling zooming on the map,
- Information tags on each ship; adjustable vectors; speed, heading, and turn ratio,
- Collision warning,
- 360-camera view and adjustable radar range,
- A specific harbour view was also suggested.

The information display graphics used in the concept testing were built on existing solutions. Testing revealed that it was necessary to develop information displays in an ROC from the ground up. A common set of principles for all graphics should be developed to ease use and understanding. Additionally, information and presentations should be tailored to changes in operational conditions and phases throughout a voyage (Kaarstad et al., 2022).

In this chapter, the conceptual design phase is explored, focusing on integrating novel technologies to support autonomous ship operations. It emphasises ensuring high SA and manageable workloads for human operators, addressing key considerations such as optimal viewpoints, essential information, and effective display design for the supervision of up to three vessels.

8.2.4 Phase D: Detailed Design

The purpose of phase D is to develop detailed design solutions for the ROC, including physical layout, information screen layout supporting operational scenarios and tailored information graphics.

The findings from the conceptual design phase showed that mirroring the layout of a traditional bridge would not be adequate for the operation of a fleet of ships from an ROC station. In the detailed design phase, the screen layout was kept as can be seen in Figure 8.9.

ISO 9241-210 provides a framework for the ROC design where Endsley's SA model is a key part, enabled by Vicente and Rasmussen's EID approach (Endsley et al., 2003). Thus, a new set of information graphics was developed from the ground up (Braseth et al., 2022). The need for efficient Human-Machine Interfaces (HMIs) when operating automated ships from an ROC is highlighted (Karvonen & Martio, 2018). A limited set of research-oriented HMI concepts was found: EID (Vicente, Rasmussen, 1992); the Open Bridge Design System (OpenBridge, 2023), which offers a design library for maritime interfaces; and Information Rich Design (Braseth, 2015), for larger overview displays. None of these is aimed explicitly at challenges related to MASS (Maritime Autonomous Surface Ships).

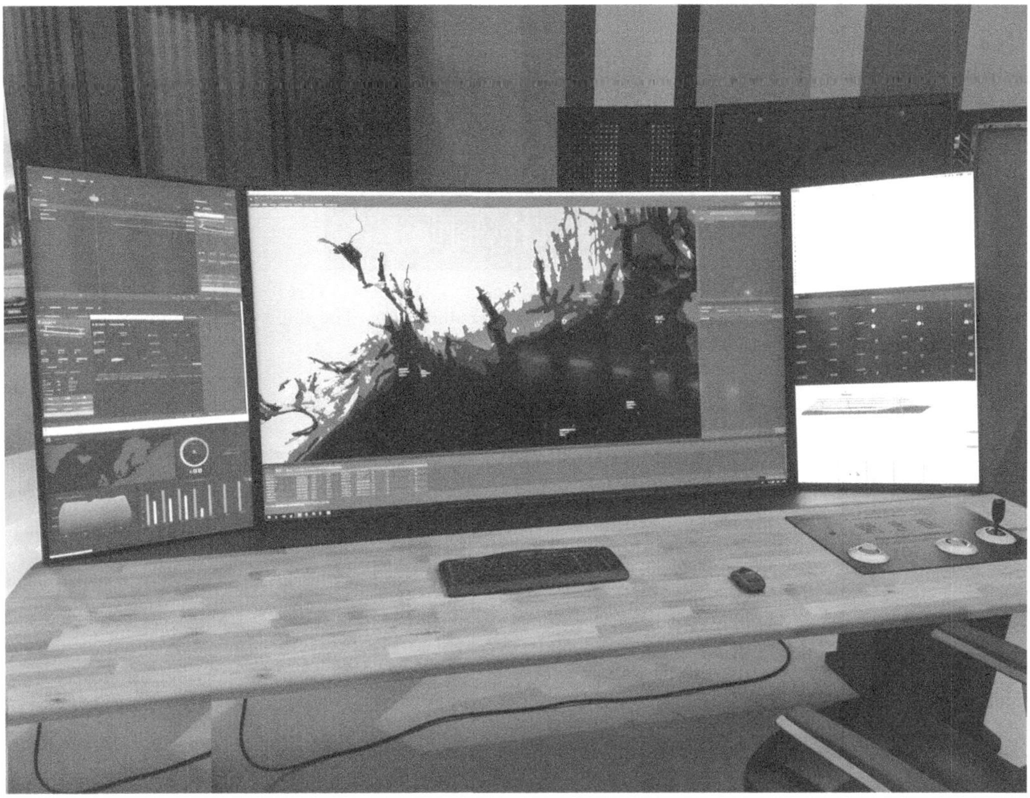

FIGURE 8.9 ROC operator station.

(With permission from author J.B Høstmark.)

This led to the development of tailoring interaction-design and graphics principles for operating MASS. The graphics were required to support SA and enable safe, efficient operation. Graphics were optimised to display information needed to supervise and operate ships. A strong focus was placed on presenting information to support an efficient, rapid perception of the operational picture. Graphics should support operators by projecting the near-future situation.

The OOW often comments on the cluttered view of the ECDIS, as more and more information has been added over the years, with limited filtering options. Figures 8.10, 8.11, 8.12, and 8.13 illustrate the approach to creating an uncluttered view that prioritises and highlights necessary information. The ship's points (metaphor) in the direction of its heading shows how the own ship, the selected ship (in command), and other ships are distinguished (Figure 8.10).

Alphanumeric information about the ship uses a standard format without units displayed (Figure 8.11).

The identification (name), automation level, direction (in degrees), and speed (in knots) were grouped for the active ship (Gestalt proximity). The example in Figure 8.12 visualises the ship AS002, which deviates slightly from the planned route. The offset from the planned route (line between waypoints) supports the detection of deviation from a route (AS002). Trailing dots show the sailed route and speed indicated by the distance between the four trailing dots.

The use of bottom-up data-driven events, such as urgent situations, was supported by strong visual pop-out effects. A gentle dynamic alarm spot (Braseth, 2015) was used to attract attention, avoiding annoying blinking and flashing graphics that can be tiring.

FIGURE 8.10 Distinguishing own ship, selected ship, and other ships. The ship symbol points in the direction of its heading.

(With permission from author A.O. Braseth.)

FIGURE 8.11 AS002, automation level A1, heading 177 degrees, and speed 5.7 knots are displayed in a standardised tag format; the units are implicitly understood.

(With permission from author A.O. Braseth.)

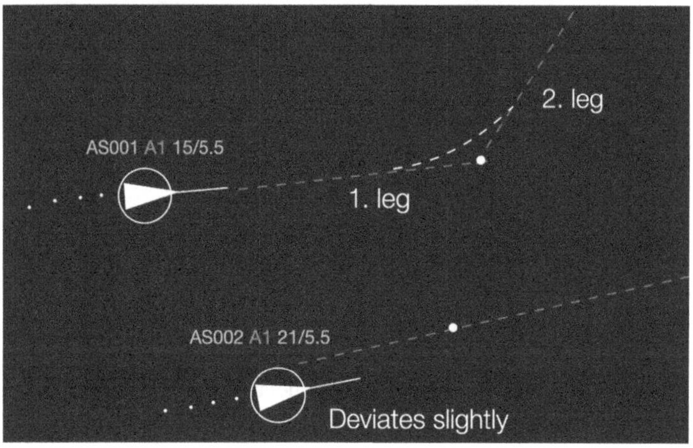

FIGURE 8.12 Visualising the planned routes to see ahead, AS002 deviates slightly.

(With permission from author A.O. Braseth.)

In Figure 8.13, the graphic is independent of whether the vessel is 'my ship' or 'other ship'. If no evasive action is taken as the BCR/TCPA values decrease, end points/red (for a black and white carbon copy the red color will not be visible in the figure) dots indicating the nearest point of contact and dynamic alarm-spot flashes with a large white bubble show how the dynamic alarm spot warns the operator of a potential collision.

Information about the nearest point/time-to-contact for each ship was also added. (Bow Crossing Range, BCR; Time to Closest Point of Approach, TCPA).

The new information graphics were then integrated into a new map view, shown in Figure 8.14. A zoomed-out view of the operational area was added to the left screen. This screen provided the operator with information about ships approaching from a distance, well in advance of their entry

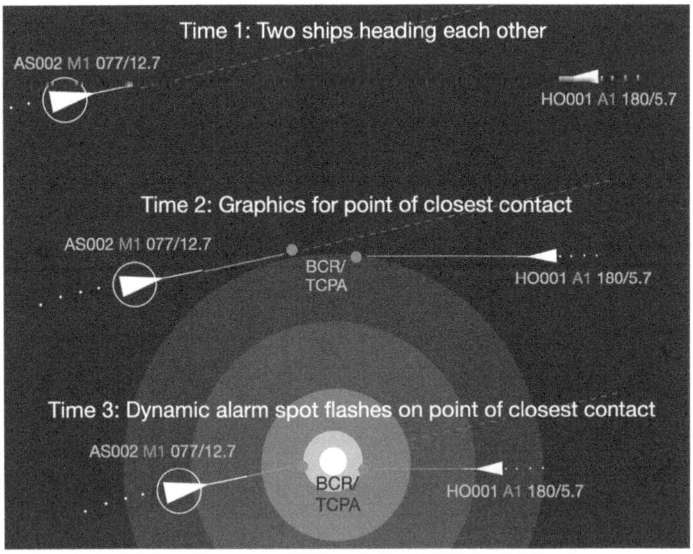

FIGURE 8.13 The graphic is independent of 'my ship', 'other ship'. Suppose no evasive action is taken as the BCR/TCPA reduces, end points/red (in black-and-white reproduction the red color will not be visible in the figure) dots showing the nearest point of contact and dynamic alarm-spot flashes with a large white bubble.

(With permission from author A.O. Braseth.)

into the operational area for own ships. To minimise fatigue and eye strain, sea areas deeper than 10 m were shown in dark grey compared to the traditional light blue (Microsoft, 2023).

The conceptual design phase revealed that replicating the traditional bridge layout was insufficient for operating multiple ships from a Remote Operation Centre (ROC). In the detailed design phase, a new set of information graphics was developed to optimise HMIs for MASS, focusing on clear, uncluttered displays that support SA and efficient operation. The graphics included features like dynamic alarm-spot flashes for urgent situations, visual cues to distinguish ships, and a zoomed-out map view to enhance operator awareness of approaching vessels and reduce fatigue.

8.2.5 Phase E: Operational Feedback LOAS

This phase focuses on validating the design objectives through operational feedback and eye-tracking data, with careful analysis to assess SA levels and provide valuable insights for future projects and evaluations.

In the evaluation phase, operational feedback was collected through simulations to assess the effectiveness of the ROC design in maintaining high SA. Eye-tracking technology was used to gather objective, non-intrusive data on where operators focused their attention during the simulation, complementing traditional feedback methods such as interviews. The analysis of eye-tracking data confirmed that the ROC design supported all three levels of SA, from perceiving essential information to understanding it and projecting future events, with varying preferences depending on the stress level of the operators.

Before completing the ROC design, operational feedback was collected to validate the detailed design. To ensure safe operation from an ROC, particular focus was placed on maintaining a high level of SA.

The feedback collected for the ROC was based on a second simulation study with an updated operational scenario and display graphics based on the bird's-eye-view solution E. Traditional methods for assessing human-machine interaction include NASA-TLX (Task Load Index) ratings, mid-test interviews, pausing the simulation to conduct interviews, and post-test interviews.

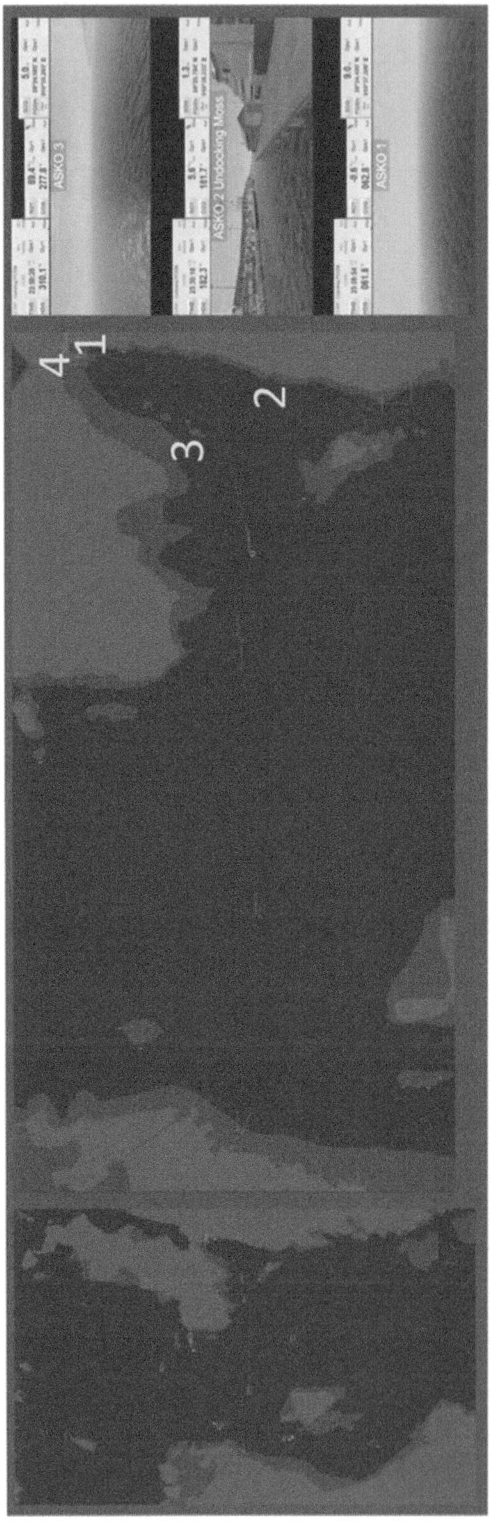

FIGURE 8.14 ROC graphical information display.

(With permission from author J.B Høstmark.)

Note that the NASA-TLX can be used both as a design tool and later to assess or validate the workload of the final ROC design. All these methods have limitations in that they interrupt and disturb the test subject, risking influencing or skewing the data. Additionally, these methods measure the test subject's conscious thoughts, thereby hiding information that could provide insight into non-consensus behaviour or aspects.

To pursue an objective, non-intrusive data-collection method, eye tracking was used. Eye-tracking measures where subjects look. With this data, one may determine where the test subjects put their visual focus at any given time.

While eye-tracking offered objective insight into attention and visual behaviour, subjective methods such as NASA-TLX could complement this by capturing perceived mental and temporal demands, providing multimethod validation.

In the operational feedback tests, the screen layout was static, with a zoomed-out view of the operational area on the left, a main map view in the centre, and a video feed from each vessel with numeric data overlays on the right. See Figure 8.14 for a detailed graphical representation.

The left, centre, and right information feeds were displayed on the three screens shown in Figure 8.9, with 43-inch screens on each side and a 65-inch screen in the centre. With the test subjects sitting approximately 1 m from the screens, they needed to move their heads to shift focus between screens.

Head-mounted eye-tracking glasses were used to enable both eye and head movement. The eye-focus point was represented in the head frame (glasses frame), meaning that the screens moved relatively in the glasses' video feed when turning the head. A computer program was developed to stabilise the eye-focus point within the display screen frame. See Figure 8.15 showing one

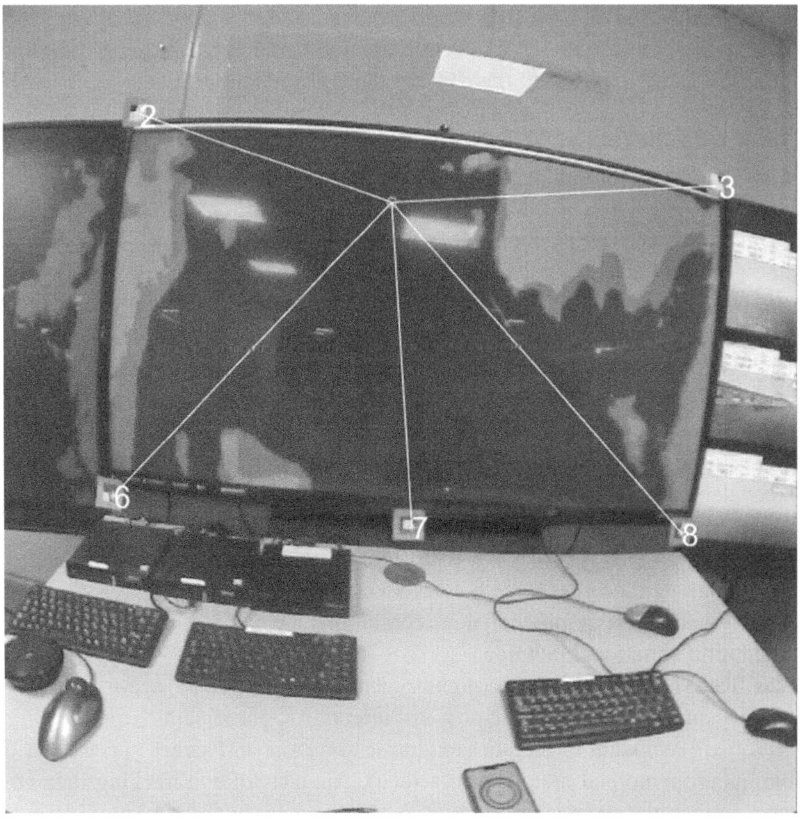

FIGURE 8.15 View from test subject with eye-tracking overlay.

(With permission from author J.B Høstmark.)

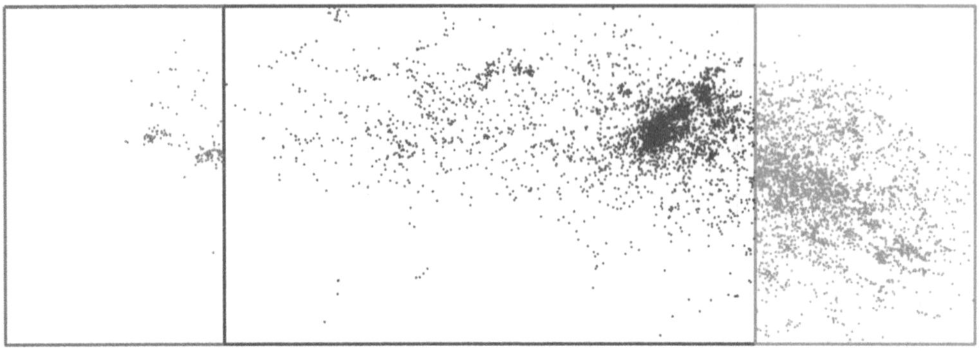

FIGURE 8.16 Eye-tracking point cloud.

(With permission from author J.B Høstmark.)

video-feed frame with tracking markers 2, 3, 6, 7, 8 for image recognition of screen location, and eye-focus point to the centre of the screen.

Eye-tracking focus points were then mapped onto a 2D image spanning the three screens' frames and stacked in time to create a 3D point cloud. Cloud dimensions are left-right, up-down, and time. See Figure 8.16, eye-tracking point cloud with point cloud for one test session; the time dimension is 'collapsed', showing all datapoints at the same time.

With this new tool, the ROC design could be evaluated based on feedback from traditional methods, along with an objective data-collection method using eye-tracking data. This enabled the tester to assess how participants used their vision during simulation tests.

8.2.5.1 Phase E: Evaluation of SA Level

Using eye-tracking data in combination with data from interviews, the effectiveness of the proposed design was evaluated.

SA Level One focuses on the perception of elements in the environment. This means sensing and identifying the necessary data to understand information elements.

Analysis of which information elements the test subject used at different times was based on the eye-tracking point cloud.

Sharp vision is concentrated in the macula lutea, covering approximately ± 9 degrees of the field of view. The fovea is a cone-shaped depression in the central retina measuring 1.5 mm in diameter, corresponding to 5° of the visual field (Wikipedia 2025). From this, we can argue that the eye-focus point and the rest time of the eyes on each information element provide the necessary condition for perceiving the element. It is unlikely that information elements outside the fovea have been perceived. With this insight, information can be classified into elements that are sensed and those that are not. However, the eye-tracking data do not give explicit proof of understanding the information elements.

SA Level Two deals with comprehension, combining the various information elements into a coherent, meaningful situational picture.

This means that SA level two is a complex mental process where eye-tracking data does not provide direct measurements. During the post-test interviews, participants explained that their initial response was to gain oversight of the big area traffic situation (left screen) first. Then, focus on the vessels in the main operational area (centre screen). Analysis of eye-tracking data confirmed this pattern. Moreover, the time spent at the zoomed-out traffic area (left screen) provides a measurement of how easily the information was perceived. Additionally, the zoomed-out view was preferred over

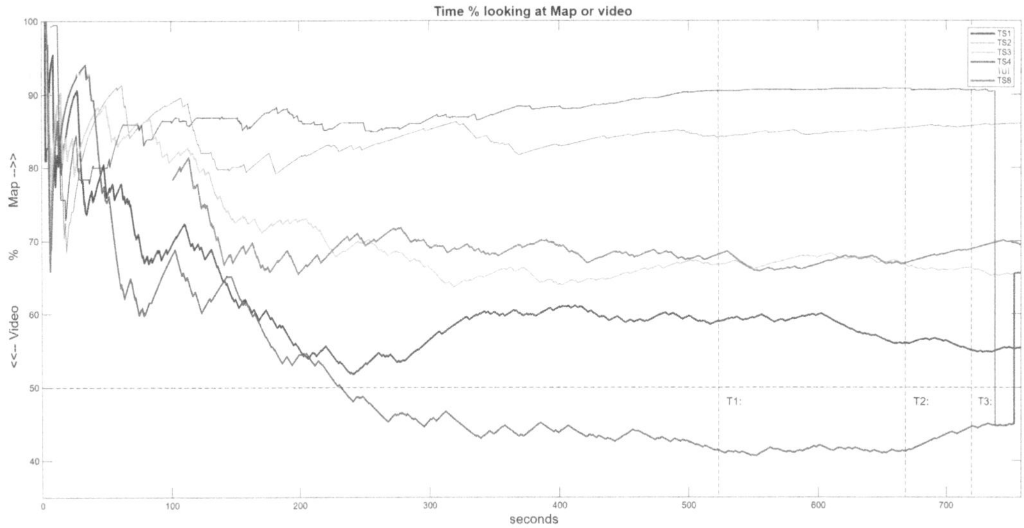

FIGURE 8.17 Preferred information source during test scenario.

(With permission from author J.B Høstmark.)

the video-feed view to create SA at the start of the operational scenario. Careful evaluation of Figure 8.17 shows that the map view is the preferred information source during the first minute of the test scenario.

The transition to focus on the centre screen indicates that the information elements have been perceived (SA level 1) and that each information element has been comprehended as a meaningful situational picture. This evidence was supported by participants in the post-test interviews, who confirmed that the data on the left screen were easy to comprehend and provided the necessary information about traffic in the operational area.

SA Level Three evaluates support for projection forward in time: how well the design supports prediction of what will happen in the near future.

This was evaluated by using indirect methods in combination with qualitative interview findings. From T1 to T2, two vessels were on a collision course. All participants tried to call the other vessel on VHF within a reasonable time; furthermore, eye-tracking showed that test subjects checked the video-feed screen to confirm the other vessel was in sight. At T3, the situation is resolved, as shown in the eye-tracking data, with visual focus shifting to the main map screen. This clearly indicates that the information displayed was sufficient to achieve SA level three (projection forward in time) when vessels were on a collision course. Furthermore, it was learned that preferred display varies across specific events and with increased stress levels. As scenarios with higher stress levels emerge, visual focus is shifted from a map-centric to a video-centric view.

Operational feedback indicates adequate SA at levels one, two, and three for ROC operators supervising ships in a MASS environment. This is supported by the consistency of operational feedback from test subjects and by evidence from eye-tracking data analysis.

Phase E focused on collecting operational feedback through simulations, with a particular emphasis on using eye-tracking technology to assess operator attention and SA. Eye-tracking provided objective, non-intrusive data on where operators focused during the simulation, complementing traditional methods like interviews and task load indices. The analysis of the eye-tracking data confirmed that the ROC design effectively supported all three levels of SA, helping operators perceive, comprehend, and project future scenarios based on their visual focus.

8.3 SUMMARY

The development of the ROC illustrates how systematic design, incorporating HF, iterative testing, and learning from near-real operations, can reduce uncertainty and support safe, efficient, and trustworthy remote operations. This process exemplifies the Safety by Design principle – proactively learning through design rather than reactively through accidents.

'Design insights for future ROCs'

- Early multidisciplinary collaboration is critical.
- Clear visualisation (bird's-eye vs. vessel-centric) improves SA for multi-ship monitoring.
- Consistent graphical principles prevent cognitive overload.
- Eye-tracking reveals attention patterns and visual fatigue.
- Workload tools (NASA-TLX, interviews) may complement physiological data at the end of the project.
- Iterative testing builds operator trust and system resilience.

Through the ROC example, we have looked at applying a structured methodology when faced with new, challenging design problems, where technology and humans need to work together to maintain a high level of safety.

Although not evident at the outset, the value of taking time to clarify which problem to solve cannot be underestimated. This forms the basis for maturing an enabling analysis and definition needed to structure and scope the work. Only then can a goal-oriented conceptual design phase begin.

During the conceptual design phase, one should emphasise the importance of exploring a broad range of possibilities, rather than leaning towards personal preferences.

It is advisable to use a structured approach, combined with proven data-collection methods, when collecting feedback on the concepts.

Strengths and weaknesses of each concept should be summarised for comparison.

Selection of the preferred concept must be based on careful evaluation of the test data of the concepts and how they fit with the design requirements defined in the analysis and definition phase.

Once a pure conceptual design is chosen, the detailed design can begin. Although design is an iterative process, the fundamental building blocks for the selected design should stay consistent.

Operational feedback is ideally a continuous effort, but most important during the commissioning of a new system or solution.

Careful selection of data-gathering tools and evaluation methods, together with a team of competent multidisciplinary experts, offers the best chance for a successful control centre design that meets the needs of human operators, safety, and efficiency.

Table 8.1 maps the five design phases in ISO 11064 to the Safety by Design Perspective and HF Design focus.

TABLE 8.1
Five Phases in ISO 11064

ISO 11064 Phase	Safety by Design Perspective	HF Design Focus
A. Clarification	Understand operational context and hazards before design	Define scope, user roles, and constraints
B. Analysis & Definition	Map socio-technical system, identify HF requirements	Perform task analysis and map information needs
C. Conceptual Design	Explore alternatives safely before implementation	Consider cognitive workload; ensure SA design support human
D. Detailed Design	Integrate documented HF principles into HMI	Ecological interface, clutter reduction
E. Operational Feedback	Learn from operation and refine design	Eye-tracking, workload validation

REFERENCES

BGPH (2022) *Bridge Procedures Guide* (6th ed.). Witherbys Publishing. ISBN 10: 1913997073. ISBN-13: 978-1913997076.

Braseth, A. O. (2015). Information-rich design: A concept for large-screen display graphics: Design principles and graphic elements for real-world complex processes, PhD thesis, NTNU, http://hdl.handle.net/11250/292990

Braseth, A. O., Kaarstad, M., & Høstmark, J. B. (2022). Remote supervision of autonomous ships: Principles for interaction display graphics. In *Proceedings of the 33rd European Safety and Reliability Conference (ESREL'23).*

DNVGL (2018). *Class Guideline, DNVGL-CG-0264, Autonomous and Remotely Operated Ships*, Downloaded on 20 April 2021, http://rules.dnvgl.com/docs/pdf/dnvgl/cg/2018-09/dnvgl-cg-0264.pdf

Endsley M. R. (2013). Situation awareness, In Lee, J.D., & Kirlik, A. (Eds.), *The Oxford Handbook of Cognitive Engineering* (p. 89). Oxford University Press.

Endsley, M. R., Bolté, B., & Jones, D. G. (2003). *Designing for Situation Awareness, An Approach to User-Centered Design*. Taylor & Francis, London & New York.

IMO (2018). *IMO Takes First Steps to Address Autonomous Ships*, accessed on 20 April 2021, https://www.imo.org/en/MediaCentre/PressBriefings/Pages/08-MSC-99-MASS-scoping.aspx

ISO (2020/11064). ISO 11064-10:2020 Ergonomic design of control centres, https://www.iso.org/standard/71558.html

ISO (2020/9241). ISO 9241-110:2020 Ergonomics of human-system interaction Part 110: Interaction principles. ISO/TC 159/SC 4. *ISO 9241-210:2019 Ergonomics of Human-System Interaction – Part 210: Human-Centred Design for Interactive Systems.*

Kaarstad, M., Braseth, A. O., Høstmark, J. B., & Strommen, G. (2022). Supervising autonomous ships – A simulator study with navigators and vessel traffic supervisors. In *Proceedings of the 32nd European Safety and Reliability Conference (ESREL'21).*

Karvonen, H., & Martio, J. (2018). HF issues in maritime autonomous surface ship systems development. In *Proceedings of the 1st International Conference on Maritime Autonomous Surface Ships (ICMASS'18)*, Busan, Korea.

Microsoft (2023). *Use the Dark Theme in Microsoft Edge*, accessed on 29 March 2023. Microsoft Support.

MUNIN (2016) *Maritime Unmanned Navigation through Intelligence in Networks*, downloaded on 28 March 2023, http://www.unmanned-ship.org/munin

Openbridge (2023). *Homepage/About*, downloaded on 28 March 2023, https://www.openbridge.no/home/about

Porathe, T. (2023) Overview of recent research. In *Human-Centred Autonomous Shipping* (pp. 158–208). CRC Press.

Sjøfartsdirektoratet (RSV 12-2020). Guidance in connection with the construction or installation of automated functionality aimed at performing unmanned or partially unmanned operations. Journal No.: 2020/37328-1 SDM/JKS.

SVAN (2018). *Safer Vessel with Autonomous Navigations. Rolls-Royce and Finferries Demonstrate World's First Fully Autonomous Ferry | Rolls-Royce.* https://www.rolls-royce.com/media/press-releases/2018/03-12-2018-rr-and-finferries-demonstrate-worlds-first-fully-autonomous-ferry.aspx

Vicente K.J., & Rasmussen J. (1992). Ecological interface design: Theoretical foundations. *IEEE Transactions on Systems, Man, and Cybernetics*, 22(4), pp. 1–18.

Wikipedia (2025). *Peripheral Vision*, accessed on 30 May 2025. https://en.wikipedia.org/wiki/Peripheral_vision; Strasburger, H., Rentschler, I., & Jüttner, M. (2011). Peripheral vision and pattern recognition: A review. *Journal of vision*, *11*(5), 13–13.

9 Human Factors, Safety, and Technology in the High Arctic Context, Svalbard

Gunhild Birgitte Sætren

9.1 INTRODUCTION

The High Arctic is becoming an increasingly important context for research on human factors, technology development, and safety, as environmental, industrial, and geopolitical changes accelerate.

Several developments contribute to the growing relevance of the High Arctic. First, rapid sea-ice decline (Yang et al., 2023) has increased interest in the Northeast Passage, creating new opportunities for shipping, fisheries, and autonomous maritime technology (Sørensen, 2021). Second, warmer conditions reveal land areas with valuable natural resources, raising the region's economic significance. Third, the area offers unique scientific opportunities for studying climate change, ecosystems, and environmental dynamics. Fourth, greater accessibility has driven significant growth in year-round tourism, including starlight and northern lights tourism (Visit Svalbard, 2025). Finally, geopolitical interest in the region has intensified, particularly among Russia, China, and the United States.

Because the High Arctic is undergoing rapid and unprecedented environmental change, understanding these developments is essential for anticipating their impact on human performance, technology reliability, and system design. It is also important because these conditions may foreshadow the impacts of climate change in other parts of the world.

The High Arctic offers a unique real-world testbed for evaluating technological solutions, as harsh weather and unpredictable conditions reveal system vulnerabilities that may remain hidden in more temperate environments. If technology works in the extreme weather conditions here, it will work in all places with harsh winter weather. For example, exploring automation and remote operations may both remove people from harm's way and stress-test the technology under harsh conditions.

Recent advances in automation and drone technology illustrate how Arctic safety is increasingly shaped by human–machine interaction. The Petroleum Safety Authority's study on drone use in northern operations (Bakken et al., 2020) emphasises that meaningful human control, robust design, and systematic risk management are essential when autonomy meets Arctic conditions.

Accidents and near misses in High Arctic operations offer valuable insights into the complex interactions between humans, technology, and extreme environmental conditions. Incorporating these insights into design processes reflects the principles of Safety by Design (Woodier, 2025).

This chapter examines key High Arctic conditions, the associated human factors challenges, and how Safety by Design principles can support the development of robust, human-centred technologies that operate reliably under extreme environmental conditions.

9.2 HIGH ARCTIC CONTEXT

The Arctic is divided into three ecological zones: the High Arctic, the Low Arctic, and the Subarctic (Figure 9.1). These zones differ primarily in climate and vegetation, which together determine which plants and animals can survive.

DOI: 10.1201/9781003741824-12

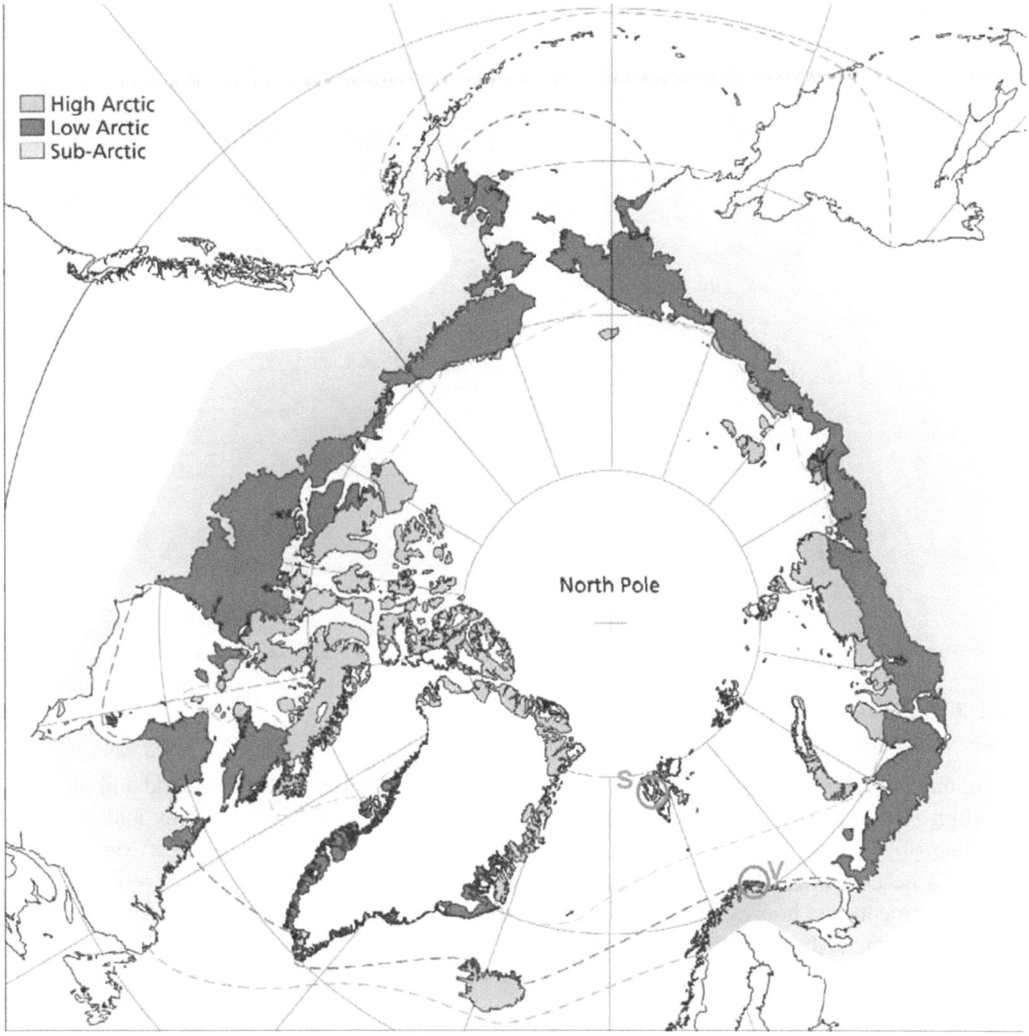

FIGURE 9.1 The three zones are the High Arctic, the Low Arctic, and the Subarctic.

(Source: COAT UiT/CAFF (Arctic Council).)

Trees grow in the Subarctic but cannot grow in either the Low Arctic or the High Arctic due to colder summer temperatures. The tree line thus marks the boundary between the Subarctic and the Low Arctic.

The Low Arctic is defined as areas where the average July temperature is below 10°C, which is too cold to support tree growth. Vegetation mainly consists of low shrubs, dwarf shrubs, and hardy plants adapted to short summers.

The High Arctic has an even harsher climate. Growing conditions are so limited that only small, resilient vegetation – such as mosses, grasses, and lichen – can survive. Almost no shrubs or tall plants are found here (Arctic Council, 2026).

9.2.1 The People in Svalbard

As of March 1, 2025, a total of 2,556 people were living on Svalbard, with approximately 60% Norwegian citizens and 40% non-Norwegian citizens. The gender balance is close to 50–50 (Statistics Norway, 2025). The central city on Svalbard is Longyearbyen, situated at 78 degrees

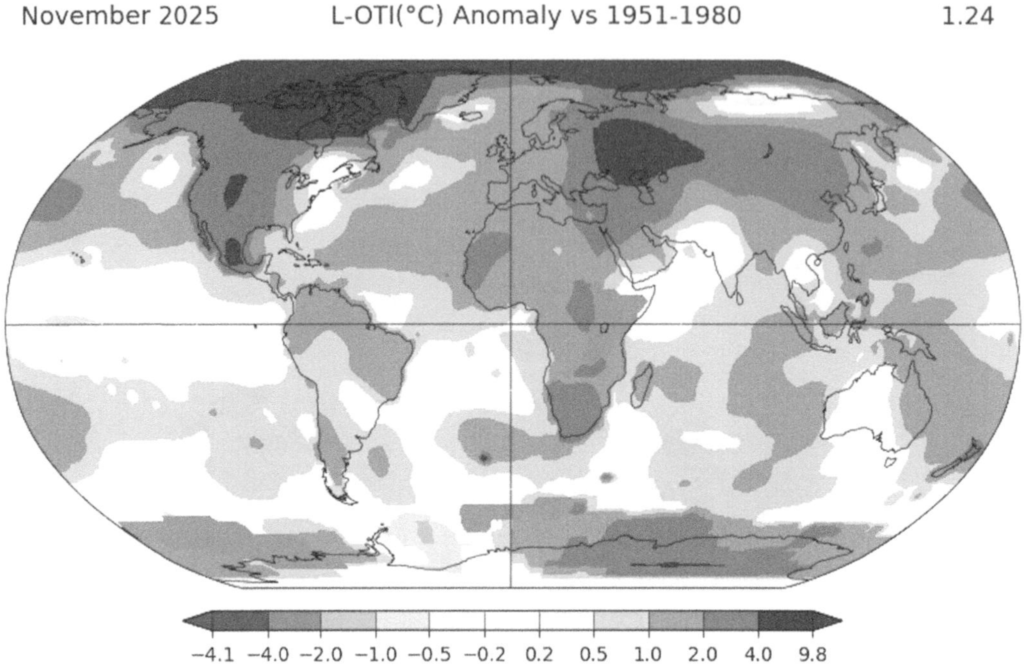

FIGURE 9.2 GISS surface temperature analysis (Global map, NASA).

north, making it the northernmost settlement with over 1,000 inhabitants in the world and placed in the High Arctic. This city, located at the base of the left side of Longyear Valley and along the coastline of Adventfjorden (Figure 9.7), is relatively small and densely populated. Consequently, there is a necessity to utilise the surrounding natural environment. The residents live in relatively small apartments and houses, and there are difficulties in providing enough housing for those living and working there (Longyearbyen Community Council, 2023a).

Moreover, Svalbard experiences significant population turnover. Over the past five years, more than 2,400 individuals have moved to the region, while over 2,300 have departed (Statistics Norway, 2024). In 2022, the median residency duration for the Norwegian population was 3.6 years (Statistics Norway, 2022). This turnover creates a diverse but highly transient community, where collective experience is limited and long-term contextual knowledge becomes fragmented. It can be described as short societal memory. This further means that citizens find it harder to detect changes in nature and society. Humans generally struggle to detect gradual environmental change (Rensink et al., 1997). In a high-turnover community like Svalbard, this challenge is amplified because residents lack long-term reference cues and personal experience with historical conditions. The limited number of long-term inhabitants further reduces the available pool of experience and contextual knowledge.

In addition to the locals, Svalbard is a popular tourist destination, with increased visits also through the dark season from October to February. In 2024, 67,656 tourists visited Longyearbyen. Additionally, 67,126 cruise passengers visited Svalbard. These numbers have had a steady increase over the past decade, apart from the corona years (Visit Svalbard, 2025). In a city with about 2,500 inhabitants, the number of tourists significantly impacts the local infrastructure, going into nature for recreation, economy, and daily life, often leading to both opportunities and challenges for the residents.

9.2.2 Changing Climate

The High Arctic is the place on earth with the most rapid changes due to climate change (NSIDC, 2025). The temperatures are rising (Rantanen et al., 2022), leading to a more unstable weather situation, permafrost melting, and a decreasing amount of sea ice. The sea ice of the High Arctic reached its record low maximum in 47 years on March 22, 2025 (NSIDC, 2025).

In the Arctic, the weather conditions are changing due to climate change. The rapid increase in Arctic mean surface temperature is known as Arctic amplification and is strongest in winter (Bintanja & van der Linden, 2013). This results in a substantial decline in sea ice, which again leads to warmer, wetter Arctic air (Vihma, 2014), and 2024 was a year with new weather records for high temperatures, glacier mass loss, and rainfall. August 2024 was the warmest month ever recorded in Longyearbyen, and the summer season was 2.9 degrees above average (European Commission, 2025). Further, for weather stations operating over 50 years on Svalbard, Bjørnøya, Hopen, Akseløya, Isfjord Radio, Svalbard Airport, and Ny-Ålesund, all had new records for average temperature for the summer season (Meteorologisk institutt, 2025). This contributed to a record-high rate of glacier melting and the largest mass loss of any glacier region globally (European Commission, 2025).

It is also a region characterised by extreme weather and high winds, where new records have recently been observed. In March 2025, gusts up to 57.9 metres per second were measured at Platåberget just outside Longyearbyen (Yr, 2025a). Schoolchildren were not allowed to walk home from school before the wind calmed down in February 2025, when the wind at the school in the city reached gusts of up to 42.3 metres per second (Yr.no, 2025b). On that day, when the wind had subsided, buses were arranged to transport the children home. Regarding precipitation, Svalbard is defined as an Arctic desert. However, precipitation is also increasing year-round, and on 8 July 2024, a new record of rain was noted with 16.6 mm (Yr.no, 2025c).

Temperatures in Longyearbyen have varied between minus 26.2 and plus 20.2°C from April 2024 to April 2025, making them highly variable over the course of one year (Yr.no, 2025c).

These extreme and rapidly shifting conditions make future states harder to predict and place additional demands on human-centred technology design, which must account for uncertainty, variability, and degraded predictability.

With increasingly harsh weather year-round, this will pose additional challenges for safety and emergency rescue efforts. It is also one of the most vulnerable areas in the world. Potential accidents could have a devastating impact on the environment, making rescue missions extremely difficult.

There is growing interest in autonomous ships operated without human crews in the Arctic. The environment provides a structured testing ground for human-centred system design specifically adapted to Arctic conditions.

9.2.3 Remoteness and Wildlife

Svalbard is geographically remote from the rest of the world, with the nearest airport in Tromsø, approximately 1.5 hours by flight from Svalbard Airport. Additionally, flight capacity fluctuates throughout the year, sometimes resulting in several days without available plane tickets to and from Svalbard. These long flights or extended shipping distances can impact both maintenance and rescue operations. This remoteness is also linked to limited infrastructure, including sparse maritime ports, roads, and power (Albrechtsen & Indreiten, 2021).

Wildlife on Svalbard is also worth mentioning. There are threats here that make fieldwork different from many other places. Foremost, the threat posed by polar bears is genuine, and it is imperative that individuals do not venture into the field without appropriate weaponry, the requisite skills, and authorisation to use it. Other, more invisible threats are rabies, which was newly discovered in May and July 2025 in two foxes on Svalbard (Veterinærinstituttet, 2025). Due to the presence of the fox tapeworm (Echinococcus multilocularis), it is advised to avoid drinking water from natural sources while hiking. Therefore, it is essential to carry sufficient water for the entire duration of field trips. It could also be possible to develop portable water purification technology or analyse water quality to help reduce this risk by enabling people in the field to safely use natural water sources.

From 26 September 2025, drinking water in Longyearbyen was considered unsafe to drink (Svalbardposten, 2025d), leading to the import of drinking water from the mainland. Additionally, the company Svalbard Bryggeri began purifying water and supplying it to Longyearbyen's residents at the local store (Svalbardposten, 2025e), in a situation that appears to be ongoing.

While the environmental context defines the physical challenges, the human element determines how these challenges translate into safety risks and technology needs.

9.3 HIGH ARCTIC HUMAN FACTORS

For those working in maintenance, guiding tourists in the field, or simply living and working in the High Arctic, human factors are highly relevant when considering technological development in the context of human-centred design. The High Arctic environment is an excellent basis for design standards and operations in demanding environments, such as ensuring safety in cold, harsh conditions. Equipment must be designed to withstand freezing and challenging contexts, and users must be able to operate it in demanding environments. Guiding standards for user-centred design, such as ISO 9241-210, should be used, suitable checklists for the working environment should be applied (Johnsen et al., 2025), and experienced users should be consulted. Old examples can be found in Roald Amundsen's polar expedition in 1903–1906, which utilised the experiences of the local Inuit, for instance, when selecting clothing and using dogs.

9.3.1 Human Performance in the High Arctic

Human performance in the High Arctic is tightly coupled to environmental, organisational, and technological conditions that shape perception, situational awareness, decision-making, and workload. This spans from the individual to the group, system, and organisation (e.g., Perrow, 1984; Reason, 1990, 1997; Rasmussen, 1997; Dekker, 2011; Weick & Sutcliffe, 2015), as well as the physical conditions such as weather and lighting. This aligns with Hollnagel's Safety-II perspective (Hollnagel, 2014): safety should not only be about preventing failure but also about ensuring that everyday work succeeds under varying conditions. The harsh Arctic environment, where trees cannot grow, is characterised by cold temperatures, strong winds, and sparse vegetation. This absence of vegetation creates a visually sparse environment with few reference cues, significantly affecting human spatial perception (Allen, 1999; Nett et al., 2025). The absence of visual anchors reduces depth perception and increases the likelihood of distance-estimation errors, a well-documented effect in low-cue environments (Murphy, 1982). Such conditions also affect cognitive processes such as orientation and navigation, as the monotony of the terrain can lead to disorientation and reduced situational awareness. Planning and risk assessment thus become more challenging because inaccurate mental models of distance and scale can lead to poor time estimations and misjudgement of hazards, such as approaching weather, polar bears, or glacier proximity.

Cold weather impairs both physical and cognitive functioning by reducing motor function and control, slowing reaction times, and decreasing alertness. Further, it increases discomfort and reduces decision-making abilities (Balindres et al., 2016; Mäkinen, 2007) and could have long-term health consequences (Ikäheimo, 2018). Therefore, cold implies an increased risk for human error (Reason, 1990; Wærø et al., 2018), which must be taken into consideration for human-centred technology design when conducting human error identification analysis (Stanton et al., 2013).

Strong winds degrade verbal communication by masking speech, interfering with auditory cues, and reducing the clarity of both direct conversation and radio communication. Wind can produce significant noise, making it difficult to communicate, such as over the phone. Wind combined with weather like snow and cold can make it hard to talk to each other, as shown in Figure 9.3. Further, wearing extensive clothing, including helmets, can make verbal communication challenging. To effectively use technology such as walkie-talkies or phones, or to hear each other when in proximity, individuals may need to remove their helmets, which could lead to freezing if no shelter is available.

FIGURE 9.3 Challenges in communication due to technology, weather, and clothing in the High Arctic.

Photo: Gunhild Birgitte Sætren.

Another factor in the High Arctic is the wildlife. To be able to spot a polar bear, you need visualisation and to be alert and attentive, in addition to carrying a weapon to, if necessary, kill it. Wearing a gun could be an extra stressful part of an excursion, and the weapon itself could lead to unwanted incidents and accidents.

Humans are also not good at staying alert for extended periods (Mackworth, 1948). Being in a group in the field is no exception to this. In February 2025, a group consisting of two guides and ten tourists embarked on a routine snow scooter excursion to Fredheim, located near Longyearbyen. Upon arrival, they initially searched for polar bears but eventually relaxed. As the group comprised tourists and guides who were unfamiliar with each other, social conformity led to polite conversation and mutual attention. While the guide was preparing food, the group became momentarily inattentive. During this brief period, a polar bear approached and placed a paw on the back of a tourist who fell while attempting to escape. Fortunately, the polar bear was more curious than hungry and did not harm the tourist, resulting in no serious injuries. A later review of the photographs taken by the tourists showed the polar bear approaching the group, unnoticed by anyone at the time (Svalbardposten, 2025a; Svalbardposten, 2025b). This incident illustrates several human factors challenges: reduced visual sight, group conformity, vigilance decrement, divided attention, and the absence of healthy scepticism when conditions feel safe. It could also be a field where technological development could help prevent future encounters with potentially dangerous wildlife.

Further, the environment lacks many distinguishing cues, for instance, no trees and similar colours, which makes it more challenging to navigate without technological equipment. Then, if the equipment is small and to be handled in the field with snow and storm and wearing a helmet without your glasses, it could be a challenge, see Figure 9.4.

Additionally, most GPS devices and other technological equipment are typically designed for use without gloves, which poses challenges in cold, windy conditions. For instance, frostbite can easily occur in harsh weather (e.g., Olsen, 2024) if the use of technology necessitates the removal of protective gear such as gloves and helmets.

FIGURE 9.4 Challenges in use of technology due to normal clothing in the High Arctic.

(Photo: Gunhild Birgitte Sætren.)

Snowmobiles are also one of the most important means of transportation. However, this is related to injuries (NRK, 2016). One example could be that in snowmobile design, placing the throttle on the handlebar introduces a control-layout design risk, which can trigger a slip error (Reason, 1990), especially under high cognitive load or unexpected events.

So, preparing for fieldwork is important. Check the weather, review satellite websites, and talk with people who have experience in the field. However, weather can change very quickly and unpredictably in the High Arctic, so technology developed for field activities must be designed for that environment. A human-centred approach is essential for ensuring safety and operational efficiency.

9.3.2 GPS and Communication Challenges in the High Arctic

One crucial dimension of technology design for safety in Arctic environments is the positioning and communication reliability. The Havtil/SINTEF report (Bakken et al., 2020) highlights that satellite-based communication and navigation systems (GPS, Galileo, Iridium) are significantly degraded north of 70°N due to low satellite elevation angles, ionospheric disturbances, and limited satellite visibility. This results in intermittent or inaccurate positioning, which directly affects navigation, situational awareness, and control for drones, crewed vessels, and remote or autonomous systems. These limitations have direct implications for human factors and system safety. In summary, GPS and communication resilience must be seen as part of human-centred safety design in the High Arctic.

9.3.2.1 Human Factors Challenges for Operations

Situational awareness (Endsley, 1995; Stanton et al., 2007) can be significantly compromised in emergency centrals or remote operations centres (ROCs) when navigation or telemetry data freezes or drifts. In such cases, operators may lose real-time awareness of the operational environment, necessitating the use of redundant navigation systems such as inertial navigation or terrain mapping sensors to ensure safe and secure control in the current situation.

Decision-making is also affected when the technologies that support it become unreliable. Inaccurate GPS data can mislead operators during mission planning or emergency response, potentially undermining trust in data-driven warnings. This may result in false alerts, such as incorrect avalanche locations or misjudged vessel drift.

Cognitive workload could increase significantly when communication dropouts occur, as operators are forced to rely on uncertain assumptions or degraded manual monitoring. This heightened mental strain can compromise decision-making under pressure and elevate the risk of human error (Reason, 1990; Tierney et al., 2025), particularly in high-stakes environments where timely and accurate responses are critical.

9.3.2.2 Safety by Design Implications

Systems must incorporate fail-safe communication-loss handling, as in graceful degradation modes. Interface design should clearly indicate when GPS or communication data is unreliable so that users can adapt. Training programmes should prepare personnel for low-communication operations, emphasising teamwork and redundancy. These factors would help prevent active failures based on latent conditions (Reason, 1990; Reason et al., 2006).

Under High Arctic conditions, navigating uncertainty is complicated by factors such as poor satellite coverage, drifting sea ice, and rapidly changing weather. This uncertainty increases cognitive workload and forces operators to make decisions with incomplete or ambiguous data. Human decision-making under stress can lead to poor choices or risk-prone behaviour, which in turn affects operational safety (Feng et al., 2025). At the same time, the system reliability, such as the robustness of the positioning systems and communication links, can either mitigate or amplify this uncertainty. If the system fails or degrades, humans must compensate, which in turn increases the workload and probabilities for errors. In contrast, reliable systems reduce uncertainty, enabling better decisions. Redundant sensors and robust communication systems can reduce uncertainty and improve decision quality.

Together, these factors create a dynamic feedback loop in which navigation uncertainty increases cognitive load, human decision making shapes operational risk, and system reliability either amplifies or mitigates uncertainty.

As seen in Figure 9.5, navigating uncertainty increases cognitive load, human decision-making influences operational risk and system demand, and system reliability mitigates or amplifies uncertainty, all in a High Arctic context.

9.3.3 Designing for Human Performance and Safety in High Arctic Conditions

Based on High Arctic factors and human factors considerations specific to the High Arctic environment, Safety by Design principles must account for these when designing technology for operations (see Figure 9.6).

Technology for High Arctic operations must incorporate principles of cognitive engineering (Norman, 1986), ensuring that systems align with human cognitive capabilities under extreme conditions. Such technology must be intuitive and easy to use, especially when extreme cold and wind, as well as stress, create high cognitive load. Additionally, systems should incorporate error-management principles, including built-in fail-safes informed by potential error modes identified through structured analysis (Kirwan, 1998).

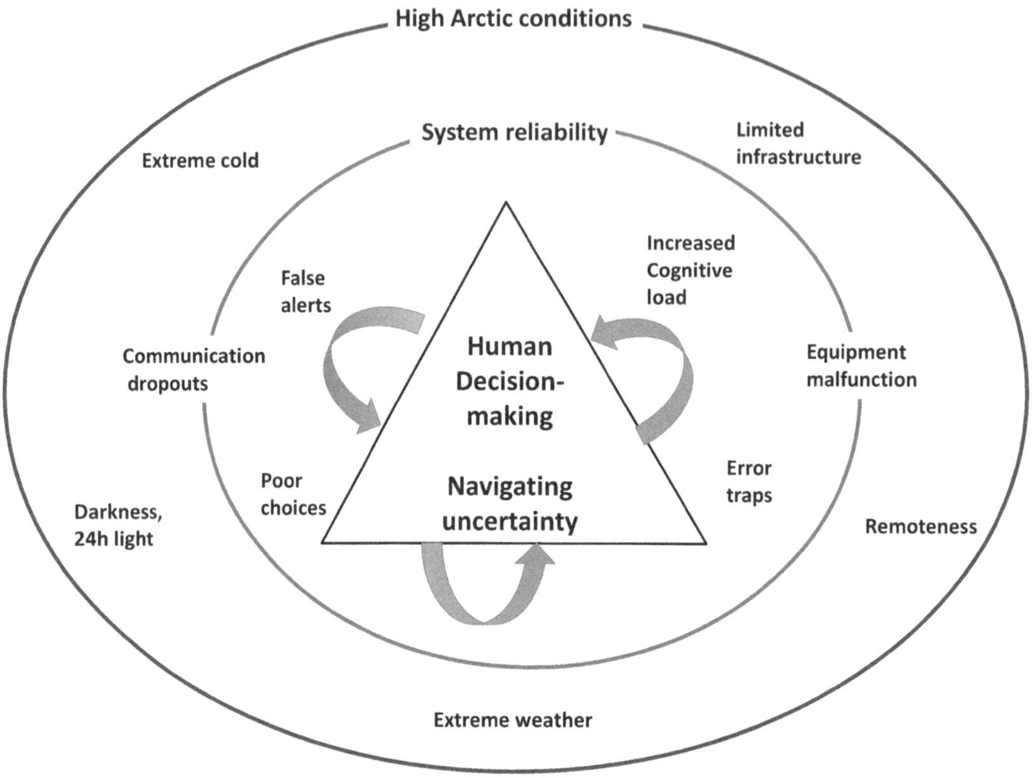

FIGURE 9.5 Interaction loop on navigating uncertainty in human decision-making and system reliability in High Arctic conditions.

High Arctic Factors
Darkness / 24-hour daylight; Remoteness and limited rescue capacity; Limited infrastructure; Wildlife threats (e.g., polar bear, rabies); Extreme weather (wind, precipitation, ice); Avalanche risk; Permafrost melt; Energy insecurity; Satellite limitations (GPS degradation, low elevation angles)
Human Factors Considerations
Visual and auditory limitations (low visibility, wind, noise); Clothing constraints (reduced dexterity, limited tactile feedback); Increased probability of human error; Elevated cognitive workload; Lack of visual reference cues; Degraded situational awareness; Physical fatigue; Alert fatigue (e.g., avalanche sensors, evacuations)
Safety by Design Principles
Intuitive interfaces with minimal cognitive load; Operability with gloves and bulky clothing; Clear feedback when GPS or positioning degrades; Large visual indicators (avoid reliance on sound); Fail-safe modes and redundancy; Real-time situational awareness support; Error-tolerant design; Trust management (credible alerts, confident indicators)

FIGURE 9.6 Human factors in High Arctic technology development on safety by design implications.

The technology must also be adaptable to fast-changing conditions and user needs. Further, there are vast possibilities for technologies to help users make informed decisions in unpredictable High Arctic conditions. This could, for instance, relate to real-time data analysis and navigation aids.

Building on the understanding of human performance, this next section explores how specific technological solutions, such as avalanche sensors and renewable energy, are being developed and tested for High Arctic conditions.

9.4 LONGYEARBYEN CHALLENGES AND TECHNOLOGY

In Longyearbyen, where most people on Svalbard live and work, they live in a multi-hazard context, including risks such as no access to locally clean drinking water (NRK, 2025), air pollution posing a direct health risk (Oliver-Serra et al., 2025; Svalbardposten, 2024), polyfluoroalkyl substances (PFAS) pollution in the ground surrounding the city of Longyearbyen (Ahrens et al., 2023; Miljødirektoratet, 2022), avalanche dangers (Hancock et al., 2024), and risks to reliable energy supply (Svalbardposten, 2025c). Two are chosen to explore in this chapter: (1) avalanches and (2) energy supply.

Svalbard faces several challenges due to natural hazards that are far more severe than those on mainland Norway. These challenges could be mitigated by advanced AI technology capable of predicting and monitoring processes to prevent unwanted incidents both in the short and long term. Additionally, the High Arctic provides an ideal context for testing technology designed to withstand extreme environments.

9.4.1 Avalanches

Climate-driven changes in weather, temperature, and ground stability have significantly increased avalanche risk for residents in Longyearbyen. The city of Longyearbyen is situated between the mountains Sukkertoppen (The Sugar Top), Gruvefjellet (The Mine Mountain), Sarkofagen (The Sarcophagus), and Platåberget (The Plateau Mountain). They are shown from left to right in Figure 9.7. The location of the old coal-mining city was determined by the establishment of mines in 1905 (Holm, 2018), rather than by considerations of safety from avalanche damage. In December 2015 and February 2017, the city was struck by avalanches from Sukkertoppen, resulting in 14 houses being hit and destroyed, and 2 fatalities (Brattlien et al., 2016; Hancock et al., 2018).

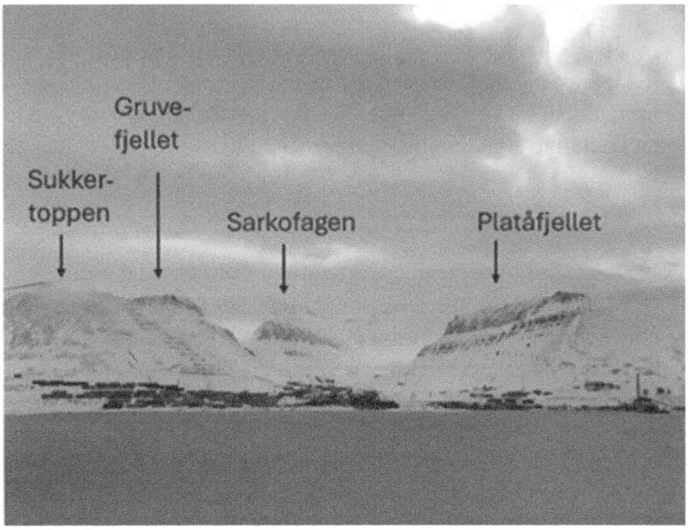

FIGURE 9.7 Longyearbyen, a city under mountains.

(Photo: Gunhild Birgitte Sætren.)

The risk of avalanches in the city has been recognised for many years. Smaller avalanches have been reported annually in the Lia area, below Sukkertoppen, according to Hestnes (1992). However, the report does not specify the size of these avalanches or the exact number of years they have been occurring. After the fatal accident, risk perception and behaviour changed in the city (Sætren et al., 2025; Turner, 1978). The increased awareness of avalanche danger led to daily avalanche forecasting, and a research team explored how technology could improve it.

Automated measurements of snow-covered properties were installed, providing key data source for avalanche forecasting (Hancock et al., 2023) as a basis for decisions on evacuations or area closing. In addition, physical installations were built after the avalanches to hinder them in different parts of Longyearbyen, at a cost of 350 million NOK, and were completed in 2025 (NTE, 2025). The cost for a city of 2500 inhabitants shows the government's interest in keeping the city safe and upholding Norwegian society to maintain sovereignty in Svalbard.

High population turnover may contribute to reduced risk awareness, as many residents lack long-term experience with local hazards and may underestimate low-frequency, high-consequence events. However, research has found that the perception of risk from natural hazards is increasing in Longyearbyen (Sætren et al., 2025). As this is a place where people tend to live two to five years (Statistics Norway, 2022), the high turnover leads to a short societal memory, which also affects the understanding of changes in the natural environment and potential dangers. People here, like in most places on earth, are mostly preoccupied with their everyday lives, which often do not include risks for avalanches or landslides if they have never experienced them.

Research further indicates that people drawn to life in the High Arctic may be motivated more by the pursuit of adventure than by engagement in community participation and local politics. Together with changes in the social structure, including the 2023 exclusion of non-Norwegians from voting and political participation (Sætren et al., 2025), this may affect the assessment of long-term decisions and the understanding of their consequences for society as a whole.

Technology development in avalanche security is now focused on snow depth sensors (Hancock et al., 2023). However, permafrost is melting (Dobiński & Kasprzak, 2022), and warmer seasons are occurring during the summer. This, in combination with a porous stone type, leads to a more unstable ground throughout the year. Increasing warning systems based on ground-sensing technology to detect instability could prevent accidents. At present, this is handled reactively, as illustrated by the following example from August 2024. Then, hikers on Sarkofagen observed a landslide and notified the Governor of Svalbard, who subsequently issued a warning via their website and Facebook page (Governor of Svalbard, 2024).

Figure 9.8 illustrates the line as one of the most frequented hiking trails near Longyearbyen, while the circled area represents the extent of the landslide.

Part of the trail has eroded, making it dangerous. Having technology that could monitor changes in ground movement could have given a warning before the event, rather than after.

9.4.2 Energy

Longyearbyen has been self-sufficient with energy supply from the coal produced in the local mines. However, in Prop. 1 S (2021–2022), the government requested that the Longyearbyen Community Council develop a concrete energy plan during 2022. The plan's premise was that renewable energy should be phased in. Thus, in 2023, coal as an energy source was shut down, and the energy supply was switched to diesel (Longyearbyen Community Council, 2023b). This is meant to be a temporary solution until green, sustainable energy can be used as a robust, reliable source. Still, there are no suggestions about how long the temporary situation will last.

The Norwegian Government asked for an investigation into energy solutions, including an additional sea cable or the possibility of using nuclear power, LNG, and LBG (liquefied natural gas and biogas), in recommendation 31 S (2023–2024). Related to this, windmills and solar panels are viewed as alternative energy sources.

Windmills could be a sustainable source of energy in Svalbard. The changing climate in Svalbard, with wetter weather, however, makes icing a challenge. Structures, such as turbine blades, could

FIGURE 9.8 Sarkofagen with a landslide.

(Photo: Governor of Svalbard, 2024.)

experience significant damage from icing. Icing can lead to different problems, including measurement errors, power losses, overproduction, mechanical failures, electrical failures, and safety hazards (e.g., Parent & Ilinca, 2011). Ice could further lead to energy loss, for instance, with turbines losing their aerodynamic shape (Parent & Ilinca, 2011; Rekuviene et al., 2024), and potentially dangerous accidents, such as ice being thrown off the rotors.

Mechanical failure and fatigue fractures lead to increased maintenance, while maintenance in the harsh Arctic environment is challenging due to weather and remoteness. Thus, the robustness of windmills needs to be suited to the High Arctic climate. To transition to long-term renewable energy systems in Longyearbyen, the Arctic context and human factors must be incorporated into the development process (Ringkjøb et al., 2020).

Energy systems in the Arctic, therefore, require robust designs, maintainable in remote conditions, and aligned with human factors considerations such as workload, diagnostic clarity, and safe maintenance access.

9.5 CONCLUSION

In conclusion, addressing human factors and safety in the High Arctic requires a multifaceted approach that integrates technological advancements with an understanding of the unique environmental challenges. The rapid climate changes, extreme weather conditions, and remote location necessitate robust, adaptable, and intuitive technologies to ensure safety and operational efficiency. By prioritising human-centred design alongside error management, situational awareness can be enhanced, and risk mitigation associated with natural hazards can be improved. The High Arctic serves as a natural testbed for Safety by Design, where extreme conditions expose system vulnerabilities and demand robust, adaptive, and human-centred solutions. By combining human factors insight, technological robustness, and continuous learning across systems, Arctic operations can serve as models for designing resilience into both human and the new autonomous/AI-based systems.

Design lessons for safe operations in the High Arctic are suggested to:

- Build *environmental resilience* into technology to mitigate risk (icing, temperature, communication failure).
- Ensure *meaningful human control* in all situations, including those involving autonomous and remote systems.
- Use *risk-based design*. Identify failure modes during design and before technology deployment.
- Foster *cross-sector competence* networks for learning and experience sharing.
- Design for *redundancy and robustness* in critical functions such as communication, navigation, power, and other prioritised areas.
- Prioritise human and ergonomic design for operators working in cold, dark, and stressful environments where GPS signals are difficult to receive.

REFERENCES

Ahrens, L., Rakovic, J., Ekdahl, S., & Kallenborn, R. (2023). Environmental distribution of per-and polyfluoroalkyl substances (PFAS) on Svalbard: Local sources and long-range transport to the Arctic. *Chemosphere*, *345*, 140463.

Albrechtsen, E & Indreiten, M (2021). Editorial: Arctic safety. *Safety Science*, *137*, 105165 https://doi.org/10.1016/j.ssci.2021.105165

Allen, G. L. (1999). Spatial abilities, cognitive maps, and wayfinding. *Wayfinding Behavior: Cognitive Mapping and Other Spatial Processes*, *4680*, 46–80.

Arctic Council (2026). *Definitions of the Arctic Region: According to Vegetation Zones*. University of Lapland. Retrieved March 23, 2026 from https://arcticcentre.org/en/arctic-region/maps/definitions-of-the-arctic-region

Bakken, T., Holmstrøm, S., Johnsen, S. O., Merz, M., Grøtli, E. I., Transeth, A., Risholm, P., & Storvold, R. (2020). *Bruk av droner i nordområdene: HMS-forhold knyttet til bruk av droner i petroleumsaktiviteten i nord* (Sintef Rapport 2020:00238). Sintef.

Balindres, A. R., Kumar, R. & Markeset, T. (2016). Effects of Arctic conditions on human performance. *Advances in Intelligent Systems and Computing*, *489*, 257–663

Bintanja, R., & van der Linden, E. (2013) The changing seasonal climate in the Arctic. *Scientific Reports*, ***3***, 1556, https://doi.org/10.1038/srep01556

Brattlien, K., Hestnes, E., & Domaas, U. (2016). *The Svalbard Fatal Accident December 2015 - Assisting Local Authorities in the Acute Phase*. Norges Geotekniske Institutt.

Dekker, S. (2011). *The Field Guide to Understanding Human Error* (2nd ed.). Ashgate.

Dobiński, W., & Kasprzak, M. (2022). Permafrost base degradation: Characteristics and unknown threat with specific example from Hornsund, Svalbard. *Frontiers in Earth Science*, *10*, 802157.

Endsley, M. R. (1995). Toward a theory of situation awareness in dynamic systems. *Human Factors*, *37*(1), 32–64. https://doi.org/10.1518/001872095779049543

European Commission (2025). European state of the climate. Report 2024. European State of the Climate 2024 | Copernicus.

Feng, T., Huang, L., Peng, X., Qiao, T., & Wu, X. (2025). An integrative review of cognitive workload assessment for safety management. *BMC Nursing*, 24(1), 1363.

Governmental Recommendation (2023). *Recommendation 31 S (2023-2024) Innstilling til Stortinget fra utenriks- og forsvarskomiteen*. Innstilling.

Governor of Svalbard (2024). Ferdsel i fjell på Svalbard. [Walking and being in mountains in Svalbard] Retrieved May 9, 2025 from Ferdsel i fjell på Svalbard | Sysselmesteren på Svalbard.

Hancock, H., Indreiten, M., & Jaedicke, C. (2024). *Avalanche Risk Management in Longyearbyen, Svalbard*. The Norwegian Geotechnical Institute. Retrieved from https://arc.lib.montana.edu/snow-science/objects/ISSW2024_O7.6.pdf

Hancock, H., Jenssen, E., Indreiten, M., & Albrechtsen, E. (2023). Development of a sensor system to support avalanche risk management in Arctic Norway. In *Proceedings of the International Snow Science Workshop*.

Hancock, H., Prokop, A., Eckerstorfer, M., & Hendrikx, J. (2018). Combining high spatial resolution snow mapping and meteorological analyses to improve forecasting of destructive avalanches in Longyearbyen, Svalbard. *Cold Regions Science and Technology*, 154, 120–132. https://doi.org/10.1016/j.coldregions.2018.05.011

Hestnes, E. (1992). *Lia Longyearbyen. Vurdering av snøskredfare og tiltak for å hindre ulykker ved vei 226-230.* Svalbard Samfunnsdrift.

Hollnagel, E. (2014). *Safety-I and Safety-II: The past and future of safety management*. Ashgate.

Holm, K. (2018). *Longyearbyen – Svalbard: historisk veiviser* (4. utg.). Kari Holm Forlag

Ikäheimo, T. M. (2018). Cardiovascular diseases, cold exposure and exercise. *Temperature*, *5*(2), 123–146. https://doi.org/10.1080/23328940.2017.1414014

Johnsen, S. O., Thieme, C., Aminoff, H., Park, J., Marius Fernander, M., Boren, M., . . . & Kirwan, B. (2025). *CRIOP A Scenario Method for Crisis Intervention and Operability Analysis*. Norway: SINTEF Technology and Society.

Kirwan, B. (1998). Human error identification techniques for risk assessment of high risk systems—Part 1: review and evaluation of techniques. *Applied Ergonomics*, *29*(3), 157–177.

Longyearbyen Community Council (2023a). Boligbehovsutdredning. Revisjon og status etter boligbehovsutredningen fra 2019. Boligbehovsutredning+2023-+med+Vedlegg+om+status+boligpotensial+i+Longyearbyen[1].pdf

Longyearbyen Community Council (2023b). 20 spørsmål om arbeidet med energiomstilling og overgang til diesel. 20 spørsmål og svar om arbeidet med energiomstilling og overgang til diesel - Longyearbyen lokalstyre.

Mackworth, N. H. (1948). The breakdown of vigilance during prolonged visual search. *Quarterly Journal of Experimental Psychology*, *1*(1), 6–21.

Mäkinen, T. M. (2007). Human cold exposure, adaptation, and performance in high latitude environments. *American Journal of Human Biology*, *19*, 155–164.

Meteorologisk institutt (2025). Været i Norge. Klimatologisk månedsoversikt sommersesongen 2024. MET-info-16-2024.pdf

Miljødirektoratet (2022) Tiltaksplan for opprydding PFAS – forurenset grunn på Svalbard Lufthavn. [Action plan for PFAS remedation – contaminated soil at Svalbard Airport] Tiltaksplan for opprydning i PFAS-forurenset grunn på Svalbard lufthavn - Miljødirektoratet ønsker innspill - miljodirektoratet.no

Murphy, G. L. (1982). Cue validity and levels of categorization. *Psychological Bulletin*, *91*(1), 174.

National Snow and Ice Data Center (NSIDC) (2025, April 2). *Arctic Sea Ice Sets a Record Low Maximum in 2025. Arctic Sea Ice Sets a Record Low Maximum in 2025*. National Snow and Ice Data Center

Nett, L., Guth, T. A., Büchel, P. K., Rungratsameetaweemana, N., & Kunz, L. (2025). Behavioral investigation of allocentric and egocentric cognitive maps in human spatial memory. *bioRxiv*, 2025-01.

Norman, D. A. (1986). Cognitive engineering. *User Centered System Design*, *31*(61), 2.

NRK (2016). En ekstrem økning i antall skuterskader. [An extreme increase in the number of snowmobile injuries – translated by AI]. Published March 8. Sterk økning i skuterskader på Svalbard – NRK Troms og Finnmark.

NRK (2025). Lokalstyret advarer mot drikkevannet i Longyearbyen. [The local council warns against drinking the tap water in Longyearbyen]. Published September 26th Link: Lokalstyret advarer mot drikkevannet i Longyearbyen – NRK Troms og Finnmark.

NTE (2025). NVE's sikringsarbeider mot flom og skred i Longyearbyen er ferdigstilt. Retrieved May 12th from NVEs sikringsarbeider mot flom og skred i Longyearbyen er ferdigstilt | Norges vassdrags- og energidirektorat (NVE).

Oliver-Serra, A., Siwik, L., Leszczyńska, N., Sikora, M., Ciesielski, T. M., Valseth, E., . . . & Paszyński, M. (2025, July). Pollution simulations and in-field measurements performed in March at Longyearbyen, Spitzbergen. In *International Conference on Computational Science* (pp. 353–360). Cham: Springer Nature Switzerland.

Olsen, A. (2024). Extreme low temperature safety. In *Ship Operations in Extreme Low Temperature Environments* (pp. 473–495). Cham: Springer Nature Switzerland.

Parent, O. & Ilinca, A. (2011). Anti-icing and de-icing techniques for wind turbines: Critical review. *Cold Regions Science and Technology*, *65*(1), 88–96.

Perrow, C. (1984). *Normal Accidents: Living with High-Risk Technologies*. Basic Books.

Rantanen, M., Karpechko, A. Y., Lipponen, A., Nordling, K., Hyvärinen, O., Ruosteenoja, K., . . . & Laaksonen, A. (2022). The Arctic has warmed nearly four times faster than the globe since 1979. *Communications Earth & Environment*, *3*, 168. https://doi.org/10.1038/s43247-022-00498-3

Rasmussen, J. (1997). Risk management in a dynamic society: A modelling problem. *Safety Science*, *27*(2–3), 183–213. https://doi.org/10.1016/S0925-7535(97)00052-0

Reason, J. (1990). *Human Error*. Cambridge University Press.

Reason, J. (1997). *Managing the Risks of Organizational Accidents*. Ashgate.

Reason, J., Hollnagel, E., & Paries, J. (2006). Revisiting the Swiss Cheese model of accidents. *EEC Note*, *13* (06).

Rekuviene, R., Saeidiharzand, S., Mažeika, L., Samaitis, V., Jankauskas, A., Sadaghiani, A. K., . . . & Koşar, A. (2024). A review on passive and active anti-icing and de-icing technologies. *Applied Thermal Engineering*, *250*, 123474.

Rensink, R. A., O'Regan, J. K., & Clark, J. J. (1997). To see or not to see: The need for attention to perceive changes in scenes. *Psychological Science*, *8*(5), 368–373.

Ringkjøb, H.-K., Haugan, P. M., & Nybø, A. (2020). Transitioning remote Arctic settlements to renewable energy systems – A modelling study of Longyearbyen, Svalbard. *Applied Energy*, *258*, 114079, https://doi.org/10.1016/j.apenergy.2019.114079

Sætren, G. B., Lund-Kordahl, I., Vaag, J. R., Lund, M. S., & Badu, J. (2025). Safety in the High Arctic. In *Proceedings of the 35th European Safety and Reliability and the 33th Society for Risk Analysis Europe Conference*.

Sørensen, A. (2021). Testination North. In Wassmann, P. (Ed.), *Whither the Arctic Ocean?: Research, Knowledge Needs, and Development en Route to the New Arctic* (pp. 165–175). Fundación BBVA.

Stanton, N. A., Salmon, P., Rafferty, L., Walker, G. H., Baber, C., & Jenkins, D. P. (2013). *Human Factors Methods: A Practical Guide for Engineering and Design* (2nd ed.). Ashgate Publishing Ltd.

Stanton, N. A., Stewart, R., Harris, D., Houghton, R. J., Baber, C., McMaster, R., Salmon, P., Hoyle, G., Walker, G., Young, M. S., Linsell, M., Dymott, R., & Green, D. (2007). Distributed situation awareness in dynamic systems: Theoretical development and application of an ergonomics methodology. *Ergonomics*, *49*(12), 1288–1311. https://doi.org/10.1080/00140130600612762

Statistics Norway (2022). Samfunnsforhold på Svalbard, befolkning, økonomi og levekår. [Social Conditions in Svalbard, Population, Economy, and Living Conditions], https://www.ssb.no/virksomheter-foretak-ogregnskap/virksomheter-ogforetak/artikler/samfunnsforhold-pasvalbard/_/attachment/inline/234f5e8c-6116-41ab-98b6-2f8a239a11d8:645812a8ec6254d8570e1ae539b1db4a6ba40103/RAPP2023-36.pdf31. 10.2024

Statistics Norway (2024). Population on Svalbard. 12622: Population 1 January (H1) or 1 July (H2) in Longyearbyen and Ny-Ålesund, by citizenship and sex 2019H1 - 2024H2. Statbank Norway. https://www.ssb.no/en/statbank/table/12622 31.10.2024

Statistics Norway (2025, May 1). Befolkningen på Svalbard – SSB.

Svalbardposten (2024). Utgjør en direkte helserisiko [Constitutes a direct health risk]. Published on 8 December.

Svalbardposten (2025a). Ingen straffes for skudd mot isbjørn [No punishemnet for firing towards polar bear]. Published on 14 March.

Svalbardposten (2025b). Turist fikk bjørnepote på ryggen [Tourist got polar bear paw on the back]. Published on 3 March.

Svalbardposten (2025c). Diesel er et nødvendig steg på veien til det grønne skiftet i Longyearbyen [Diesel is a necessary step on the path toward the green transition in Longyearbyen]. Published on 11 June.

Svalbardposten (2025d). LL advarer: Ikke drikk vannet fra springen [Local Council warns: Do not drink the tap water – translated by AI]. Published on 25 September. Link LL advarer: Ikke drikk vannet fra springen.

Svalbardposten (2025e). Dette var dråpen [This was the last straw – translated by AI]. Link: Dette var dråpen.

Tierney, M., Adeyemi, E. O., & Bommer, S. (2025). A systematic literature review: Cognitive workload assessment in human factors research. *Advances in Human-Computer Interaction*, *2025*, 9313239.

Turner, B. A. (1978). *Man Made Disaster*. London: Wykeham Publications.

Veterinærinstituttet (2025, 7. juli). *Rabies påvist hos fjellrev på Svalbard – andre påvising i 2025*. https://www.vetinst.no/nyheter/rabies-pavist-hos-fjellrev-pa-svalbard-andre-pavising-i-2025 [Rabies detected in Arctic fox on Svalbard – second case confirmed in 2025 - translated by AI].

Vihma, T. (2014). Effects of Arctic sea ice decline on weather and climate: A review. *Surveys in Geophysics*, *35*, 1175–1214. https://doi.org/10.1007/s10712-014-9284-0

Visit Svalbard (2025). Svalbard-statistikken 2024 [Svalbard statistics 2024]. Årsstatistikk2024.pdf

Wærø, I., Rosness, R., & Kilskar, S. S. (2018). *Human Performance and Safety in ARCTIC Environments*. Report number 102016139. SINTEF.

Weick, K. E., & Sutcliffe, K. M. (2015). *Managing the Unexpected* (3rd ed.). John Wiley & Sons.
White Paper. (2021). Prop. 1 S (2021-2022). For budsjettåret 2022 — Statsbudsjettet (Gul bok) [For the Fiscal Year 2022 — The National Budget (Yellow Book)]. Retrieved from Prop. 1 S (2021–2022) - regjeringen.no.
Woodier, N. (Ed.). (2025). *Learning from Near Misses: Cross-Sector Reflections to Support Safety Management*. CRC Press.
Yang, M., Qiu, Y., Huang, L., Cheng, M., Chen, J., Cheng, B., & Jiang, Z. (2023). Changes in sea surface temperature and sea ice concentration in the Arctic Ocean over the past two decades. *Remote Sensing*, *15*(4), 1095.
Yr.no (2025a). Historikk vær Platåberget 23. mars 2025 Yr - Platåberget - Historikk som graf - 23. mars 2025.
Yr.no (2025b). Historikk vær Longyearbyen 7. februar 2025 Yr - Longyearbyen - Historikk som graf - 7. februar 2025.
Yr.no (2025c). Historikk Longyearbyen siste 13 måneder. Yr - Longyearbyen - Historikk som graf - Siste 13 måneder.

10 Safety-Critical Task Analysis in Automation and Remote Operations

Marius Fernander

10.1 INTRODUCTION

In today's increasingly automated world, the interaction between humans and automation is crucial for maintaining system safety. When properly designed, automation can improve consistency, increase efficiency, and greatly enhance safety results. However, poorly designed automation can create new risks, sometimes with catastrophic consequences.

Notable examples include the crash of Air France Flight 447 and the two tragic Boeing 737 Max accidents, which together claimed 346 lives. Investigations into the Boeing incidents revealed several key issues, among them critical flaws in the design of the Maneuvering Characteristics Augmentation System (MCAS), which relied on input from a single angle-of-attack sensor. Compounding the issue was insufficient pilot training and a lack of awareness about the system's behaviour (Hopkins, 2025). These events underscore the dangers of over-reliance on automation in the absence of robust design safeguards, automation transparency, system resilience, and adequate human–system integration.

To mitigate risks, industries that operate safety-critical systems, such as aviation, maritime, and oil and gas, apply structured risk analysis methods. Among these, Safety-Critical Task Analysis (SCTA) is one method that can be used to understand and improve human (and system) performance in complex operations. By systematically analysing tasks, SCTA can reveal where automation must provide timely cues, feedback, and support to ensure that operators can successfully operate and monitor critical processes. This is particularly vital in high-stakes environments such as aviation, autonomous driving, and industrial process control. SCTA can further identify tasks where human error could lead to serious consequences and guide the design of systems that reduce the probability of error while increasing system tolerance to inevitable human variability.

Designing for safety involves anticipating how humans interact with systems before incidents occur. SCTA, utilizing the Analysis of Pre-accident Operator Actions (APOA) framework, provides a systematic approach to identify potential task failures and ways to redesign systems for resilience and error tolerance. This chapter explains how APOA integrates human factors, task analysis, and design standards to ensure safety is an inherent part of engineered systems. It introduces SCTA based on the recent development of the APOA method (DNV, 2023). Through a real industry example, it will be demonstrated how SCTA can be applied to improve system safety by enhancing human performance and increasing error tolerance in safety-critical systems. The chapter aims to offer practical, step-by-step guidance on conducting SCTA, along with background information on task analysis.

We will begin by exploring some central concepts that are important to be familiar with before explaining the different steps in the underlying process, that is, the APOA method. The last part of this chapter will focus on how SCTA can be used to improve automation.

DOI: 10.1201/9781003741824-13

10.2 ERROR TOLERANCE

Errors, *or human variability*, are both universal and inevitable and are part of being human (Reason & Hobbs, 2003). Errors can therefore be expected, and it is essential that we create systems that are safely designed and flexible enough to handle errors and variability in task execution (Conklin, 2019). In APOA, three principles of error tolerance are outlined, namely:

1. Error pathway blocking
2. Error recovery enhancement
3. Error consequence reduction

If human errors (or human mistakes) cannot be prevented, it may be possible to implement measures to block error pathways, so they don't have any adverse effects on the system. One way of blocking an error pathway is to implement interlocks that prevent unsafe actions from propagating, ensuring that an incorrect operator action cannot lead to a hazardous consequence.

If blocking the error pathway is not possible, one option could be to increase the likelihood of error recovery. An example of an error recovery measure could be to provide alarms or require peer checks to increase the probability of detecting an error before control of a hazard is lost.

The last error-reduction approach is to reduce the consequences of an error if it cannot be prevented, blocked, or recovered. Consequence reduction here refers to minimizing the consequences a human error may have in terms of safety. One example of consequence reduction is to implement some logic in the control system that will bring the system to a safe state if an operator forgets to shut off the supply to a tank. Upon high level, the control system could for instance automatically shut off the supply by stopping a pump and closing a valve on the supply line.

APOA and other methods for SCTA provide tools for designing error-tolerant systems. APOA also stipulates other principles to improve system safety by identifying error traps and error-producing conditions so that they can be eliminated. Both approaches will be explained and exemplified further in this chapter.

10.3 HIERARCHY OF CONTROLS

The principles referred to above are also in line with the hierarchy of controls, a risk reduction principle that is mandatory in many industries (e.g., NIOSH, 2023).

This principle asserts that measures at the top of the hierarchy are most effective for risk reduction (Figure 10.1). Although eliminating or substituting hazards may be challenging—since they often constitute core business activities—priority should be given to engineered controls over administrative measures or personal protective equipment (PPE). Accordingly, systems should be designed to be inherently safe or error-tolerant, rather than relying primarily on training and procedural compliance.

10.4 SAFETY-CRITICAL TASK ANALYSIS

Task analysis is the study of what a person (or group) is required to do in terms of physical behaviour and/or cognitive processes when it comes to achieving a system goal (Kirwan & Ainsworth, 1992). SCTA is a set of methods that extends basic task analysis to assess how and why a task can fail. In SCTA, the goal is not only to understand what operators do but also why specific actions may fail and how system design can reduce the likelihood or consequences of such failures.

Basic task analysis is a foundational Human Factors Engineering (HFE) tool and underpins several essential activities, including the following:

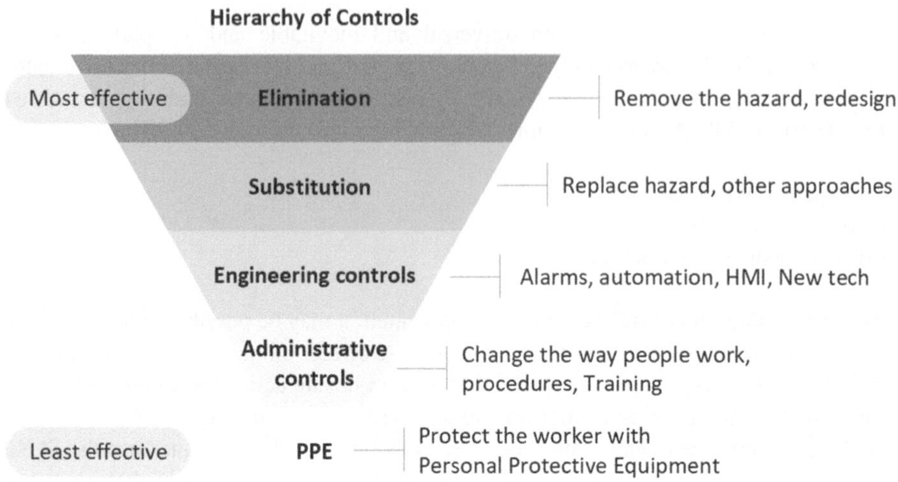

FIGURE 10.1 Hierarchy of controls.

- System and interface design
    - o Human–Machine Interface (HMI): Designing user-friendly interfaces for control rooms, industrial systems, and software by understanding user tasks and workflows.
    - o Product design: Improving usability and user experience by analysing how users interact with products.
- Training and competence development
    - o Training program development: Creating effective training modules by breaking down complex tasks into teachable steps.
- Procedure and documentation development
    - o Standard Operating Procedures (SOPs): Developing clear, step-by-step instructions for critical tasks.
    - o Emergency response plans: Structuring emergency procedures to ensure all necessary actions are covered and easily understood.
- Safety and risk assessment
    - o Human Reliability Analysis (HRA): Identifying potential for human error and designing safeguards (covers both qualitative and quantitative methods).
    - o Workload and Operability Assessment: Evaluating whether tasks can be performed safely and efficiently under various conditions.
    - o Health risk assessment and ergonomic risk analysis: Analysing various exposures and strains.

We know that people are fallible, and by systematically looking for potential errors, SCTA can increase safety by improving human performance and making systems more error tolerant.

Safety-critical tasks (SCTs) can be separated into several categories. Table 10.1 describes the different types of tasks and explains how human errors can negatively affect safety.

APOA was primarily developed to analyse pre-initiator and initiator errors, but the tools included in the guideline can be used to analyse all types of SCTs. For detailed analyses of post-initiator tasks (i.e., responding to emergencies), it is recommended to use other methods such as Petro-HRA (Blackett et al., 2022) or similar methods.

TABLE 10.1
Task Types and Errors

Task Type	Negative Contribution	Example
Inspection, testing, and maintenance of critical systems and equipment (including barriers).	Introduce (or fail to reveal) latent failures. Referred to as *pre-initiator errors*.	Wrong line-up of valves after e.g., maintenance that may cause unexpected pressure build-up at a later stage.
Monitor and control normal, yet safety-critical operations.	Initiate an accident sequence, solely or in combination with technical failures. Referred to as *initiator errors*.	An operator mistakenly starts loading to the wrong vessel or tank, causing unintended pressure buildup.
Respond to safety-critical failures and accident scenarios.	Allow a hazardous event to escalate into an accident which results in losses. Referred to as *post-initiator* errors.	Operators fail to shut down loading pumps or close valves upon high-pressure alarm in a tank.

Source: Adopted from APOA.

10.4.1 (Quantitative) Human Reliability Analysis

HRA has often been used to quantify the likelihood of human error. However, such probability estimates are criticized to be uncertain and varying among analysts. In addition, quantitative analyses can be resource-demanding with diminishing returns compared to pure qualitative analyses (like APOA). The selection of methodology should be based on the study's required output. If a quantitative Human Error Probability (HEP) is not required, a qualitative method such as APOA is generally preferable. In some cases, one may, however, need a HEP to feed into a broader risk/reliability analysis.

Regardless of the method applied to identify and assess the likelihood of errors, it is important that the analysis builds upon a solid task analysis foundation. Otherwise, the HRA risks reducing complex human behaviour to abstract numbers that are divorced from operational reality. Experience from real-world applications also shows that workers do highly relate to the qualitative part of the analysis, and less to the error probabilities (HEPs), especially when these HEPs are low (humans are in general insensitive to low probabilities).

Task Analysis and Human Error Identification (HEI) are required steps in several HRA methods. It is important that these activities are performed as they establish the factual, design-specific understanding of the following:

- What operators actually do (work as done)
- What cognitive, physical, and collaborative demands exist
- Which cues, controls, and feedback mechanisms support decision-making
- How the interface and procedures enable or constrain performance

When this understanding exists, HRA can be applied more meaningfully – or, as in APOA, it can be replaced by qualitative, design-sensitive analysis that focuses on *why* errors occur and *how* systems can be redesigned to prevent or recover from them. In short, the overall outcome of an HRA is quantification of human reliability (task failure), but the task analysis and the HEI explain it – making task analysis an essential foundation for human-centred safety design.

10.5 SUPPORT FROM BEST PRACTICES STANDARDS

The techniques and methods described in the following are also supported by best-practice methods for human-centred design, such as ISO 11064 (2000–2023), ISO 9241-210 (2019), and IEC 63303 (2024).

In the control-room ergonomics standard (ISO 11064), task analysis is the essential starting point for designing control-room layouts, displays, and alarm systems. The analysis ensures that spatial, cognitive, and organizational design choices reflect actual operational tasks. APOA directly addresses these needs by mapping operator activities, interactions, and workload, forming the analytical foundation for ISO-compliant human-centred design.

The user-centred design standards, such as ISO 9241-210, should use task analysis or SCTA in specific activities: (1) understanding and defining the context of use, (2) specifying user and organizational requirements, (3) creating design solutions, and (4) evaluating them against requirements. SCTA supports Steps 1–2 and provides feedback for Steps 3–4.

HMI design, as outlined in IEC 63303, stresses that process information shown to the operator should be simple, relevant, and directly aligned with task requirements. Task analysis forms the basis for this by determining what operators do, the information they need, and how they interact with controls and displays. This HMI standard highlights that interface structure, information hierarchy, and alarm logic should directly follow task sequences and cognitive demands.

10.6 THE DIFFERENT STEPS IN APOA

APOA (DNV, 2023) is a complete method for SCTA and consists of the following seven steps:

1. Establish context
2. Task identification and screening
3. *Task analysis*
4. *Error analysis*
5. Model accident causation (optional)
6. *Improve human performance and error tolerance*
7. Prioritize and communicate recommendations

This chapter introduces all the steps required to perform the analysis. The main focus will, however, be spent on Steps 3, 4, and 6. Step 5 (error modelling) is an optional step that requires some prior knowledge of fault tree modelling and will not be covered in detail. More detailed guidance can be found in the APOA guideline (DNV, 2023).

APOA can facilitate a clear feedback loop by providing results from Steps 3–6 of APOA to designers, engineers, and procedure developers before final implementation. This promotes learning and design iteration in line with ISO 9241-210.

10.6.1 Step 1 – Establish Context (Project Definition/MoC – Management of Change)

In general, risk assessments begin with establishing a clear understanding of the context in which the operation and task(s) are performed (ISO, 2009). This is important because it can have a direct influence on how the people involved understand the objectives, process, and results from the analysis. Figure 10.2 is not exhaustive but presents some of the contextual factors which could be relevant when establishing context.

This first step also aligns with best practices in project definition and change management, ensuring that the scope and boundaries of the analysis are well understood.

When performing an analysis, it is advisable to include a system diagram that illustrates the physical process. For process systems, a process flow diagram (PFD) or a high-level piping and instrumentation diagram (P&ID) should be sufficient. This can also be used when defining the boundaries or battery limits of the analysis.

10.6.2 Step 2 – Task Identification and Prioritization

The second step of APOA is to identify which tasks or parts of an operation are considered safety-critical and should be selected for further analysis. Performing tasks and error analysis

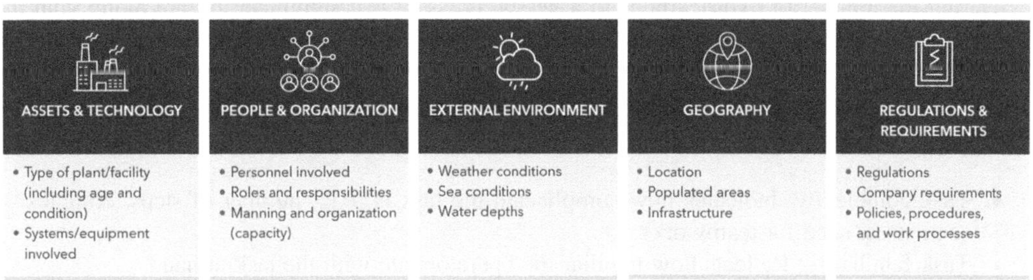

FIGURE 10.2 Contextual factors that should be described.

(Adopted from APOA.)

(Steps 3 and 4) are time-consuming activities, so the main purpose of this step is to identify those tasks that are regarded as most critical so that effort can be directed where it provides the highest value.

Words like *operation*, *task*, *sub-task*, *task step*, and *action* will be used in the subsequent sections, and understanding the relationship and difference is important for further understanding.

An operation is a high-level description of an industrial activity (e.g., 'drill well', 'operate plant'). Each operation consists of several tasks, which in turn can be decomposed into task steps or sub-tasks. Sub-tasks can be broken down further into observable physical actions and cognitive processes carried out by people (see Figure 10.5 for an example).

When identifying potentially safety-critical tasks, it is possible to use different hazard taxonomies, such as the energy wheel (ref. Figure 10.3), to check whether one or more of the eight different hazards/energy forms are, or may be, present during a specific part of an operation (a task).

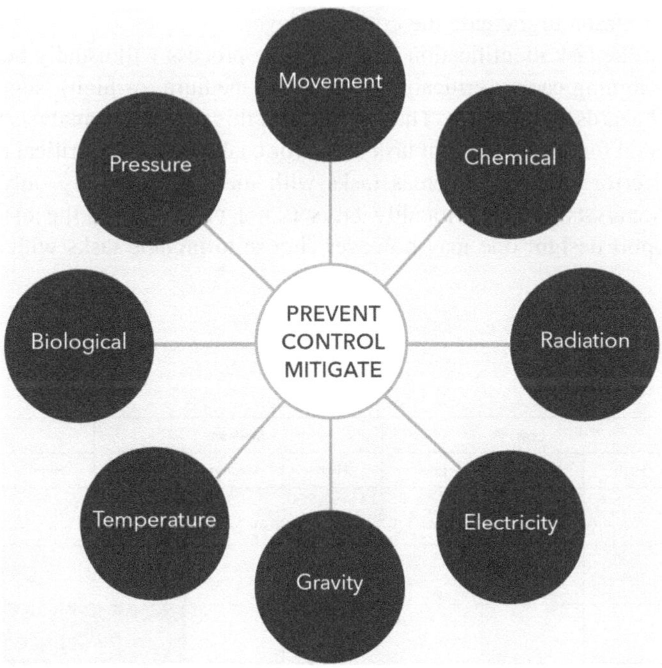

FIGURE 10.3 Energy wheel.

(Adopted from NSA, 2014.)

During the transfer of hydrocarbons to a vessel, pressure is a significant hazard along with the hazardous hydrocarbons themselves. It should therefore be further examined whether human actions can lead to a loss of control over these hazards.

APOA also suggests using task characteristics when assessing task criticality. The three criteria used in APOA are as follows:

- Task complexity: Indicates how complicated the task is (e.g., number of steps, complex sequences, need for teamwork)
- Task familiarity: Reflects how familiar the operators are with the task at hand
- Error consequence: The severity of the consequence related to an operator error

The matrix presented in Figure 10.4 should be understood as a coarse screening aid for determining the depth of subsequent analysis, rather than as a definitive measure of safety significance. Analysts should therefore consider not only complexity, familiarity, and consequence, but also the degree of human reliance and the vulnerability of apparently simple but high-consequence tasks to slips, lapses, and omissions, as such tasks may still require detailed SCTA. Each criterion is rated as high, medium, or low, following the definitions in the APOA guideline. Initial task criticality is then determined using the matrix in Figure 10.4. Other factors may also be relevant when assessing task criticality. To make the process more flexible and allow the analyst to differentiate better, APOA suggests some adjustment criteria that can be used, namely:

- Equipment/plant condition
- Frequency of hazard exposure
- History of near-misses and incidents

The analyst can adjust the criticality by one level (i.e., from medium to high) if one or more of these criteria are relevant. If incident data shows that most events occur during execution of some specific tasks, this can be a reason to increase the criticality level.

The outcome of the task identification and screening process will usually be a table listing the main tasks and assigning each a criticality score (low, medium, or high), supported by descriptions of potential hazards and failures. The results from this part of the analysis will determine the level of detail needed for the subsequent task and error analysis. Highly critical tasks will require a detailed task and error analysis, whereas tasks with medium criticality only require a coarse analysis. Further analysis of low-criticality tasks is not necessary. If the main purpose of the analysis is to support design, one may however choose to include tasks with low criticality for further analysis.

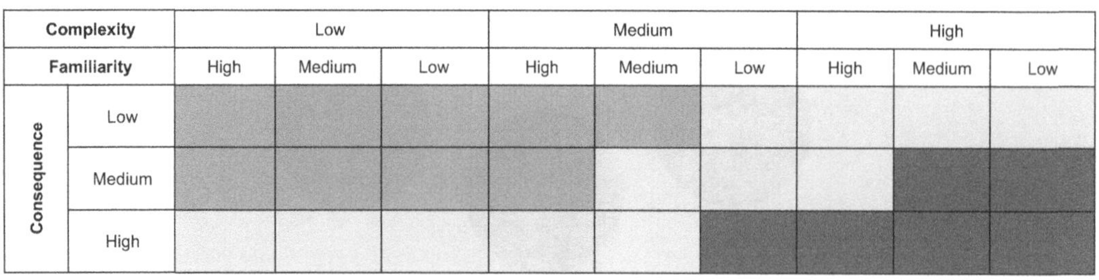

FIGURE 10.4 Matrix for ranking task criticality.

(Inspired by Energy Institute, 2020 (Annex A).)

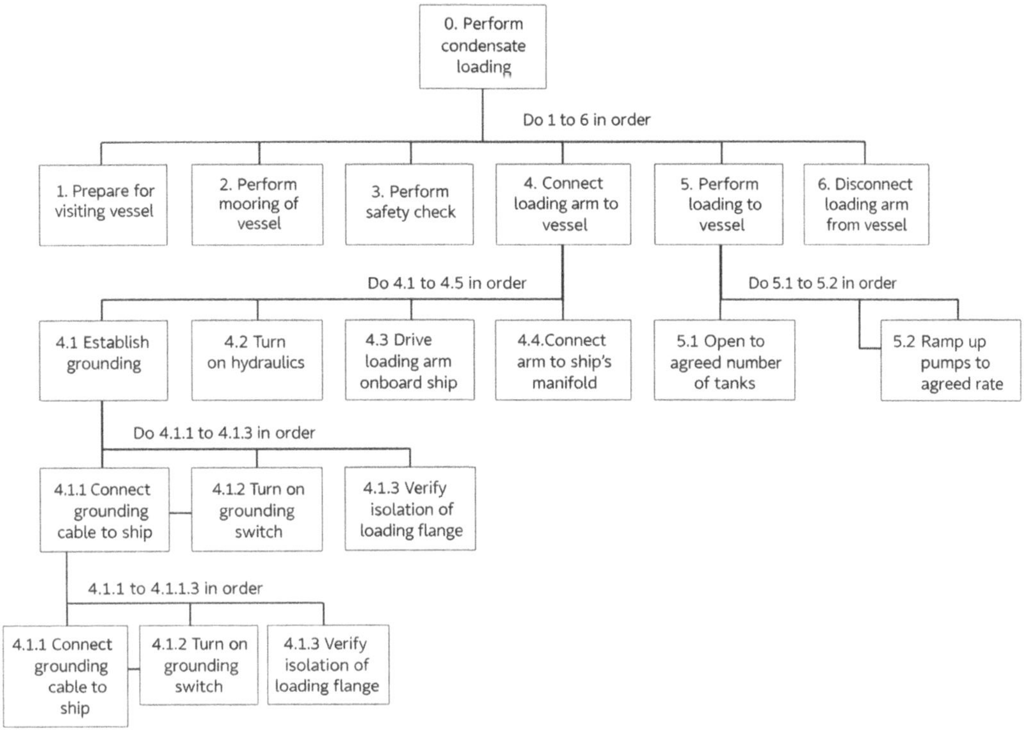

FIGURE 10.5 Hierarchical task analysis of 'Perform condensate loading' (simplified version).

10.6.3 Step 3 – Task Analysis

In the field of Human Factors, mastering task analysis is probably the most important skill because it has many application areas. Several advanced Human Factors methods, such as workload analysis (including cognitive workload), safety-critical task analysis, and HRA, are based on fundamental task analysis techniques. By mapping task structure and cognitive demands, task analysis supports targeted workload evaluation using the NASA-TLX, allowing assessment of how different task elements contribute to perceived mental, physical, and temporal load.

To accurately identify what operators need to do to achieve a system goal, data collection is essential. Briefly described in this step, data collection is an ongoing activity throughout the entire analysis process. Some data is also required when defining context and conducting the screening. Both written instructions, like procedures and checklists, and visual observations are considered valuable data sources. A combination of these is often beneficial because it allows for recording data that procedures alone do not cover. It also helps prevent analysis based on false assumptions – such as assuming that work as imagined (WAI) equals work as done (WAD) (see, i.e., Hollnagel & Clay-Williams, 2022). Industry experience shows this is not always the case. When performing task analysis, it is common practice to conduct both hierarchical task analysis (HTA) and tabular task analysis (TTA). Conducting an HTA involves following four steps:

1. Define the main goal of the task: Explain what the task intends to achieve. This is the primary objective and should be placed at the top of the hierarchy.
2. Identify the tasks (and sub-tasks/actions) needed to reach the goal: Break down each part of the task by asking how it is accomplished. Repeat the breakdown until each sub-task is expressed at a level where actions, decisions, and cues can be clearly identified.

3. Use task criticality to decide which steps need further breakdown: Use screening results to determine the level of detail. Break down critical tasks thoroughly. For medium critical tasks, detailed breakdowns are optional. For low-criticality tasks, breakdowns might not be necessary.
4. Organize the HTA diagram based on task relationships and order: Arrange the HTA to clearly show how tasks, sub-tasks, and actions connect and the sequence in which they are performed (the plan). See Figure 10.5.

It is recommended to include sequences/plans directly within the HTA diagram. There are many different types of plans. To name a few:

- Fixed sequence plans (i.e., 'do Steps 1, 2, and 3' in sequence)
- Optional sequence plans (i.e., 'do Steps 1, 2, and 3' in any order)
- Contingent sequence plans (i.e., 'do Step 1, then Step 2, if pressure level is above 20 PSI, then continue with Step 3')
- Concurrent (i.e., 'do Steps 1 and 2 simultaneously')
- Cyclical ('do Steps 1 to 5, then repeat')

More advanced plans could also include elements of all the above.

Figure 10.5 shows an example of an HTA performed for a transfer operation. The goal of the operation (*perform condensate loading*) is displayed on top of the hierarchy. The answer to the question 'how' the goal is achieved is given by tasks 1–6 in a fixed sequential order (the plan). For the tasks that are considered most critical according to the screening (tasks 4 and 5), the same process is repeated.

Further breakdown of sub-task 4.1 shows how one can decompose a task into a concrete set of recognizable actions and cognitive processes. The level of detail shown here may be required when performing HEI (see Step 4 – Error analysis) as the guidewords are typically defined to identify potential errors at this level (see error taxonomy in Table 10.5).

Creating an HTA is a good starting point for understanding how the task is performed. It is recommended to establish the HTA before moving forward with the analysis. The HTA is a flexible format, and it is easy to move boxes around. A disadvantage of the HTA format is that it cannot record much task-related data directly in the diagram. To capture more information about tasks, it is recommended to convert the HTA into a tabular format (TTA).

For tasks with medium criticality, the following information should be recorded:

- ID: The unique identifier for the task, task step, or action
- Task description: Brief description of the main sub-tasks for each task
- Performed by/personnel: Role(s) performing the sub-tasks/actions
- Performance shaping factors (PSFs): Factors that can influence human performance (see Table 10.4)
- Comments/actions: Comments, actions, and follow-up items (not shown in Table 10.2)

Additional information to be collected for a detailed task analysis includes the following:

- Procedure reference: Reference to a specific step in a procedure.
- Control panel/equipment/HMI: A list of the equipment, displays and/or controls used to perform the task. If there are known issues with these, this should be noted in the 'performance shaping factors' column.
- Cue: A brief description of the cue (instruction) for the operator to carry out a particular action. This could be a step in a procedure, a message from a co-worker, equipment status, operator knowledge, etc.
- Feedback: A brief description of the feedback that the operator receives to confirm that the action has been correctly performed.[1] (See Notes) This could include visual confirmation from equipment directly or from the HMI.

TABLE 10.2
Example of a Coarse Task Analysis

ID	Task Description	Performed By	Performance Shaping Factors
4	Connect loading arm to vessel • Establish grounding • Turn on hydraulics • Drive loading arm onboard ship • Connect arm to ship's manifold • Perform leak test	– Field operator (ship) – Field operator (dock) – Loading master	– Loading checklists – Buddy-check is included in the checklist – Loading arm control panel/joystick – Control-room HMI and communication systems – Courses have been developed for loading operations

Table 10.3 shows an example of how a detailed task analysis may look.

SCTA is a method used to identify potential human errors and the factors or conditions that increase or decrease the likelihood of errors. Dekker (2006) states that human error is not random but systematically linked to features of people's tools, tasks, and operating environment (i.e., errors are inherent to the overall system and often arise from inadequate design). To identify these factors or conditions, most SCTA guidelines include a taxonomy that can be used. During the development of APOA, a comprehensive review of Human Factors and human performance literature was conducted, including sources such as the U.S. Department of Energy (2009a, 2009b), IAEA (2013), IMO (2018), Franciosi et al. (2019), and the Energy Institute (2008, 2020). The results of the literature review are presented in Table 10.4. This list can serve as a basis for discussions to identify potential factors that can positively or negatively influence human performance. During the task analysis, the analyst should note the PSFs that can positively or negatively influence operator performance.

In the analysis, it is not necessary to evaluate how PSFs affect every individual sub-task or action. Doing so can be resource-intensive and overly time-consuming with diminishing returns, especially when efforts are spent on low-criticality task steps/actions. Therefore, a more detailed review of PSFs should be integrated into the error analysis (Step 4) for tasks with medium and high criticality. It should also be highlighted that the overall objective of the analysis can dictate which PSFs to focus on. If the methodology is applied to design an automation system, function allocation and HMI may be PSFs that are more relevant than those listed under *environmental factors*.

10.6.4 Step 4 – Error Analysis

Error analysis is a method used to understand how human errors (slips, lapses, and mistakes) can lead to task failure and accidents (DNV, 2023). This includes identifying how contextual (workplace) and individual PSFs might create error-producing conditions and error traps. Error analysis should be performed after the task analysis is complete or sufficiently developed; that is, no additional task steps or actions are expected to be identified, and the task structure (plans) will not change. This is especially important when conducting a detailed error analysis. Like task analysis, error analysis can be time-consuming, so it is recommended to base the analysis on task criticality (ref. Step 2).

An important part of the process is identifying potential errors that may occur during the execution of an action (or a cognitive process). There are different taxonomies available, and APOA uses the Structured Human Error Reduction and Prediction Approach (SHERPA), developed by

TABLE 10.3
Example of a Detailed Task Analysis

ID	Task Description	Procedure Ref.	Personnel	Control Panel/ Equipment/ HMI	Cue	Feedback	Performance Shaping Factors
5	Perform loading to vessel						
Plan: Do 5.1 to 5.X in sequence							
5.1	Open to agreed number of tanks according to loading agreement	4.0	Ship crew	Valves on ship manifold operated from bridge	Procedure/loading agreement	Valves change status from closed to open on HMI.	*Task preparation and planning* Loading rate and how many valves should be open are covered in the pre-loading meeting between 1st officer and loading master.
5.2	Ramp up pumps to agreed rate	4.1	Control-room operator	HMI	Informed by field operator on ship	Pumps symbol changes shape/colour. Rate will change from 0 to agreed rate.	*Communication; task complexity* Critical period when adjusting up and down the rates. It is important that crew on ship and in the terminal control room can communicate freely during this period, but radio noise is quite common.

TABLE 10.4
List of Performance Shaping Factors (PSFs)

Task/Job Factors

- Task characteristics, incl. complexity, difficulty, familiarity, ambiguity
- Time available vs. required to perform task
- Divided attention, different tasks competing for attention
- Distractions and interruptions
- Control panels/human–machine interfaces (HMI)
- Automation design and allocation of functions
- Interlocks and protection systems (error tolerance)
- Alarm system, e.g., alarm floods, prioritization/categorization
- Clarity of signs/signals/instructions/labels/other information
- Tools appropriate for task, incl. condition and availability/accessibility
- Procedures, e.g., inadequate, unavailable, incomplete, or inappropriate
- Routines for preparing, planning, and handing over a task
- Communication with colleagues, supervision, contractors, others
- Physical working environment – noise, heat, space, lighting, ventilation, humidity, vibrations, access

Individual/person factors

- Formal competence, e.g., coursework and certificates
- Informal competence, i.e., knowledge, skills (incl. social), and experience
- Work overload/underload (physical or cognitive)
- Fatigue, e.g., due to long working hours
- Physical capability and condition
- Stress, incl. experience threats towards own or others' safety
- Job satisfaction, morale, and attitudes
- Motivation and attention vs. other priorities
- Mismatch between perceived and actual risk

Organizational Factors

- Work pressures and conflicting goals, e.g., production vs. safety (incl. peer pressure)
- Level and nature of supervision/leadership, lines of authority
- Drills, table-top exercises, and on-the-job training
- Teamwork and collaboration, coaching of others
- Manning/staffing levels, availability/capacity/selection of human resources
- Organization and structure, e.g., clarity of roles and responsibilities
- Consequences of failure to follow rules/procedures (compliance)
- Effectiveness of organizational learning (learning from experiences)
- Organizational or safety culture, e.g., everyone breaks the rules
- Management system, e.g., policies, work processes, and change management
- Psychosocial working environment, e.g., support from supervisor and peers

Environmental Factors

- Outdoor temperature
- Weather (rain, snow, hail, wind)
- Time of day/natural light

Embrey (1986), with some adjustments (e.g., Blackett et al., 2022). Table 10.5 presents categories of actions (and cognitive processes) and their associated error modes.

For task 4, one sub-task is to connect a grounding cable. One specific error mode associated with this action is that the operator forgets to connect the grounding cable. This error can hence be linked to the error mode A11 – operation omitted. It is neither realistic nor necessary to apply

TABLE 10.5
Human Error Taxonomy

Diagnosis Errors	Information Communication Errors
Dx1-Diagnosis not made	I1-Information not communicated
Dx2-Diagnosis incorrect/incomplete	I2-Wrong information communicated
Dx3-Diagnosis too early/too late (mistimed)	I3-Incomplete/incorrect communication of information
Planning Errors	I4-Information communication unclear
P1-Plan omitted	**Information Retrieval Errors**
P2-Plan incorrect	R1-Information not obtained
P3-Plan incomplete	R2-Wrong information obtained
Decision Errors	R3-Information retrieval incomplete
D1-Correct decision based on wrong/missing information	R4-Information obtained incorrectly (misheard/wrongly interpreted)
D2-Incorrect decision based on right information	R5-Information not available/accessible
D3-Incorrect decision based on wrong/missing information	**Action Errors**
D4-Failure to make a decision (impasse)	A1-Operation too long/short
Checking Errors	A2-Operation too early/late (mistimed)
C1-Check omitted	A3-Operation in wrong direction
C2-Check incomplete	A4-Operation too little/much
C3-Right check on wrong object	A5-Operation too fast/slow
C4-Wrong check on right object	A6-Misalign
C5-Wrong check on wrong object	A7-Misplacement
C6-Check too early/late mistimed	A8-Right operation on wrong object
Selection Errors	A9-Wrong operation on right object
S1-Selection omitted	A10-Wrong operation on wrong object
S2-Wrong selection made	A11-Operation omitted
Intentional Acts	A12-Operation incomplete
IA1-Intentional (but unsafe) variability in task execution[a]	A13-Operation in wrong order

[a] Refers to cases where there is variability among workers in how a task is performed, and this has a potential of being an unsafe work practice. Can in principle be many of the other human errors, but with the difference being that they are intentional due to some performance shaping factor (e.g., time pressure and workload).

every guideword to every action; analysts should instead focus on the most credible error modes. Still, the analyst should be familiar with the taxonomy so that they can identify the most credible error modes.

Data for error analysis can be gathered through various activities such as document reviews, incident record reviews, field observations, and workshops. Using multiple sources usually yields the most comprehensive information. It is strongly recommended to hold a workshop with participating operational personnel, technical/system experts, and safety representatives to conduct a structured error analysis. No matter how the error analysis is carried out, the goal is to examine each sub-task or action to identify potential human errors, describe their consequences, and evaluate PSFs that could adversely affect human performance.

For the task analysis, data should be recorded at various levels of detail corresponding to the initial task criticality assessment (see Step 2). APOA recommends recording data in the following format for tasks with medium criticality:

- Potential errors and consequences: Errors that can either initiate hazardous events or introduce latent failures into the system. Both errors and consequences are described in the same column.
- Performance shaping factors: If done at separate points in time, revisit the PSFs captured as part of the coarse task analysis to determine whether any error traps or error-producing conditions are present.
- Comments and actions: Record any comments needed for additional information or clarification, as well as recommended actions to improve safety.

The coarse error analysis is an extension of the task analysis performed. As you can see in Table 10.6, potential errors and their consequences are included. There could also be a separate column in the table for comments and actions, but this will be covered under Step 6 (Error reduction).

As for the task analysis (see Step 3), APOA recommends recording more detailed and organized data for highly critical tasks. The following columns should be included for these tasks:

- Potential errors: Identify and describe potential errors related to the various sub-tasks and actions.
- Likely consequence: Describe the likely consequences of each error.
- Consequence level: Rank the consequence as 'high', 'medium', or 'low' based on consequence severity. Use the organization's risk matrix, or the one included in APOA.
- Recovery opportunity: Describe if there are any recovery opportunities. Recovery means detecting and fixing the error before any negative consequences materialize.
- Recovery level: Rank recovery level (likelihood of recovery) using the categories 'high', 'medium', or 'low'.
- Criticality: Assess the errors' criticality levels based on their consequences and recovery opportunities. This is performed by applying the matrix as shown in Figure 10.6.

TABLE 10.6
Example of a Coarse Error Analysis

ID	Task Description	Performance Shaping Factors	Potential Errors and Consequences
4	Connect loading arm to vessel – Establish grounding – Turn on hydraulics – Drive loading arm onboard ship – Connect arm to ship's manifold – Perform leak test	– Loading checklists available but not consistently filled out on work site by all shifts. – Buddy-check is included in the checklist. – No visual indication on the loading arm joystick confirming whether grounding is established. – No signal/indication in the control room regarding the status of grounding. – Crew has not received adequate training in performing cleaning of isolation flanges.	1. Omits to connect grounding cable or connects cable to a non-conductive surface (e.g., coated by paint). May cause static electricity that could lead to ignition of a gas cloud. 2. Omits to turn on grounding switch. May cause static electricity, which may lead to ignition of a gas cloud. 3. Incorrect operation of loading arm is causing crushing injuries to personnel. 4. Leak test incorrectly performed, increasing the risk of condensate leak during loading.

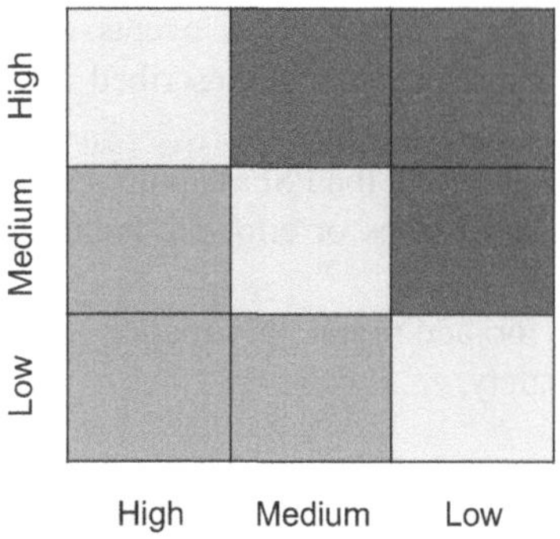

Criticality level	
High	Risk reducing measures must be implemented
Medium	Risk reducing measures are recommended
Low	Risk reducing measures can be considered

FIGURE 10.6 Error criticality matrix and requirements for error reduction.

- Worst-case outcome: Map what the worst-case outcomes could be, assuming the incident would escalate. This could be valuable input when performing error modelling (ref. Step 5).
- PSFs: Examine and record how performance shaping factors may represent error traps and error-producing conditions.

As for the coarse error analysis, a column to record any additional comments and risk-reducing measures (actions) should also be included. Table 10.7 shows an example of how a detailed error analysis may look.

Criticality is classified using a simple function of consequence severity and recovery potential, as illustrated in the matrix in Figure 10.6. The need for risk-reducing measures should likewise be determined based on the criticality of the error. See Step 6 for further discussion.

10.6.5 Step 5 – Model Accident Causation

Error analysis (Step 4) examines how individual human errors impact outcomes. Understanding how combinations of human errors (and other events or conditions) can lead to accidents may sometimes offer valuable insights for risk reduction. Various techniques are available to model human factors in accident causation (see e.g., Kirwan & Ainsworth, 1992 and Kirwan, 1994). However, for APOA, which focuses on errors during the pre-initiator and initiator stages, fault tree analysis (FTA) is recommended.

As FTA is considered an advanced topic, this step will not be covered in detail in this chapter. From Figure 10.7, you can see that the top event (fire/explosion on tank deck) is the result of several 'human basic events' (errors) occurring. On the left side of the fault tree, it is shown that pressure build-up in the system requires at least one of two potential human errors to happen (represented by an OR-gate). The loading rate must be higher than specified (1.1.1), or the ship's crew opens too few manifold valves (1.1.2). Since the operation is monitored from both the shore side (control room) and the ship side, there are multiple opportunities to take mitigating actions. These actions may also fail and are represented by the 'human basic events' 1.2.1 to 1.2.3. For ignition to occur, an ignition source must also be present. Errors related to this intermediate event are not shown in this FTA.

The actions associated with the intermediate event 1.2 are defined in Table 10.1 as post-initiator errors, which APOA was not designed to analyse. When performing a FTA, new actions may be identified. Users can utilize the tools in APOA to analyse these actions further or choose other methods, such as Petro-HRA (Blackett et al., 2022).

TABLE 10.7
Example of a Detailed Error Analysis

ID	Task Step	Potential Errors	Likely Consequence	Consequence Level (H/M/L)	Recovery Opportunity	Recovery Level (H/M/L)	Criticality (H/M/L)	Worst Case Outcome	Potential Error Traps or Error Producing Conditions (PSFs)
5	Perform loading								
Plan: Do 5.1 to 5.X in sequence									
5.1	Open to agreed number of tanks according to loading agreement.	Ship crew is open to fewer tanks than agreed.	Pressure build-up in the loading line. Risk of over-pressuring the system (PVs on tank deck may open).	H	Pressure/rate will be monitored both from CCR and from the bridge. Mitigating action can be taken from both sides.	H	M	Ignited HC leak If ignition source control fails	Miscommunication between the 1st officer and loading master can lead to incorrect number of tanks being opened and prepared
5.2	Ramp up pumps to agreed rate	Pumps are ramped up to a higher rate than the ship can receive.	Pressure build-up in the loading line. Risk of over-pressuring the system (PVs on tank deck may open).	H	Pressure/rate will be monitored both from CCR and from the bridge. Mitigating action can be taken from both sides.	H	M	Ignited HC leak If ignition source control fails	Distractions and radio noise during ramp up of the pumps can lead to too high loading rate being applied.

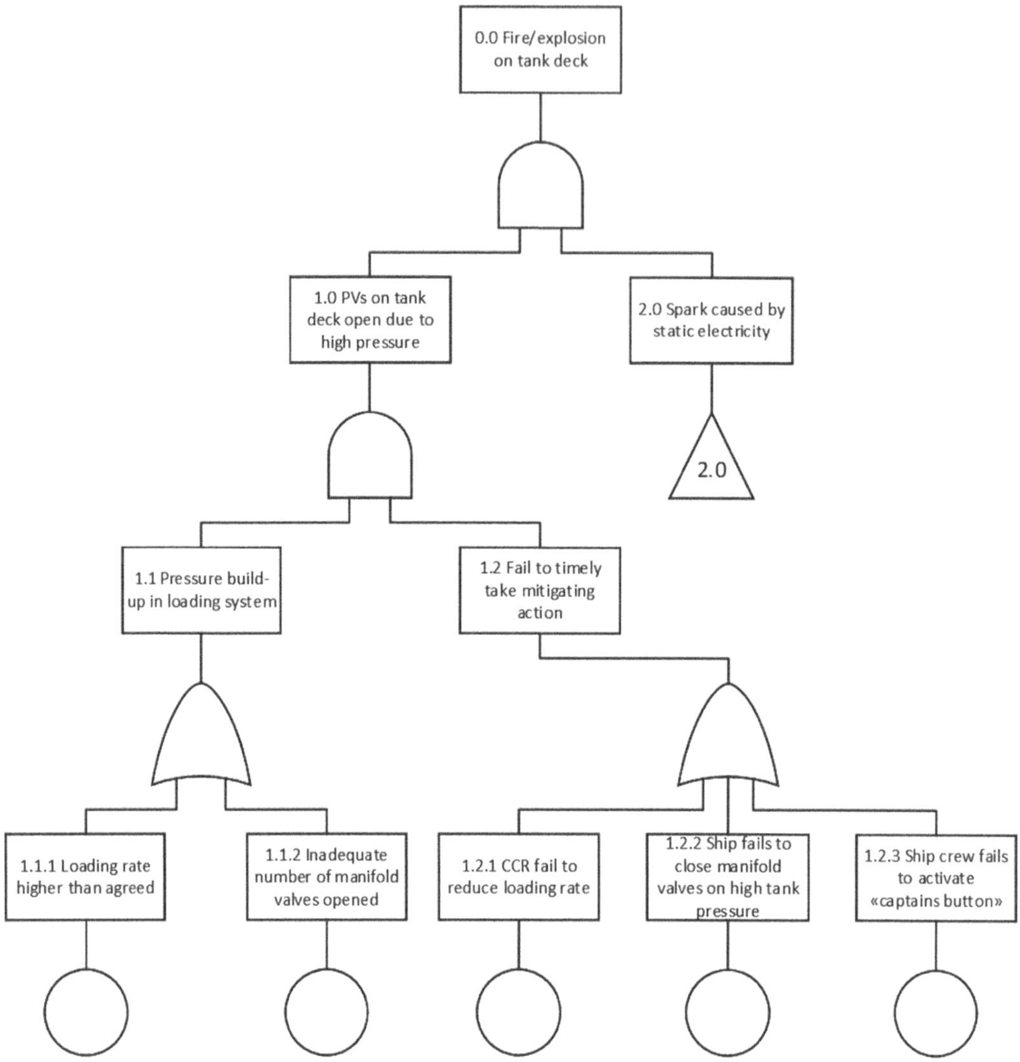

FIGURE 10.7 Example of a fault tree analysis.

10.6.6 Step 6 – Improve Human Performance and Error Tolerance

In APOA, developing measures aimed at improving human (and system) performance are as follows:

- Examine the task analysis and consider how the task and PSFs can be improved to facilitate successful task performance.
- Examine the error analysis to identify error traps and error-producing conditions and determine how these can be eliminated or mitigated through design improvements.

In addition, (system) error tolerance can be improved by recommending designs which make sure that:

- Error doesn't have an adverse effect on the system (error pathway blocking).
- Errors can be easily recovered before they have a negative effect on the system (error recovery enhancement).
- Errors' consequences are minimized to the greatest extent possible (error consequence reduction).

This approach and these principles are very useful when designing and improving automation systems as recommendations can target both how the system behaves (logic) and how systems interact with their users (HMI).

Table 10.8 shows an example of how a task and error analysis may look like when including risk-reducing measures. Note that some of the table columns have been removed to make space for the additional 'Recommended actions' column.

Since this chapter focuses on how SCTA can enhance automation, let us consider the error related to ramping up the pumps (item 5.2). In this case, the system is not very tolerant of a plausible human error. Because such errors are likely to occur, one solution is to incorporate greater error tolerance into the system. This can be achieved by modifying or improving the control system.

By implementing an early warning alarm for high pressure (PAH), the system can improve error recovery, enabling the operator to quickly reduce the supply. If the operator cannot shut off the supply for some reason, the control system can be further enhanced to automatically shut off the supply once a certain pressure threshold (PAHH) is reached. This will minimize or even eliminate any negative safety outcomes if the operator is unable to control the pressure. Other safety features, such as pressure relief valves (PRVs), are also important measures to consider.

This analysis showed that ensuring proper electrical grounding is crucial, as static electricity could serve as a potential ignition source if a leak occurs. One recommended measure from this study was to add a signal to the control system that prevents any loading operations from continuing

TABLE 10.8
TTA and Error Analysis Updated to Include Recommended Actions

ID	Task Step	Potential Errors	Likely Consequence	Recovery Opportunity	Potential Error Traps or Error-Producing Conditions	Recommended Actions
5	Perform loading Plan: Do 5.1 to 5.X in sequence					
5.1	Open to agreed number of tanks according to loading agreement.	Ship crew open to fewer tanks than agreed.	Pressure build-up in the loading line. Risk of over-pressuring the system (PVs on tank deck may open).	Pressure/rate will be monitored both from CCR and from the bridge. Mitigating action can be taken from both sides.	Miscommunication between the 1st officer and loading master can lead to incorrect number of tanks being opened and prepared.	Ensure that the ship's UHF has priority on the network to secure that communication is not corrupted/ interfered by other communication on the network.
5.2	Ramp up pumps to agreed rate	Pumps are ramped up to a higher rate than the ship can receive.	Pressure build-up in the loading line. Risk of over-pressuring the system (PVs on tank deck may open).	Pressure/rate will be monitored both from CCR and from the bridge. Mitigating action can be taken from both sides.	Distractions and radio noise during ramp up of the pumps can lead to too high loading rate being applied. Alarms and logic not designed to assist operator.	Implement a PAH alarm in the control system and automatic shut-down of loading pumps on PAHH.
...	...	...	...	...	...	...

unless grounding is verified. This creates a software interlock that blocks the error pathway, preventing the system from allowing an error to occur.

10.6.7 Step 7 – Prioritize and Communicate Recommendations

Step 6 often results in a detailed list of observations and recommendations that can be resource-intensive for the risk owner to follow up on and implement. Therefore, Step 7 in the APOA process involves prioritizing and communicating these recommendations to stakeholders. Management and operators can participate in relevant APOA workshops – this aligns with the participation principle of ISO 9241-210 (and action research), increasing ownership and execution of safety improvements.

Ensuring that actions are assigned the correct priority is essential to guarantee they will be followed up on. APOA guides the assignment of appropriate priority levels to actions.

- Use error criticality as a guide; errors with high consequences and low recovery potential should be prioritized over less critical errors.
- Cost-benefit analysis: Evaluate the expenses of risk reduction against the benefits gained.
- Examine whether the current solution or design complies with regulatory requirements. Any non-compliance may necessitate the implementation of risk-reducing measures regardless of other considerations.
- Hierarchy of controls, as presented in the introduction, is often a mandatory risk reduction principle that should be used when working with risk reduction.

Visualizing results using diagrams like bow tie diagrams, Swiss cheese models, and fault trees can be an effective way to present the findings, as these visualizations may give the reader a more comprehensive understanding of system safety and layered defences. However, creating these visualizations can be resource-intensive and is generally recommended only for critical errors or top events.

10.7 SUMMARY

APOA demonstrates how SCTA supports user-centred design. By integrating SCTA into the iterative processes outlined by ISO 9241-210, ISO 11064, and IEC 63303, organizations can verify that HMIs, procedures, and work environments align with real user tasks, cognitive demands, and organizational realities – turning analysis into actionable design enhancements and embedding safety as a fundamental system characteristic. APOA outcomes can promote organizational learning through change initiatives, competence management, and training.

EXAMPLE OF A SIMPLE TASK ANALYSIS (BOILING WATER FOR TEA)

To illustrate the logic of task analysis before applying it to complex industrial operations, consider the everyday activity of *boiling water for tea*.

The example below shows how a simple task can be decomposed hierarchically and then converted into a tabular task description. Even trivial actions reveal how cues, feedback, and potential errors can be identified through structured analysis.

HIERARCHICAL TASK ANALYSIS (HTA) 'MAKE A CUP OF TEA'

Goal: Prepare a cup of tea with boiling water.

Task ID	Task/Sub-Task	Plan
0	Make a cup of tea	Do 1–4 in sequence
1	Fill kettle with water	Ensure water level between min & max marks
2	Switch kettle on	Continue when indicator light shows 'on'
3	Prepare cup with teabag	Add sugar or milk if desired
4	Pour boiled water into cup	Check for boiling, do not leave the area, Wait until kettle switches off before pouring
5	Switch kettle off	Check off light

TABULAR TASK ANALYSIS (TTA) 'BOIL WATER FOR TEA'

ID	Task Step	Cue	Feedback	Performance Shaping Factors	Potential Errors
1	Fill kettle with water	Visual check of water level	Water visible in kettle	Lighting, distractions Check distance between kettle and water tap; avoid long distance	Forget to add water; overfill
2	Switch kettle on	Plugged in, press switch	Indicator light, sound	Power supply, design of switch	Fail to switch on, switch wrong device
3	Wait for water to boil	Sound of boiling, auto shut off	Steam, light turns off	Time pressure, noise	Kettle does not switch off when boiling; user may remove the kettle too early.
4	Pour water into cup	Kettle off, lid secure	Cup fills	Grip, weight, distractions	Check close distance, avoid long trip with hot water – spill water; scalding
5	Check power off on kettle	Switch off	No light, no connection	Design of power off	Fire, destroyed kettle due to high temperature

KEY ISSUES

- Even simple activities contain *sequential, contingent, and feedback loops* that can be mapped systematically.
- *Cues and feedback* show where the design supports the user (indicator lights, sound, tactile confirmation).
- *Potential errors* illustrate where design improvements could help (auto shut-off, ergonomic handle, fill-line markings).
- The same logic scales directly to complex industrial operations such as loading hydrocarbons or controlling a reactor.

NOTE

1 Recording information about task cues and feedback can be very helpful later in the analysis when performing error analysis and when working with improvement measures as missing task cues and feedback often have negative effect on human performance and system safety.

REFERENCES

Blackett, C., Farbrot, J. E., Øie, S., & Fernander, M. (2022). *The Petro-HRA Guideline*, rev. 1. Retrieved from: https://ife.no/en/project/the-petro-hra-project

Conklin, T. (2019). *5 Principles of Human and Organizational Performance (HOP) with Dr. Todd Conklin*. Vector Solutions. Retrieved from https://www.vectorsolutions.com/resources/blogs/5-principles-of-human-and-organizational-performance-hop-with-dr-todd-conklin

Dekker, S. (2006). *The Field Guide to Understanding 'Human Error'*. CRC Press.

DNV (2023). *Analysis of Pre-Accident Operator Actions*. Retrieved from https://www.dnv.com/Publications/analysis-of-pre-accident-operator-actions-247857

Embrey, D. E. (1986). *SHERPA: A Systematic Human Error Reduction and Prediction Approach*. International Meeting on Advances in Nuclear Power Systems.

Energy Institute (2008). *Guidance on Investigating and Analyzing Human and Organisational Factors Aspects of Incidents and Accidents*. London: Energy Institute.

Energy Institute (2020). *Guidance on Human Factors Safety Critical Task Analysis*. Energy Institute. Retrieved from https://www.hpog.org/assets/documents/Guidance-on-human-factors-safety-critical-task-anlaysis-jk.pdf

Franciosi, C., Di Pasquale, V., Iannone, R., & Miranda, S. (2019). A taxonomy of performance shaping factors for human reliability. *Journal of Industrial Engineering and Management*, 12(1), 115–132. https://doi.org/10.3926/jiem.2702

Hollnagel, E., & Clay-Williams, R. (2022). Work-as-imagined and work-as-done. In *Implementation Science* (1st ed.). Routledge.

Hopkins, A. (2025). *Boeing, the 737 MAX Crisis and Aviation Safety: The Perils of Profit-Driven Engineering*. CRC Press.

IAEA (2013). *Managing Human Performance to Improve Nuclear Facility Operation* (No. NG-T-2.7). Vienna: International Atomic Energy Agency. Retrieved from https://www-pub.iaea.org/MTCD/Publications/PDF/Pub1623_web.pdf

IEC 63303 (2024). *Human–Machine Interfaces for Process Automation Systems*. Geneva, Switzerland: International Electrotechnical Commission (IEC).

IMO (2018). *Revised Guidelines for Formal Safety Assessment (FSA) - For use in the IMO Rule-Making Process*. MSC-MEPC.2/Circ.12/Rev.2. London: International Maritime Organization.

ISO 11064 (2000–2023). *Ergonomic Design of Control Centres (Parts 1–7)*. Geneva, Switzerland: International Organization for Standardization (ISO).

ISO 31000 (2009). *Risk Management - Principles and guidelines*. Geneva, Switzerland: *International Organization for Standardization (ISO)*.

ISO 9241-210 (2019). *Ergonomics of Human-System Interaction – Part 210: Human-Centred Design for Interactive Systems*. Geneva, Switzerland: International Organization for Standardization (ISO).

Kirwan, B. (1994). *A Guide to Practical HRA*. London: Taylor and Francis.

Kirwan, B., & Ainsworth, L. K. (1992). *A Guide to Task Analysis*. London: Taylor & Francis.

NIOSH (2023). *Centers for Disease Control and Prevention*. Retrieved from https://www.cdc.gov/niosh/topics/hierarchy/default.html

NSA (2014). *Barrier Management in Operation for the Rig Industry - Good Practices*. Norwegian Shipowners Association.

Reason, J., & Hobbs, A. (2003). *Managing Maintenance Error: A Practical Guide*. CRC Press.

U.S. Department of Energy (2009a). *Human Performance Improvement Handbook - Volume 1: Concepts and Principles*. Retrieved from https://www.standards.doe.gov/standards-documents/1000/1028-BHdbk-2009-v1/@@images/file

U.S. Department of Energy (2009b). *Human Performance Handbook - Volume 2: Human Performance Tools for Individuals, Work Teams and Management*. Retrieved from https://www.standards.doe.gov/files/doe-hdbk-1028-2009-human-performance-improvement-handbook-volume-2-human-performance-tools-for-individuals-work-teams-and-management

11 Situation Awareness in Aviation

An Ecological Approach to Support Human Operators at Different Levels of Automation

Max Mulder, Clark Borst, and Marinus M. (René) van Paassen

11.1 INTRODUCTION

With the advent of ever more powerful computers and algorithms across various domains, it seems that full automation requiring no human involvement will be achieved soon. For many, the rapid rise of artificial intelligence (AI), primarily machine learning oriented, marks the end of any need for human interaction in controlling systems.

There are at least two arguments against this claim. First, the 'intelligence' of algorithms is debatable. Most progress is made in well-defined problems like playing chess or go, controlling 'closed' systems like home appliances (thermostat, washing machine), and pattern recognition based on statistical inference. In safety-critical domains, such as aerospace or automotive, automation designers face more fundamental challenges. Systems are 'open'; they interact with their environment in extensive, complicated, unpredictable ways. The designer cannot *anticipate*, or even imagine, all possible events in advance. An intelligent agent is needed to cope with unpredictable, unanticipated events. That agent must be able to 'think about', reason and deal with, ideally, all possible *situations* – defined here as nested, convoluted combinations of events and states – often based on noisy, uncertain, incomplete information.

Second, the pace of introducing higher levels of automation (LOA) varies for different systems. In cases where risks are small and safety concerns can be easily contained, such as in the washing machine, automation has been and will continue to be introduced at a rapid pace. For safety-critical domains, however, requirements for mitigating risks are extremely high. Commercial aviation, for example, accepts risk levels lower than 10^{-9}. With advanced, mostly Large Language Models (LLM)-based, AI algorithms reported to be 'hallucinating' significantly, in some benchmarks, in the range of 30–45%, (Huang et al. 2025), it will take considerable time and effort before these algorithms are useful and certifiable in high-risk environments.

For the foreseeable future, the only available intelligent agent that, after education, training, and proper support, is capable of adapting and creatively dealing with unforeseen situations is the human operator. Before systems can work fully autonomously, our main safety net, our last line of defence, will be human operators. The question then is how to support this operator in the best possible way, using all the means provided by current-day automation tools (including AI) and interface technologies. This chapter aims to address this question, starting with defining the main design requirement.

DOI: 10.1201/9781003741824-14

First and foremost, when humans are responsible, the automation and human-machine interface should help the human operator to be *aware* of what is happening in the system under control, and the environment in which it functions. She should also have sufficient means to act on that system in a timely, efficient, and effective manner. Hence, in Section 11.2 of this chapter, we start by discussing the concept of *situation awareness* (SA), its definition(s) and meaning. Whereas many investigations focus on determining how aware operators are of situations, our work focuses on discovering what situations the operator should be aware of in the first place. What *are* situations, exactly? And how to present these situations to the operator in a meaningful, supportive manner? (Flach et al. 2004; van Paassen et al. 2018)

In our research, we adopted the 'ecological' approach to interface design, proposed by Vicente and Rasmussen in the early 1990s (Vicente and Rasmussen 1990, 1992). Section 11.3 briefly discusses the ecological approach to supporting human operators, which combines Gibson's principles of ecological perception, centred around the concepts of affordances and constraints (Gibson 1979), with Rasmussen's methods for analysing human cognition (Rasmussen 1983; Rasmussen et al. 1994).

The remainder of the chapter then describes an example of applying the main principles of ecological interface design (EID) to support operators in aviation. The Conflict Detection and Resolution (CD&R) task will be discussed in air traffic control (ATC). Section 11.4 first describes the current tasks, interface, and automation support of ATC operators. It will also explain why most attempts to introduce higher LOA in this task have failed. Section 11.5 then presents an advanced ecological decision-support tool, the Solution Space Display (SSD). The SSD, driven by clever algorithms, shows air traffic controllers the full 'solution space' of CD&R tasks.

In Section 11.6, we discuss how the SSD can be combined with automation advice. Sheridan and Verplank's (1978) early taxonomy of automation levels will be used to explain how solution space-based decision-support tools can facilitate smooth transitions between various LOA. The chapter ends with a discussion in Section 11.7 of our findings and possible ways forward.

11.2 SITUATION AWARENESS

Designers of automation and the human-machine interface generally focus on achieving four goals. First and foremost, the human must remain in charge and have the means available to do the task at hand in an efficient and effective manner (Billings 1997). Second, the resulting human-automation team should *perform* well, meeting external demands and achieving the main purposes for which the system under control exists (Lee and Seppelt 2009). Third, operator *workload* should be tuned well, not too high to avoid fatigue but also not too low, to prevent boredom and keep the operator engaged, preferably in a meaningful, satisfying way. And fourth, the SA of the operator (and, when working together in a team, also the automation, as we will argue) should be high at all times.

Whereas responsibility, performance, and workload are generally more straightforward to define and measure, the question of what SA exactly means, how it can be measured, and how one can 'design for it' is more difficult to answer. In aviation, SA studies, together with workload studies, have dominated the literature since the mid-1980s, after the rapid advance of automation in aircraft cockpits. A literature survey published only 15 years later (Breton and Rousseau 2001) showed that by that time already more than 27 proposed definitions of SA were published. Clearly, a topic of scientific debate.

The pioneering work of Endsley (1995a, 1995b) led to a three-level SA model – perception, comprehension, projection – resembling information-processing theory, and a first definition of SA: 'the perception of environmental elements and events with respect to time or space, the comprehension of their meaning, and the projection of their status into the future'. We will use this definition here and will elaborate on our interpretation of words like 'elements and events in time or space', 'meaning', and 'projection' using an operational example.

The past 30 years have shown a plethora of studies on SA in many domains. Typical of many of these studies is that they first define what the operator 'should be aware of', then try to experimentally measure the level of that awareness. The underlying idea is that a 'good' interface leads to a 'high' SA (and vice versa), and as a result, various techniques, ranging from questionnaires to neuroscientific tools, to measure SA have been developed. Although it is not our intention to argue against these studies, many of which have indeed been very valuable in the design and evaluation of human-machine systems, we *do* issue a warning about oversimplifying the problem at hand here, that is, being able to actually determine 'how well' human operators, or any intelligent agent, are truly aware of what is happening (Flach et al. 2004).

Consider a typical pilot simulator training scenario, in which the aircraft they control (here, manual control is assumed, using elevator and thrust) approaches the runway, which normally occurs along a three-degree downwards and straight reference path, Figure 11.1.

At some point, a situation is created where the aircraft is flying much higher than the reference approach trajectory ($H > H_{ref}$) with a velocity that is too low ($V < V_{ref}$), situation (a) in Figure 11.1. Then we can ask the pilot what the aircraft's altitude and velocity are, to check whether she is aware of the off-nominal states. 'What is your altitude? What is your speed?' It remains to be seen, however, whether this pilot is aware of the full 'situation'.

Both aircraft states can be brought back to their nominal values, situation (b) in Figure 11.1, by *exchanging* the excess in potential energy (altitude) with the deficit in kinetic energy (speed), through commanding the aircraft to pitch-down using elevator control. In dealing with this situation, it could well be that some pilots first increase engine thrust to accelerate the aircraft to the right speed, then later find out, when using elevator control to expedite the aircraft descent to the reference path, that they are flying much too fast for landing. Then, how 'aware' was the pilot when she correctly identified the off-nominal states and then decided to increase thrust? These states are *coupled*, and only *together* define the actual state of the aircraft in this particular task, which can be formulated as a controlled reduction of the aircraft's *total energy*, while keeping within constraints on kinetic energy (speed) and potential energy (altitude) before touchdown (Amelink et al. 2005). How aware was the pilot of the situation?

So, what is the right question to ask this pilot? What *is* the situation the pilot should be aware of? Clearly, it is not just altitude and speed of the aircraft. While these are indeed essential states for pilots to monitor throughout the flight, especially when flying close to the ground, their true *meaning* depends on the task context. Answering the question '*what* question(s) to ask operators to test their situation awareness?' can sometimes be relatively easy but in operationally relevant cases often quickly becomes rather complicated. Especially in open systems, where many of the system's interactions with the environment are difficult to understand, model, control, and predict beforehand.

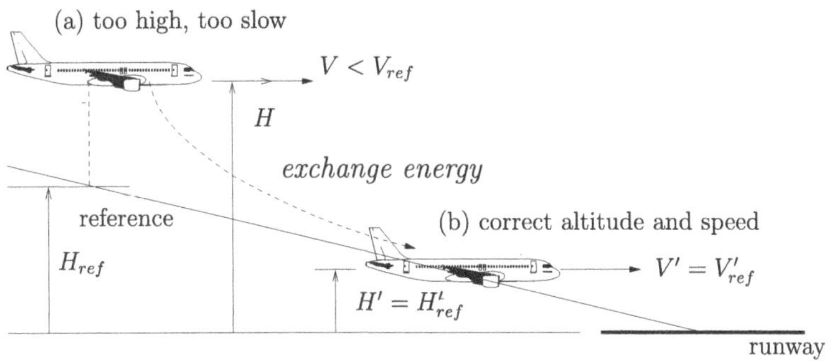

FIGURE 11.1 Aircraft approaching the runway: (a) Flying too high ($H > H_{ref}$) and too slow ($V < V_{ref}$), then exchanging energy to arrive on the trajectory with correct altitude and velocity. (b) Figure is not to scale.

The question of what the operator should be aware of may therefore be more important to answer than the question of what the operator is aware of (Flach et al. 2004). It precedes the latter question. Hence, much of the work in our group focuses on investigating 'situations', that is, what should operators be aware of? Rather than attempting to look 'inside' the head of the operator, to see what she is aware of, our research focuses on the *environment*, inside of which the operator must perform her task. What is the operator 'inside of'? What does the environment and system under control allow the operator to do to fulfil her goals? These questions are reminiscent of James Gibson's ecological approach to understanding human perception and action, centred around the concept of *affordances* (Gibson 1986, 1979).

In the next sections, we will elaborate on our work, which ultimately aims to design *ecological* interfaces to support human operator decision-making and control, mostly in aviation.

11.3 ECOLOGICAL INTERFACE DESIGN

EID is a framework to design the human-machine interface and accompanying automation. It emerged in the late 1980s when Vicente and Rasmussen investigated how interfaces can be created for complex, open work domains, such that the operator is supported at all three levels of Rasmussen's 'skills, rules, knowledge' (SRK) taxonomy of human cognitive control (Rasmussen 1983; Vicente and Rasmussen 1990, 1992). They argued that whereas humans can easily explore and exploit the (patterns in the) *natural* environment with their powerful, versatile perceptual mechanisms, for operators of complex (industrial) systems, like nuclear power plants, this is not possible.

The long response times, the many subsystems, and their intricate couplings and interconnections make complex systems react more slowly and much more difficult to predict than easier systems or tools. In addition, the aforementioned 'open-ness' of these systems means that not all interactions with the environment, all situations and events, can be accounted for beforehand. It is the operator who must have an accurate knowledge of this system, its structure, functioning and interactions with the environment, supported by comprehensive information, enabling her to act on signals, signs, and symbols at all three SRK levels. For an extensive explanation of the EID approach to interface design, its working principles and analysis techniques, the reader is referred to Vicente's book *Cognitive Work Analysis* (1999). We will briefly summarise the main thoughts and principles here.

According to Gibson, in his ecological approach to human visual perception, perception and action are coupled. When acting in the environment, humans can 'directly perceive' what (elements in) the environment allow them to do, or not, to fulfil their goals, the concept of 'affordances'. Affordances are the possibilities and constraints for actions and achieving goals, specified by the environment. Consider for example a hiker who walks through a hilly environment with some larger and smaller trees. Depending on the situation, a large tree allows a lost traveller to climb into it and serve as a lookout for familiar places or paths. For a cold traveller, the smaller trees' branches can be used to make a fire. If the weather turns bad, the branches and leaves can serve as shelter. Clearly, trees have different meanings for the hiker, depending on what she needs, and these are all specified by the environment. Some trees are too high, though, and are useless. Hence, there are constraints to what trees 'afford' the traveller.

Vicente and Rasmussen took this as a starting point for EID. Basically, in EID the designer should aim to transfer the cognitive process of understanding and interacting with the complex system to a *perceptual* process. Since the human-machine interface is the medium for interacting with the system, that interface should reveal and show the structure of the work domain in such a way that it is more compatible with human natural perception. Ideally, this structure is mapped onto the geometry of the interface in such a way that it can be directly *perceived*. In short, 'to make visible the invisible'. The interface acts as an external mental model and should support the operator on all three levels of cognitive control.

Rather than showing the operator what to do in a particular situation, a typical ecological interface shows the *constraints for action*. This leaves operators with the flexibility to act (and adapt) within the remaining space of action possibilities. In doing so, ideally, ecological interfaces show operators what is happening, what their options for actions are, and what the consequences of an action could be (what if?).

Identifying the constraints for action often requires rigorous analysis of the work domain, that is, the system to be controlled, its environment, and potential interactions. This Work Domain Analysis (WDA) is conducted on multiple levels of functional abstraction; see Vicente (1999) for more details. Ultimately, the goals at the highest level, typically the reason for the system to exist in the first place (its functional purpose), the means to achieve these goals, typically the action possibilities for the operators, and the way these means and goals are related, the so-called means-ends relationships, should all be directly visible from the display.

One of the benefits of performing this functional analysis is that, in doing so, the interface and automation designers become *experts* in understanding the system at hand. Being able to define the constraints of the system and its means-ends relationships also means that a route towards automation is opened-up. The constraints and means-ends relationships are typically *actor-independent*; these not only hold for the *human* operator, but for *any* other intelligent, automated agent. The analysis focuses on the system and the environment in which it needs to function, the constraints and (often physical) laws that underlie the system's dynamic behaviour – in Gibson's terms: what the environment affords – and what control actions are possible. Creating an appropriate (visual) representation of that 'universal action space' yields a representation that is *independent of which agent, human or automation, is controlling the system* (Van Paassen et al. 2013). Human and automation agents *can work with the same representation*, paving the way for human-automation teaming.

To illustrate these points in a concrete and (hopefully) understandable way, the next three sections will discuss an example, in ATC, a safety-critical domain where human operators have been, and will continue to be for a long time, the ultimate decision maker.

11.4 AIR TRAFFIC CONTROL, AN INTRODUCTION

The example domain chosen in this chapter to explain our approach to interface design is ATC. This is a complex socio-technical system, and for the sake of illustrating our thoughts, we describe it in highly simplified terms. Generally speaking, while flying en-route from A to B at high altitudes, aircraft cross airspace *sectors*, controlled by (teams of) Air Traffic Control Officers (ATCOs), responsible for managing a safe and efficient operation within their sector. In this section, we will introduce the main en-route ATCO tasks, the current interface and automation tools.

11.4.1 Conflict Detection and Resolution

We focus on the task of guaranteeing the safety of air traffic within a sector. Aircraft are not allowed to fly too close to other aircraft; they need to be *separated* either horizontally (by at least 5 Nautical Miles (NM)) or vertically (by at least 1,000 feet). These conflict boundaries can be interpreted as each aircraft surrounded by an 'ice-hockey puck', a flat circular disk. When the separation limits are crossed, a Loss-of-Separation (LOS) event happens, which should be avoided. In ATC, a *conflict* is defined as a potential future LOS. One of the main tasks of an ATCO is to avoid any conflicts, a task often assisted by automation. To maintain safety, the ATCO must *detect* potential conflicts and *resolve* them on time, a task known as Conflict Detection and Resolution (CD&R) (Ribeiro et al. 2020).

11.4.2 The ATCO Interface

A typical ATCO working position is illustrated in Figure 11.2.

The main visual interface is the electronic radar display known as the PVD. It shows the sector from above, with the sector boundaries and sector entry- and exit-waypoints (e.g., 'COOKY'), and

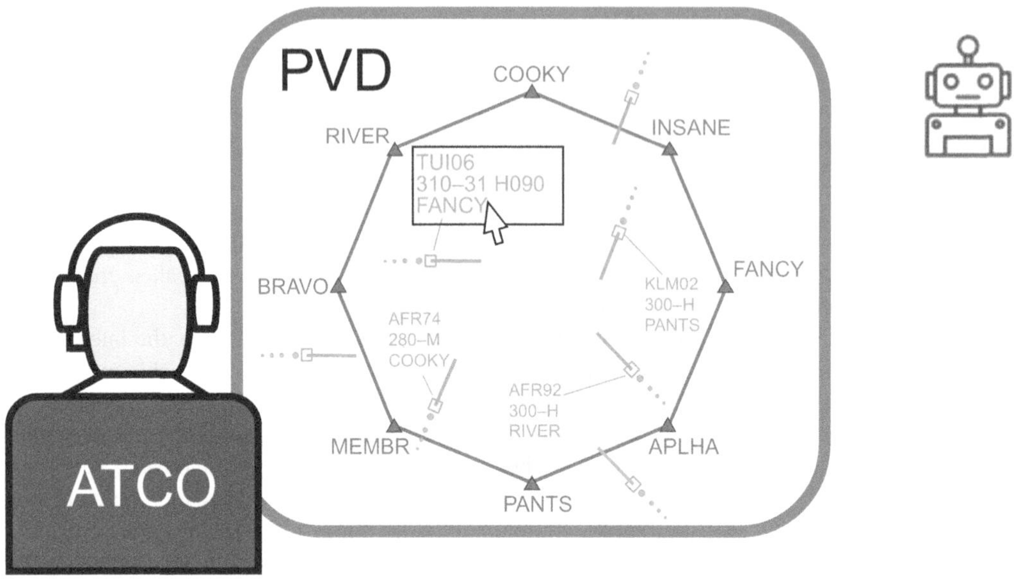

FIGURE 11.2 ATCO working on her sector, shown on the Plan View Display (PVD) (not to scale, for illustration purposes only). Automation is represented by the robot icon (top right).

also the aircraft flying in or close to the sector under control. Using some device, like a computer mouse or trackball, the ATCO can click on an aircraft label; in Figure 11.2, flight 'TUI06' has been clicked upon. The label shows information about the selected aircraft, such as its callsign 'TUI06', weight class (here: 'H' from 'heavy'), its flight level (here: '310', this aircraft flies at an altitude of 31,000 feet), its desired exit waypoint (here: 'FANCY', at the eastern sector boundary). This particular aircraft has received a heading command (heading 090, i.e., 'fly east') and will not exactly pass its exit waypoint without further ATCO action.

ATCOs communicate with aircraft pilots through radiotelephony, that is, voice. Since the ATCO is responsible for safety, their commands *must* be followed up by the pilot. The ATCO works on one aircraft at a time. The ATCO can command pilots to change the aircraft's flight in numerous ways. Considering en-route air traffic only, we limit the discussion to the ATCO issuing heading (HDG) and speed (SPD) commands.

11.4.3 Automation Safety Nets

Several safety nets are in place to assist the ATCO in detecting conflicts. Only one will be discussed here, the Short-Term Conflict Alert (STCA). While the ATCO is managing the traffic in the sector, computer algorithms are continuously checking, based on trajectory predictions, whether conflicts may occur in the near future. When the automation detects a conflict to occur within the next 2 minutes, it issues an alert to the ATCO, the STCA.

This is illustrated in Figure 11.3, where the automation has detected a conflict between aircraft 'KLM02' and 'AFR92', requiring immediate attention of the ATCO. Note that in real-life operations, ATCOs consider an STCA alert issued in their sector a sign of poor performance.

The internal workings of conflict alerting algorithms will become clear in the next section, where we introduce the SSD. For now, it is important to note that an analysis of STCA alerts issued in everyday operations showed that close to 50% of these alerts were *caused by operator's responses to previous alerts* (Lillo et al. 2009). That is, with the current radar-like electronic interface, when air traffic controllers 'resolve' a conflict, their resolution actions often trigger new conflicts later.

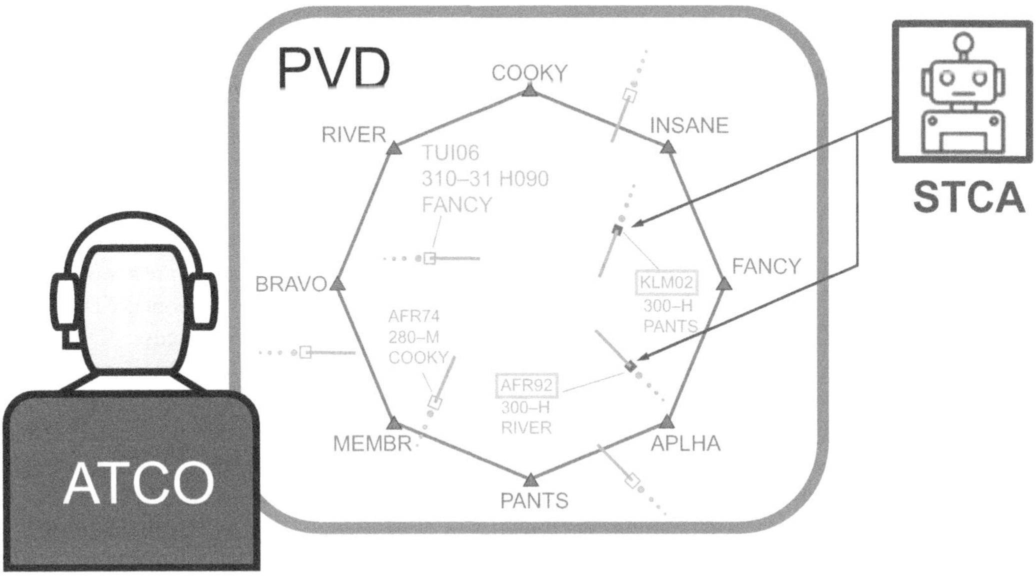

FIGURE 11.3 An STCA alert is issued by automation, a conflict is detected, and the ATCO needs to resolve it.

An inefficient and ineffective process where much can be gained using, as we will see, relatively minor changes to the current interface.

In addition, automation is currently limited to detecting conflicts, not resolving them. Decades of research have resulted in many algorithms also capable of conflict resolution, but these have rarely been used in real operations. Issues with ATCO trust and acceptance of these tools have prohibited their widespread use (Westin et al. 2016). Imagine a situation, illustrated in Figure 11.4, where automation would advise the ATCO to select aircraft 'AFR92' and change its heading to, for example, 270 degrees.

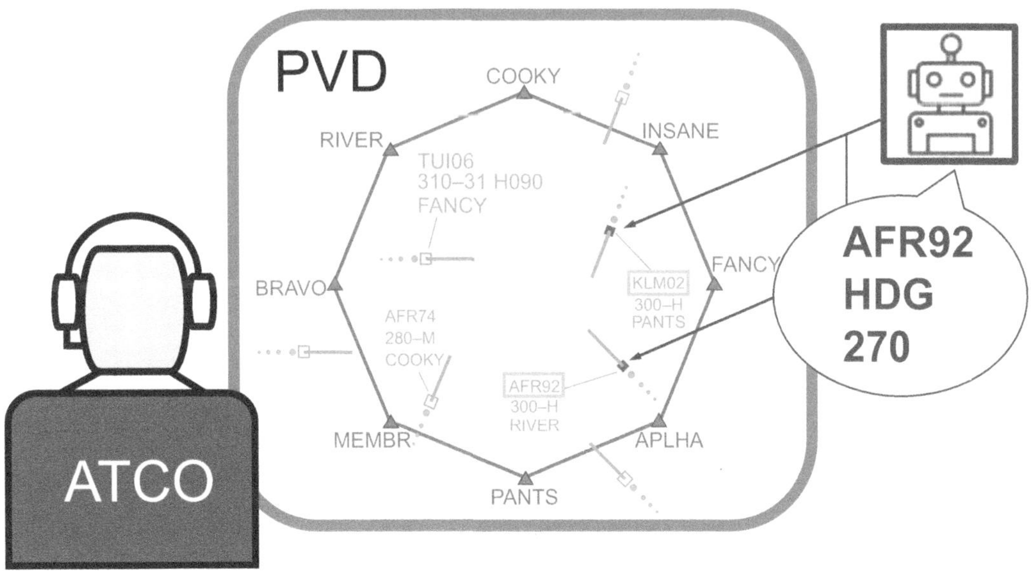

FIGURE 11.4 Automation advises the ATCO to resolve the conflict through issuing an HDG command to 'AFR92'.

Then, *how* does the ATCO know whether the automation is correct? How is the conflict resolved; for instance, which aircraft will pass in front of the other aircraft? Does the advice align with the ATCO's own plans for the sector? Why not work with the other aircraft, 'KLM02', and change, for example, the velocity of this aircraft to resolve the conflict? Answering these questions is not easy, yet again, much can be gained through minor changes to the interface design.

11.5 THE SOLUTION SPACE DISPLAY

In this section we introduce the SSD, its background, how it works, and what it shows, first for two aircraft conflicts, then for multi-aircraft conflicts. The discussion focuses on horizontal CD&R, that is, two-dimensional (2D) conflicts. At the end of the section, we briefly discuss 3D and 4D prototypes. We focus on *tactical* CD&R, that is, working on short-term conflict (few minutes) resolution, as opposed to strategic CD&R, which treats the problem on a longer timescale (hours), often involving trajectory optimisation. In addition, only *state-based* CD&R is discussed, where aircraft are all assumed to fly along a straight trajectory and can turn instantaneously. Note that the SSD visualisation is not constrained by any of these assumptions; these are made here only to keep the discussion and explanation simple.

11.5.1 CD&R IN THE HORIZONTAL PLANE

Consider two aircraft on crossing trajectories, Figure 11.5. Without loss of generalisation, we consider the conflict geometry from the perspective of the ATCO working with one aircraft, called 'own-ship', and consider the other aircraft the 'intruder'. Both aircraft are assumed to have fixed velocities and headings; wind effects are neglected, meaning that both aircraft fly along straight trajectories.

Suppose we 'bring the intruder aircraft to a stand-still', then we need to subtract the velocities of *both* aircraft with intruder's velocity vector V_{int}. Own-ship (*Own*) then travels with the *relative* velocity V_{rel} towards the intruder (*Int*), along the relative path illustrated in Figure 11.6. We draw a circular 'protected zone', with the stationary intruder aircraft position as its centre and radius R_{PZ} equal to the horizontal separation boundary of 5 NM, the 'ice-hockey puck' introduced before.

In this situation, the relative path crosses the circle, and a loss of separation would occur without any further action, a conflict. It starts at position *a* and ends at position *b* in Figure 11.6. The smallest

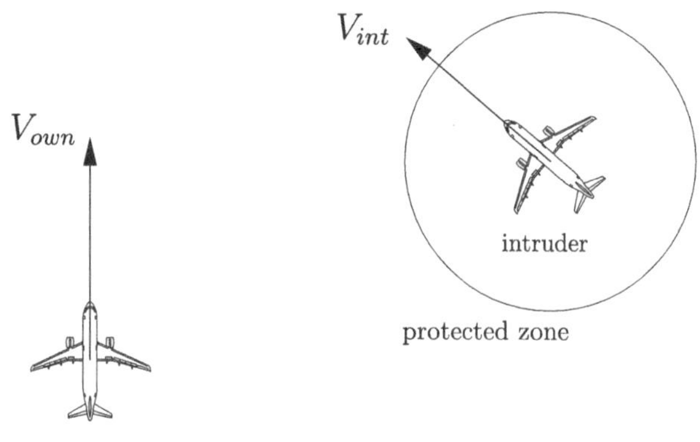

FIGURE 11.5 Bird's-eye view of two aircraft flying at the same altitude, with the own-ship aircraft bottom left and the intruder aircraft top right, with the 5NM circular protected zone. The velocities and headings of both aircraft are illustrated with vectors. Figure is not to scale.

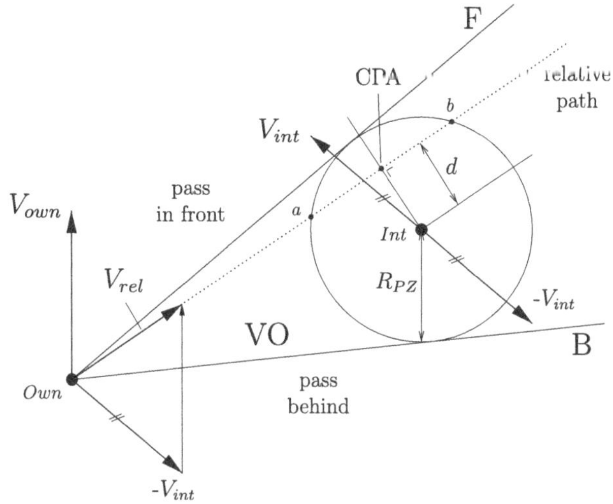

FIGURE 11.6 Conflict geometry variables.

distance d between the two aircraft is called the Closest Point of Approach (CPA). When $d < R_{PZ}$, a LOS occurs. This is how CD&R algorithms generally operate (Ribeiro et al. 2020).

The ATCO can resolve the conflict by changing the relative velocity vector, such that it does not lie in-between the two lines tangent to the separation circle, which form a triangular shape, with its origin located at the position of own-ship. This triangle is referred to as the Velocity Obstacle (VO) (Degré and Lefèvre 1981; Tychonievich et al. 1989; Fiorini and Shiller 1998; Chakravarthy and Ghose 1998). The VO represents all combinations of own-ship speed and direction that would result in a conflict, while velocities outside the VO ensure safe separation. When changing the relative velocity such that it points to the left of line F in Figure 11.6, the own-ship will pass *in front* of the other aircraft. When changing the relative velocity such that it points to the right of line B in Figure 11.6, the own-ship will pass *behind* the other aircraft.

Changing the relative velocity of the aircraft pair forms the basis of resolving most, if not all, conflicts in the horizontal plane. However, it is not a very convenient variable for the ATCO to change. After all, the ATCO can only issue *absolute* velocity commands to *either* the own-ship *or* the intruder aircraft. As here the ATCO is assumed to be interacting with own-ship; a clever way to show how the *absolute* velocity of own-ship V_{own} can be changed to resolve the conflict is through translating the VO's origin with the velocity of the intruder aircraft, resulting in Figure 11.7 (Van Dam et al. 2008). In line with the main tenets of EID, we have 'made visible the invisible', as the ATCO can now *see at a glance* four crucial state variables.

First, she can observe the relative velocity through the line connecting the VO origin (the 'tip of the triangle') with the tip of own-ship velocity vector. Second, conflict detection comes down to checking whether the tip of the own-ship velocity vector lies within the VO (= conflict), or not (= no conflict). Third, she can also see *how* the velocity of own-ship (the 'tip of the velocity vector') can be changed to bring the relative velocity outside of the VO. Fourth, she can see the consequences of changing the velocity of own-ship for the path of own-ship relative to the intruder; that is, will it pass in front or behind the other aircraft?

Then, how to resolve the conflict in our example? As Figure 11.7 illustrates, one possibility would be to either increase or significantly decrease the *velocity* of own-ship V_{own}. Aircraft velocities, however, are constrained as well. This particular aircraft cannot fly faster than V_{max}; it cannot fly slower than V_{min}. These limits, depending on the type of aircraft and the altitude at which it flies, are shown using the two concentric circles. The tip of the own-ship velocity vector should always stay within the two circles as it then satisfies the velocity constraint.

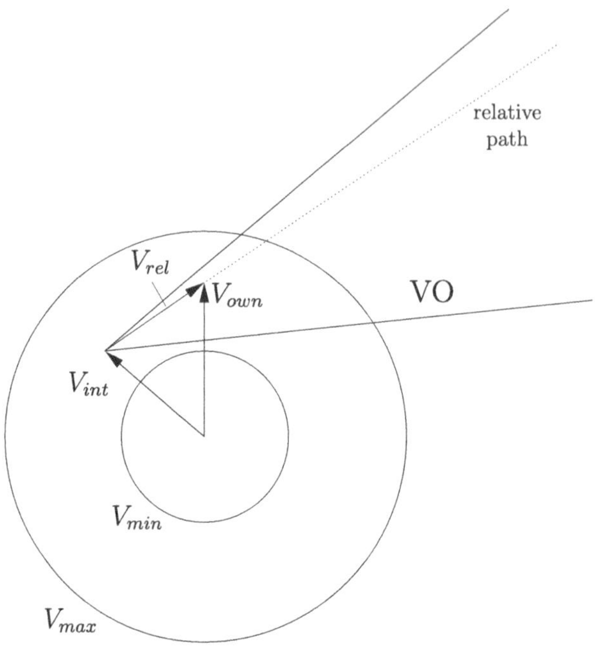

FIGURE 11.7 Moving the velocity obstacle (VO) to see the conflict geometry from the perspective of own-ship.

Another possibility would be to change the direction of the own-ship velocity, that is, changing the own-ship *heading*. Again, as long as the tip of the own-ship velocity vector remains within the two circles (satisfying the velocity constraints) and moves outside of the VO (avoiding a conflict with the intruder aircraft), the ATCO can issue any heading command. Changing the own-ship velocity vector such that it becomes similar to the intruder aircraft velocity (in Figure 11.7, turning the own-ship to the left 45 degrees and decelerating it slightly) would mean that the relative velocity vector becomes very small. Both aircraft will then fly *parallel* to each other, and resolving the conflict will take a very long time. Hence, these 'solutions' are to be avoided.

From Figure 11.7, it is a small step to the SSD, Figure 11.8.

When an ATCO selects an aircraft, the automation gathers all information to draw the SSD, which then 'pops up' on the PVD, to show all constraints (own-ship velocity, avoiding conflicts with other aircraft) the ATCO should adhere to when resolving a conflict, or when there is another need to change the own-ship aircraft trajectory.

In the following sub-sections, we will explain a bit more about how the SSD works. But before we do so, let it be clear that it is the ATCO, and *not* automation, who selects an aircraft to work on. In this simple example with just two aircraft, the ATCO could also have selected the *intruder* aircraft to work on. Then automation draws the SSD from the perspective of the intruder aircraft, which is then the own-ship.

This is illustrated in Figure 11.9, showing some of the 'symmetry' characteristics of the conflict geometry (Van Dam et al. 2008).

11.5.2 Showing the SSD on the ATCO Interface

Before explaining some of the properties of the SSD in more detail, it will first be shown how the interface can be integrated in the ATCO working position. Figure 11.10 returns to the situation depicted in Figure 11.3, where the automation issued an STCA alert. The ATCO has reacted to this

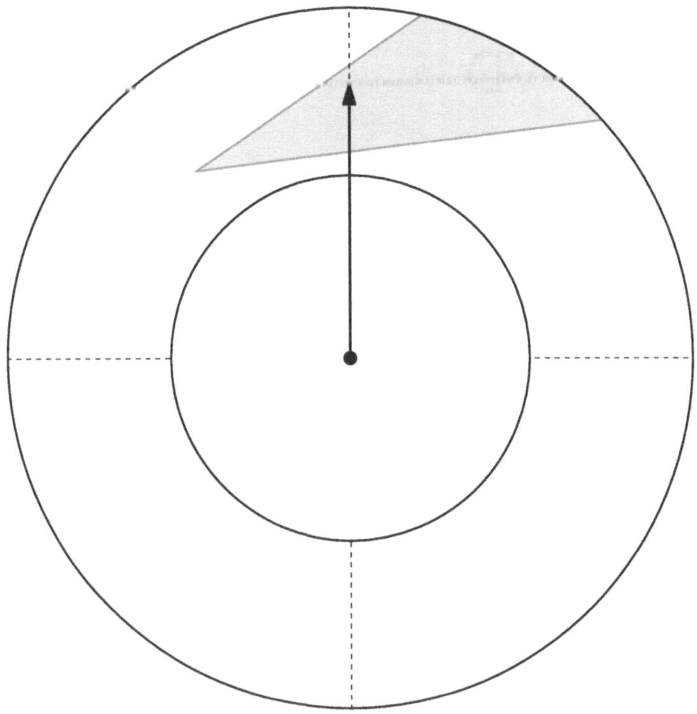

FIGURE 11.8 The Solution Space Display (SSD) for the conflict example in the text.

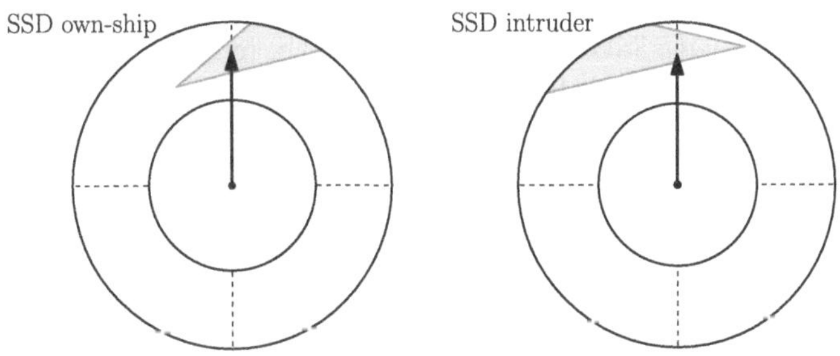

FIGURE 11.9 The SSD when selecting the own-ship (left), and the SSD when selecting the intruder aircraft (right), for the conflict example in the text. Note the symmetry.

alert by clicking on one of the aircraft involved, 'AFR92', and selected a 'show SSD' option in its label.

Then automation, based on the information available on the conflict (the aircraft positions, velocities, headings), visualises the conflict through the SSD, such that the ATCO can 'see' the situation and act on it. Here, the conflict can be resolved by turning 'AFR92' to the west, for example, by issuing command 'fly HDG 270'. The ATCO can monitor the consequences of giving this command through the SSD, or on the PVD, when having disengaged the SSD presentation. Note that Figure 11.10 shows the SSD in a cartoon-like way, obscuring and cluttering the PVD considerably. Obviously, these are to be avoided in the real implementation of the SSD.

When the aircraft crew followed up on this command, as they should, at some point, the aircraft 'AFR92' will need to be re-directed again to fly to its exit waypoint 'RIVER', which is at the ATCO's discretion. The SSD can also be used to time this manoeuvre (not shown).

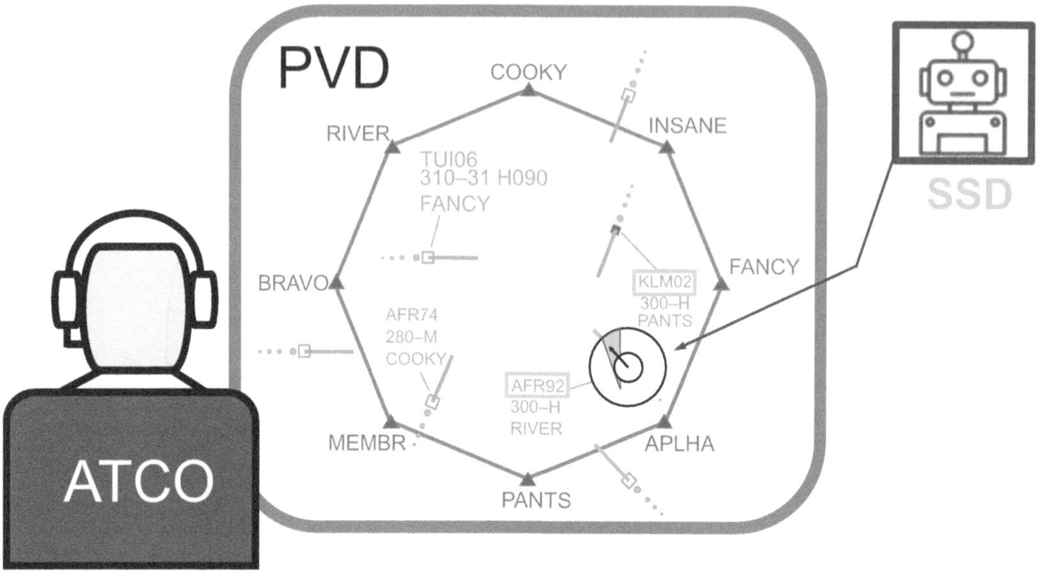

FIGURE 11.10 Showing the SSD on the PVD (not to scale, for illustration purposes only).

11.5.3 Resolving Conflicts with the SSD

From the above, it is clear that there are infinitely many ways to resolve conflicts. This section will first explain how the SSD shows some common solutions (heading command, speed command, combined heading, and speed command) and how the 'best' resolution can be directly observed from the SSD. We will continue to work with the conflict geometry defined in Section 11.5.1. To reduce paper length, we only show the top-half of the SSD. After all, ATCOs will be reluctant to 'send an aircraft back', that is, turn it more than ± 90 degrees. We then briefly explain multi-aircraft conflict resolution situations.

11.5.3.1 Heading Resolutions

The SSD directly shows all possible heading resolutions. In Figure 11.11a, the (red) shaded area indicates that, given the current velocity of own-ship, headings between *H1* and *H2* are blocked, as they would lead to a conflict with the intruder.

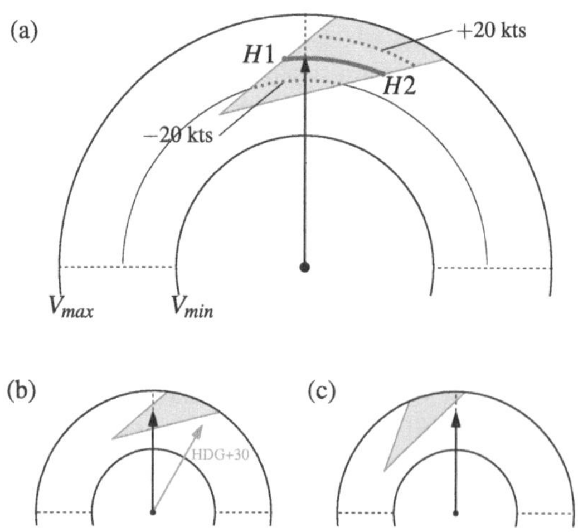

FIGURE 11.11 Resolving the conflict using heading command. (a) The SSD shows blocked heading range (*H1–H2*) where loss of separation occurs for the current speed. (b) Turning own-ship right 30 deg. (c) Conflict resolved.

To safely avoid the conflict, the own-ship should either: turn left to a heading < *H1*, passing in front of the intruder, or turn right to a heading > *H2*, passing behind the intruder. The size of the blocked heading band (*H1–H2*) *changes* depending on the aircraft's velocity, shown by the dashed semi-circles for ±20 knots adjustments. The VO can be considered a 'family of semi-circles', showing all speed-heading combinations that lead to a conflict with the other aircraft. An ATCO familiar with the SSD will immediately *see* that first commanding the aircraft to slow down will, after it has decelerated, result in a situation where a smaller heading change can resolve the conflict.

Although any heading outside the blocked area resolves the conflict, controllers generally prefer resolutions that minimise disruption to the aircraft flight path and minimise changes in the flow of aircraft in their sector. Figure 11.11b shows that this ATCO has chosen a different resolution, the (blue) vector, the own-ship is turned 30 degrees to the right and will pass behind the other aircraft. Figure 11.11c shows the SSD after the own-ship has completed its turn manoeuvre.

11.5.3.2 Speed Resolutions

The SSD also shows all possible speed resolutions. In Figure 11.12a, the (red) shaded area indicates that, given the current heading of own-ship, all speeds between *S1* and *S2* are blocked, as they would lead to a conflict with the intruder. To safely avoid the conflict, the own-ship should either: slow down to a speed < *S1*, passing behind the intruder, or accelerate to a speed > *S2*, passing in front of the intruder. The blocked speed-band (*S1–S2*) changes when changing the aircraft heading, illustrated with the +15 deg and -15 deg line segments. One could see the VO as a 'family of line-segments', showing all speed-heading combinations that lead to a conflict with the other aircraft. An ATCO familiar with the SSD will recognise that first commanding the aircraft to turn left will, after some time, result in a situation where a smaller speed change is required to resolve the conflict. However, this manoeuvre has a trade-off as the relative velocity (visible on the SSD as the difference between the tip of the VO triangle and the tip of the own-ship velocity vector) decreases, meaning that the own-ship will fly more parallel to the intruder, and it will take longer to return to its intended flight path.

Figure 11.12b shows the resolution that this ATCO has chosen, the (blue) vector; the own-ship will decelerate with 20 knots and will pass behind the other aircraft. Figure 11.12c shows the SSD after the own-ship has completed its deceleration.

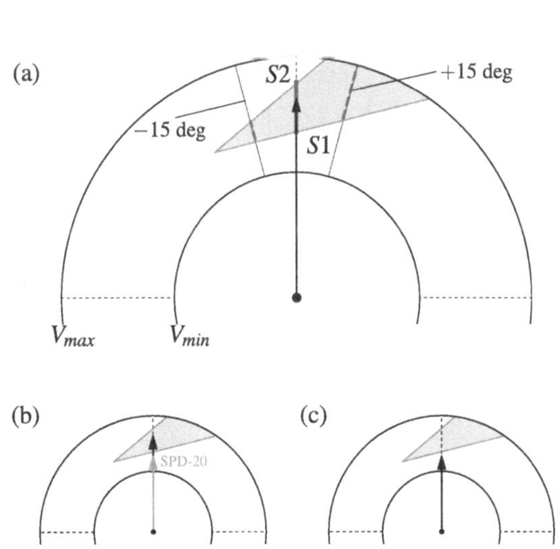

FIGURE 11.12 The same conflict situation as in Figure 11.11, now resolving the conflict using a speed command. (a) The SSD shows blocked speed range (*S1–S2*) where loss of separation occurs for the current heading. (b) Reducing own-ship speed 20 kts. (c) Conflict resolved. (Colors on WEB version)

11.5.3.3 Combined Heading and Speed Resolutions, Optimal Resolution

The example conflict can be resolved by either changing the own-ship heading, its speed, or a combination of both. ATCOs can issue multiple commands in one go, making combined heading and speed commands a feasible option. This raises the question of whether an *optimal* resolution exists. This question is not easy to answer, however, as aircraft operations are highly optimised to save fuel, increase comfort, and maintain a certain flight plan.

From a purely geometrical standpoint, the optimal conflict resolution involves changing the own-ship heading and speed in the smallest possible way that still avoids the conflict. This minimises disruption to both the own-ship flight path and the overall air traffic flow in the sector. The SSD provides an immediate visual indication of this optimal manoeuvre, as illustrated in Figure 11.13. Basically, moving the tip of the own-ship velocity vector along a line *perpendicular* to the nearest VO 'leg' (here, pass in front (leg F), i.e., increase speed with 7 kts, turn left 3 degrees) is the geometrically optimal resolution (Van Dam et al. 2008).

Figure 11.13b shows the resolution that this ATCO has chosen, the (blue) vector, the own-ship will accelerate with 10 knots, turn left 5 degrees, and will pass *in front* of the other aircraft. Figure 11.13c shows the SSD after the own-ship has completed its turn manoeuvre and is flying at its commanded increased velocity.

11.5.4 Conflicts with Two or More Aircraft

Potential conflicts with two or more aircraft are easily shown on the SSD. For all aircraft flying in the vicinity of the own-ship, a velocity obstacle is included. Figure 11.14 shows the example conflict again, now with two additional intruder aircraft (intruders 2 and 3) positioned to the left and behind the own-ship, respectively.

Both aircraft fly faster than own-ship, as the tips of their VO triangles are closer to the maximum speed of own-ship than the current own-ship velocity. Although the current path of own-ship only leads to a conflict with intruder 1, resolving this conflict now leaves the ATCO with fewer options, as she must also take the constraints on own-ship heading and speed imposed by intruders 2 and 3 into account. Basically, each nearby aircraft blocks a part of the available heading-speed commands that can be issued to own-ship.

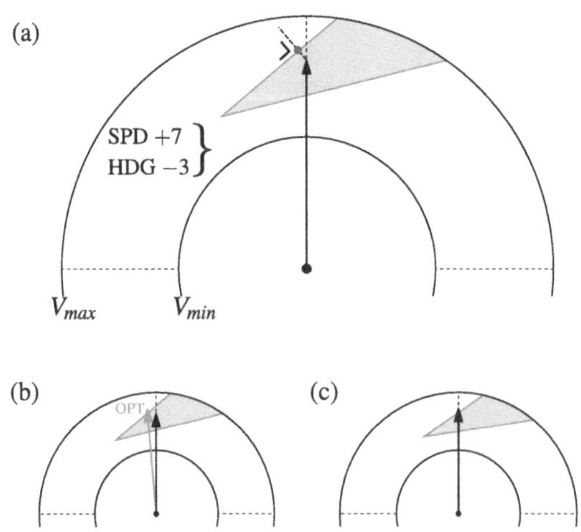

FIGURE 11.13 The same conflict situation as in Figure 11.11, now resolving the conflict using a combined heading and speed command. (a) Optimal resolution (OPT) of the conflict, a combination of a heading (turn own-ship left 3 deg) and speed (increase own-ship speed 7 kts) command. (b) Resolution selected. (c) Conflict resolved (Colors in WEB version of the book)

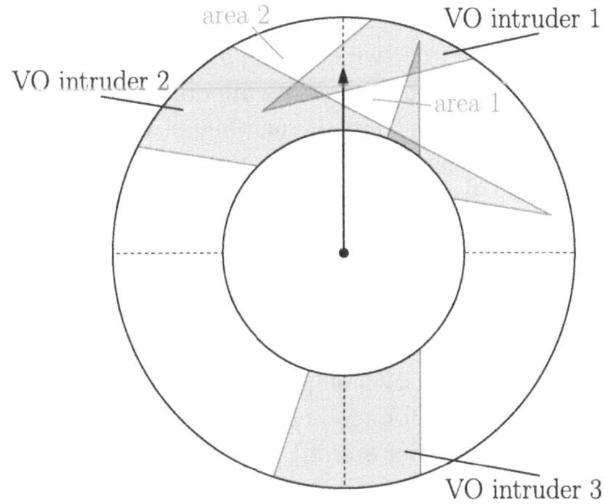

FIGURE 11.14 The SSD with three 'intruder' aircraft flying near the own-ship, showing their velocity obstacles and two regions (area 1, 2) where the own-ship velocity vector can be moved to resolve the example conflict.

Out of infinitely many options, the ATCO could in this case choose to decelerate own-ship and turn it right (area 1). Then the conflict with intruder 1 is resolved, the own-ship will pass this aircraft from behind. Intruders 2 and 3, flying faster than own-ship, will pass own-ship from behind. The ATCO could also choose to accelerate own-ship and turn it left (area 2, closer to the optimal solution discussed above). Then own-ship will pass intruder 1 in front, intruders 2 and 3 will pass own-ship from behind.

11.5.5 More Realistic Cases, 2D, 3D, and 4D Solution Spaces

The SSD discussed in this chapter was the first iteration of the 'solution space in ATC' concept in our group, originally developed to support self-separation by pilots (Van Dam et al. 2008; Heylen et al. 2008; Ellerbroek et al. 2011, 2013a, 2013b). Since then, many more iterations have been made, leaving the state-based assumption to include full future trajectory (intent) constraints and state-transition dynamics (Mercado-Velasco et al. 2015), including the altitude dimension (Lodder et al. 2011), merging tasks (Mercado-Velasco et al. 2021), and four-dimensional aircraft trajectory control (Klomp et al. 2016). Effects of sensor faults on working with the SSD have been analysed (Borst et al. 2017), as well as its integration with AI-based automated advisory systems (Van Rooijen et al. 2020).

11.5.6 Design for Situation Awareness?

Before starting our discussion of whether and how solution-space-based interfaces can help to facilitate introducing higher LOA, let's first return to the question raised in Section 11.2 of whether we succeeded to 'design for situation awareness'. Perhaps not surprisingly, we are inclined to answer this question affirmatively. We argue that the SSD contributes meaningfully to SA. By providing an immediate visualisation of the own-ship's position and motion relative to nearby aircraft, the SSD offers operators a clear view of:

- the task goals (e.g., maintaining safety by efficiently detecting and resolving conflicts),
- the available means (e.g., heading and speed adjustments of own-ship),
- the constraints imposed by the environment (here: other aircraft), and
- the means-end relationships that link actions (speed and heading commands) to their consequences (collision avoidance, resolution manoeuvre, flow of air traffic).

The SSD shows the 'situation' of the own-ship position and motion relative to all aircraft flying in its vicinity at a glance. It provides a true 'window' on all the goals and the means to achieve these goals while satisfying all relevant constraints. Most importantly, it shows the means-end relationships ('what can we do about a situation?') in an unambiguous, *visual* way. The SSD makes visible the invisible. The 'elements of time and space', their 'meaning', and their 'projection' are captured all together in one interface.

With the SSD (and its later iterations mentioned in Section 11.5.5), an ATCO can 'directly perceive' what the aircraft-under-control and its environment afford her to do. Once the SSD is shown, an ATCO experienced with working with the SSD can directly see whether there is a conflict or not (detection), whether that conflict is imminent or not (time, distance), what the options are to resolve the conflict (resolution, change own-ship heading, velocity, or both), and what the consequences will be of selecting one option for action (e.g., the own-ship will turn right and pass behind intruder aircraft 1 and in front of intruder aircraft 2), even in complicated multi-aircraft situations which are very difficult to solve without support. As evidenced by the finding that ATCOs are currently resolving conflicts, of which 50% have resulted from *their own actions* in the past (Lillo et al. 2009).

The SSD presents all possible control actions to the operator that can be issued to own-ship, given the task-relevant constraints caused by own-ship (e.g., acceptable velocities, desired trajectory, exit waypoint) and other aircraft flying nearby (prevent collisions). The reader should note that these constraints *are valid for any agent* working on own-ship, human, or automation. Hence, it is a small step to design an interface that can show automation advice (or even fully automated control actions) to the operator in such a way that she can rapidly 'see' what the automation agent is proposing, what the consequences are of adopting/approving the automation advice, such as whether it aligns with the ATCO's own intentions in managing the traffic in her sector. After all, what the automation suggests should also meet the restrictions set by the task and situation; these do not differ from the constraints the ATCO must adhere to herself. Given that the SSD reveals all system constraints and that these constraints are agent-independent, it lends itself well to improve transparency when presenting action suggestions produced by an automated agent (Borst et al. 2015). In the next section, we will briefly explain our view on how to *integrate* automated advice systems, possibly driven by AI, in the SSD.

11.6 WORKING WITH THE SSD ON DIFFERENT LEVELS OF AUTOMATION

Involving automation in the operator's task means that, in some way, the division of roles and responsibilities between human and automation (and their delegation) must be (re-)defined. To serve this purpose, at least 12 (!) 'Levels of Automation' (LOA) taxonomies have been described in the past (Vagia et al. 2016). We will use Sheridan and Verplank's (1978) early LOA taxonomy, as it is relatively simple and served as the basis for several more recent, more extensive taxonomies discussed at the end of this section.

11.6.1 Sheridan and Verplank's LOA Taxonomy

In 1978, Sheridan and Verplank were the first to define a taxonomy of 'Levels of Automation' in human-computer decision-making, focusing on the tele-operation of underwater robots. It is a one-dimensional scale with ten levels, where at higher levels automation does more, and the operator does less. At Level 1, the human tells automation what action to implement, at Levels 2 to 6 the automation suggests action possibilities to the operator, while at Levels 7 to 9 the automation only informs the operator about its decision *after* its action has been implemented. At Level 10, the automation acts fully autonomously. Note that the wording of the taxonomy levels, and even the number of levels, have changed a few times over the years since its earliest publication (Parasuraman et al. 2000; Sheridan 2011; Vagia et al. 2016).

In the following, we will concentrate on discussing the potential of human-automation teamwork in ATC, at Levels 1 to 8, with the SSD as human-machine interface. Levels 9 to 10 imply little or no human involvement and are considered less relevant in the context of the example discussed throughout this chapter. The wording of the LOAs will be adapted to better reflect the terms used in ATC, the possible roles of automation, and the allocation of work between the ATCO and automation agent.

11.6.2 Human-Automation Teamwork in ATC

Before discussing how some of the LOA can be combined with the SSD, let's first briefly discuss the general concept and what the interface may look like, see Figure 11.15. Here, we have the ATCO working on her sector, assuming that she has selected an aircraft to work with (perhaps because of an STCA issued by automation), and the automation agent, apart from constructing and visualising the SSD, includes a *suggestion* for the operator to resolve the conflict. Possible interactions between the ATCO and automation, using the SSD, will be discussed at the lower eight levels of Sheridan and Verplank's LOA taxonomy. We use the example conflict introduced in Section 11.5, with two aircraft in conflict, *after* an STCA was issued by automation. That is, the conflict has been detected by an algorithm working completely autonomously at Level 10 ('Computer does whole job if it decides it should be done, and if so tells human, if it decides he should be told'). We focus on the ATCO task of *resolving* this conflict at various LOA. We first discuss Levels 1 to 4, then Levels 5 and 6, and then 7 and 8.

11.6.3 Working with the SSD at Levels 1 to 4

Level 1 ('The computer offers no assistance: human must take all decisions and actions') is interpreted here such that the automation does not provide any support to the operator for *resolving* the conflict, after issuing an STCA alert. The SSD is *not* shown; the operator must do all the work by herself, without any support from automation. At Level 2 ('The computer offers a complete set of decision/action alternatives'), *only* when the ATCO clicks on the label of one of the aircraft involved

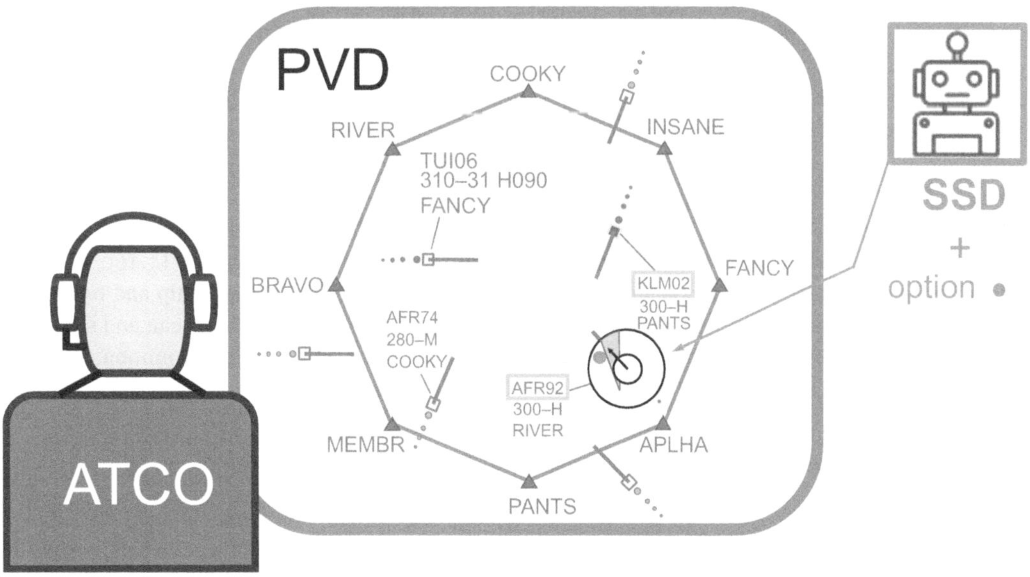

FIGURE 11.15 Showing the SSD on the PVD, with automated resolution suggestion (option, (in blue)) presented.

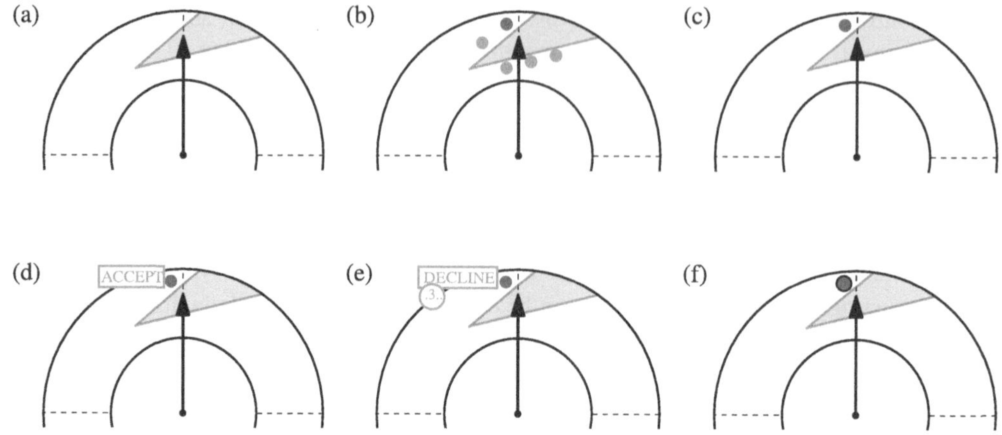

FIGURE 11.16 Showing the SSD at LOA Levels 2 (a) to 7/8 (f) in Sheridan and Verplank's taxonomy. See the main text for explanations.

in the conflict, and requests the SSD to be shown, the automation constructs and shows the SSD, Figure 11.16a. Based on information on the positions, headings, altitudes, and velocities of all aircraft flying nearby, the relevant velocity obstacles are shown. In line with Level 2, the SSD has the unique property (relative to many other CD&R interfaces known to the authors) of showing not just one, but *all* options for resolving the conflict. With the SSD, automation helps determine *all* the options in one go.

At Level 3 ('Computer narrows the selection down to a few, and suggests one, which human need not follow'), the automation selects and shows some possible resolutions ((blue) circles) and one suggestion ((red) circle), on the SSD, see Figure 11.16b. The resolutions could, for instance, include the optimal resolution (OPT, Section 11.5.3) and a resolution that would disturb the traffic flow the least. The suggested resolution could be one that may be preferred by the ATCO, for example, based on automation 'remembering' how that ATCO resolved similar conflicts (in terms of conflict geometry, the traffic density) before (Van Rooijen et al. 2020).

At Level 4 ('Computer suggests one solution, which human may or may not follow'), the automation only shows one suggestion ((red) circle), Figure 11.16c. The ATCO can then follow up on this advice or think of and implement a different action to resolve the conflict.

Clearly, with the SSD, the ATCO can immediately 'see' whether the automation advice provided at Levels 3 and 4 is correct (will it move the tip of the own-ship velocity vector out of the VO?) and also whether the advice makes any sense. For example, in a conflict with a slower-moving intruder aircraft, automation could suggest controlling the own-ship such that it passes that aircraft in front, which is a common ATC strategy often adopted by ATCOs. Indeed, in Figure 11.16c, the (red) circle indicates that the automation suggests that the ATCO accelerate the own-ship and turn it left, such that it will pass the (slower) intruder aircraft in front. These 'ATC practices' can and should be included in the design of decision-support systems, as these often reflect common, 'deeper' relationships in the work domain, in this case, relative motion affordances.

11.6.4 Working with the SSD at Levels 5 and 6

From Levels 5 to 6, the collaboration between human and automation shifts in authority. At Level 5 ('Computer selects that solution and executes it if human approves'), also known as *management by consent*, the operator is still in charge, but the automation is ready to implement the action all by itself, when the operator agrees on the proposed solution and clicks the 'ACCEPT' button, Figure 11.16d. While the operator can safely ignore the automation advice at this Level, this changes at Level 6.

At Level 6 ('Computer selects the solution, allows the human restricted time to veto it before execution'), the automation presents the resolution action. It will proceed with implementing that action unless the operator cancels it within a specified time. This is called *management by exception*. Figure 11.16e illustrates this principle. Automation shows a 'DECLINE' option, as well as a timer ((blue) circle) which shows, for example, how many seconds the ATCO has to decline the proposed solution. When the ATCO does not respond, or not respond in time, automation will execute the proposed solution. Implementing automation to resolve conflicts at Levels 5 and 6 (or higher) has not been very successful in ATC, for many reasons, but primarily because ATCOs do not accept automation to 'interfere too much' with their work (Westin et al. 2016). The high workload associated with mentally 'checking' the automation advice, the (perceived) lack of automation's understanding of the ATCO's plans for the sector, etc., all make introducing higher automation levels difficult. The SSD can play a useful role in this, however, as the ATCO can immediately 'see' what the automation advice 'means', that is, whether it is safe and how the traffic pattern will change, reducing the need for checking the advice given and the workload associated with that.

In addition, perhaps with the help of AI, the suggested resolution could be tailored to the individual ATCO (Hilburn et al. 2014), for familiar conflicts, which then together with the familiar conflict 'symbol' presented by the SSD, that is, the shape of the velocity obstacle triangle corresponding to some typical, well-known, conflict geometry, may lead to rapid understanding and earlier acceptance, reducing workload and increasing SA.

11.6.5 Working with the SSD at Levels 7 or 8

At these higher LOA, the authority of resolving the conflict lies fully with automation, with the only difference in whether automation informs the human or not.

At Level 7 ('Computer selects and executes the solution, and necessarily tells human what it did'), the automation resolves the conflict and, through the SSD, shows the ATCO how it changed the motion of own-ship. Figure 11.16f shows the selected resolution (red circle) and that it has been implemented (black outside edge).

At Level 8 ('Computer selects and executes the solution and tells the human what it did only if asked by the human'), the automation resolves the conflict and, *only* at the request of the ATCO, shows how it changed the motion of the own-ship. Its presentation may look very much like what is suggested in Figure 11.16f.

It goes without saying that these two LOA are very high and, when implemented, automation should be able to deal with many, if not all, situations. Apart from significantly changing the work of ATCOs, in many ways, the chance that automation fails to resolve the conflict should be extremely low. In these high-risk, safety- and time-critical situations, it will be extremely difficult for even a trained human operator to act as a last-resort safety net. Putting the final responsibility for system performance and safety with the human operator would be unfair and should be prevented.

11.6.6 Final Thoughts, Other LOA Taxonomies

We argue that the SSD, providing a 'window' on the world of conflict detection and resolution, can serve to facilitate smooth transitions, from lower to higher LOA levels, and vice versa, in a way that the ATCO will be involved, engaged, and aware of the situation at all times. Instead of the automation merely advising what to do to resolve a conflict, without opportunities for the operator to check the advice, Figure 11.4, combining the advice with the SSD, Figure 11.16, *does* enable for rapid checks of resolution feasibility, correctness, and whether the advice 'fits the plan' of the ATCO. Discussing how the SSD can work together with automation, at various levels, also revealed the difficulty of applying this (and other) LOA taxonomy. And that is that automation has more faces than just giving advice, the 'output' of automation. After all, before being able to construct and show the SSD, other automation subsystems have already conditioned, sampled, filtered, and processed the data from many different sensors. Some other automation subsystem(s) then derive more meaningful information from the data, and again other subsystems show this data on the electronic PVD. These are all working at the 'input' of the human-automation system and are often fully automated.

More recent LOA taxonomies have therefore extended Sheridan's taxonomy to also include the 'input'. The result is again similar to the common four 'information processing' stages of any system, of any human-automation team (Parasuraman et al. 2000), from information selection to processing, to making decisions, to activating the control action. Not surprisingly, this closely resembles Endsley's model for SA which is based on the same universal information-processing scheme.

Clearly then, many of the subsystems, especially at the information selection and processing sides, have often been programmed to work fully autonomously (Levels 8–10), without *any* human intervention, based on clever designs and elaborate failproof algorithms from human-automation designers. It is especially the decision-making and control-activation stages which operate at lower LOA levels, for reasons of safety, trust, risk, all related to the fact that in many open systems automation designers cannot foresee, or even imagine, all possible situations beforehand, and build in a heuristic to automatically deal with these situations. It leaves the final decision, and with that the responsibility for that decision, to the human operator.

11.7 DISCUSSION

In this chapter, we have tried to explain to the reader what our view is on how to design human-machine systems. Our main argument is that, for complex, open, safety-critical systems, automation, and interface designers *cannot foresee and account for every possible situation*. When considered as a distribution of chances that common, less common, and rare events occur, in open systems the tail of this distribution is extremely long and has many (unknown) dimensions. It is inevitable that the automation we build is *incomplete*; we are not able to 'close all loops', therefore, active intelligence is needed in the actual operation to resolve the unforeseen events.

Similar to more conventional types of automation, machine learning-based systems are likely to remain brittle, in some cases inevitably operating at the 'edges' of their data set used for learning. In fact, it is doubtful whether current AI, merely statistical inference systems that are trained to obtain the best possible 'overall' accuracy in detecting patterns, will *over*-emphasise being correct in dealing with the common events, which often occur, *at the cost* of being correct in the less common, rare events where we need 'intelligence' the most. In high-risk, safety-critical systems, there will always be a requirement for including some intelligent agent that can adapt, be creative, or simply can 'muddle through' in an imperfect world (Flach, 2012). We strongly believe that, for a long time to come, the *only* technically viable and societal acceptable option is trained, professional human operators. This view is reflected in the European Commission's 2024 Artificial Intelligence Act, Chapter III, Article 14, on *Human Oversight of AI*, which states that '…high-risk AI systems must be designed in a way that allows humans to effectively oversee them' (EU, 2024).

The million-dollar question is then *how* to prepare the human operator for the unknown, unanticipated events? And how to design clever automation tools and human-machine interfaces to support the human in dealing with these events? Clearly, operator education, training, and continuously extending the capabilities of support systems will remain extremely helpful. On top of that, we propose that the support systems help the operator to rapidly become an *expert* in performing the task, understanding the means-ends relationships, and dealing with all constraints that come with satisfying the system's purposes. We argue that the design of *ecological* interfaces can be a feasible path to create both interface and automation that helps human operators to become experts. Through the rigorous functional analysis of the work domain that stands at the basis of the automation and interface design in the ecological approach, the *designer becomes a domain expert* herself. The visual interface and the automation tools required to process and visualise the information in the most meaningful way act as an 'external mental model', a window on the world. Ecological interfaces provide richer, more meaningful information to operators than traditional ones. They mediate signals, signs, and symbols in the same interface, supporting cognition on all levels of the SRK taxonomy. Operators can directly perceive what the system and environment afford them to do and why that is the case.

Given these benefits, the Maastricht Upper Area Control Centre (MUAC) – one of Europe's most innovative ATC centres – has adopted our 'solution space' approach in its ongoing ARGOS automation project and initiated operational testing of a variant of the SSD (MUAC, 2024). MUAC's version, known as the Lateral Obstacle and Resolution Display (LORD), is a somewhat simplified SSD that visualises available heading options within a ± 40-degree range relative to the flight track, assuming constant speed. Basically, the *H1-H2* heading band is shown in Figure 11.11.

The SSD discussed in this chapter, as well as many of its derived interface variants, presents all possible actions and relevant constraints in a visually integrated fashion, allowing operators to perceive and act in 'one go'. It shows multi-aircraft conflict situations as they are, the 'elements and events in time and space', their meaning, how the situation will evolve, projection, both in cases where the operator does not intervene and when acting on the situation. On purpose, here we used the words Endsley uses in her *definition* of SA (Endsley 1995a, 1995b) to emphasise that, in our view, the SSD *is* the ultimate design *for* SA, in this particular task.

The reader should note that, in contrast to what many believe, ecological interfaces may *not be intuitive*; they often require extensive experience while working with them in dynamic settings to discover the richness and many dimensions of what they visualise. Also, EID is not an easy trajectory; it requires the designer to obtain a deep understanding of the problem at hand, which may be difficult and often requires inputs from several disciplines, often involving engineering. It may also force the designer to look for an alternative representation for a control problem that lends itself better for visualisation (van Paassen et al. 2018), like the use of velocity obstacles in the SSD, compared to traditional CD&R interfaces, which merely show operators the CPA.

Another advantage of taking an ecological approach to interface and automation design is that the WDA is *actor-independent*. That is, the constraints and means-ends relationships that follow from the analysis are true for *any* operator, human or machine. This in turn means that the ecological interface can act as the perfect intermediary between human and machine, because they work on the basis of the same *representation* of the task to be done. Sharing an understanding of the constraints involved, and working on them together, creates a true joint cognitive system (Woods and Hollnagel 2006). Operator and automation share the same 'view' on the situation. Hence, ecological interfaces and the ecological design approach do not prohibit the use of automation. On the contrary, we argue that ecological interfaces can serve as transparent windows into what automation proposes or does (Borst et al. 2015).

In this chapter, we therefore also showed an example of how an ecological interface can be used to integrate the advice provided by an automated agent at various LOA. Depending on the situation, and perhaps even set by the operator at hand, some LOA could temporarily serve the operator better than another. For instance, in high workload situations, the operator could task automation to take over part of the work. The ecological interface allows automation to share its advice in a meaningful way and operators to quickly check the validity and suitability of the suggestions. What level of automation is most appropriate, how and when it should change, and what role the operator should play in these adaptations remain subjects for future research.

What has perhaps also become clear from our chapter is that automation and interface design are *not* an exact science. There is a fierce debate on several of its key tenets, with, for example, over 27 definitions for SA (Breton and Rousseau 2001), and more than 12 LOA taxonomies (Vagia et al. 2016). Also, there is not one particular entity which we can label as '*the* automation'. Neither do systems contain a box labelled 'safety device'. Rather, automation consists of many subsystems working together, often working at different LOA levels. Statements such as 'this process is operating at LOA level 4' are a huge simplification of reality. In most cases, many sub-processes have already been fully automated, whereas on the other hand it could well be that in some events automation completely fails and the operator must step in to save the day. As long as this is the case, and as long as humans are ultimately responsible, *humans must remain in command, and automation can assist.* Which is the first and central tenet of Billings' guidelines for human-centred automation (Billings 1997), based on lessons learned from the surge in automation in cockpits of commercial aircraft during the 1980s–1990s.

In conclusion, human operators should be able to do their job in the best possible way, supported by clever interfaces and capable, reliable automation tools. Chances are high that, no matter what we do, the ironies of automation (Bainbridge 1983) will continue to exist. Nonetheless, we do believe that we can do a much better job in designing automation and interfaces to facilitate real teamwork between humans (clever problem-solvers with intrinsic values) and machines (faster, never tired, able to optimise in many dimensions at a time, but need guidance *what* to do). Perhaps AI techniques can be embedded within novel, more clever tools so that we can put these tools in the hands of capable human operators. To *augment* their intelligence, rather than replace it.

ACKNOWLEDGEMENTS

The authors would like to thank their students and PhD candidates who, during the past decades, worked with them on this topic, contributing many inventions and new ideas to support human operators.

REFERENCES

Amelink, M. H. J., M. Mulder, M. M. Van Paassen, and J. M. Flach (2005). Theoretical Foundations for a Total Energy-Based Perspective Flight-Path Display. *The International Journal of Aviation Psychology 15*(3), 205–231.

Bainbridge, L. (1983). Ironies of Automation. *Automatica 19*(6), 775–779.

Billings, C. E. (1997). *Aviation Automation – The Search for a Human-Centered Approach.* Mahwah, NJ: Lawrence Erlbaum Associates, Inc.

Borst, C., V. A. Bijsterbosch, M. M. Van Paassen, and M. Mulder (2017). Ecological Interface Design: Supporting Fault Diagnosis of Automated Advice in a Supervisory Air Traffic Control Task. *Cognition, Technology, and Work 19*, 545–560.

Borst, C., J. M. Flach, and J. Ellerbroek (2015). Beyond Ecological Interface Design: Lessons from Concerns and Misconceptions. *IEEE Transactions on Human-Machine Systems 45*(2), 164–175.

Breton, R. and R. Rousseau (2001). *Situation Awareness: A Review of the Concept and its Measurement.* Technical report DRDC TR-2001-220, Defence Research and Development.

Chakravarthy, A. and D. Ghose (1998). Obstacle Avoidance in a Dynamic Environment: A Collision Cone Approach. *IEEE System, Man and Cybernetics - Part A: Systems and Humans 28*(5), 562–574.

Degré, T. and X. Lefèvre (1981). A Collision Avoidance System. *Journal of Navigation 34*(2), 294–302.

Ellerbroek, J., K. C. R. Brantegem, M. M. Van Paassen, N. de Gelder, and M. Mulder (2013a). Experimental Evaluation of a Coplanar Airborne Separation Display. *IEEE Transactions on Human-Machine Systems 43*(3), 290–301.

Ellerbroek, J., K. C. R. Brantegem, M. M. Van Paassen, and M. Mulder (2013b). Design of a Coplanar Airborne Separation Display. *IEEE Transactions on Human-Machine Systems 43*(3), 277–289.

Ellerbroek, J., M. Visser, S. B. J. van Dam, M. Mulder, and M. M. Van Paassen (2011). Design of an Airborne Three-Dimensional Separation Assistance Display. *IEEE Transactions on Systems, Man & Cybernetics, Part A 41*(5), 863–875.

Endsley, M. (1995a). Measurement of Situation Awareness in Dynamic Systems. *Human Factors 37*(1), 65–84.

Endsley, M. (1995b). Toward a Theory of Situation Awareness. *Human Factors 37*(1), 32–64.

EU. (2024). *EU Artificial Intelligence Act*, Chapter III, Article 14. https://artificialintelligenceact.eu/article/14/

Fiorini, P. and Z. Shiller (1998). Motion Planning in Dynamic Environments Using Velocity Obstacles. *International Journal of Robotics Research 17*, 760–772.

Flach, J. M. (2012). Complexity: Learning to Muddle Through. *Cognition, Technology & Work 14*(3), 187–197.

Flach, J. M., M. Mulder, and M. M. Van Paassen (2004). The Concept of the Situation in Psychology. In S. Banbury and S. Tremblay (Eds.), *A Cognitive Approach to Situation Awareness: Theory and Application*, 42–60. Oxon, UK: Ashgate Publishing.

Gibson, J. J. (1979). *The Ecological Approach to Visual Perception*. Houghton: Mifflin and Company.

Gibson, J. J. (1986). *The Ecological Approach to Visual Perception*. Hillsdale, NJ: Lawrence Erlbaum Associates, originally published in 1979.

Heylen, F. M., S. B. J. Van Dam, M. Mulder, and M. M. Van Paassen (2008). Design of a Vertical Separation Assistance Display. In *Proceedings of the AIAA GNC Conference*, Honolulu, HI, August 18–21 (AIAA 2008-6969).

Hilburn, B. G., C. A. L. Westin, and C. Borst (2014). Will Controllers Accept a Machine That Thinks Like They Think? The Role of Strategic Conformance in Decision Aiding Automation. *Air Traffic Control Quarterly 22*(2), 115–136.

Huang, L., W. Yu, W. Ma, W. Zhong, Z. Feng, H. Wang, Q. Chen, W. Peng, X. Feng, B. Qin, and T. Liu (2025). A Survey on Hallucination in Large Language Models: Principles, Taxonomy, Challenges, and Open Questions, *ACM Transactions on Information Systems 43*(2), Article 42, 1–55.

Klomp, R. E., C. Borst, M. M. Van Paassen, and M. Mulder (2016). Expertise Level, Control Strategies, and Robustness in Future Air Traffic Control Decision Aiding. *IEEE Transactions on Human-Machine Systems 46*(2), 255–266.

Lee, J. D. and Seppelt, B. D. (2009). Human Factors in Automation Design. In S. Y. Nof (Ed.), *Springer Handbook of Automation*, 417–436. Springer.

Lillo, F., S. Pozzi, A. Tedeschi, G. Ferrara, G. Matrella, F. Lieutaud, B. Lucat, and A. Licu (2009). Coupling and Complexity of Interaction of STCA Networks. In *Proceedings of the EUROCONTROL Innovative ATM Research Workshop*, 1–12, Brétigny-sur-Orge, France, December 1–3.

Lodder, J., J. Comans, M. M. Van Paassen, and M. Mulder (2011). Altitude-Extended Solution Space Diagram for Air Traffic Controllers. In *Proceedings of the 16th International Symposium on Aviation Psychology (ISAP)*, May 2–5, 345–350, Wright State University, Dayton, OH.

Mercado-Velasco, G. A., C. Borst, J. Ellerbroek, M. M. Van Paassen, and M. Mulder (2015). The Use of Intent Information in Conflict Detection and Resolution Models Based on Dynamic Velocity Obstacles. *IEEE Transactions on Intelligent Transportation Systems 16*(4), 2297–2302.

Mercado-Velasco, G. A., C. Borst, M. M. Van Paassen, and M. Mulder (2021). Solution Space Decision Support for Reducing Controller Workload in Route Merging Task. *Journal of Aircraft 58*(1), 125–137.

MUAC. (2024, 18 March). LORD adds Speed and Flexibility to Decisions by Maastricht Controllers, *EUROCONTROL Skyway Magazine*, (79), 54–56. https://www.eurocontrol.int/article/lord-adds-speed-and-flexibility-decisions-maastricht-controllers

Parasuraman, R., T. B. Sheridan, and C. D. Wickens (2000). A Model for Types and Levels of Human Interaction with Automation. *IEEE Transactions on Systems, Man, Cybernetics, Part A 30*(3), 286–297.

Rasmussen, J. (1983). Skills, Rules, and Knowledge; Signals, Signs, and Symbols, and other Distinctions in Human Performance Models. *IEEE Transactions on Systems, Man, and Cybernetics 13*, 257–266.

Rasmussen, J., A. Pejtersen, and L. Goodstein (1994). *Cognitive Systems Engineering*. New York: Wiley.

Ribeiro, M., J. Ellerbroek, and J. M. Hoekstra (2020). Review of Conflict Resolution Methods for Manned and Unmanned Aviation. *Aerospace 7*(79), 1–37.

Sheridan, T. B. (2011). Adaptive Automation, Level of Automation, Allocation Authority, Supervisory Control, and Adaptive Control: Distinctions and Modes of Adaptation. *IEEE Transactions on Systems, Man, and Cybernetics - Part A: Systems and Humans 41*(4), 662–667.

Sheridan, T. B. and W. L. Verplank (1978). *Human and Computer Control of Undersea Teleoperators*. Technical report, Massachusetts Institute of Technology, Cambridge, MA.

Tychonievich, L., D. Zaret, J. Mantegna, R. Evans, E. Muehle, and S. Martin (1989). Maneuvering-Board Approach to Path Planning with Moving Obstacles. In *International Joint Conference on Artificial Intelligence*, 1017–1021.

Vagia, M., A. A. Transeth, and S. A. Fjerdingen, (2016). A Literature Review on the Levels of Automation During the Years. What are the different taxonomies that have been proposed? *Applied Ergonomics 53* (A), 190–202.

Van Dam, S. B. J., M. Mulder, and M. M. Van Paassen (2008). Ecological Interface Design of a Tactical Airborne Separation Assistance Tool. *IEEE Transactions on Systems, Man & Cybernetics, Part A 38*(6), 1221–1233.

Van Paassen, M. M., C. Borst, J. Ellerbroek, M. Mulder, and J. M. Flach (2018). Ecological Interface Design for Vehicle Locomotion Control. *IEEE Transactions on Human-Machine Systems 48*(5), 541–555.

Van Paassen, M. M., C. Borst, R. Klomp, M. Mulder, P. Van Leeuwen, and M. Mooij (2013). Designing for Shared Cognition in Air Traffic Management. *Journal of Aerospace Operations 2*(1), 39–51.

Van Rooijen, S. J., J. Ellerbroek, C. Borst, and E. Van Kampen (2020). Toward Individual-Sensitive Automation for Air Traffic Control Using Convolutional Neural Networks. *Journal of Air Transportation 28*(3), 105–113.

Vicente, K. J. (1999). *Cognitive Work Analysis – Toward Safe, Productive and Healthy Computer-Based Work.* Mahwah, NJ: Lawrence Erlbaum Associates.

Vicente, K. J. and J. Rasmussen (1990). The Ecology of Human-Machine Systems II: Mediating "Direct-Perception" in Complex Work Domains. *Ecological Psychology 2*(3), 207–249.

Vicente, K. J. and J. Rasmussen (1992). Ecological Interface Design: Theoretical Foundations. *IEEE Transactions on Systems, Man, and Cybernetics 22*(4), 589–606.

Westin, C. A. L., C. Borst, and B. G. Hilburn (2016). Strategic Conformance: Overcoming Acceptance Issues of Decision Aiding Automation? *IEEE Transactions on Human-Machine Systems 46*(1), 41–52.

Woods, D. D. and E. Hollnagel (2006). *Joint Cognitive Systems: Patterns in Cognitive Systems Engineering.* Boca Raton, FL: Taylor and Francis.

12 Alarm Management

Towards a Human-Centred Approach

Houda Briwa, Maria Chiara Leva, Stig Ole Johnsen, and Anders L. Madsen

12.1 UNDERSTANDING HUMAN–SYSTEM INTERACTION IN ALARM MANAGEMENT

In modern control rooms, defined by ISO 11064-5:2008 as centralised facilities for remote monitoring and control of safety-critical processes, operators rely on human–machine interfaces (HMIs), including displays, alarms, and communication tools, to build and maintain situational awareness and to coordinate corrective actions (ISO, 2008). Alarm management is a practical expression of Safety by Design, as it integrates human capabilities, system constraints, and operational feedback to prevent degraded system states from escalating into incidents. An alarm is an audible or visible means of indicating to the operator an equipment malfunction, process deviation, or abnormal condition requiring a timely response (*EEMUA 191*). These signals, typically a combination of tones, flashing lights, and text messages, activate when measurements cross predefined thresholds (*EEMUA 191*). Their main purpose is to focus operator attention on emerging hazards before higher-level safety systems, such as emergency shutdown functions, intervene. However, technological developments in control rooms have often been driven more by engineering possibilities than by operator needs, creating potential mismatches between system behaviour and human capabilities (Brazier, 2010).

To understand alarm management from a human-factors perspective, three questions guide this chapter: (1) What role do alarms play in human–system interaction? (2) How have failures shaped current standards and practices? and (3) How do automation and AI reshape operator workload, trust, and oversight? These questions provide the structure for the remainder of the chapter.

Human information-processing capacity is limited (Dehais et al., 2020). As a result, excessive alarm rates, referred to as alarm flooding, can overwhelm operators and impair their performance. When task demands exceed attentional capacity, mental overload may occur, leading to inattention, fatigue, and reduced vigilance. Conversely, when operators face too few demands, they can experience underload. Underload induces boredom and disengagement and impairs performance (Dehais et al., 2020). To address these challenges, system and procedural designers must optimise operator workload, balancing task demands with human capabilities to maintain effective performance in complex environments (Stanton et al., 2009).

One approach to aligning alarm-handling tasks with operators' cognitive capacities is alarm management. This set of practices ensures that alarm systems remain effective throughout their operational life. According to ANSI/ISA-18.2, alarm management follows six phases: philosophy, identification, rationalisation, detailed design, implementation, and maintenance. Each phase is grounded in human-factors engineering principles (ISA & ANSI, 2016). The goal of these practices is to ensure alarms are meaningful, prioritised, and actionable, eliminating spurious and nuisance alerts and continuously improving system performance (*EEMUA 191*).

DOI: 10.1201/9781003741824-15

Still, even well-designed systems can fail under certain conditions. If alarms are too frequent or poorly prioritised, operators may ignore critical alerts or become desensitised. As such, ongoing monitoring, review, and optimisation of alarm-system performance are essential. In the following section, we examine real-world incidents in which alarm-system failures have contributed to major accidents, highlighting the need for sustained, human-centred alarm management.

12.2 WHY ALARM MANAGEMENT MATTERS: LESSONS FROM THE PAST

Alarm management plays a crucial role in ensuring the safety and efficiency of industrial operations. However, ineffective alarm management increases demand on safety systems and, because no protective layer is infallible, this ultimately raises safety risk, potentially leading to unsafe situations and, in extreme cases, fatal accidents.

The history of alarm-system failures in the oil and gas industry and other sectors provides essential lessons for alarm-management practices today. Several catastrophic events, such as the Piper Alpha disaster in 1988, the Milford Haven explosion in 1994, the BP Texas Refinery explosion in 2005, the Deepwater Horizon blowout in 2010, and the Pryor Trust Well gas well blowout in 2018, were all exacerbated by alarm-system failures or or alarm-management weaknesses (Briwa et al., 2022). Investigations following these events consistently identified poor alarm design and management, alarm floods, and a lack of proper prioritisation as the primary contributing factors. These cases illustrate that alarm failures are fundamentally design and organisational problems rather than operator failings.

Table 12.1 summarises the alarm issues from several significant accidents, offering insights into common system flaws that contributed to these events. These major incidents illustrate recurring weaknesses in alarm-related processes. Despite occurring across different industries and decades, they exhibit common patterns: alarm floods, poor prioritisation, ignored alarms, and design mismatches between system behaviour and operator needs.

The recurring failures demonstrate why later standards such as EEMUA 191, ISA-18.2, and IEC 62682 emphasise alarm prioritisation, acceptable operator workload, and human-centred design

TABLE 12.1
Overview of Critical Alarm-System Failures in Major Process Safety Accidents

Year	Accident	Key Alarm Issues	Consequences
1988	Piper Alpha (Cullen, 1990)	Alarm-system failure, uncertain alarm status, and loss of public-address/general-alarm capability contributed to confusion and impaired emergency response during the Piper Alpha disaster	167 fatalities, the destruction of an offshore platform, and financial losses of approximately £2 billion (Cullen, 1990).
1994	Milford Haven Refinery (HSE, 1997)	Control board operator was overwhelmed by alarm flood in an emergency situation. Alarms poorly prioritised.	26 injuries and financial losses of £48 million (HSE, 1997).
2005	BP Texas Refinery (USCSB, 2007)	Alarm-system failure, delayed responses, and missed mitigation opportunities	15 deaths, 180 injuries, and over $1.5 billion in losses (USCSB, 2007).
2010	BP Deepwater Horizon (United States Coast Guard, 2011)	Bypassing of alarms and shutdown systems, allowing hazardous conditions to escalate	11 fatalities, 16 injuries, and the largest oil spill in U.S. history (United States Coast Guard, 2011).
2010	Kalamazoo River Oil Spill (National Transportation Safety Board, 2012)	Ignored alarms and failure to follow procedures, leading to delayed response	320 people with exposure symptoms, contamination of Kalamazoo River and wetlands, and cleanup costs exceeding $767 million.
2012	Columbia Gas Transmission (NTSB, 2014)	Failure to act on critical leak alarm	Release and combustion of 76 million cubic feet of natural gas with financial losses of $8.7 million.

(Continued)

TABLE 12.1 (Continued)

Year	Accident	Key Alarm Issues	Consequences
2018	Pryor Trust Well Gas Well Blowout (USCSB, 2019)	The drillers deactivated the entire alarm system, causing them to miss critical signs of the gas influx and impending blowout. The alarm system, poorly designed for different operational phases, generated numerous non-critical alarms in the 14 hours prior, likely leading the drillers to disable it to avoid overload.	Five workers lost their lives due to alarm-related failures (USCSB, 2019).
2018	Husky Superior Refinery (USCSB, 2024)	Board operators were dealing with an alarm flood for nearly 12 hours before the incident occurred, with 3,712 alarms going off during this period. Excessive alarms contributed to delays and errors in response and mitigation of the overflow of naphtha into the fuel gas system.	2 fatalities, release of 23,000 pounds of naphtha, and property damage of $597 million (USCSB, 2024).
2019	Equinor – Statfjord A (Oplenskedal & Bjørheim, 2019)	Frequent alarms were caused by electrical noise from a new battery pack. Over 20 DP alarms triggered in 45 minutes were deemed non-critical.	1 injury, 218 people evacuated, and extensive damage to vessel and lifeboat station (Oplenskedal & Bjørheim, 2019).
2023	*Viking sky* (Norwegian Safety Investigation Authority (NSIA), 2024)	The ECR operators on *Viking Sky* were overwhelmed by a large number of ungrouped alarms with no prioritisation. Design and configuration issues likely impaired the effectiveness of the engineering officers. At the time of construction, no specific standards existed for designing engine room alarm systems in the maritime industry.	19 injuries, structural damage from wave impacts, including lifeboat systems, and evacuation of passengers (Norwegian Safety Investigation Authority (NSIA), 2024).

Source: Adapted from Briwa et al. (2022) and Goel et al. (2017).

principles, and formalise expectations for management reporting and oversight. Across decades, recurring problems include poor alarm design, alarm flooding, inconsistent prioritisation, and insufficient prioritisation of human factors.

Alarm-related issues continue to surface in major incidents, suggesting that existing knowledge and recommendations are not consistently applied. Over the years, several technical documents have been developed, ranging from industry best practices to formal standards, to guide the design and performance of alarm systems. The next section reviews how these documents have evolved and how they define, structure, and measure the effectiveness of alarm systems.

12.3 CURRENT PRACTICES AND STANDARDS IN ALARM MANAGEMENT

12.3.1 Evolution of Alarm-Management Frameworks

Following the Milford Haven refinery explosion (1994), the UK Health and Safety Executive published the Alarm Management Process Report (1998), the first industry-wide synthesis of good practice. Rather than prescribing strict rules, it distilled lessons on alarm configuration, prioritisation, operator response, and performance monitoring based on real plant experience.

This work laid the foundation for EEMUA 191 (1999), which became the conceptual backbone for modern alarm management. The guideline translated human-factors principles and incident learnings into structured recommendations for alarm philosophy, rationalisation, performance targets, and lifecycle governance. Although not mandatory, EEMUA 191 quickly gained global influence due to its strong empirical grounding.

In 2009, these principles were formalised into the ANSI/ISA-18.2 standard, introducing defined lifecycle phases, key performance indicators (KPIs) (e.g., alarm rates, standing alarms), and audit requirements for U.S. industries. The IEC 62682 standard subsequently harmonised with ISA-18.2 for international use. Sector-specific standards followed include ISO 11064-5 for control-room ergonomics, API RP 1167 for pipeline SCADA alarms, Germany's NAMUR guidelines, and Norway's PSA YA-711 and NORSOK I-002/I-005. These standards are periodically revised to incorporate technological developments and field experience.

Industry consortia further reinforced the evidence base. The ASM Consortium demonstrated that high-performing sites can reliably meet or exceed EEMUA 191 performance targets. In parallel, human-factors engineering frameworks (e.g., Chandrasegaran et al., 2020) embedded alarm management into broader considerations of control-room layout, cognitive workload, and organisational resilience.

Taken together, guidelines, standards, and technical norms form a dynamic and increasingly harmonised framework for alarm management in high-risk sectors such as offshore oil and gas. Table 12.2 summarises the principal documents formalising alarm management and user-centred design and revision histories.

TABLE 12.2
Regulatory and Guideline Documents Formalising Alarm Management and User-Centred Design in Offshore Oil & Gas

Document	Issuing Body	Synopsis	Year/ Rev. History
Standards			
IEC 62682:2022, Management of Alarm Systems for the Process Industries (IEC 62682, 2022)	International Electrotechnical Commission (IEC)	Defines principles and processes for managing alarm systems via control systems and HMIs in process industry facilities. It applies to alarms from BPCS, annunciators, packaged systems, and SIS across continuous, batch, and discrete processes. Alarm systems notify operators of abnormal conditions and support response, using process measurements and logic. It includes HMIs for communication, as well as alarm logs, historians, and performance metrics.	2014 2022
ISO 11064-5:2008 Ergonomics of human-system interaction. (ISO, 2008)	International Organization for Standardization (ISO)	This part of ISO 11064 presents principles and gives requirements and recommendations for displays, controls, and their interaction, in the design of control-centre hardware and software.	2008
ISO 6385:2016 – Ergonomics Principles in the Design of Work Systems (ISO, 2016)	International Organization for Standardization (ISO)	It provides the philosophy for designing the work systems with ergonomics principles, including alarm-system design, from establishing the goals to evaluation of the designed work systems. It also stipulates involvement of ergonomists during the design development process.	2004 2016

(*Continued*)

TABLE 12.2 (Continued)

Document	Issuing Body	Synopsis	Year/ Rev. History
NORSOK S-002N:2004 working environment (Standard Norge (SN), 2004)	Standards Norway (Standard Norge)	It sets ergonomic and environmental design requirements for new and modified offshore petroleum installations, including drilling, production, pipelines, and accommodation units. It mandates systematic management of working-environment issues throughout project development to ensure a safe, healthy, and efficient workplace during operations.	1997 2004 2018 2018 + AC2018 2018 + AC2021
Guidelines and best practices			
EEMUA 191, Alarm systems: a guide to design, management and procurement (Engineering Equipment Material Users Association, 2013)	The Engineering Equipment and Materials Users' Association. (EEMUA)	It provides guidance on designing, managing, and procuring an effective alarm system. It is intended to help in improving existing systems and in developing new facilities during plant construction or during alarm-system refurbishments	1999 2007 2013 2024
YA-710/YA-711 – Principles for design of alarm systems (NPD, 2001)	Norwegian Ocean Industry Authority (Havtil) previous PSA	It provides basic principles and guidelines on alarm generation, structuring, prioritisation, and presentation for offshore installations on the Norwegian Continental Shelf.	2001

Source: Adapted from Briwa et al. (2022) and Chandrasegaran et al. (2020).

12.3.2 Alarm-Management Principles and Performance Criteria

Based on the frameworks described in Section 12.3.1, the fundamental design principles and quantifiable criteria that define an effective alarm system are distilled. Although alarms are critical for safety, several design flaws are frequently encountered and can impair operator performance during disturbances. These include nuisance alarms, where a single initiator generates many alarms within a short period; false alarms, where an alarm activates even though no abnormal or unsafe condition is present; alarm flooding, where more alarms are received than a single console operator can physically address; confusing messages; poor organisation and placement; inadequate visual and auditory prominence; and signals that blend with normal indicators (O'Hara & Fleger, 2020).

According to EEMUA 191, a well-designed alarm must focus attention only on truly significant conditions, convey its urgency through clear prioritisation, present an unambiguous message, and guide the operator towards the correct response. In practice, these principles are formalised in an alarm-management philosophy, a site-specific document that defines the purpose of alarms, design and prioritisation rules, operator response expectations, and roles and responsibilities across the alarm lifecycle. The philosophy provides the reference against which individual alarms are rationalised and ongoing performance is assessed.

In the ANSI/ISA-18.2 lifecycle, three main steps are particularly important for continuous improvement of alarm systems: rationalisation, monitoring and assessment, and audit.

Rationalisation is carried out for both new and existing systems. In new systems, it is performed before detailed design; in existing systems, it is revisited during operation and maintenance. In each case, alarms are evaluated against predefined criteria, and their setpoints, consequences, required responses, priorities, and classifications are all documented to yield a risk-aligned, prioritised alarm database.

Monitoring and assessment is performed through routine analysis of alarm performance data and learning from incidents, to identify deviations from the established alarm philosophy and to determine when setting adjustments or advanced techniques are warranted. In this phase, industry standards and guidelines specify KPIs against which alarm-system performance is measured. A minimum of 30 days of operational data is generally recommended, although actual metric values may vary with process characteristics, operator expertise, HMI design, automation level, environmental conditions, and the nature of the alarms themselves.

Audit encompasses a review of all lifecycle phases, to confirm compliance, pinpoint deficiencies, and guide further system refinement.

Complementing these metrics, human-factors guidance (O'Hara & Fleger, 2020) recommends reviewing alarm systems along seven dimensions:

- How are alarms defined and treated (processed and prioritised)?
- How are alarms displayed?
- How users interact with and control alarms?
- How reliable is the alarm system, and how is it tested and maintained?
- How do operators respond to alarms?
- How are alarm control and display elements integrated?
- How are alarms integrated with the other Human–System Integration (HSI) elements, such as general controls and displays?

However, metrics alone cannot fully capture how operators actually interact with alarm systems. For this reason, complementary human-factors-based evaluations like Crisis Intervention and Operability (CRIOP) remain essential to verify key metrics (Table 12.3).

TABLE 12.3
Comparison of Alarm-Performance Metrics and Their Target Values across Key Industry Standards and Guidelines

Evaluation Criteria	Guideline Documents	Recommended Value
Criteria 1: Average Alarm Rate		
Long-term average alarm rate in steady-state operation	EEMUA 191	≤1 per 10 minutes
Average alarm rate	IEC 62682	≤2 per 10 minutes
Average alarms per control room operator per hour	ISO 11064-5	≤1 per 6 minutes
Average annunciated alarm rate per operating position	ISA-18.2	<2 per 10 minutes
Alarm rate in steady operational conditions	PSA YA-711	1 per 5 minutes
Criteria 2: Peak Alarm Rate		
Alarms in 10 minutes after plant upset	EEMUA 191	<10 per 10 minutes
Maximum peak alarm rate during upset conditions	ISO 11064-5	3–5 per minute (and <6 in 10 minutes)
Peak annunciated alarm rate per operating position	ISA-18.2	≤10 per 10 minutes

(Continued)

TABLE 12.3 (Continued)

Evaluation Criteria	Guideline Documents	Recommended Value
Criteria 3: Proportion of Time Alarm Rates Are Outside of Acceptability Target		
% of time outside average alarm rate	EEMUA 191	10%
Criteria 4: Average Standing Alarms		
Average number of standing alarms	EEMUA 191	<10
Stale alarms	ISA-18.2	<5 per day
Number of standing alarms associated with 'on-line' equipment	ISO 11064-5	<5
Criteria 5: Priority Distribution (Low/Med/High)		
Priority distribution	EEMUA 191	80%/15%/5%
Annunciated priority distribution	ISA-18.2	80%/15%/5%

Source: Adapted from Briwa, Leva, and Turner (2022).

12.3.3 CRIOP Methodology

Alongside performance metrics and standardised guidelines, broader human factors methodologies have been applied to assess the usability of alarm systems in control rooms. One such method is the CRIOP framework, which was developed to support the verification and validation of control-centre design in line with ISO 11064 and HF best-practice standards. It is based on participatory HF design to explore 'work as imagined' versus 'work as done' together with management and the workforce, fostering organisational learning. CRIOP includes a section on control and safety systems that covers alarm-related aspects, including prioritisation, suppression of non-essential alarms, clarity of presentation, KPI reporting, and alignment with operator tasks (Johnsen et al., 2025). The method combines checklist-based reviews with scenario walkthroughs to examine how alarm functions perform in real settings, that is, 'work as done'. This process can help identify potential mismatches between alarm-system behaviour and operator expectations in the early design phases and in operations. CRIOP has been used in the Norwegian oil and gas industry to support control-room assessments, including alarm-related considerations, as part of a broader focus on human–system integration. Its use in conjunction with conventional performance criteria provides a complementary perspective on alarm-system effectiveness from a human factors standpoint (Johnsen et al., 2025). CRIOP reveals interactional breakdowns, gaps between design intent and operational reality, that standards alone cannot detect.

Despite the availability of mature standards, guidelines, and documented best practices, alarm-related failures continue to contribute to recent industrial incidents. This recurrence indicates a persistent gap between regulatory knowledge, operational implementation, and management of 'work as done'. Notably, much of the progress in advancing alarm-management practices continues to be driven by regulators rather than industry leadership. The next section explores the underlying barriers to effective implementation and discusses strategic pathways to support more sustainable and proactive alarm management.

12.4 EFFECTIVE ALARM MANAGEMENT IN PRACTICE

12.4.1 Implementation Barriers

In 2022, the Petroleum Safety Authority Norway (Havtil) conducted a multi-site audit of alarm-system monitoring and control-room operations across 54 facilities (Ueland et al., 2022). The Havtil audit identified three critical gaps in the implementation of alarm-management strategies.

First, neither new nor standing alarms were being adequately reviewed, resulting in alarm overload that masks genuine process deviations and hazards. Second, companies lacked documented analyses of the control-room working environment, indicating insufficient integration of technical, operational, and organisational safeguards. Third, the data generated by alarm monitoring and ergonomic assessments were not being processed or fed back into operational and organisational controls, preventing targeted improvements in alarm performance and control-room conditions.

Further, organisations' reliance on highly granular, efficiency-focused KPIs measured by the second and reviewed daily increases time pressure on operators while failing to demonstrate how these metrics drive technological improvements or reduce risk. This gap underscores the need to link performance data to actionable process refinements to enhance both safety and learning (Bergh et al., 2024).

These operational shortcomings reflect deeper obstacles to learning from the past. The CIEHF white paper (CIEHF, 2020) identifies key elements for why organisations struggle to learn from adverse events:

12.4.1.1 Failure to Accept the Need for Change

Near misses and weak signals often go unaddressed because teams lack the 'requisite imagination' to foresee adverse outcomes. Success in averting harm can dull urgency, and assigning a 'responsible manager' to the immediate workgroup can limit recommendations to that manager's remit. When events are dismissed as rare coincidences, preventive action is easily deferred.

12.4.1.2 Cognitive Biases

Normalisation of deviance occurs when unsafe conditions repeat without consequence and become accepted as routine. Distancing by assigning each deviation's attributes to a unique anomaly obscures systemic weaknesses. Both biases cause critical warning signs to be overlooked. The CIEHF illustrates this with UK rail Automatic Warning System (AWS), where a single audible alarm is used for events ranging from speed restrictions to a more critical one, the need to stop. This has contributed to trains passing red signals, but because drivers and managers have grown up with it, the design has become accepted as normal and is often overlooked as a contributing factor in incidents.

12.4.1.3 Overly Constraining the Scope of Change

Investigations often focus narrowly on the individuals, equipment, or procedures directly involved, neglecting wider systemic factors. This 'what you look for is what you find' mindset reinforces a human-error narrative and prevents organisation-wide remedies. For instance, as highlighted in Table 12.1, alarm design and management problems were repeatedly identified as root causes or contributing factors. Yet the narrative around several of these events still centres on proximal operator actions (e.g., ignored alarms, failure to follow procedures, deactivation of alarm systems), illustrating how a focus on 'human error' or direct causes can obscure the underlying weaknesses in alarm-system design and management.

12.4.1.4 Relying on Quick Fixes

Under time pressure, superficial remedies, such as adding procedural checks, are favoured over addressing latent system vulnerabilities. While easy to implement, these fixes impose new burdens and fail to capture the full scope of necessary change.

12.4.1.5 Failure to Embed the Reason for Change

Without robust knowledge-sharing processes, the link between identified risks and controls fades as organisational memory erodes. Immediate post-incident enthusiasm gives way to old habits unless continuous reminders, training refreshers, and accessible documentation are maintained.

12.4.1.6 A Compliance Mindset

When the focus shifts to 'ticking the regulator's box', genuine risk understanding is sidelined. Multinational operations may comply with local regulations at the expense of internal best practices, hindering the transfer of critical safety lessons across borders.

Together, these factors explain why, despite well-established standards and guidelines, sustained learning and lasting improvement remain elusive.

12.4.1.7 Failure to Monitor and Evaluate the Effectiveness of Change

Sustained learning is not achieved simply by implementing actions and closing them on time. Organisations often measure whether recommendations have been completed rather than whether they have actually reduced risk or addressed underlying causes. Effective learning requires performance-oriented recommendations and KPIs that test whether new arrangements improve system performance (e.g., whether personnel can reliably determine the status of safety-critical systems) and also detect unintended consequences in other parts of the socio-technical system.

A further barrier to effective alarm management is that improvement efforts often concentrate on technical properties of alarms while paying limited attention to how control-room operators (CROs) actually use them in practice. Industry guidance such as EEMUA 191 reflects this, largely framing the CRO as part of the engineered alarm chain and emphasising alarm response rather than the wider context in which decisions are made (EEMUA 191, p.19). However, as noted by The Energy Institute (2016), alarms form just one element within a much richer information environment that includes shift handovers, process graphics, communications with colleagues, CCTV, weather data, and the operator's own understanding of the plant. Which alarms are noticed and how they are treated depends heavily on prior experience, expectations, and beliefs about the current plant state. Alarm interpretation is therefore strongly shaped by operators' mental models and pattern-recognition strategies. Experienced CROs often use familiar alarm patterns to identify what is most important and to cope with heavy alarm loads, while novices, with less-developed mental models, typically need longer to interpret the same information. These cognitive shortcuts can be helpful but also introduce vulnerabilities: mental models may be inaccurate or incomplete, and confirmation bias can lead operators to seek and prioritise information that supports an initial hypothesis while overlooking conflicting cues (The Energy Institute, 2016). This has direct consequences for alarm rationalisation. Alarms that appear superfluous from a technical standpoint, for example, because similar information is available elsewhere, may nonetheless act as critical prompts within operators' mental models, such as early indicators of specific failure modes. Removing or downgrading such alarms without understanding their role in everyday sensemaking risks unintentionally weakening situation awareness (SA) (The Energy Institute, 2016).

12.4.2 Towards Sustainable Alarm-Management Practices

Understanding these barriers reveals that overcoming them requires better practices and the resources and structured change mechanisms to embed those practices effectively. Effective alarm-system management requires ongoing investment, yet securing that funding is difficult without a clear way to quantify its value. Because alarms rarely appear as direct causes of incidents, organisations lack a recognised method to calculate the financial impact of poor alarm performance. Establishing a reliable metric for these costs, covering both operational losses and increased safety risks, would be essential for making a compelling business case to senior management (Schliemann-Haug, 2024).

Introducing better alarm management requires not only upfront investment but also a structured change process to ensure lasting adoption. Despite two decades of Human Factors in Control network work in Norway, major operators such as Equinor and others have shown limited engagement; pilot projects and awareness campaigns struggle to gain traction amid competing priorities. To overcome this inertia, the network recommends a six-step framework based on change management principles to institutionalise alarm-management practices (Table 12.4; Kotter, 2012):

TABLE 12.4
Six-Step Framework for Sustainable Alarm Management Practices

Steps	Implementation Guidance
1. Build a Compelling Business Case	
1.1. Quantify the risks and costs of poor alarm practices.	- Document key issues from the Havtil audits. - Use real-world examples (accidents, near misses, inefficiencies) where poor alarms contributed to problems.
1.2. Show positive return on investment (ROI).	- Highlight how better alarms can reduce downtime, improve operator performance, and increase system reliability. - Frame alarm management not as a cost but as risk mitigation and performance enhancement.
2. Get Executive Sponsorship	
2.1. Identify a champion at the leadership level who understands the risk and can advocate for the project.	
2.2. Frame it in leadership language.	- Focus on strategic risks (e.g., safety, compliance, reputation) and strategic gains (e.g., operational excellence).
2.3. Tie it to regulatory or corporate goals.	- If possible, connect it to compliance standards (e.g., ISA-18.2, EEMUA 191, IEC 62682) or corporate initiatives like operational excellence, safety culture, or digitalisation or CRIOP.
3. Start Small: Pilot Programme	
3.1. Select a critical area (e.g., a single control room, a critical production unit) to improve alarms first.	
3.2. Demonstrate quick wins	- Show measurable improvements (reduced nuisance alarms, faster response times). - Use the pilot to gather success stories and internal credibility.
4. Involve Operators and Key Stakeholders Early	
4.1. Engage the people affected by alarms: operators, engineers, maintenance staff.	
4.2. Co-create the alarm philosophy.	- Involve them in defining what constitutes a good alarm (e.g., sound, prioritisation, and quantity). - Listen to their pain points and show how better alarms solve their real problems (not just management's).
5. Make the New Alarm Philosophy Mandatory	
5.1. Formalise the practice.	- Introduce an Alarm Philosophy Document that defines design principles, alarm prioritisation, alarm response expectations, etc. - Integrate into engineering standards: Make alarm management part of control-system lifecycle (design, testing, operation) and CRIOP checklists.
5.2. Audit and monitor.	- Set KPIs for alarm-system performance (e.g., alarm rate, % standing alarms, operator load). - Include alarm-system reviews in regular audits, especially easy issues such as the number of standing alarms.
6. Communicate, Communicate, Communicate	
6.1. Tell the story.	- Repeatedly explain why alarm management matters. Use real incidents (anonymised if needed) to remind people what's at stake.
6.2. Celebrate success.	- Publicise early wins, like reduced alarm rates or improved operator performance.

Source: Adapted from Kotter (2012).

While structured frameworks and best practices lay a foundation for effective alarm management, the growing integration of automation and artificial intelligence (AI) into process industries introduces new dimensions and challenges for human factors. As systems evolve towards higher autonomy, alarm design and management must account for shifting roles, changing cognitive demands, and the risk of reduced operator engagement. As alarm systems increasingly incorporate AI-driven diagnostics and predictive analytics, new challenges arise related to model brittleness, transparency, and shifting cognitive demands. The next section explores these emerging dynamics, examining how AI and automation reshape alarm-handling tasks and what this means for human reliability, awareness, and safety.

12.5 INTEGRATING AUTOMATION AND AI IN ALARM MANAGEMENT

Automation and AI do not eliminate the need for alarm management. They change it as algorithms take over sensing, filtering, and recommending actions; the operator's role shifts from direct control to supervisory oversight. This underscores the importance of transparency, meaningful human control, and cognitive alignment between humans and machines.

This section begins by clarifying the concepts of automation, autonomy, and artificial intelligence and their implementation from an HF perspective. It then reviews the progression of alarm management, from foundational Distributed Control Systems (DCS) logic through successive automation improvements to modern AI-integrated systems, and concludes with an assessment of the technical limitations and human-factors implications introduced by these advances.

12.5.1 Automation from HF Perspectives

To clarify the roles of technology in alarm systems, three terms are distinguished (National Academies of Sciences, Engineering and Medicine, 2022):

Automation refers to the execution of tasks by a system without continuous human involvement and may range from mechanical routines to rule-based programmable logic.

Autonomous systems, on the other hand, have the capacity to make context-sensitive decisions in unanticipated conditions without explicit pre-programming, and can serve as human proxies in complex environments.

Artificial Intelligence (AI) extends these capabilities by embedding learning and reasoning mechanisms into automated systems, allowing them to perform cognitive tasks such as interpretation, classification, and planning through symbolic or data-driven approaches.

From a human-factors perspective, automation is defined by the cognitive stages it replaces or augments. Parasuraman et al. (2000) identify four such stages: information acquisition, information analysis, decision and action selection, and action implementation. For instance, automation (and AI) may acquire raw data from sensors (information acquisition), process and interpret that data (information analysis), recommend or select actions (decision and action selection), or execute those actions (action implementation). Each function can also be automated to varying levels, from fully manual to fully automated, depending on how much authority is delegated to the system (McLeod, 2022).

12.5.2 Role of Automation and AI in Alarm-Management Practices

Automation has fundamentally reshaped alarm-handling practices in process industries since the introduction of DCS. The shift began in the 1970s when the DCS replaced analogue panels with centralised digital interfaces.[1] This transition marked the integration of alarm functions into software logic, moving away from physical relays. Through the 1980s and 1990s, DCS technology evolved to support greater system integration and improved visualisation, with growing interoperability with platforms such as SCADA (Poştovei et al., 2020). With the onset of Industry 4.0, DCSs have incorporated real-time analytics, cloud connectivity, and AI capabilities. These developments have extended to alarm management, where AI is increasingly used to support alarm prioritisation, anomaly detection, and operator decision-making (Segovia & Theorin, 2012).

12.5.3 Limitations of Automation and AI

(Johnsen, 2021) identified a chain of human–machine integration challenges in recent automation projects. First, interfaces grew overly complex when systems from multiple suppliers were combined, which in turn meant that legacy alarms were not reassessed during upgrades. This oversight, coupled with technology-driven designs developed without user involvement, eroded 'meaningful human control' and left operators uncertain about automated functions. Consequently, inadequate planning and training led to costly, late-stage modifications.

Moreover, as AI-based functions are introduced into complex operations, the literature highlights several reliability and adaptability risks that need to be managed, which can be grouped into four main categories.

12.5.3.1 Brittleness of AI

AI models tend to perform well only within the scope of their training data or predefined logic. When faced with novel or unforeseen conditions, their performance may degrade, requiring retraining, recalibration, or updated validation (DNV Energy Systems, 2024).

12.5.3.2 Overfitting and Model Degradation

Overfitting occurs when models become overly specialised to training data, resulting in reduced performance when applied to new inputs. As operating conditions change, due to equipment ageing, process adjustments, or environmental shifts, model degradation may occur if not properly monitored and updated (DNV Energy Systems, 2024).

12.5.3.3 Perceptual Challenges and Hidden Bias

Many AI models still struggle to accurately interpret objects in noisy environments or process natural language inputs. These models also operate as opaque 'black boxes', concealing biases introduced by skewed or incomplete training datasets and complicating efforts to detect or correct them. Furthermore, current AI lacks a model of causality; it recognises patterns but cannot reason about cause-and-effect relationships. As a result, it struggles to predict consequences, adapt to dynamic situations, or respond effectively when conditions deviate from past experience (National Academies of Sciences, Engineering and Medicine, 2022).

12.5.3.4 Security and Reliability Risk

Alarm systems depend on reliable software and hardware infrastructure. Software bugs and hardware failures can undermine system availability and accuracy. Environmental factors, such as elevated temperatures or electromagnetic interference, may exacerbate these failures. Additionally, malicious actions, including cyberattacks or signal disruption, pose serious threats to data integrity and system continuity (DNV Energy Systems, 2024). Given these vulnerabilities, safety-critical systems require independent layers of protection. Where AI is employed to optimise alarm management, conventional BPCS (Basic Process Control System) alarms must remain fully operational, and independent safety systems (e.g., SIS) must continue to function as separate protection layers.

These technical constraints not only undermine alarm accuracy but also reshape the cognitive demands placed on operators. Taken together, these technical vulnerabilities translate directly into new human-factors challenges, particularly related to SA, trust, and workload. The next subsection examines the human-factors challenges that arise from these limitations.

12.5.4 Human Factors Considerations

SA, defined as the perception of relevant information, comprehension of its meaning, and anticipation of future states (Endsley, 1995), is critical for safe and effective control-room performance. However, automation (and AI) can undermine SA by providing insufficient feedback or obscuring system behaviour, making it difficult for operators to comprehend evolving conditions or anticipate future states, particularly during abnormal events (e.g., alarm floods) (Endsley, 2018).

These SA vulnerabilities are particularly evident in oil and gas process control environments, where automation and remote monitoring have replaced direct physical engagement with digital interfaces. By eliminating sensory inputs that traditionally supported sensemaking in critical situations, operators may become 'out of the loop', less engaged, and slower to respond when automation fails or unexpected scenarios arise (Johnsen & Winge, 2023).

A key concern in this context is over-trust in automation. When systems perform reliably, operators may divert attention to other tasks, leading to neglected monitoring. In complex, multi-tasking environments, this coping strategy further reduces vigilance and weakens situational awareness. Additionally, high false alarm rates can erode trust, prompting operators to disregard or disable alarms, even when they are relevant (Endsley, 2018).

These challenges are amplified with the adoption of AI and machine learning. While these systems can reduce physical workload during normal operations, they can shift and sometimes increase cognitive demands, especially when they lack transparency or fail to align with human mental models (Saghafian et al., 2025). Regulatory audits have echoed these concerns, highlighting how embedded AI systems frequently lack interpretability, making it harder for operators to anticipate or understand system behaviour. Under conditions of fatigue, passive monitoring, or cognitive overload, this opacity further degrades trust and situational awareness (Bergh et al., 2024).

12.5.5 Human–AI Conflicts

To translate the human factors vulnerabilities identified above into a formal risk framework, Arunthavanathan et al. (2024) employ a fault-tree case study to formalise how mismatches between AI and human operators, referred to as Human Intelligence (HI), can escalate into hazardous control actions. In their model, AI continuously interprets real-time sensor data using algorithms trained on historical records, while HI applies experiential knowledge and situational awareness to oversee the same data stream. They identify three distinct stages at which conflicts may arise:

- **Observation conflicts**: misreads or omissions in sensor data by AI or HI.
- **Interpretation conflicts**: divergent conclusions drawn from the same observations.
- **Action conflicts**: conflicting commands issued to the same actuator.

The fault-tree analysis employed by Arunthavanathan et al. (2024) traces failure paths from initial observation errors, whether due to sensor faults, cyberattacks, data-quality issues, or human reading mistakes, through successive interpretation discrepancies and ultimately to conflicting control actions. An early observation conflict may propagate into an interpretation conflict if AI and HI reach different decisions; if unaddressed, this divergence leads to action conflicts that threaten safe operation.

This analysis highlights that conflicts between AI and human operators arise from their interactions rather than isolated component failures. Addressing these interaction-driven risks requires a comprehensive, human-centred alarm-management approach, which is presented in the following section.

12.6 TOWARDS A HUMAN-CENTRED ALARM-MANAGEMENT APPROACH

While the successful integration of technology, including AI, into control-room operations requires robust data and model governance, regulatory and ethical compliance, and organisational readiness (Enemosah, 2021), AI-based applications can add specific challenges to each of these domains.

Drawing on a review of the relevant literature, the specific requirements and recommended practices for each domain are examined in the subsections that follow.

12.6.1 Data and Model Governance

Robust data and model governance ensure that AI outputs remain accurate over time. High-quality sensor data must be cleaned, synchronised, and labelled, supplemented by synthetic examples when fault records are sparse, and models require continuous drift monitoring with semi-automated retraining pipelines that incorporate operator feedback (Enemosah, 2021; Janssen et al., 2020).

The stakeholder roadmap in Enemosah (2021) prescribes a three-stage approach related to data and model governance in their roadmap for stakeholders in implementing intelligent decision-support systems in oil and gas operations: first, prioritise assets by assessing sensor coverage, data quality, and system maturity; next, modernise the data pipeline through standardised protocols, edge preprocessing, and secure data feeds; then, develop and pilot AI models using historical data, expert validation, and human-in-the-loop safeguards; and finally, establish continuous monitoring and regular retraining to detect drift and incorporate operator feedback, ensuring sustained model accuracy.

12.6.2 Regulatory and Ethical Compliance

Regulatory and ethical compliance demands that AI systems offer transparent audit trails, explainable decision logic, and built-in bias detection. These safeguards satisfy safety regulations, maintain accountability, and enable operators to critically assess AI recommendations in safety-critical scenarios (Enemosah, 2021).

12.6.2.1 EU Regulation on AI

The EU AI Act (Madiega, 2021) establishes a harmonised legal framework for all AI systems placed on the EU market or used within the Union, requiring conformity assessments under the New Legislative Framework (Articles 114 and 16 TFEU) and defining ‘AI system’ broadly to include machine-learning, logic-based, and statistical approaches. A key feature is its risk-based classification, which distinguishes four tiers of AI applications:

- **Unacceptable Risk**: Practices banned outright, including subliminal manipulation, exploitation of vulnerable groups, social scoring by authorities, and real-time biometric identification in public spaces.
- **High Risk**: Systems serving as safety components under EU directives (e.g., Machinery Directive, ATEX) or operating in Annex III sectors (e.g., gas delivery, digital infrastructure), subject to documented risk management, rigorous testing, human-in-the-loop oversight, transparency, and cybersecurity requirements.
- **Limited Risk**: Applications such as chatbots, emotion-recognition, biometric categorisation, and generative media, which must clearly disclose their AI nature.
- **Minimal Risk**: Low-impact systems are exempt from additional rules but encouraged to adopt voluntary codes of conduct.

Each Member State designates one or more national supervisory authorities to monitor compliance and enforce the AI Act, while a European Artificial Intelligence Board provides pan-EU coordination and guidance. Operators in breach of these regulations face administrative fines proportional to the severity of their non-compliance.

To apply these provisions within the petroleum sector, DNV Energy Systems' (2024) analysis of Article 6 draws two principal distinctions:

- **High-risk systems**, including AI embedded in products governed by Annex I safety directives or serving as safety components in Annex III critical-infrastructure areas (e.g., gas delivery), must meet full pre-market requirements for risk management, robust testing, human-in-the-loop oversight, transparency, and cybersecurity.
- **Grey-area applications**, such as advisory, planning, or condition-monitoring tools limited to simple procedures, fall outside explicit high-risk definitions and require case-by-case assessment.

12.6.2.2 Global Frameworks for AI Safety in Process Industries

DNV Energy Systems (2024) references several key international standards and guidelines for AI safety in process industries. These documents, covering topics from lifecycle management to functional safety and risk governance, are summarised in the Appendix (Table 12.A.1) and grouped into three categories: lifecycle, safety, and data quality.

In addition to these standards and guidelines, it is important to acknowledge the DNV report itself as the primary source from which these references were adapted. Commissioned by Havtil to assess AI-related major-accident risks in the petroleum sector, the report not only identifies relevant lifecycle, safety, and governance standards but also provides the strategic context and industry-specific recommendations that underpin their application. It urges the industry to jointly establish unified best-practice guidelines for AI applications, thereby harmonising safety levels, streamlining compliance with Havtil's regulations, and aligning with the EU AI Act. It also recommends exploring AI-driven methods for automated detection of unsafe conditions to mitigate 'out-of-the-loop' risks introduced by advanced automation and calls for novel qualification and maintenance strategies, such as regression-testing frameworks, to manage AI's non-deterministic behaviour and preserve ongoing software integrity in safety-critical environments.

Building on these industry-wide recommendations Enemosah (2021) further emphasises the need to embed successful AI pilots into enterprise workflows under a robust governance framework. This requires seamless integration of decision-support tools with systems such as computerised maintenance management system (CMMS), SCADA, and Enterprise Resource Planning (ERP), complemented by formal audit trails, clearly defined roles, and comprehensive security controls. Compliance is strengthened by mapping these governance measures to the EU AI Act's risk-based requirements and applicable international standards (e.g., ISO/IEC lifecycle and safety directives), thereby ensuring that transparency, traceability, and data-retention policies fulfil both regulatory and ethical obligations.

12.6.3 Organisational Readiness

Organisational readiness and change management cement AI's role in daily practice. Role-specific training programmes, clear governance for alert validation and automated actions, and leadership-driven communication of pilot successes create the cultural and procedural infrastructure needed for sustained AI adoption (Enemosah, 2021).

According to Enemosah (2021), the roadmap's concluding phases address organisational readiness by embedding AI into daily practice. This begins with role-specific training, delivered via hands-on workshops and simulation exercises, to familiarise operators, engineers, and managers with new decision-support tools. Concurrently, change management led by executive sponsors and cross-functional teams fosters stakeholder engagement and iterative system refinement based on real-world feedback. Finally, periodic model retraining and performance audits ensure the AI solution evolves alongside operational changes, establishing a self-improving asset that maintains alignment with business objectives and safety requirements.

Johnsen (2021) highlights three key practices for successful adoption: appointing a single, accountable vendor to simplify scope and responsibilities; securing early, visible management support and strong user involvement to foster positive engagement; and applying user-centric design standards through iterative, user-driven testing to enhance usability and operator confidence.

Roessler (n.d.) emphasises that effective alarm management depends on sustained organisational readiness through robust change management and leadership engagement. Companies must maintain a documented Management of Change (MOC) process, capturing, reviewing, and approving all alarm modifications, while ensuring ongoing operations, maintenance, and auditing to preserve lifecycle integrity.

More concretely, when considering automation or AI for remote control and alarm management, one might look to operational models that incorporate graded levels of operator readiness, as described by S. T. Porathe (2023) on the MUNIN project for autonomous ships. Passive monitoring involves operators being on call when the system detects a problem, with each readiness level defined by a specific response time based on the operational context. A key challenge in this setup is the time-to-get-in-the-loop, since immediate takeover is not always possible. To address this, a grace period is needed during which the automation manages the situation. If the operator does not respond in time, the system shifts to a fail-to-safe mode suited to the situation (T. Porathe et al., 2020).

Such an approach requires a high level of automation explainability and transparency to ensure timely and effective human intervention. In this context, it has been recommended that Human Readiness Levels (HRLs) be used alongside Technology Readiness Levels (TRLs) to better assess system maturity from both technical and human perspectives. To support this integration, Human Reliability Analysis (HRA) can be applied to inform HRL determination by quantifying human performance and identifying potential sources of error. These approaches allow a scalable supervision and sustainable integration of automation into operational practice (Boring, 2022). Ultimately, embedding HF principles from the concept phase onwards ensures alarm systems are designed not only for compliance but for resilient human performance.

12.7 CONCLUSION AND DIRECTIONS FOR FUTURE RESEARCH

This chapter has traced the evolution of alarm management from the human-alarm system interactions revealed by accident investigations, through established frameworks and standards, to the persistent implementation barriers that continue to undermine best practices of alarm management. The integration of automation and AI introduces further complexity, challenging traditional assumptions about human oversight, data integrity, and system transparency and demands a coordinated response across three domains: data and model governance, regulatory and ethical compliance, and organisational readiness. Five key takeaways emerge from this analysis:

1. Alarm systems must be human-centred.
2. Standards provide structure but not complete solutions.
3. Organisational learning and change management are essential for sustained improvement.
4. AI integration requires transparency and robust governance.
5. Future safety depends on maintaining meaningful human control.

Looking forward, industry-wide collaboration will be essential to bridge the remaining gaps. In their state-of-the-art assessment of AI risks, DNV Energy Systems (2024) recommends three collective initiatives for the petroleum sector:

Joint AI-focused Guidelines. Develop sector-specific best-practice standards to harmonise safety levels and streamline compliance with the EU AI Act.

Automated Unsafe-Condition Detection. Invest in AI-driven monitoring tools to compensate for reduced operator visibility in high-automation environments.

AI Qualification and Maintenance Frameworks. Establish validation and regression-testing protocols tailored to non-deterministic AI outputs to ensure software integrity over time.

These recommendations should be pursued in parallel with the domain-specific measures outlined in Sections 12.4.2 and 12.6. Forming cross-company working groups to draft unified AI guidance, conducting R&D on real-time hazard detectors, and piloting new AI lifecycle assurance methods will together support the transition to truly sustainable, human-centred alarm management.

ACKNOWLEDGEMENT

This work has been supported by a Marie Skłodowska-Curie Innovative Training Network Fellowship of the European Commission's Horizon 2020 Programme under contract number 955901 CISC.

APPENDIX

TABLE 12.A.1
Key International Standards and Guidelines for AI Safety in Process Industries

Document	Issuing Body	Synopsis	Reference
Lifecycle			
AI Standards & Assurance Roadmap	National Standards Authority of Ireland (NSAI)	Developed as part of Ireland's National AI Strategy and informed by the Top Team on Standards in AI, this document sets out actions to align Ireland's AI standards leadership with the requirements of the EU AI Act and strengthen the country's readiness for its implementation.	National Standards Authority of Ireland (NSAI), (2023)
DNV-RP-0671 Assurance of AI-enabled systems	DNV	Defines a governance framework for AI-enabled systems, offering lifecycle guidance to ensure they remain trustworthy and responsibly managed. It forms part of the UK Centre for Data Ethics and Innovation's portfolio of AI assurance techniques.	Haugen et al. (2024)
ISO/IEC 22989 Information technology – Artificial intelligence – Concepts and terminology of artificial intelligence	International Organization for Standardization (ISO) and International Electrotechnical Commission (IEC)	Defines key AI terminology and concepts to support consistent understanding across standards development and stakeholder communications in any organisation.	ISO & IEC (2022), p. 22989
ISO/IEC 23894 Information technology – Artificial intelligence – Guidance on risk management	International Organization for Standardization (ISO) and International Electrotechnical Commission (IEC)	Offers organisations a framework for identifying, assessing, and integrating AI-specific risks into their broader risk-management processes, detailing steps for the effective implementation and continual integration of risk management throughout AI product and service lifecycles.	(ISO/IEC 23894, 2023, p. 23894)

(*Continued*)

TABLE 12.A.1 (Continued)

Document	Issuing Body	Synopsis	Reference
ISO/IEC 42001 Information technology – Artificial intelligence – Management system	International Organization for Standardization (ISO) and International Electrotechnical Commission (IEC)	Establishes a framework for an Artificial Intelligence Management System (AIMS), guiding organisations in the responsible development, deployment, and continual improvement of AI by addressing ethics, transparency, risk management, and governance.	ISO & IEC (2023), p. 42001
Management System Support for Trustworthy Artificial Intelligence	Fraunhofer Research Institute	Assesses the AIMS (ISO/IEC 42001) Draft's ability to help organisations embed trustworthiness in AI lifecycles by comparing its governance, risk, and compliance requirements against existing regulatory and standards frameworks.	(Mock et al., 2021)
Safety			
AMLAS Assurance of Machine Learning for Use in Autonomous Systems	Assuring Autonomy International Programme (AAIP) University of York	Defines a safety-assurance framework for machine learning across its lifecycle, covering scoping, safety requirements elicitation, data management, model learning, verification, and deployment, and provides safety-argument patterns that connect system-level hazard analyses to ML safety claims, explicitly detailing assumptions, trade-offs, and uncertainties.	Richard Hawkins et al. (2021)
IEC 61508 Functional safety of electrical/ electronic/programmable-electronic safety-related systems	International Electrotechnical Commission (IEC)	Defines generic functional-safety requirements for electrical, electronic, and programmable-electronic safety-related systems across all applications, serving as the principal reference for control-system safety design. Unlike its sector-specific counterpart EN 62061, EN 61508 itself is not harmonised.	(IEC, 2010, p. 61508)
IEC 61511 Functional safety – Safety instrumented systems for the process industry sector	International Electrotechnical Commission (IEC)	Specifies requirements for the specification, design, installation, operation, and maintenance of Safety Instrumented Systems (SIS) in the process industry, serving as the sector-specific implementation of IEC 61508:2010.	IEC (2016), p. 61511
Safe AI – how is this possible	Fraunhofer Research Institute	Outlines key challenges to AI safety, including unpredictable model behaviour, specification gaps, and dynamic operating environments, and proposes a rigorous engineering framework for uncertainty reduction, verification, and continuous monitoring to ensure AI systems behave safely within defined confidence limits.	Rueß & Burton (2022)

(Continued)

TABLE 12.A.1 (Continued)

Document	Issuing Body	Synopsis	Reference
The AI Act and The Agile Safety Plan	SINTEF and NTNU	This book provides guidance on developing high-risk AI systems under the EU AI Act (2024/1689), focusing on safety planning aligned with IEC 61508. It covers regulatory context, standardisation, agile safety planning, and both technical and human-organisational considerations.	Myklebust et al. (2025)
Data Quality			
ISO/IEC 5259-1 Artificial intelligence – Data quality for analytics and machine learning (ML) – Part 1: Overview, terminology, and examples	International Organization for Standardization (ISO) and International Electrotechnical Commission (IEC)	Introduces and contextualises the ISO/ IEC 'Artificial Intelligence – Data Quality for Analytics and ML' series, establishing foundational concepts and illustrating associated technologies through use cases and usage scenarios.	(ISO & IEC, 2024a, pp. 5259–1)
ISO/IEC 5259-2 Artificial intelligence – Data quality for analytics and machine learning (ML) – Part 2: Data quality measures	International Organization for Standardization (ISO) and International Electrotechnical Commission (IEC)	Defines a data-quality model, associated metrics, and reporting guidelines for analytics and machine learning, and is intended for use by any organisation seeking to meet its data-quality objectives	ISO & IEC (2024b), pp. 5259–2
ISO/IEC 5259-3 Artificial intelligence – Data quality for analytics and machine learning (ML) – Part 3: Data quality management requirements and guidelines	International Organization for Standardization (ISO) and International Electrotechnical Commission (IEC)	Specifies requirements and guidance for an organisation-wide data-quality management process in analytics and machine learning, covering its establishment, implementation, maintenance, and continual improvement, without mandating specific procedures, metrics, or tools	ISO & IEC (2024c), pp. 5259–3
ISO/IEC 5259-4 Artificial intelligence – Data quality for analytics and machine learning (ML) – Part 4: Data quality process framework	International Organization for Standardization (ISO) and International Electrotechnical Commission (IEC)	Outlines organisation-wide practices for ensuring data quality across the entire lifecycle, acquisition, preparation, labelling, evaluation, and use, in supervised, unsupervised, semi-supervised, and reinforcement learning, without mandating specific tools.	ISO & IEC (2024d), pp. 5259–4
Towards a Standard for Identifying and Managing Bias in Artificial Intelligence	National Institute of Standards and Technology (NIST)	Provides guidelines focusing on identifying and limiting biases in AI algorithms and data, which is essential to ensure fairness and reliability in safety-critical applications.	Schwartz et al. (2022)

Source: Adapted from DNV Energy Systems (2024).

Note

1 Source: https://www.yokogawa.com/eu/solutions/products-and-services/control/control-and-safety-system/distributed-control-systems-dcs/#Details

REFERENCES

Arunthavanathan, R., Sajid, Z., Khan, F., & Pistikopoulos, E. (2024). Artificial intelligence–Human intelligence conflict and its impact on process system safety. *Digital Chemical Engineering*, *11*, 100151.

Bergh, L. I. V., Teigen, K. S., & Dørum, F. (2024). Human performance and automated operations: A regulatory perspective. *Ergonomics*, *67*(6), 744–758. https://doi.org/10.1080/00140139.2024.2321457

Boring, R. (2022). Implications of human reliability analysis for human readiness levels. *Human Error, Reliability, Resilience, and Performance*, *33*(33). https://doi.org/10.54941/ahfe1001573

Brazier, A. (2010). *Operations: A Control Room Is Only a Component in a Complex System*. Article of AB Risk Limited Managing Risks of Control Room Operations.

Briwa, H., Leva, M. C., & Turner, R. (2022). Alarm management for human performance. Are we getting better?. *Book of Extended Abstracts for the 32nd European Safety and Reliability Conference*, 3242–3251. https://doi.org/10.3850/978-981-18-5183-4_S33-05-584-cd

Chandrasegaran, D., Ghazilla, R., & Rich, K. (2020). Human factors engineering integration in the offshore O&G industry: A review of current state of practice. *Safety Science*, *125*, 104627.

CIEHF - Chartered Institute of Ergonomics and Human Factors. (2020). *Learning from Adverse Events*. https://ergonomics.org.uk/resource/learning-from-adverse-events.html

Cullen, L. W. (1990). *The Public Inquiry into the Piper Alpha Disaster* (Vols. 1–2, pp. 146–148, 174). HMSO.

Dehais, F., Lafont, A., Roy, R., & Fairclough, S. (2020). A neuroergonomics approach to mental workload, engagement and human performance. *Frontiers in Neuroscience*, *14*. https://www.frontiersin.org/journals/neuroscience/articles/10.3389/fnins.2020.00268

DNV Energy Systems. (2024). *"State of the Art" for Responsible Use of Artificial Intelligence in the Petroleum Sector* (p. 55). Norwegian Ocean Industry Authority (Havtil).

Endsley, M. R. (1995). Toward a theory of situation awareness in dynamic systems. *Human Factors*, *37*(1), 32–64.

Endsley, M. R. (2018). Automation and situation awareness. In *Automation and Human Performance* (pp. 163–181). CRC Press.

Enemosah, A. (2021). Intelligent decision support systems for oil and gas control rooms using real-time AI inference. *International Journal of Engineering Technology Research & Management*, *5*(12), 236–244.

Engineering Equipment Material Users Association. (2013). *EEMUA 191: 2013 Alarm Systems - Guide to Design, Management and Procurement*. EEMUA.

Goel, P., Datta, A., & Mannan, M. S. (2017). Industrial alarm systems: Challenges and opportunities. *Journal of Loss Prevention in the Process Industries*, *50*, 22–36.

Haugen, O. I., McGeorge, D., Myhrvold, T., Agrell, C., & Hafver, A. (2024). Assurance of AI-enabled systems. *Proceedings of the 2024 8th International Conference on Advances in Artificial Intelligence*, 100–106. https://doi.org/10.1145/3704137.3704153

Hawkins, R., Paterson, C., Picardi, C., Jia, Y., Calinescu, R., & Habli, I. (2021). *Guidance on the Assurance of Machine Learning in Autonomous Systems (AMLAS)*. Assuring Autonomy International Programme (AAIP). University of York. https://www.york.ac.uk/media/assuring-autonomy/documents/AMLASv1.1.pdf

HSE. (1997). *The Explosion and Fires at the Texaco Refinery, Milford Haven, 24 July 1994: A Report of the Investigation by the Health and Safety Executive Into the Explosion and Fires on the Pembroke Cracking Company Plant at the Texaco Refinery, Milford Haven on 24 July 1994*. Incident Report Series.

IEC. (2010). *IEC 61508-1:2010—Functional Safety of Electrical/Electronic/Programmable Electronic Safety-Related Systems—Part 1: General Requirements (see Functional Safety and IEC 61508)*. International Electrotechnical Commission (IEC).

IEC. (2016). *IEC 61511-1:2016 Functional Safety—Safety Instrumented Systems for the Process Industry Sector—Part 1: Framework, Definitions, System, Hardware and Application Programming Requirements*. International Electrotechnical Commission (IEC).

IEC. (2022). *IEC 62682:2022 Management of Alarm Systems for the Process Industries*. International Electrotechnical Commission (IEC).

ISA & ANSI. (2016). *ISA-18.2:2016 Management of Alarm Systems for the Process Industries*. International Society of Automation.

ISO. (2008). *ISO 11064-5:2008 Ergonomic Design of Control Centres Part 5: Displays and Controls*. International Organization for Standardization.

ISO. (2016). *ISO 6385:2016 Ergonomics Principles in the Design of Work Systems*. International Organization for Standardization.

ISO & IEC. (2022). *ISO/IEC 22989:2022 Information Technology—Artificial Intelligence—Artificial Intelligence Concepts and Terminology*. International Organization for Standardization (ISO) & International Electrotechnical Commission (IEC).

ISO & IEC. (2023). *ISO/IEC 42001:2023 Information Technology—Artificial Intelligence—Management System*. International Organization for Standardization & International Electrotechnical Commission.

ISO & IEC. (2024a). *ISO/IEC 5259-1:2024 Artificial Intelligence—Data Quality for Analytics and Machine Learning (ML) Part 1: Overview, Terminology, and Examples*. International Organization for Standardization & International Electrotechnical Commission.

ISO & IEC. (2024b). *ISO/IEC 5259-2:2024 Artificial Intelligence—Data Quality for Analytics and Machine Learning (ML) Part 2: Data Quality Measures*. International Organization for Standardization & International Electrotechnical Commission.

ISO & IEC. (2024c). *ISO/IEC 5259-3:2024 Artificial Intelligence—Data Quality for Analytics and Machine Learning (ML) Part 3: Data Quality Management Requirements and Guidelines*. International Organization for Standardization & International Electrotechnical Commission.

ISO & IEC. (2024d). *ISO/IEC 5259-4:2024 Artificial Intelligence—Data Quality for Analytics and Machine Learning (ML) Part 4: Data Quality Process Framework*. International Organization for Standardization & International Electrotechnical Commission.

Janssen, M., Brous, P., Estevez, E., Barbosa, L. S., & Janowski, T. (2020). Data governance: Organizing data for trustworthy Artificial Intelligence. *Government Information Quarterly*, *37*(3), 101493. https://doi.org/10.1016/j.giq.2020.101493

Johnsen, S. O. (2021). Mind the gap between automation and meaningful human control, through standards. *Proceedings of the 31st European Safety and Reliability Conference (ESREL)* (pp. 2260–2267). https://doi.org/10.3850/978-981-18-2016-8_499-cd

Johnsen, S. O., Thieme, C., Aminoff, H., Park, J.; Fernander, M., Boren, M., Kjøl, Å. A., & Kirwan, B. (2025). *CRIOP A Scenario Method for Crisis Intervention and Operability Analysis*. Norway: SINTEF Technology and Society.

Johnsen, S. O., & Winge, S. (2023). Human Factors and safety in automated and remote operations in oil and gas: A review. *ESREL 2023-Proceedings of the 33rd European Safety and Reliability Conference: The Future of Safety in the Reconnected World*, 3–7 September 2023, University of Southampton, Southampton, UK.

Kotter, J. P. (2012). *Leading Change*. Harvard Business Press.

Madiega, T. (2021). *Artificial Intelligence Act*. https://www.dirittobancario.it/wp-content/uploads/2024/09/Documento-riassuntivo-dellAI-Act-Parlamento-europeo-settembre-2024.pdf

McLeod, R. (2022). *Human Factors in Highly Automated Systems* [White Paper]. Chartered Institute of Ergonomics and Human Factors.

Mock, M., Schmitz, A., Adilova, L., Becker, D., Cremers, A. B., & Poretschkin, M. (2021). *Management System Support for Trustworthy Artificial Intelligence*. https://publica.fraunhofer.de/entities/publication/5421f1f0-7768-4b90-90c2-475374957bed/details

Myklebust, T., Stålhane, T., & Vatn, D. M. K. (2025). *The AI Act and The Agile Safety Plan*. Springer Nature Switzerland.

National Academies of Sciences, Engineering and Medicine. (2022). *Human-AI Teaming: State-of-the-Art and Research Needs*, vol. 10, p. 26355. Washington, DC: The National Academies Press.

National Standards Authority of Ireland (NSAI). (2023). *AI Standards & Assurance Roadmap*. National Standards Authority of Ireland. www.nsai.ie

National Transportation Safety Board. (2012). *Pipeline Accident Report: Enbridge Incorporated Hazardous Liquid Pipeline Rupture and Release*. https://www.ntsb.gov/investigations/AccidentReports/Reports/PAR1201.pdf

Norwegian Safety Investigation Authority (NSIA). (2024). *Report Marine: Loss of Propulsion and Near Grounding of Viking Sky, Hustadvika, Norway*, 23 March 2019. https://iims-media-library.s3.eu-west-2.amazonaws.com/wp-content/uploads/2024/04/03152300/NSIA-Viking-Shy-cruise-ship_2024_03.pdf

NPD. (2001). *YA-711—Principles for Alarm System Design*. The Norwegian Petroleum Directorate (NPD).
NTSB. (2014). *Columbia Gas Transmission Corporation Pipeline Rupture Sissonville, West Virginia*. National Transportation Safety Board.
O'Hara, J. M., & Fleger, S. (2020). *NUREG-0700 Revision 3: Human-System Interface Design Review Guidelines*. https://www.nrc.gov/docs/ML2016/ML20162A214.pdf
Oplenskedal, A., & Bjørheim, L. G. (2019). *Investigation of Collision Between Sjoborg Supply Ship and Statfjord A*. Pet Saf Authority Norway.
Parasuraman, R., Sheridan, T. B., & Wickens, C. D. (2000). A model for types and levels of human interaction with automation. *IEEE Transactions on Systems, Man, and Cybernetics-Part A: Systems and Humans*, *30*(3), 286–297.
Porathe, S. T. (2023). Alarm and hand-over concepts for human remote operators of autonomous ships. *Proceedings of The 33rd European Safety and Reliability Conference (ESREL'23).*
Porathe, T., Fjortoft, K., & Bratbergsengen, I. L. (2020). Human factors, autonomous ships and constrained coastal navigation. *IOP Conference Series: Materials Science and Engineering*, *929*(1), 012007. https://doi.org/10.1088/1757-899X/929/1/012007
Poştovei, D. A., Bulac, C., Triştiu, I., & Camachi, B. (2020). The evolution and challenges of modern distributed control systems. *IEEE 14th International Symposium on Applied Computational Intelligence and Informatics (SACI)*, 000089–000094.
Roessler, D. (n.d.). *Implementing Alarm Management: Business Value Beyond Compliance* [White Paper]. TiPS Incorporated. https://tipsweb.com/wp-content/uploads/Alarm-Management-Business-Value-Beyond-Compliance.pdf
Rueß, H., & Burton, S. (2022). Safe AI – How is this possible? (arXiv:2201.10436). https://doi.org/10.48550/arXiv.2201.10436
Saghafian, M., Vatn, D. M. K., Moltubakk, S. T., Bertheussen, L. E., Petermann, F. M., Johnsen, S. O., & Alsos, O. A. (2025). Understanding automation transparency and its adaptive design implications in safety–critical systems. *Safety Science*, *184*, 106730.
Schliemann-Haug, A. C. (2024). *Alarm Management in the Maritime Industry Volume 1*. Lloyd's Register.
Schwartz, R., Schwartz, R., Vassilev, A., Greene, K., Perine, L., Burt, A., & Hall, P. (2022). *Towards a Standard for Identifying and Managing Bias in Artificial Intelligence*, vol. 3. US Department of Commerce, National Institute of Standards and Technology.
Segovia, V. R., & Theorin, A. (2012). History of control history of PLC and DCS. *University of Lund*, *44*, 45.
Standard Norge (SN). (2004). *NORSOK S-002N:2004—Working Environment*. Standard Norge.
Stanton, N. A., Salmon, P., Jenkins, D., & Walker, G. (2009). *Human Factors in the Design and Evaluation of Central Control Room Operations*. CRC Press.
The Energy Institute. (2016, July). *Guidance for Optimising Operator Situational Awareness by Rationalising Control Room Alarms*. The Energy Institute.
U.S. Chemical Safety and Hazard Investigation Board (USCSB). (2007). Investigation report: Refinery explosion and fire (15 killed, 180 injured). BP Texas City, Texas, March 23, 2005 (Report No. 2005-04-I-TX).
Ueland, A., Teigen, K. S., Bergh, L. I. V., & Embergsrud, A. H. (2022). *Ansvar, kompetanse og vedlikehold av alarmhåndteringssystemer i kontrollrom*. (Responsibilities, Competence, and Maintenance of Alarm Management Systems in Control Rooms). Petroleum Safety Authority Norway. https://www.havtil.no/contentassets/5649e607f64f482cb89ad487984b5343/2021_93-rapport-om-ansvar-kompetanse-og-vedlikehold-av-alarmhandteringssystemer-i-kontrollrom.pdf
United States Coast Guard. (2011). *Report of Investigation into the Circumstances Surrounding the Explosion, Fire, Sinking and Loss of Eleven Crew Members Aboard the Mobile Offshore Drilling Unit Deepwater Horizon in the Gulf of Mexico* (Vol. 1). United States Coast Guard.
USCSB. (2019). *Pryor Trust Fatal Gas Well Blowout and Fire*. U.S. Chemical Safety and Hazard Investigation Board (CSB). https://www.csb.gov/pryor-trust-fatal-gas-well-blowout-and-fire/
USCSB. (2024). *Fatal Naphtha Release and Fire at BP-Husky Toledo Refinery* [Investigation Report]. U.S. Chemical Safety and Hazard Investigation Board (CSB). https://www.csb.gov/assets/1/6/final_report_-_20241.pdf

Part 4

Workload Assessment and Training

13 Workload Assessment for Modern Industrial Systems

Gerald Matthews, Dual Herzog, and Jinchao Lin

13.1 INTRODUCTION

Mental workload is most simply defined as the magnitude of cognitive demands imposed by a task. More sophisticated definitions specify that workload reflects the extent to which the person can manage demands through allocation of attention and effort, as further discussed in Section 13.1.1.

High workload is a familiar human experience, especially within industrialized societies. Meeting tight deadlines at work, driving in heavy traffic, and keeping up with a high volume of calls and emails are typical examples. People experience workload in a graded fashion along a spectrum ranging from idleness through comfortable levels of activity to overload. The extremes of overload and underload are often experienced as aversive and stressful, but not always. High workload may be accepted or even welcomed when the task performed is challenging and motivating. In general, though, people are happiest and most effective at moderate workload levels.

In industrial settings, workload assessment can contribute to enhancing productivity, safety, and wellbeing. If we can identify when workers are overloaded, we can redesign the task, allocate additional personnel or artificial support systems to relieve the burden, or train workers to manage task demands better. These are activities that should be done as early as possible to support safety by design. It can also be important to mitigate underload, especially in an age of industrial automation in which employees are tasked with monitoring automated systems that rarely fail.

To manage workload, we must measure it reliably and validly, which has proved to be surprisingly difficult. We can measure workload in different ways, from subjective reports and questionnaires, from psychophysiological measures of brain activity, and from changes in performance. As discussed below, each of these approaches has value, but different measures may fail to converge. A person might report high workload, but their brain activity shows nothing unusual, for example. It is also important to interpret workload measures within a broader psychological and organizational context. A team leader working to complete an exciting, innovative project that will transform a company's prospects might experience high workload. So too might a worker on a fast-paced assembly line, aiming only to hold onto their job for a few months until they retire. These two individuals might get the same score on a workload scale, but the consequences of that workload for the person and for the organization are likely to be very different.

This chapter provides an overview of how psychologists define and measure mental workload in applied settings. We focus especially on applications to industry, broadly defined to include activities such as commercial transportation and "knowledge work," as well as traditional factory and plant operations. We aim to provide a high-level overview of the principal issues for researchers and practitioners; for a more detailed account, see Matthews and Reinerman-Jones (2017). The development of workload assessment methods to manage the various challenges of new technologies, such as Artificial Intelligence (AI), merits special attention. We will conclude by emphasizing the need for contextual interpretation of workload scores, as part of a multidimensional response to operational demands.

DOI: 10.1201/9781003741824-17

13.1.1 Defining Workload

To define workload, we must distinguish the external demands of the task performed from the person's responses to demands. Defining "demands" can itself be challenging, especially in complex operational environments, but some elements can be defined objectively, such as the number of items to be held in short-term memory, or the number of subtasks to be performed concurrently. Workload, however, refers to a relationship between demands and the person's capacity to manage those demands and maintain effective task performance (Cain, 2007; Matthews & Reinerman-Jones, 2017; Young et al., 2015). A simple, intuitive definition is that workload is the difference between the processing capacity needed to process task stimuli effectively and the amount of capacity actually available. Even more simply, workload could be defined as the amount of effort applied to the task.

13.1.1.1 Attentional Resource Theory and Workload

Some researchers have advanced beyond intuitive definitions by grounding the definition of workload in attentional resource theory (Navon & Gopher, 1979). This theory assumes that certain types of information processing, especially those that are conscious and deliberate, require a metaphorical pool of energy or fuel to operate, experienced subjectively as mental effort (Matthews et al., 2000). We can think of mental processes as being like light bulbs powered by a battery. As we attach more and more lights (or replace low-wattage bulbs with high-wattage ones), eventually the battery cannot provide enough power, and the lights start to dim. That is, capacity required exceeds capacity supplied, and workload represents the difference between needed and available capacity.

The strength of resource theory is that it allows us to specify how people respond to changing demands, even with complex, operational tasks for which it is difficult to specify in detail how information is processed. Resource theory explains how individual task difficulty, needs for multitasking, and compromised operator functioning can all lead to resource insufficiency and performance failures. It also explains the superior performance of experts relative to novices, which in part reflects the automatization of consistent elements of processing and reduced needs for processing capacity (Ackerman, 1988). The theory also highlights the importance of allocation as well as availability of resources. Resources can be allocated voluntarily to different processing activities, but a poor strategy for allocation may result in performance impairments. Although beyond our present scope, there has also been extensive neuropsychological research on brain circuits for attention that may underlie resource availability.

13.1.1.2 Challenges to Resource Theory

There are two basic challenges to resource theory, which in turn affect definitions of workload. One issue is that lack of resources is not the only explanation for performance breakdown when tasks are demanding (Matthews et al., 2000). Breakdown in the form of errors or slowed response may reflect overload of a single component of the cognitive architecture for processing, rather than a general mechanism for energization. For example, verbal short-term memory has a limited capacity. Failure to keep a long shopping list in memory might reflect overload of this specific memory store rather than any general resource. Similarly, slowed color-naming on the color-word Stroop test reflects response competition between the two-color stimuli rather than lack of resources. When performing two complex tasks simultaneously such as having separate conversations with two different people, performance may be impaired due to failure to segregate the two streams of processing. Collectively, performance failures of these kinds are attributed to *structural interference*, competition for a specific processing unit rather than competition for a limited resource. Subjectively, we know when we are struggling to perform a task, but we do not necessarily know what the source of our difficulties is. Subjective workload is unlikely to be a pure measure of resource allocation.

A second issue is that there may not be a single general-purpose resource for processing. Instead, there may be multiple resources, each of which energizes a different type of processing. For example, people can often combine performance of a visual and an auditory task quite successfully, suggesting that tasks in different sensory modalities require different multiple resources. The most elaborate multiple resource theory (Wickens, 2008) categorizes resources by modality, by stage of processing (perception, central processing, and response), and by type of processing (verbal or spatial). Multiple resource theory implies that workload too is a multidimensional construct (Boles & Dillard, 2015).

A compromise view is that there are multiple resources, but complex tasks require control of processing by a supervisory executive system which draws on a general resource (Cooper & Shallice, 2000). From an applied perspective, it is often useful to define workload as unitary and identify levels of workload that may be problematic. However, for some applications, such as interface design, multiple resource theory is more useful. Poor design might variously lead to difficulties in perceiving displays (input), in making decisions based on displayed information (central processing), or in responding using the available controls (output). In such cases, breaking down the workload by the stage of processing facilitates design solutions to performance issues (Hart & Wickens, 1990).

13.1.2 Applications of Workload Assessment

Historically, workload research is often traced to concerns about industrial health and efficiency in the years following World War 1. High workload, along with long work hours and poor working environments, was one of a number of factors perceived as threats to productivity and occupational health. Renewed focus on the cognitive limitations of pilots and other military personnel during World War 2 was a further stimulus to research. Workload assessment remains important for traditional domains of inquiry, including manufacturing industry, power generation, military operations, and civilian ground and air transportation.

13.1.2.1 Workload Assessment in the Digital Age

Table 13.1 illustrates some new trends for workload assessment across a range of applied domains. Some general themes are as follows.

13.1.2.1.1 Interaction with Automated and Autonomous Systems

Both overload and underload are concerns for interactions with systems (Parasuraman & Riley, 1997). Designing systems and interfaces that do not impose excessive workload is a longstanding concern of human factors practitioners. Autonomous systems may elevate workload in various ways, including

TABLE 13.1
Emerging Issues for Workload Assessment in the Digital Age

Domain of Application	Emerging Issues
Aerospace	Unmanned autonomous systems Support systems, including augmented reality and decision support
Surface transportation	Vehicle automation In-car systems and distraction
Industry	Training for – and with – new technology Human-robot interaction
Health	Laparoscopic surgery Health information systems and technologies
Education	Multimedia and virtual reality training Instructional design for online learning
Cybersecurity	Compliance with security protocols High-volume cyber defense
Military applications	Human-machine teaming AI for Command and Control (C2)

lack of transparency and explainability. In some contexts, such as cyber defense, the capabilities of the technology to generate a high volume of alerts may overload the operator. By contrast, the widespread introduction of automation has also revitalized concerns about underload, as workers shifted from being skilled machine operators to supervisors of automation. We return to these issues in Sections 13.3 and 13.4.

13.1.2.1.2 Workload Transitions

Sudden shifts in task demands, in both upward and downward directions, can be destabilizing for operators. Systems that require abrupt shifts in level of operator control, such as partly autonomous vehicles that require periodic manual takeover, are vulnerable to these effects. Transitions may also be driven by external events and emergencies such as industrial accidents and disasters.

13.1.2.1.3 Training and Technology

In general, moderate demands that challenge but do not overload the trainee are optimal for learning. As training is increasingly facilitated by technologies including immersive simulation and extended reality (XR), it remains important to assess the workload imposed by new methods. Cognitive load theory for education and training distinguishes intrinsic load that reflects the complexity of learning materials from extraneous load that results from how the learner receives and interacts with materials. The theory has been applied to various forms of instructional design (Chen, Paas & Sweller, 2023).

13.1.2.1.4 Teaming

Team performance imposes demands associated with accomplishing the goals of the task (taskwork) and with coordinating team activities (teamwork). Most workload research addresses demand on individual workers, but teams as well as individuals can be taken as the unit of analysis for assessment. We can evaluate the ability of the team as a whole to manage task demands. For example, a team may fail to meet performance challenges if it fails to communicate effectively, even if individual team members are highly proficient. Scales have been developed to assess the collective workload of teams (Liu et al., 2024).

Teaming with autonomous systems raises novel workload challenges (van de Merwe, Mallam & Nazir, 2024). The demands of taskwork may be increased by factors such as poor design of interfaces and displays, suboptimal allocation of subtasks to humans and autonomy, and failures to trust the system appropriately. Similarly, teamwork may be demanding when humans and autonomy fail to communicate and coordinate effectively. Mitigating demands requires design solutions such as transparency of machine operation that support shared situational awareness, as well as training in management of autonomy. Workload assessment during the system design process is important both for safety and to support effective human-machine collaboration. AI has the potential to massively increase the capabilities of teams, but it can also amplify the challenges of collaboration, as further discussed in Section 13.4.1.

13.1.2.2 Goals of Applied Workload Assessment

Workload assessment typically addresses one or more of the following goals across the different domains of application: *safety*, *productivity*, and *wellbeing* (Matthews & Reinerman-Jones, 2017). The first, and often the most important goal, is to prevent excessive workload from causing human errors that threaten safety. What level of workload is "safe" needs to be determined with the application context concerned by investigating how error probabilities track workload. Workload impacts can be determined from multiple lines of evidence, including experimental studies using work simulation, observation of the operational environment, and input from subject matter experts (SMEs). Threats to performance associated with underload such as loss of vigilance and task engagement may also need to be assessed.

A second goal is to optimize productivity. Extremes of workload can impair the efficiency of task performance even if there is no threat to safety. In general, people perform best at moderate levels of workload (Hancock & Warm, 1989), but workload-performance associations vary with context. Expertise, intrinsic task motivation, perceptions of the task as a manageable challenge, and strategy for resource allocation all contribute to maintaining performance under high workloads (Matthews, Wohleber & Lin, 2019).

A third goal is to promote wellbeing. Prolonged periods of very low or high workload in organizational settings tend to be stressful (e.g., Ganster & Rosen, 2013). Chronic work stress is associated with mental and physical issues, including depression and burnout. As for performance, contextual factors such as task engagement moderate the impact of workload on wellbeing, but even highly motivated employees can be vulnerable to burnout from excessive chronic workload. Occupational health depends on multiple factors, and workload impacts should be considered in the context of factors such as control over work activities, supervisor and organizational support, and the individual's resilience.

13.2 METHODS OF WORKLOAD ASSESSMENT

Assessing cognitive workload is essential in high-stakes industries such as oil, gas, and nuclear energy, where operators must transition rapidly from low to high cognitive demands. Effective measurement helps inform system design, training, and safety protocols by capturing how individuals experience and manage cognitive load in dynamic environments. A range of assessment methods exists – subjective, behavioral, and physiological – each offering a different lens on operator performance and mental effort. Because workload is a multidimensional construct, these methods may not always align, and each presents unique strengths and limitations. Understanding how they function in practice – and how they complement one another – can support more informed implementation in industrial settings.

13.2.1 Evaluating Workload Measures

Before reviewing different assessment methods, we will discuss how researchers and practitioners evaluate the different measures available. Table 13.2 shows a widely used set of criteria developed by Eggemeier et al. (1991). They can be used as a checklist for determining the suitability of a measure for a particular application. The first three criteria establish the validity of the measure.

TABLE 13.2
Eggemeier et al.'s (1991) Criteria for Evaluating Workload Metrics

Criterion	Description
Sensitivity	Capacity of the instrument to detect changes in task difficulty or cognitive demands
Diagnosticity	Capacity of the instrument to differentiate distinct sources of workload, such as specific capacities or multiple resources
Selectivity	Sensitivity of the instrument only to differences in cognitive demands, not to changes in other factors (e.g., physical workload, stress response)
Reliability	Consistent assessment of mental workload
Intrusiveness	Lack of interference with primary-task performance
Implementation requirements	Practical constraints associated with instrumentation, software, and training
Operator acceptance	Operator perception of the validity and usefulness of the procedure

Sensitivity is the most basic criterion; changes in task demands should "move the needle" on the workload gauge. Measures differ in *diagnosticity*. Some simply provide a workload value, where others are informative about the factors that produce workload. *Selectivity* is often challenging. Workload measures may be influenced by other factors. For example, stress can elevate subjective workload, and physical activity raises heart rate. *Reliability* is a basic requirement for any measurement instrument. If we measure workload repeatedly, on the same task under the same conditions, the instrument should give similar values each time. Reliability can also be assessed with statistics for internal consistency, such as Cronbach's α. The remaining criteria – *intrusiveness*, *implementation*, and *operator acceptance* – refer to practical issues that arise in assessment. They are especially important in applied settings.

The measures we will review all have demonstrated sensitivity and reliability, at least in some contexts. They tend to differ more with respect to the remaining criteria, whose respective importance varies with the context for use. For example, diagnosticity may not be critical if the aim is simply to establish a "redline" or maximum value of acceptable workload for an operational task.

13.2.2 Subjective Measures

Subjective measures rely on self-report from the operator, typically after task completion. These tools are cost-effective, easy to administer, and well-suited to environments where physiological or behavioral monitoring is impractical. In operational conditions, where physiological and behavioral tools may interfere with task performance or prove difficult to deploy, subjective ratings may be the most feasible option for capturing workload under routine and emergency conditions.

Subjective ratings often reflect the perceived effort, task difficulty, emotional strain, and other internal experiences not easily observable from the outside. However, their interpretability can vary based on timing, context, and individual differences in introspective accuracy or willingness to report difficulty. For example, high immersion in a task ("flow") may encourage effort but reduce the person's awareness of workload (Gründling et al., 2023).

13.2.2.1 Unidimensional Scales

These include simple numeric scales intended to capture overall workload with minimal diagnostic detail. The Modified Cooper-Harper Scale (MCH) and Overall Workload (OW) scale fall into this category (Hill et al., 1992; Vidulich & Tsang, 1987; Wierwille & Casali, 1983). They are efficient and fast to administer but may lack sensitivity to specific workload components such as mental effort versus time pressure.

13.2.2.2 Multidimensional Scales

The most widely used multidimensional tool is the NASA Task Load Index (NASA-TLX), which assesses six dimensions: mental demand, physical demand, temporal demand, performance, effort, and frustration (Hart & Staveland, 1988). Other tools, such as the Subjective Workload Assessment Technique (SWAT: Reid & Nygren, 1988) and the Dundee Stress State Questionnaire (DSSQ) (Matthews et al., 2002), provide broader insight into task engagement, distress, and cognitive-affective states relevant to workload.

13.2.2.3 Strengths and Limitations

Subjective measures have strong face validity and are widely accepted in both research and practice. Tools like the NASA-TLX are commonly used across a variety of tasks and domains. Their key strengths are simplicity, adaptability, and ease of use, making them especially valuable during post-task debriefings, after-action reviews, or training evaluations. However, they also come with notable limitations as measurement instruments (de Winter, 2014). Retrospective bias can distort responses, especially if ratings are influenced by task outcomes. These measures also lack the temporal resolution needed for real-time monitoring. Additionally, responses are inherently subjective and can vary across individuals experiencing similar task demands. Because subjective tools rely on personal interpretation and delayed reporting, they are better suited for capturing overall impressions of workload rather than precise, moment-to-moment changes.

13.2.3 Behavioral Measures

Behavioral measures assess workload by tracking either primary or other task performance, often using built-in system data like alarm responses or error rates (Butmee, Lansdown & Walker, 2019). Evaluating the changes in these metrics can provide an indirect measure of an individual's workload. However, skilled operators may maintain performance through compensatory strategies, masking workload unless multiple indicators are used.

13.2.3.1 Primary-Task Performance

These measures reflect observable aspects of task performance, including metrics like error rates, task completion times, and operational accuracy. In practice, they may involve delays in detecting alarms, inefficient navigation behaviors such as excessive interface switching or repeated commands, or degraded responsiveness during routine procedures. Signs like delayed reactions to critical events or failure to promptly address system fluctuations can indicate increased cognitive workload.

13.2.3.2 Secondary Task Performance

Secondary tasks are added alongside a primary task to assess spare attentional capacity, offering an indirect yet sensitive measure of cognitive workload (Ogden, Levine & Eisner, 1979). For example, an individual might be asked to press a button in response to a tone or monitor a peripheral display while managing another task. Declines in secondary task performance typically indicate that cognitive resources are being taxed by the primary activity. While effective in controlled environments, secondary tasks are often impractical in applied settings. They can divert attention from essential responsibilities and implementing them without disrupting performance requires careful design (Young & Stanton, 2007). As a result, their use is generally limited to simulations, training contexts, or research applications.

13.2.3.3 Strategic Behavior and Adaptation

Under high workload, individuals often adapt their interaction patterns in ways that reflect cognitive strain (Matthews et al., 2000). These adaptations may include longer decision times, increased reliance on automated defaults, reduced frequency of manual inputs, or omission of non-essential checks. While such strategies can help preserve core task performance under pressure, they may also signal depleted attentional resources and reduced situational awareness. To capture these subtle shifts, behavioral workload measures often track indicators such as the time it takes to resume a task after an interruption, latency in detecting anomalies or errors, and deviations from typical task sequences. These metrics provide insight into the operator's capacity to monitor and manage system demands, especially when mental resources are stretched. However, interpreting such behaviors requires careful context analysis, as some adaptations may reflect efficient prioritization rather than degradation.

13.2.3.4 Strengths and Limitations

Behavioral measures are typically non-intrusive and seamlessly integrated into existing systems, enabling continuous data collection during live operations. They can be automatically captured through human-machine interfaces, making them scalable across industrial sites. However, their usefulness depends on context: performance is task-specific and may not generalize, compensatory strategies can mask workload, and interpretation often requires baseline comparisons – such as evaluating whether a technician's current response time deviates from their usual pattern.

These measures are most informative when combined with physiological or subjective data. For example, a pattern of precise task execution paired with delayed reactions to unexpected events or prolonged pauses between actions may indicate elevated cognitive load, even in the absence of overt errors.

13.2.4 Physiological Measures

Physiological measures capture involuntary bodily responses linked to cognitive effort, stress, and arousal, offering insight into workload changes that may not be evident through performance or self-report (Charles & Nixon, 2019). In operational environments, they support real-time monitoring, fatigue management, and post-incident analysis. These measures are typically categorized by system type: Central Nervous System (CNS) activity, Autonomic Nervous System (ANS) responses, eye movements, and speech and facial cues.

13.2.4.1 Central Nervous System (CNS) Measures

CNS-based methods assess brain activity related to attention and cognitive control. The electroencephalogram (EEG), the most common in laboratory settings, provides high temporal resolution and links changes in brain-wave activity to greater mental effort (Gevins et al., 1995). Wearable EEG systems now allow limited use in operational settings. Functional near-infrared spectroscopy (fNIRS), which measures prefrontal blood oxygenation, is another popular measure. However, fNIRS has lower temporal resolution than EEG and is more sensitive to motion artifacts, limiting its effectiveness in highly dynamic or mobile operational environments.

13.2.4.2 Autonomic Nervous System (ANS) Measures

ANS measures reflect physiological arousal and regulation via the sympathetic and parasympathetic systems. Common indicators include heart rate, heart rate variability, respiratory rate, blood pressure, and electrodermal activity. These signals, often recorded using wearables like chest straps or wristbands, can indicate rising workload but are also susceptible to confounding factors such as movement, emotion, and environmental conditions.

13.2.4.3 Oculometrics

Oculometric measures track visual attention and workload through indicators such as pupil diameter, blink rate and duration, fixation patterns, and scan path variability (Marquart, Cabrall & De Winter, 2015). The Index of Cognitive Activity (ICA) isolates pupil changes tied to mental effort and is less sensitive to lighting than raw pupil diameter (Marshall, 2002). These metrics are especially useful in settings where visual monitoring is central, like control rooms or inspection terminals. However, most systems require an individual to be stationary to perform accurately.

13.2.4.4 Speech and Facial Metrics

Vocal and facial responses offer unobtrusive indicators of workload. Increased speech pauses and changes in pitch and loudness can signal cognitive strain (Brenner et al., 1994), while micro-expressions and facial electromyography (EMG) detect subtle muscle activity linked to stress or focus. These measures are particularly useful in settings where communication is closely tied to task performance, such as alarm coordination or team-based operations. However, implementation is difficult, making these measures uncommon.

13.2.4.5 Strengths and Limitations

Physiological measures provide continuous, objective data that can detect workload changes not visible through performance or self-report, making them especially useful for identifying early signs of fatigue, stress, or overload. However, they come with several limitations. Physiological responses often lack specificity, as similar patterns may result from stress, mental effort, or environmental influences. For example, elevated heart rate or increased skin conductance might reflect high cognitive demand, but they could just as easily be triggered by heat, anxiety, or physical exertion, making it difficult to isolate cognitive workload as the source. Many methods also require baseline calibration, trained personnel, and extensive data processing. Additionally, some sensors remain too intrusive or fragile for rugged industrial conditions.

Despite earlier challenges, advances in wearable and contactless technologies have made physiological assessment more practical and accessible. These improvements have led to growing use in dynamic fields like defense and industrial operations, where such data supports adaptive systems and helps manage operator workload. However, there are concerns that wearables may elicit the Hawthorne effect; that is, awareness of being monitored may influence the behavior of research participants (Lastoria et al., 2025).

13.2.5 Divergence among Measures

A key challenge in workload assessment is that different types of measures often produce divergent results – a phenomenon known as dissociation (Hancock & Matthews, 2019). This occurs when subjective ratings, behavioral performance, and physiological signals do not align consistently. Alternate physiological measures can also fail to converge with one another (Matthews et al., 2015). In industrial settings, such discrepancies can complicate diagnosis of operator strain, evaluation of training, or validation of system changes.

For example, a technician might report high workload on the NASA-TLX while showing stable heart rate variability and unaltered performance metrics. Alternatively, physiological indicators may reveal elevated arousal even when self-reported workload is low, possibly due to confounds such as underreporting influenced by overconfidence or social desirability.

Dissociation does not necessarily indicate measurement failure; rather, it may highlight workload's multidimensional nature (Matthews & Reinerman-Jones, 2017). Each method captures different aspects: subjective measures reflect perceived effort or emotional strain, behavioral measures track outcomes that may remain stable despite rising effort, and physiological signals detect internal states that may not be outwardly visible. This complexity is especially relevant in high-stakes environments, where workload involves constant shifts in task demands, vigilance, and environmental pressures.

Recognizing dissociation has practical value. Sole reliance on performance metrics may miss hidden strain, while subjective ratings alone may underestimate overload during critical tasks. A multi-method approach helps interpret discrepancies not as errors, but as meaningful indicators of different dimensions of the operator's experience.

13.2.6 Practical Considerations for Implementation

Implementing workload assessment in operational settings requires balancing technical capabilities with practical realities. Even well-validated tools must align with on-the-ground workflows and constraints to be effective. Understanding how these factors shape implementation can help ensure that workload monitoring supports rather than disrupts operations.

13.2.6.1 Intrusiveness and Disruption

The degree to which a tool interferes with task execution is a major concern. Physiological sensors like EEG or facial EMG may be impractical where safety gear, mobility, or confined spaces limit usability. Similarly, secondary task methods, though informative, can divert attention from primary duties if poorly integrated. Subjective assessments are least intrusive but require time and focus, making them difficult to use during shifts or emergencies. As a result, many organizations reserve these for post-task debriefs, sacrificing real-time accuracy.

13.2.6.2 Technical and Environmental Constraints

Industrial environments introduce noise, vibration, lighting changes, and electromagnetic interference, all of which can degrade signal quality – especially for sensitive sensors like eye trackers or voice analyzers. Devices must be rugged, waterproof, and safety-compliant. For wearable tech, battery life and wireless reliability are critical. Tools that require controlled lighting or stable postures may perform poorly in the field.

13.2.6.3 Cost and Infrastructure Requirements

Costs vary widely. Subjective tools are low-cost, while advanced physiological systems require expensive equipment, trained staff, and integration with existing IT systems. Cost-benefit tradeoffs must consider the long-term value of incident prevention and fatigue management. For example, while wearable EEG or eye-tracking systems may entail substantial upfront investment, their ability to detect early signs of overload could help prevent costly errors or downtime. Conversely, relying solely on inexpensive self-report tools may miss subtle workload shifts that precede performance degradation, particularly in dynamic or high-pressure environments. Selecting the right balance depends on available resources, system criticality, and the level of precision required.

13.2.6.4 Data Management and Interpretation

Collecting data is only the first step – extracting meaningful insights is just as important. Physiological data require preprocessing, baseline comparisons, and attention to confounds such as caffeine or emotional state. Behavioral and subjective data are easier to work with but may still need normalization. Effective data management involves organizing streams from multiple sources, ensuring time alignment, and handling missing or noisy data. Interpretation depends not only on statistical patterns but also on understanding task context and individual variability.

13.2.6.5 Operator Acceptance and Organizational Culture

Successful implementation hinges on user acceptance. Operators may resist physiological monitoring if it feels invasive or they worry that subjective ratings could be used against them. Transparency, privacy protection, and clear communication about data use are essential. Trust is easily eroded if data are ignored or used punitively. When framed as tools for safety and resilience – not surveillance – workload measures are more likely to be embraced.

13.2.7 Sensitivity of Workload Assessment

A successful workload assessment should be sensitive to workload variations and have its value changes as cognitive demands change (Matthews & Reinerman-Jones, 2017). In other words, if we consistently see higher values for the metric in various high-demand conditions, we can be confident of sensitivity. By contrast, if even large changes in demand produce little movement in the needle, then the metric is insensitive.

In practice, results from empirical studies (Hughes et al., 2023) showed that workload responses induced by experimental manipulations vary in magnitude, depending on factors such as sample and other manipulations, suggesting that sensitivity variation of the metrics detecting workload changes varies in different circumstances.

Psychophysiological workload measures are validated based on evidence of associations between variations in task demands and changes in physiological response. Psychophysiological measures are irreplaceable as they allow researchers to administer continuous measurement of participants' responses to experimental manipulations.

In a series of simulated studies in the nuclear power plant (NPP) operation context, the metrics of the psychophysiological workload measures (e.g., EEG and ECG) were generally sensitive in detecting the workload changes induced by the more demanding tasks performed by operators. Although psychophysiological measures may be more precise in some circumstances, subjective measures are more practical and easier to use. In addition, subjective measures reflect how participants feel about the experimental workload manipulations. In other words, subjective measures can reveal participants' perceived workload levels. Compared to the simple and quick tasks, subjective measures (e.g., NASA-TLX) appear to be more sensitive in distinguishing the workload changes induced by more complex tasks that require sustained attention (Hughes et al., 2023).

13.2.8 Summary of Measurement Issues

Workload is a complex construct that requires equally diverse assessment strategies. Careful assessment of measures for their suitability for any specific research or practical application is essential. Eggemeier et al.'s (1991) criteria provide a useful scheme for evaluating measures. A more detailed guide to designing workload assessment studies can be found in Matthews and Reinerman-Jones (2017).

This section reviews three primary categories – subjective, physiological, and behavioral – each offering distinct insights into operator experience (see Table 13.3). Subjective measures are simple and reflect perceived effort; behavioral measures can be integrated into existing systems but may miss hidden strain due to compensatory strategies; physiological measures provide continuous, objective data but involve confounding factors and implementation challenges.

A recurring theme is the potential for dissociation among measures – instances where reported experience, performance, and physiological response do not align. Rather than indicating flaws, these discrepancies may reflect the complex, multidimensional nature of workload. Acknowledging such differences can offer valuable insights for diagnostics, design, and system support.

Implementation must consider factors such as cost, usability, and organizational readiness. No single method suits all scenarios. Effective strategies combine multiple indicators, iterate on implementation, and remain grounded in the operational realities of high-risk industries. For industrial engineers, the goal is not just measurement but turning that insight into lasting, practical improvements.

TABLE 13.3
A Summary of Three Methods for Workload Assessment

	Subjective	Physiological	Behavioral
Methods	Post-task questionnaires and ratings	Central and autonomic nervous system, hemodynamic, and ocular responses	Primary and secondary task performance
Strengths	Quick and cheap, typically sensitive	Objective metrics based on cognitive neuroscience	Objective metrics linked to operational error probabilities
Weaknesses	Unsuitable for continuous monitoring, biases in self-reports	Potentially intrusive, often fail to converge with each other	Vulnerable to changes in operator strategy
Sample application	Identifying a "redline" at which workload is unacceptably high	Evaluating alternative interfaces for a system	Determining variation in demands during vehicle driving

13.3 APPLICATIONS TO MODERN INDUSTRY

Workload assessment has origins in attempts to manage and optimize factory work, and industrial applications remain important. This section provides an overview of research and methods in three contemporary contexts: manufacturing industry, transportation, and power generation. Workload assessment supports each of the three human factors goals previously described – to maintain safety during operations, to maximize productivity and quality of work, and to support worker wellbeing. While safety is the top priority, at least in well-managed organizations, the priorities for workload assessment and mitigation vary across different industrial contexts. In transportation and power generation, a primary aim is to identify and mitigate workload issues that might cause human errors that lead to accidents and disasters. Productivity is naturally important for manufacturing and service industries.

Physical safety has become less important in industries in which people are interacting only with digital systems rather than with potentially dangerous machinery. However, the dangers of interacting with the digital world, such as exposure to cyberattacks, are becoming more common. Maintaining vigilance for cyber threats such as phishing emails and compliance with sometimes onerous security protocols is becoming a significant source of workload and stress in the contemporary workplace (Matthews et al., 2025a).

13.3.1 Manufacturing Industry

Historically, human factors practitioners in industry have been more concerned with physical than with cognitive ergonomics. Physical demands predominated in traditional factory work, and they were also easier to assess objectively than cognitive demands. The widespread introduction of computer technology, together with advancements in workload measurement, revitalized interest in cognitive demands and workload in manufacturing. As in other domains, there are multiple types of demand factors that influence cognitive workload.

13.3.1.1 Equipment and Task Design

Poor design of equipment, screen displays, and system interfaces elevates workload. The multiple resource model discussed previously suggests that specific resources can become overloaded by poorly designed displays and interfaces (Hart & Wickens, 1990). For example, poorly legible text and visual clutter elevate workload resulting from visual input demands. Conversely, engaging multiple resources may mitigate overload issues, for example, by using a mixture of text and graphic display elements so as not to overload either verbal or spatial/imaging resources. Structuring work activities can also help. In a study of car assembly, Brolin, Thorvald, and Case (2017) showed that organizing components within a structured kit lowered workload and enhanced performance, relative to requiring workers to search for components within a rack. Human-centered design, coupled with detailed task analysis, has been proposed as a method for identifying steps in vehicle assembly that may be vulnerable to overload (Brunzini et al., 2024). Workload scales can be used throughout the system design process from early prototyping to advanced testing and evaluation.

13.3.1.2 Volume of Work

Even routine tasks can overload the operator if there is high time pressure to execute tasks or if extensive multi-tasking is required. Impacts of high-volume working can be mitigated by assigning workers some level of personal control over how to complete assignments.

13.3.1.3 Underload

The automation of much factory work has changed the role of workers from skilled agents to passive monitors of automation, which, typically, will only rarely require human intervention. That is, the prime requirement of the job becomes vigilance for occasional automation failures.

Many laboratory studies of vigilance show that people are prone to increasing errors in signal detection over time (the vigilance decrement: Warm, Parasuraman & Matthews, 2008). Paradoxically, although the overt demands of vigilance may be low, the effort required to sustain the focus of attention requires attentional resources and imposes subjective workload. Various psychophysiological measures, especially EEG, can be used to track declining resources and increasing workload during vigilance tasks, including those that require monitoring automation.

13.3.1.4 Dynamic Factors

In most jobs, work demands vary over time. Xie and Salvendy (2000) distinguished different temporal aspects of workload, including instantaneous, peak, average, and accumulated workload, which may jointly influence post-task workload ratings in enigmatic ways. Abrupt changes in workload, interruptions, and switching between tasks requiring different mental sets can all elevate workload over and above immediate task demands. Handling sudden increases in workload is critical for managing emergency events.

13.3.1.5 Interaction of Physical and Cognitive Workload

Measurement efforts typically aim to separate physical and cognitive workload, but one may influence the other (e.g., DiDomenico & Nussbaum, 2011). Physical demands, especially if prolonged, produce distraction and fatigue that can interfere with cognitive processing. Similarly, cognitive demands can lead to neglect of the physical environment. In designing controls for interfaces, it is especially important to assess both cognitive and physical demands. For example, tactile and haptic feedback that makes a control easier to use minimizes the need for attentional resources that would otherwise be diverted from concurrent cognitive task processing.

13.3.1.6 Work Environment

Factors such as loud noise, uncomfortable ambient temperatures, and poor lighting can interfere with work quality. These factors operate primarily as stressors rather than demand factors (though lighting affects visual perception), but they may be experienced as elevating subjective workload also. Ignoring distraction is itself a subtask; processing irrelevant speech can divert resources from the primary task performed.

13.3.1.7 Organizational Factors

From a sociotechnical standpoint, the level of support and training the organization provides can affect workload beyond immediate task demands. Specific demands of this kind include difficulties in obtaining technical support when systems fail, requirements to use physical or digital systems that are unsuitable for the work performed, and social demands imposed by coworkers and supervisors. Like environmental factors, these influences tend to elicit stress and perceptions of workload concurrently.

13.3.2 Transportation and Vehicle Operation

Aviation was an important focus for early studies of cognitive workload, leading to development of specialized measures of piloting workload. Research concerns included identifying phases of flying in which workload was especially elevated, such as take-off and landing, designing cockpits to reduce workload, and continuous psycho-physiological monitoring of workload (Kantowitz & Casper, 2017).

Emergency events are, of course, especially concerning. Workload is elevated because the pilot has to diagnose the source of the event, which may well be unexpected and unusual. High levels of stress are also likely. In the Air France 447 (2009) and Boeing 737 MAX (2018/2019) crashes, pilots were overloaded by the demands of understanding unexpected autopilot behavior, and their cognitive capacity was reduced by stress. The more demanding phases of flight require extensive multi-tasking. Workload is mitigated by checklists for executing procedures, but pilots nevertheless require skills in knowing which subtasks to prioritize, especially during unusual events. Effective crew resource management also serves to manage workload. Aviation studies also highlight the importance of workload factors in the wider infrastructure that supports commercial flight, especially air traffic control and airplane construction and maintenance.

Workload issues are also important in other areas of transportation, including surface vehicle driving, maritime operations, and control of unmanned vehicles. Reviews of workload in vehicle driving (e.g., Paxion, Galy & Berthelon, 2014) typically distinguish environmental, vehicle, and driver factors in workload. The demands of vehicle driving are influenced by environmental factors such as weather and lighting, as well as design of roadways and intersections. Vehicle factors include dashboard design, ease of use of controls, and visibility (e.g., via mirrors). Current research investigates workload impacts of vehicle automation (Shahini & Zahabi, 2022).

Driver factors partly reflect the driver's cognitive attributes; for example, inexperienced and older drivers may experience higher workload. These various factors commonly interact. Older drivers may be especially impacted by poor visibility and glare at night, due to age-related impairment of the visual system. Driver workload also reflects the ability of drivers to regulate demands voluntarily, for example, by changing vehicle speed or headway to a vehicle in front, so that driver workload is typically moderate. Driver error is not just the product of overload but also failing to match resource allocation to demands appropriately (Matthews & Desmond, 2002). Indeed, increases in environmental demands or secondary tasking can sometimes produce improvements in driver performance as the driver brings up resource allocation to the level needed (Paxion et al., 2014).

Research has addressed the impact of driver distraction associated with phone use and in-car information and entertainment systems (Horrey, Lesch & Garabet, 2009). Drivers are typically confident that they can manage the additional workload from using a device concurrently with driving. They may also use compensatory strategies such as slowing down while taking a call. We noted that sometimes secondary task activities enhance performance, especially when the workload is low initially. However, phone use more typically impairs performance and safety, especially when drivers take their eyes off the road. Horrey et al. (2009) found that drivers' subjective workload ratings were insensitive to the added demands of dual tasking.

13.3.3 Process Control and Power Generation

Process control tasks have some commonalities with the manufacturing industry as they require people to manipulate physical systems (via an interface) and some with vehicle operation, in that operators work within a dynamic external environment, which is mostly but not entirely predictable. Features of process control include complexity of operations (often supported by automation), continuous supervision and monitoring of operations, and high variability in demands depending on the mode of operation (Pütz et al., 2024).

Nuclear power generation is one of the most extensively studied instances of process control in relation to workload (see Braarud, 2024, for a review). Operators in the Main Control Room (MCR) for an NPP have the primary task of supervising and controlling operations (Hughes et al., 2023). They may also perform additional activities, including interface management, maintaining equipment, and coordinating plant support functions. Primary tasking is potentially demanding due to the number of instruments and controls (I&C) required for operations, which may be several thousand. Specific tasks include checking I&C, detecting system state changes, assessing situations, planning responses, and response implementation. There are additional demands associated with operator teaming and communication. Both routine and emergency operations make extensive use of well-defined procedures, which typically keep workloads moderate. However, workload may spike, especially during unusual events that procedures do not cover, so that operators must use their knowledge of plant operations to diagnose and correct the anomaly.

The United States Nuclear Regulatory Commission (NRC) has since 2012 maintained a research program on operator workload that exemplifies how research in this domain is conducted (Hughes et al., 2022). It utilizes simulations of NPP operation that are used to manipulate demand factors systematically and to test their impacts on subjective workload, physiological metrics, and errors during performance. For practical reasons, it is challenging to recruit a sufficient number of current operators for research studies. The NRC research has recruited novice samples to perform proceduralized scenarios such as executing an emergency shutdown due to loss of AC power, following appropriate training. Because workload issues result from the built-in processing limitations of the brain, results are expected to generalize to expert operators. (Generalization to experts has been verified by NRC using former operator samples.) However, the methodology is not suitable for knowledge-based activities such as fault diagnosis, for which expertise is essential.

The NRC research has shown that different operator subtasks differ in their workload on both subjective and psychophysiological measures (Hughes et al., 2023). The most demanding element of operations appears to be detection, for example, monitoring a gauge to detect when the value goes out of range. Basic research also shows that monitoring tasks of this kind can impose high workload (Warm et al., 2008). Research has also investigated interface effects, such as differences in workload in using a conventional computer workstation versus a touchscreen, for example. This research is important for guiding modernization of Human Systems Integration (HSI). Study findings also support the licensing mission of the NRC. Plant operators applying for a license to build or modify a plant must demonstrate that MCR systems will not impose excessive workload on operators in different scenarios. Evidence on workload factors can be incorporated into Human Reliability Analysis (HRA: Bye, 2023) to determine if a system modification will unduly increase human error probability.

13.3.4 Recommendations for Practitioners in Engineering and Safety

Table 13.4 shows a summary checklist for practitioners. More detailed guidelines for workload assessment may be found in Matthews and Reinerman-Jones (2017). Design applications support the safety by design principle (Hale, Kirwan & Kjellén, 2007), that is, designers should seek to anticipate workload issues proactively throughout the different stages of the design process.

13.4 DIGITALIZATION AND WORKLOAD ASSESSMENT

Industry, along with other human factors domains, is being transformed by digital technology for autonomous operations and AI. We will review the workload issues raised by advancements in technology and the assessment challenges for future industrial and transportation applications.

Industrial innovations and advancements are often linked to a series of "industrial revolutions" marked by major shifts in the technology and social organization of production. The origins of workload assessment lie in the Second Industrial Revolution, which introduced mass production, electrification, and standardization of components and products, beginning in the late 19th century. These developments emphasized efficiency, coordination, and the human role within mechanized systems, prompting early interest in measuring mental and physical workload. The Third Industrial Revolution – often referred to as the Digital Revolution – began in the mid-to-late 20th century and

TABLE 13.4
Checklist for Applying Workload Assessment in Industry

Define workload assessment goals: safety, performance, or wellbeing
Select measure(s) using Eggemeier et al. (1991) criteria and evidence on validity
Ensure minimal task disruption through balancing intrusiveness vs. fidelity
Calibrate baseline workload for comparison
Combine multiple data streams for validation
Use results to redesign interfaces, task allocation, or training
Verify effectiveness of changes prompted by results as part of a continuous cycle of evaluation and improvement

introduced the widespread use of personal computers, business software, digital communication, and the internet. This transition shifted many work demands from physical to cognitive, increasing the need for refined workload assessment techniques in information-rich and automated environments.

As we have seen, theories of workload and methodologies for assessment have proved well-suited to investigating the demands of "knowledge work" performed by individuals performing at a computer workstation. Indeed, a shift in focus from the factory to the office counters some of the practical challenges of assessing workload in a physically dynamic environment. However, the radical changes to the nature of work that are currently unfolding raise new challenges for understanding and measuring workload.

13.4.1 Novel Challenges for Workload Research

The latest industrial revolution – sometimes referred to as Industry 4.0 – is based on advanced digital automation and communication that enables AI, cloud computing, and the Internet of Things (IoT). That is, digital systems increasingly mediate numerous activities at work and other contexts. Industry 5.0, which envisages increased collaboration between humans and AI, will accelerate this trend (Trstenjak et al., 2025).

The digitalization of life poses some challenges for workload research and applications (Matthews et al., 2025a, 2025b):

13.4.1.1 Novel Cognitive Demands

Systems powered by AI, especially deep learning, often lack transparency and explainability, which elevates demands on human users, especially when the system behaves unexpectedly or commits an error. In some contexts, working with intelligent systems may be characterized by uncertainty, increasing the workload of human operators. Such issues arise in automated systems, for example, in using the autopilot in aviation. They will be particularly salient in working with autonomy, that is, a system that has agency over its decisions, because of such systems' greater degrees of freedom of action.

13.4.1.2 Teaming Demands

Teams made up of humans experience workload from teamwork as well as immediate taskwork, where teamwork refers to behaviors such as communication and providing backup support that help the team meet its goals as a cohesive unit. Team workload may be measured as an emergent property of the team and not just the mean workload of individuals (Zhang, She & Li, 2023). Manufacturing and service industries may increasingly require humans to team with intelligent robots and virtual agents. These work scenarios will impose novel demands as the person learns how to team effectively with virtual systems. The advent of humanlike systems and natural language communication will, in some respects, reduce workload but may also be challenging for some people. Accurate calibration of trust to objective agent capabilities is critical for team performance, but deciding what advanced systems can be trusted to do is itself a form of demand, especially in high-stakes or safety-critical situations. Appropriate organizational support for human-autonomy teams is also important for mitigating the demands of teaming. Managers should promote the team membership of the autonomous system together with human decisional ownership of the team's work products (Zercher, Jussupow, and Heinzl, 2025).

13.4.1.3 Self-Regulative Demands

Beyond task performance and teaming, intelligent technology may also challenge the person's sense of self. Being directed, monitored, and evaluated by an algorithm rather than a human manager may be perceived as dehumanizing. Rideshare drivers already work under these conditions. Challenges to self-regulation will be expressed primarily as stress, but finding one's identity and resolving decision authority in a machine world may also add to the workload.

13.4.2 Power Plant Digitalization and Operator Workload

As a case study of the challenges of new technology, we consider digitalization of the nuclear energy industry (Boring, Ulrich & Lew, 2023). *Digitalization* is distinguished from *digitization* in that it adds novel functionality (and potentially novel task demands) to the control room environment. Digitization is merely the replacement of analog with digital I&C, which is now a mature technology. Boring et al. (2023) defined multiple levels of digitalization (LODs) as shown in Table 13.5. Higher LODs provide increasing scope for automation of plant operations, such as information gathering on plant state to support the alarms delivered to the operator. They are also integral to novel plant designs, such as Small Modular Reactors (SMRs), that can be operated remotely with extensive support from automation.

Two reviews (Bye, 2023; Porthin, Liinasuo, & Kling, 2020) have addressed the implications of digitalization for understanding operator performance in the context of HRA and anticipating human error. Table 13.6 identifies some common features of modernized control rooms that may impact operator workload. In general, the additional information and functionality of digitalized systems are beneficial to the extent that they enhance operator situational awareness of the state of the plant and guide operators toward effective control actions.

Especially in highly automated systems such as SMRs, additional workload may actually benefit performance if it maintains operator engagement and interest in the task. However, there are also risks of information leading to overload and confusion. Good design of HSI and operator training (and periodic retraining) are essential to mitigating these risks. Bye (2023) highlights the potential of digitalization to introduce new and unexpected error mechanisms. He recommends that the reactor crew and the modernized control room should be evaluated together as a unit. Workload assessment would be an important element of such an evaluation.

TABLE 13.5
Levels of Digitalization Defined by Boring et al. (2023)

LOD	Level	Description
0	No digitalization	Entirely analog I&C or digital I&C without added functionality
1	Hybrid digitalization	Some digital instrumentation and control maintain basic functionality, while some features offer enhanced functionality
2	Full digitalization	A fully digital system that uses advantages of digital technology to enhance performance
3	Integrated digitalization	Multiple distal digitalized systems are integrated to allow synthesized functionality

TABLE 13.6
Workload Impacts of Features of Digitalized NPP Systems

Feature	Workload Impact
Systems provide additional information	Mixed: information may reduce demands of understanding system state. It may also overload the operator
User interface adds new secondary tasks	Likely to increase workload
Operator loses awareness of activities of other operators	Increases workload associated with teamwork
Digital systems updated frequently through software upgrades	Increases workload due to changes in system operation

Source: Based on Porthin et al., 2020.

There is also a relatively small but growing literature on empirical studies of workload in simulations of digitalized plants (see Lin et al., 2023, for a review). Empirical studies that have assessed workload, most commonly with subjective measures, highlight both the benefits and potential downsides of digitalization. On the positive side, studies have shown that advancements such as the introduction of computerized operating procedures can reduce workload and enhance situation awareness. However, studies have also identified additional workload that results from needs to manage the interface for control and handle the volume of information delivered by digitalized systems. Under high workload, operators may be at risk of adopting a faulty mental model for plant functioning. These examples highlight the importance of using workload assessments to evaluate new systems, along with measures of situation awareness, trust in automation, and objective performance.

13.5 CONCLUSIONS

Workload assessment is a cornerstone of Human Factors Engineering (ISO 9241-210, ISO 11064). It embodies the safety by design philosophy by integrating human performance data into iterative system design, ensuring that technology adapts to human capacity rather than the reverse. Valid measurement of workload is also important for enhancing productivity and wellbeing in multiple industrial contexts. Workload assessment has a basis in attentional resource theory. However, the psychology of how people manage task demands is complex, which produces challenges for workload assessment.

Subjective, behavioral, and psycho-physiological measures of workload are used, but alternate measures may dissociate. Choosing appropriate workload measures requires a thorough understanding of the research or application context. Explicit criteria for validity and practical applicability of measures should be referenced. It is often useful to assess workload together with other constructs such as stress, situation awareness, and trust.

Industrial applications for workload measurement have branched out from traditional manufacturing industry to include knowledge-based work, transportation, process control, and other contexts. Automated systems can produce underload rather than overload. Workload studies can contribute to interface design, readiness for emergency events, training procedures, organizational policies and procedures, and improving the work environment.

The digital revolution is ushering in new opportunities and challenges for workload assessment. Working with intelligent systems, including those with humanlike capabilities, produces novel task demands, which may require new assessment methods. Digitalization of NPP operation illustrates how workload measurement contributes to evaluation of new technologies and systems.

REFERENCES

Ackerman, P. L. (1988). Determinants of individual differences during skill acquisition: Cognitive abilities and information processing. *Journal of Experimental Psychology: General 117*(3), 288–318.

Boles, D. B., and M. B. Dillard (2015). The measurement of perceptual resources and workload. In R. R. Hoffman, P. A. Hancock, M. W. Scerbo, R. Parasuraman, and J. L. Szalma (Eds.), *The Cambridge handbook of applied perception research*, pp. 39–59. Cambridge University Press.

Boring, R. L., T. A. Ulrich, and R. Lew (2023). Levels of digitization, digitalization, and automation for advanced reactors. *Proceedings of the Human Factors and Ergonomics Society Annual Meeting 67*, 207–213.

Braarud, P. Ø. (2024). Measuring cognitive workload in the nuclear control room: A review. *Ergonomics 67*(6), 849–865.

Brenner, M., E. T. Doherty, and T. Shipp (1994). Speech measures indicating workload demand. *Aviation, Space, and Environmental Medicine 65*(1), 21–26.

Brolin, A., P. Thorvald, and K. Case (2017). Experimental study of cognitive aspects affecting human performance in manual assembly. *Production and Manufacturing Research 5*(1), 141–163.

Brunzini, A., F. Grandi, M. Peruzzini, and M. Pellicciari (2024). An integrated methodology for the assessment of stress and mental workload applied on virtual training. *International Journal of Computer Integrated Manufacturing 37*(9), 1088–1106.

Butmee, T., T. C. Lansdown, and G. H. Walker (2019). Mental workload and performance measurements in driving task: A review literature. In *Proceedings of the 20th Congress of the International Ergonomics Association (IEA'18)*, pp. 286–294. Springer International Publishing.

Bye, A. (2023). Future needs of human reliability analysis: The interaction between new technology, crew roles and performance. *Safety Science 158*, 105962.

Cain, B. (2007). *A review of the mental workload literature.* Technical report, Defence Research and Development Toronto, Canada.

Charles, R. L., and J. Nixon (2019). Measuring mental workload using physiological measures: A systematic review. *Applied Ergonomics 74*, 221–232.

Chen, O., F. Paas, and J. Sweller (2023). A cognitive load theory approach to defining and measuring task complexity through element interactivity. *Educational Psychology Review 35*(2), 63.

Cooper, R., and T. Shallice (2000). Contention scheduling and the control of routine activities. *Cognitive Neuropsychology 17*(4), 297–338.

De Winter, J. C. (2014). Controversy in human factors constructs and the explosive use of the NASA-TLX: A measurement perspective. *Cognition, Technology & Work 16*(3), 289–297.

DiDomenico, A., and M. A. Nussbaum (2011). Effects of different physical workload parameters on mental workload and performance. *International Journal of Industrial Ergonomics 41*(3), 255–260.

Eggemeier, F. T., G. F. Wilson, A. F. Kramer, and D. L. Damos (1991). Workload assessment in multi-task environments. In D. L. Damos (Ed.), *Multiple-task performance*, pp. 207–216. Taylor and Francis.

Ganster, D. C., and C. C. Rosen (2013). Work stress and employee health: A multidisciplinary review. *Journal of Management 39*(5), 1085–1122.

Gevins, A., H. Leong, R. Du, M. E. Smith, J. Le, D. DuRousseau, J. Zhang, and J. Libove (1995). Towards measurement of brain function in operational environments. *Biological Psychology 40*, 169–186.

Gründling, J. P., N. Feld, D. Zielasko, and B. Weyers (2023). Correlations of flow, usability, workload. In *Proceedings of Virtual Reality and Mixed Reality: 20th EuroXR International Conference, EuroXR 2023 14410*, pp. 153–165. Cham, Switzerland: Springer Nature.

Hale, A., B. Kirwan, and U. Kjellén (2007). Safe by design: Where are we now? *Safety Science 45*(1–2), 305–327.

Hancock, P. A., and G. Matthews (2019). Workload and performance: Associations, insensitivities, and dissociations. *Human Factors 61*(3), 374–392.

Hancock, P. A., and J. S. Warm (1989). A dynamic model of stress and sustained attention. *Human Factors 31*, 519–537.

Hart, S. G., and L. E. Staveland (1988). Development of NASA-TLX (Task Load Index): Results of empirical and theoretical research. In P. A. Hancock and N. Meshkati (Eds.), *Human mental workload*, pp. 139–183, North-Holland.

Hart, S. G., and C. D. Wickens (1990). Workload assessment and prediction. In H. R. Booher (Ed.), *MANPRINT, an approach to systems integration*, pp. 257–296. Van Nostrand Reinhold.

Hill, Susan G., H. P. Iavecchia, J. C. Byers, A. C. Bittner Jr, A. L. Zaklade, and R. E. Christ (1992). Comparison of four subjective workload rating scales. *Human Factors 34*, 429–439.

Horrey, W. J., M. F. Lesch, and A. Garabet (2009). Dissociation between driving performance and drivers' subjective estimates of performance and workload in dual-task conditions. *Journal of Safety Research 40*(1), 7–12.

Hughes, N., J. Lin, G. Matthews, D. Barber, & K. Dickerson (2023). *Human Performance Test Facility (HPTF). Volume 3 – Supplemental exploratory analyses of sensitivity of workload measures.* Research Information Letter RIL 2022-11. Washington, DC: US Nuclear Regulatory Commission.

Hughes, N. L. Reinerman-Jones, J. Lin, G. Matthews, D. Barber, and K. Dickerson (2022). *Human Performance Test Facility (HPTF). Volume 2 - Comparing operator workload and performance between digitized and analog simulated environments.* Research Information Letter RIL 2022-11. Washington, DC: U.S. Nuclear Regulatory Commission.

Kantowitz, B. H., and P. A. Casper (2017). Human workload in aviation. In R. K. Dismukes (Ed.), *Human error in aviation*, pp. 123–153. Routledge.

Lastoria, D. A. A., S. Keynes, M. Nurek, C. Kellett, and C. B. Hing (2025). The Hawthorne effect on wearable technology: A systematic review. *International Journal of Psychophysiology 214*, 113207.

Lin, J., G. Matthews, J. Schreck, K. Dickerson, and N. H. Green (2023). Evolution of workload demands of the control room with plant technology. *Human Factors and Simulation 83*, 57–66.

Liu, Y., J. Koopmann, V. Alterman, M. Wang, S. Liu, and J. Shi (2024). Duality of workload in teams: A Daily investigation of team workload and team functioning. *Journal of Management 52*(2), 614–645.

Marquart, G., C. Cabrall, and J. De Winter (2015). Review of eye-related measures of drivers' mental workload. *Procedia Manufacturing 3*, 2854–2861.

Marshall, S. P. (2002). The index of cognitive activity: Measuring cognitive workload. In *Proceedings of the IEEE 7th Conference on Human Factors and Power Plants*, p. 7. IEEE.

Matthews, G., S. E. Campbell, S. Falconer, L. A. Joyner, J. Huggins, and K. Gilliland (2002). Fundamental dimensions of subjective state in performance settings: Task engagement, distress, and worry. *Emotion 2*, 315–340.

Matthews, G., R. Cumings, E. P. De Los Santos, I. Y. Feng, and S. A. Mouloua (2025a). A new era for stress research: Supporting user performance and experience in the digital age. *Ergonomics 68*(6), 913–946.

Matthews, G., D. R. Davies, S. J. Westerman, and R. B. Stammers (2000). *Human performance: Cognition, stress and individual differences*. Psychology Press.

Matthews, G., and P. A. Desmond (2002). Task-induced fatigue states and simulated driving performance. *The Quarterly Journal of Experimental Psychology: Section A 55*(2), 659–686.

Matthews, G., P. A. Hancock, J. L. Szalma, J. Lin, and A. R. Panganiban (2025b). Individual differences in teaming with artificial intelligence, robots, and virtual agents in the workplace. In A. Tuncdogan, O. A. Acar, H. W. Volberda and K. de Ruyter (Eds.), *The Oxford handbook of individual differences in organizational contexts*, pp. 345–367. Oxford University Press.

Matthews G., and L. Reinerman-Jones (2017). *Workload assessment: How to diagnose workload issues and enhance performance*. Human Factors and Ergonomics Society.

Matthews, G., L. E. Reinerman-Jones, D. J. Barber, and J. Abich (2015). The psychometrics of mental workload: Multiple measures are sensitive but divergent. *Human Factors 57*, 125–143.

Matthews, G., R. Wohleber, and J. Lin (2019). Stress, skilled performance, and expertise: Overload and beyond. In P. Ward, J. M. Schraagen, J. Gore, and E. M. Roth (Eds.), *Oxford handbook of expertise: Research and application*, pp. 490–524. Oxford University Press.

Navon, D., and D. Gopher (1979). On the economy of the human processing system. *Psychological Review 86*(3), 214–255.

Ogden, G. D., J. M. Levine, and E. J. Eisner (1979). Measurement of workload by secondary tasks. *Human Factors 21*(5), 529–548.

Parasuraman, R., and V. Riley (1997). Humans and automation: Use, misuse, disuse, abuse. *Human Factors 39*(2), 230–253.

Paxion, J., E. Galy, and C. Berthelon (2014). Mental workload and driving. *Frontiers in Psychology 5*, 1344.

Porthin, M., M. Liinasuo, and T. Kling (2020). Effects of digitalization of nuclear power plant control rooms on human reliability analysis–A review. *Reliability Engineering and System Safety 194*, 106415.

Pütz, S., A. Mertens, L. Chuang, and V. Nitsch (2024). Physiological measures of operators' mental state in supervisory process control tasks: A scoping review. *Ergonomics 67*(6), 801–830.

Reid, G. B., and T. E. Nygren (1988). The subjective workload assessment techniques: A scaling procedure for measuring mental workload. In P. A. Hancock and N. Meshkati (Eds.), *Human mental workload*, pp. 185–218. Elsevier.

Shahini, F., and M. Zahabi (2022). Effects of levels of automation and non-driving related tasks on driver performance and workload: A review of literature and meta-analysis. *Applied Ergonomics 104*, 103824.

Trstenjak, M., A. Benešova, T. Opetuk, and H. Cajner (2025). Human factors and ergonomics in industry 5.0 – A systematic literature review. *Applied Sciences 15*(4), 2123.

van de Merwe, K., S. Mallam, and S. Nazir (2024). Agent transparency, situation awareness, mental workload, and operator performance: A systematic literature review. *Human Factors 66*(1), 180–208.

Vidulich, M. A., and P. S. Tsang (1987). Absolute magnitude estimation and relative judgement approaches to subjective workload assessment. In *Proceedings of the Human Factors Society Annual Meeting 31*, pp. 1057–1061.

Warm, J. S., R. Parasuraman, and G. Matthews (2008). Vigilance requires hard mental work and is stressful. *Human Factors 50*, 433–441.

Wickens, C. D. (2008). Multiple resources and mental workload. *Human Factors 50*, 449–455.

Wierwille, W. W., and J. G. Casali (1983). A validated rating scale for global mental workload measurement applications. In *Proceedings of the Human Factors Society Annual Meeting*, vol. 27, pp. 129–133.

Xie, B., and G. Salvendy (2000). Review and reappraisal of modelling and predicting mental workload in single-and multi-task environments. *Work and Stress 14*(1), 74–99.

Young, M. S., K. A. Brookhuis, C. D. Wickens, and P. A. Hancock (2015). State of science: Mental workload in ergonomics. *Ergonomics 58*, 1–17.

Young, M. S., and N. A. Stanton (2007). Miles away: Determining the extent of secondary task interference on simulated driving. *Theoretical Issues in Ergonomics Science 8*(3), 233–253.

Zercher, D., E. Jussupow, and A. Heinzl (2025). Team climate in team-AI collaboration: Exploring the role of decisional ownership and perceived AI team membership. In *European Conference on Information Systems (ECIS): Research Papers*, vol. 8, pp. 1–17. AISeLibrary.

Zhang, Y., M. She, and Z. Li (2023). Team workload: Review and conceptualization. *International Journal of Industrial Ergonomics 95*, 103.

14 Subjective Workload Assessment

Understanding and Applying the NASA-TLX in Safety-Critical Work

Martin Rasmussen Skogstad

14.1 INTRODUCTION

Subjective workload assessment (SWA) evaluates workload through an individual's own perceptions and experiences, capturing aspects of mental, emotional, and physical demand that objective measures may overlook. Unlike objective assessments (to the extent they are truly objective), which rely on measurable metrics such as response times, error rates, or physiological and neurological indicators, subjective methods prioritise the individual's own evaluation of task demands. SWA recognises that workload is inherently multidimensional, encompassing cognitive, emotional, and physical aspects that may not always be apparent from objective metrics alone.

Subjective assessments typically involve self-reporting tools or questionnaires in which individuals reflect on and rate their experiences of mental effort, stress, fatigue, frustration, temporal pressure, and satisfaction with their performance. These assessments provide valuable insights into how different tasks, systems, or working conditions affect individuals' psychological and physical states, and how these states in turn influence performance, error rates, and overall well-being.

SWA is particularly valuable because it captures individuals' perceived stress and strain. By revealing aspects of workload that are difficult to capture through direct measurement alone, it can help identify challenging features of a work setting, map individual differences in how situations are perceived, and evaluate the effects of introducing new technology. Such feedback is essential in domains where cognitive overload, fatigue, or reduced situational awareness may contribute to unsafe conditions, including aviation, maritime operations, healthcare, and, increasingly, automated or AI-driven environments.

This chapter will focus on the most frequently used tool for SWA – the NASA Task Load Index (NASA-TLX) (Hart, 1986; Hart & Staveland, 1988). However, most of the chapter is devoted to SWA in general. It should also be relevant to many of the other methods that have similar areas of use, underlying factors, and data collection methods (e.g., the Subjective Workload Assessment Technique, SWAT (Reid & Nygren, 1988), and the modified Cooper-Harper Scale – a tool specifically developed for pilot workload evaluation (Wierwille & Casali, 1983)). Each method brings unique advantages and limitations, necessitating careful selection based on context, objectives, and specific user groups.

In exploring SWA, this chapter will further examine the importance of subjective evaluations relative to objective measures, highlight practical implications through relevant case studies such as the collision between the frigate HNoMS *Helge Ingstad* and the oil tanker *Sola TS* outside the Sture Terminal in the Hjeltefjord in Hordaland County on 8 November 2018 (Alsos et al., 2023; Norwegian Safety Investigation Authority, 2019, 2021) and autonomous shipping scenarios, and

DOI: 10.1201/9781003741824-18

discuss complementary approaches for achieving comprehensive workload assessment. Ultimately, this discussion underscores the critical role of SWA in ensuring safety, efficiency, and trustworthiness in automation and artificial intelligence systems.

This chapter introduces NASA-TLX, discusses its administration and use, reviews key critiques, and considers its continued relevance in AI-integrated systems.

14.1.1 Understanding Subjective Workload

Subjective workload is fundamentally about individuals' perceptions of how demanding a task or set of tasks is, considering their capabilities and available resources. These perceptions are influenced by various factors, including task complexity, time pressure, sustained attention requirements, and emotional stressors. Unlike objective workload measures, subjective assessments allow individuals to articulate their internal experiences directly, providing nuanced information about their mental state and cognitive load.

The concept of subjective workload is rooted in cognitive psychology and human factors engineering, where researchers recognise that perceived task difficulty can significantly impact performance, motivation, and fatigue levels. When workload exceeds an individual's capacity, it leads to increased stress, reduced situational awareness, impaired decision-making, and heightened error rates, all of which pose critical safety risks, particularly in high-stakes settings.

Understanding subjective workload is critical for designing tasks and technologies that match human cognitive capacities and avoid overload. Effective SWA helps identify when tasks or environments may overload individuals, enabling interventions to improve performance and reduce errors. Furthermore, understanding workload through subjective assessment aids in developing targeted training programmes, designing more effective human-machine interfaces, and informing policies and procedures that promote safety and efficiency across diverse operational contexts.

14.2 NASA TASK LOAD INDEX (NASA-TLX)

The Human Performance Group developed NASA-TLX at NASA's Ames Research Centre and it has remained a popular method since the first version of the Pencil and Paper Package in 1986 (Hart, 1986), and the following description in a chapter in the book *Human Mental Workload* from the *Advances in* Psychology book series in 1988 (Hart & Staveland, 1988). The latter of these has become the standard citation for those who use and discuss the method. When Hart revisited NASA TLX and its use in *NASA-task load index (NASA-TLX); 20 years later* (Hart, 2006), she found that it had been translated into more than a dozen languages; 'NASA TLX' returned 1,670 hits on Google Scholar, and the original source had 446 citations. In 2025, this increased to 40,400 hits on Google Scholar and 20,083 citations. While caution should be taken in the direct use of these numbers or in direct comparisons between them (e.g., as Google expands its databases or changes its search algorithms), it is safe to say that NASA-TLX has remained a very popular method as it approaches its 40-year anniversary.

Part of the popularity is likely due to the method's easy access and administration. The method itself is freely available as an Open Access (OA) tool on NASA's website (https://humansystems.arc.nasa.gov/groups/tlx/downloads/TLX_pappen_manual.pdf), where you can still access the slightly skewed scanned image of a printed and folded-out version of the 'Paper and Pencil Package v.1.0'. There is also an official free NASA-TLX iOS app (https://humansystems.arc.nasa.gov/groups/tlx/tlxapp.php) that can make administration even easier. Many unofficial versions exist; however, NASA warns against these apps as they might collect Personally Identifiable Information (PII). The paper-and-pencil version remains widely used, and a creative study of NASA-TLX found that workload experienced by the administrators was lower in the paper-and-pencil version than in the app version (Noyes & Bruneau, 2007).

The method involves assessing workload across six dimensions. Three dimensions reflect task-related demands (mental, physical, temporal), and three reflect how the individual engages with the task (performance, effort, frustration). The user is first introduced to the task, preferably by performing it. However, as this is not always practical nor possible, it can be a visual presentation (e.g., video, animation, or demonstration of the task), having it described, or asking respondents to close their eyes and visualise performing the task for one to two minutes, which can be effective when physical task demonstrations are impractical. The introduction of the task will naturally have a key role when the subject is performing a novel task. However, it should not be overlooked even when the subject is performing a task well-known to them. There needs to be a clear understanding of the task definitions, including what constitutes the beginning and end of the task and the conditions under which it is being performed (e.g., is it being performed on a normal Tuesday or during a fire in the next room).

Users are presented with a short description of the dimensions. Hart includes a suggested standard list of descriptions (1988, p.13, example in Figure 14.1); however, these can be adapted to the specific task being assessed.

The subject rates each dimension individually on a 0–100 scale. Notably, Hart and Staveland (1988) intentionally do not number the scale, rather having the participant mark their response on a 21-point scale (where each point represents a difference of 5 in the 0-100 scale; Figure 14.2). Several unofficial versions use numbered scales, and a number is requested in the official verbal version. These are considered the 'Raw Rating'.

The next step is to determine the weight of each dimension. This is done through the subject selecting one in each of the 15 possible pair-wise comparisons between the dimensions, answering the question 'Which was the more important contributor to workload for the specific task(s)?' If the subject answers consistently (they do not always), this will lead to the dimensions being selected 0, 1, 2, 3, 4 and 5 times, ranking them from the least important (never chosen) to the most important (chosen five times).

The weight (0–5) is then multiplied by the Raw Rating (0–100), leading to an 'Adjusted rating' (0–500) for each dimension. The 'Adjusted Ratings' are summarised and divided by 15, leading to the final NASA-TLX score, the 'Weighted Rating' (also known as the overall workload index; 0–100).

Descriptions (Mental Demand)
How much mental and perceptual activity was required (e.g., thinking, deciding, calculating, remembering, looking, searching, etc.)? Was the task easy or demanding, simple or complex, exacting or forgiving?

FIGURE 14.1 Rating scale definition of the dimension Mental Demand.

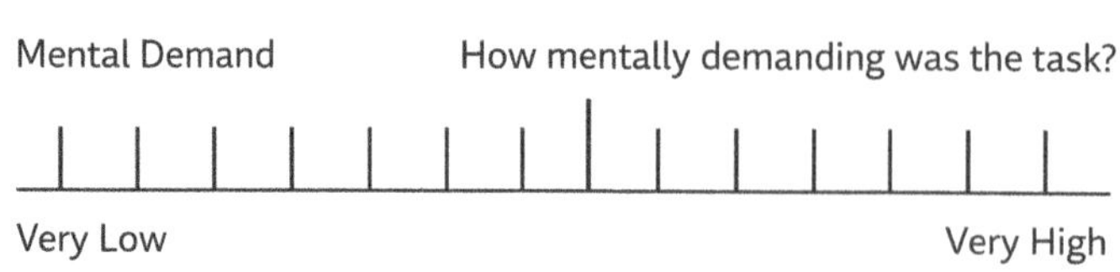

FIGURE 14.2 Rating sheet for Mental Demand.

(Source: humansystems.arc.nasa.gov)

14.3 WEAKNESSES AND CRITIQUE OF SUBJECTIVE WORKLOAD ASSESSMENT AND NASA-TLX

Despite their widespread use and valuable insights, SWAs, including NASA-TLX, have notable limitations and have been subject to critique (see de Winter, 2014, for a comprehensive review of the critique of NASA-TLX).

One significant weakness of subjective assessments is their reliance on self-reporting, which can, even though I would mainly see it as a positive, introduce biases and inconsistencies. Self-report measures are vulnerable to factors such as social-desirability bias, memory limitations, and individual differences in interpreting rating scales, all of which can affect response accuracy. For instance, individuals may underreport their workload to appear competent or overreport if the results determine which section will get more staff or if you are looking into why something went wrong.

The NASA-TLX has specifically faced criticism for its sensitivity and granularity. Some researchers note that the six NASA-TLX dimensions may partially overlap, potentially complicating interpretation if users are not given clear guidance. Additionally, the weighted scoring system, while intended to provide personalised workload profiles, may introduce complexity and confusion for respondents, particularly when administered without proper instruction or oversight.

Moreover, subjective measures like NASA-TLX generally provide limited insight into the precise cognitive processes or physiological mechanisms underlying workload experiences. Without objective data, it can be challenging to pinpoint specific aspects of task demands requiring intervention.

Finally, subjective assessments typically capture only snapshots of workload experiences rather than continuous measurements. This limitation may miss important fluctuations or dynamics in workload that occur during complex or prolonged tasks, potentially underestimating workload peaks or troughs.

Recognising these limitations, researchers and practitioners often recommend combining subjective assessments with objective measures to obtain more comprehensive and reliable evaluations of workload.

14.4 THE CONTINUED RELEVANCE OF SUBJECTIVE WORKLOAD ASSESSMENTS: NASA-TLX TODAY

NASA-TLX is an almost 40-year-old method that can be administered through paper and pencil. Looking back at the changes over the past 40 years in the technology we use and how we work, it is almost unreasonable that the same method would be relevant today. I would, however, argue that it is now as relevant as it was in 1986. Humans have not changed all that much, and we are facing (as we always have) changes and new technology that impact workload, and we need methods to capture that workload.

Subjective workload assessments remain highly relevant today because they capture human perceptions that other metrics and measurements frequently overlook. Despite significant advancements in technology and analytics, the importance of understanding the human experience of workload is growing, not diminishing.

14.4.1 Flexibility across Diverse Contexts

Today's workplaces have become increasingly varied, encompassing traditional office environments, remote work scenarios, and hybrid models. The NASA-TLX demonstrates flexibility and utility across these diverse settings, enabling consistent evaluation of subjective workload irrespective of the specific work context. Its simplicity and broad applicability make it suitable for assessing a range of tasks, from its traditional use in controllers, operators, and pilots to new tasks such as remote virtual teamwork or AI-Human teaming.

14.4.2 Workload and the *Helge Ingstad* Crash

The sinking of the Norwegian frigate *Helge Ingstad* in 2018 highlighted that workload is a current issue with a critical impact on maritime safety. Although NASA-TLX cannot be retroactively applied without strong hindsight bias, its dimensions map closely to the types of workload pressures documented in the accident, making it a useful pedagogical tool for illustrating how SWA can inform system design and crew training. Investigation reports identified high levels of cognitive and temporal workload experienced by the bridge crew as major contributing factors (many other contributing factors were present) (Norwegian Safety Investigation Authority, 2019, 2021). Alsos et al. (2023) provide a comprehensive evaluation of the human, technical, and organisational conditions. Increased workload, stemming from complex navigational conditions, multiple simultaneous tasks, and communication challenges, overwhelmed the crew's cognitive capacities. Consequently, key situational awareness cues were missed, resulting in delayed and inadequate responses to collision threats. Proactive SWA could have identified early signs of cognitive overload, prompting mitigation strategies such as redistributing tasks, improving communication protocols, or enhancing crew training, potentially preventing the incident. SWA could also have used it after the incident to show that the crew was under an unreasonable amount of workload; however, NASA-TLX is not well-suited for post-incident assessment because knowledge of the outcome may distort subjective ratings. Its structure highlights the kinds of workload pressures investigators should consider. Despite investigations revealing many systemic issues, breaches of safety rules, and barriers, the courts ultimately sentenced the officer of the watch (vaktleder in Norwegian, also referred to as conning officer) after the incident. Had the workload been systematically monitored using tools such as NASA-TLX, design challenges and the potential for early warning signs of cognitive overload might have been identified.

14.4.3 SWA with Automation and AI

The increasing integration of autonomous systems and AI into safety-critical environments – such as autonomous ships, vehicles, aviation, and healthcare – has introduced new dimensions of workload assessment and management. Although automation was initially expected to reduce workload, research shows that it often shifts workload – changing its timing, character, and distribution rather than eliminating it. Many researchers have already explored this using NASA-TLX (e.g., Bell et al., 2022; Javernik et al., 2023; Jiang et al., 2025; Shin & Prabhu, 2018), generally showing lower workload but also discussing scenarios where it could increase or change. Operators frequently experience changes in the nature and timing of demands, leading to new cognitive challenges such as maintaining vigilance during periods of automation, effectively managing system malfunctions or unexpected situations, and rapidly transitioning between passive monitoring and active intervention (Hayes & Skogstad, 2025).

One prominent issue, known as 'automation-induced complacency', arises from prolonged periods where operators passively monitor automated systems. This passivity can degrade situational awareness, slow down response times, and impair decision-making abilities during critical incidents, paradoxically increasing cognitive workload when rapid intervention is suddenly required. Similarly, automation can introduce increased demands related to trust calibration, as operators must continuously evaluate system reliability, understand its decisions, and recognise conditions that warrant overriding automated controls (Hayes & Skogstad, 2025).

In autonomous shipping, for example, the autonomous control of the ship will in most cases remove human tasks and thereby also workload. However, autonomous still means someone needs to monitor the operation, often from a control centre onshore. Operators monitoring multiple autonomous vessels remotely may face periods of low apparent workload, interspersed with intense and unpredictable spikes in cognitive demands when unexpected events occur. SWA in such contexts is crucial for identifying potential cognitive overloads, ensuring that automation complements rather than undermines human capabilities.

AI supported methods may also enhance workload assessment by combining subjective reports with physiological, neurological, or behavioural indicators. Hybrid approaches could offer continuous monitoring and early detection of overload, extending NASA-TLX's value in increasingly automated workplaces. This strategy could potentially capture the nuanced workload dynamics of advanced automated environments, ultimately ensuring that AI and automation enhance rather than compromise human performance, safety, and trust. As automation continues to evolve, hybrid approaches combining NASA-TLX with physiological or behavioural monitoring will likely define the next generation of workload assessment.

14.5 CONCLUSION

Workload remains a central determinant of human performance in safety-critical operations. SWA, and NASA-TLX in particular, provides an accessible and robust way to capture user experience – supporting safer system designs, better training, and improved decision-making.

APPENDIX: EXAMPLE OF USE OF THE NASA-TLX

NASA-TLX Workload Assessment – Example

Task: Nighttime navigation in a congested fjord.

Below is the NASA-TLX example with:

1. A scenario to be assessed.
2. Description of NASA-TLX dimensions.
3. Pairwise weighting step.
4. Filled-in ratings for each dimension.
5. Weighted workload score (final NASA-TLX result).

Scenario Describing the Task to Be Assessed

A maritime navigator must manually monitor radar, the automatic identification system (AIS), and visual cues while entering a congested fjord at night under moderate rain. The autopilot is disengaged; the officer must maintain course and speed, avoid nearby vessels, and respond to Vessel Traffic Service (VTS) calls.

The example reflects a realistic perceptual-cognitive safety-critical task.

NASA-TLX Dimensions Filled Out (Brief)

1. Mental Demand (MD)
2. Physical Demand (PD)
3. Temporal Demand (TD)
4. Performance (P) – perceived success
5. Effort (E)
6. Frustration (F)

Ratings are given on a 0–100 scale. Each dimension is filled out.

Step 1: Pairwise Comparison Weighting

Participant compares each pair (15 total) and selects the more important contributor to the workload. Example results from user.

Pair	Selection
MD vs PD	MD
MD vs TD	TD
MD vs P	MD
MD vs E	E
MD vs F	MD
PD vs TD	TD
PD vs P	P
PD vs E	E
PD vs F	F
TD vs P	TD
TD vs E	TD
TD vs F	TD
P vs E	E
P vs F	F
E vs F	E

Count how many times each dimension was chosen:

Dimension	Weight (0–5)
Mental Demand (MD)	3
Physical Demand (PD)	0
Temporal Demand (TD)	5
Performance (P)	1
Effort (E)	4
Frustration (F)	2
Sum of weights	**15**

Step 2: Raw 0–100 Ratings for the Task

Participant gives numerical ratings:

Dimension	Rating (0–100)	Interpretation
Mental Demand (MD)	80	Monitoring multiple cues; high cognitive load
Physical Demand (PD)	20	Primarily seated work
Temporal Demand (TD)	90	High time pressure due to traffic & changing conditions
Performance (P)	60	Believes performance was moderate
Effort (E)	85	Strong effort needed throughout
Frustration (F)	40	Mild frustration; not overwhelming

Step 3: Weighted NASA-TLX Score

Dimension	Rating	Weight	Weighted Score
MD	80	3	240
PD	20	0	0
TD	90	5	450
P	60	1	60
E	85	4	340
F	40	2	80
SUM		**15**	**1170**

Calculate the Workload as the sum of the Dimension weight*rating divided by the sum of weights – for this case NASA-TLX: 1170/15 = 78; Final NASA-TLX Workload Score: 78

This is consistent with a high-cognitive-load, high-temporal-pressure maritime navigation task in challenging conditions.

REFERENCES

Alsos, O. A., Johnsen, S. O., & Bjørneseth, F. B. (2023). *Menneskelige, tekniske og organisasjonsmessige forhold knyttet til Helge Ingstad ulykken*. NTNU Report. ISBN: 978-82-91917-40-5.

Bell, S. W., Kong, J. C., Clark, D. A., Carne, P., Skinner, S., Pillinger, S., Burton, P., & Brown, W. (2022). The National Aeronautics and Space Administration-task load index: NASA-TLX: evaluation of its use in surgery. *ANZ Journal of Surgery*, *92*(11), 3022–3028.

de Winter, J. C. (2014). Controversy in human factors constructs and the explosive use of the NASA-TLX: a measurement perspective. *Cognition, Technology & Work*, *16*(3), 289–297.

Hart, S. G. (1986). *NASA Task Load Index (TLX): Paper and Pencil Package-Volume 1.0 (p. 20)*. NASA Ames Research Center.

Hart, S. G. (2006). NASA-task load index (NASA-TLX); 20 years later. In *Proceedings of the Human Factors and Ergonomics Society Annual Meeting* (vol. *50*, no. 9, pp. 904–908).

Hart, S. G., & Staveland, L. E. (1988). Development of NASA-TLX (Task Load Index): Results of empirical and theoretical research. In *Human Mental Workload* (vol. 52, pp. 139–183). Elsevier.

Hayes, J., & Skogstad, M. R. (2025). From rigid orderliness to barely controlled chaos: Sociotechnical risk and AI in aviation. In *Proceedings of the 35th European Safety and Reliability & the 33rd Society for Risk Analysis Europe Conference*, Stavanger, Norway. https://doi.org/10.3850/978-981-94-3281-3_ESREL-SRA-E2025-P6532-cd

Javernik, A., Buchmeister, B., & Ojsteršek, R. (2023). The NASA-TLX approach to understand workers workload in human robot collaboration. *International Journal of Simulation modelling*, *22*(4), 574–585.

Jiang, N., Zhou, W., Hasanzadeh, S., & Duffy, V. G. (2025). ExpertGen: A comparative analysis of user performance, cognitive workload, and trust in domain-tailored generative AI. In *International Conference on Human-Computer Interaction* (pp. 178–190).

Norwegian Safety Investigation Authority. (2019). Part one report on the collision on 8 November 2018 between the frigate HNoMS Helge Ingstad and the oil tanker Sola TS outside the Sture Terminal in the Hjeltefjord in Hordaland County [Marine report].

Norwegian Safety Investigation Authority. (2021). Part two report on the collision between the frigate HNoMS 'Helge Ingstad' and the oil tanker Sola TS outside the Sture Terminal in the Hjeltefjord in Hordaland County on 8 November 2018.

Noyes, J. M., & Bruneau, D. P. (2007). A self-analysis of the NASA-TLX workload measure. *Ergonomics*, *50*(4), 514–519.

Reid, G. B., & Nygren, T. E. (1988). The subjective workload assessment technique: A scaling procedure for measuring mental workload. I. In *Advances in Psychology* (vol. *52*, pp. 185–218). Elsevier.

Shin, H., & Prabhu, V. V. (2018). Evaluating impact of AI on cognitive load of technicians during diagnosis tasks in maintenance. In *IFIP International Conference on Advances in Production Management Systems* (pp. 27–34).

Wierwille, W. W., & Casali, J. G. (1983). A validated rating scale for global mental workload measurement applications. In *Proceedings of the Human Factors Society Annual Meeting* (vol. *27*, no. 2, pp. 129–133).

15 Training and Assessment during Digitalisation

Salman Nazir

15.1 INTRODUCTION

Over the past three decades, Human Factors research has made significant advances in understanding safety, efficiency, and resilience in socio-technical systems (Endsley et al., 2025; Hollnagel, 2017). However, compared to design-oriented work such as human-centred design, usability, and standards, training and assessment have not received the same emphasis. Accident investigations across domains repeatedly show that performance breakdowns are rarely purely technical; they are typically linked to insufficient preparation, degraded situation awareness, poor non-technical skills, and a failure to transfer trained competencies into real operations (Reason, 1997; Mumaw et al., 2000). In high-risk domains such as aviation, nuclear power, maritime operations and process industries, simulation-based training has become a key strategy for preparing operators to manage rare, high-consequence scenarios that cannot be practised safely on the real system (Mallam et al., 2019; Kim et al., 2021; Sellberg, 2025; Dewan et al., 2023).

Today's technological landscape is characterised by digitalisation, advanced automation, AI-supported decision systems, remote operations, and integrated control environments (Dimitrakieva et al., 2025). These developments not only change what must be trained but also how we verify that system design actually supports human performance. In a Safety-by-Design perspective, training functions not only as a learning activity but also as a practical validation mechanism for system interfaces, alarms, procedures, and automation behaviours. Well-designed training scenarios reveal whether the design truly supports human capabilities and operational goals. These developments further elevate the importance of robust training and assessment. Automation shifts operators from active controllers to supervisory managers whose infrequent interventions during abnormal situations become more cognitively demanding and safety-critical (van de Merwe, 2024; Hollnagel, 2017).

The author's involvement in several studies across the maritime and process industries, conducted in collaboration with Norwegian universities and international partners, demonstrates that well-designed simulation environments and structured assessment approaches can enhance industrial safety, strengthen distributed situation awareness, and improve preparedness for abnormal events (Nazir et al., 2015; Kim et al., 2021; Tusher et al., 2023; Badu et al., 2025; Sharma et al., 2019). Yet persistent challenges remain, such as simulators that fail to align with user needs, overemphasis on procedural checklist training, and limited evidence on what trainees actually learn (Sellberg, 2025; Karahalil et al., 2024). This gap between technological potential and pedagogical practice is precisely where Human Factors perspectives on simulator-based training and performance assessment are most needed. Even after decades of simulator use, many training programmes across industries still lack validated, transparent, and reliable evaluation tools. Assessment practices often rely heavily on subjective instructor judgement, vary widely between centres, and are not always grounded in empirical validation. This gap continues to limit comparability, reproducibility, and, at times, the reliability of performance assessment.

On a macro level, global investment patterns emphasise the increasing strategic importance of workforce development. The U.S. organisations invested approximately USD 125.9 billion in employee training in 2009, rising to around USD 171.5 billion in 2010 (ASTD, 2010; Green and McGill, 2011).

DOI: 10.1201/9781003741824-19

Training magazine's 2023 report shows that U.S. organisations allocated USD 101.8 billion to employee learning in 2023, with notable growth in spending on external tools, digital systems, and outsourced content, indicating a shift towards technology-driven learning ecosystems (Freifeld, 2023; Grensing-Pophal, 2024; LearnExperts, 2024).

European data show that employers invest approximately 0.7% of labour costs in continuing vocational training, rising to over 1% in countries like France and the Netherlands (Cedefop, 2019). Across maritime and process sectors, these investments signal a shared recognition that a skilled, adaptable workforce is a critical resource, and that evidence-based training and assessment are essential for safety, reliability, and competitiveness (Aguinis & Kraiger, 2009; Mallam et al., 2019; Dewan et al., 2023).

15.2 ROLE OF SIMULATORS IN SAFETY-CRITICAL INDUSTRIES

Simulation has long been an essential tool for preparing individuals and organisations to operate safely in complex environments. A simulator is generally defined as a system or device that replicates real-world conditions for training, experimentation, or analysis (White and Ingalls, 2015), providing a controlled environment in which behaviours and decision-making can be practised without real-world risk. Although the Link Trainer, developed by Edwin Link in 1929 and widely used for pilot instruction during World War II, is often recognised as the first industrialised flight simulator (Jeon, 2015; Myers et al., 2018), the origins of simulation-based training go much further back. Military organisations had long used structured war gaming, map exercises, and sand table simulations to enhance tactical decision-making. Following the wartime research boom of the mid-20th century, the emergence of digital computing accelerated the sophistication of simulators, enabling broader adoption across aviation, nuclear energy, maritime operations, and healthcare (Bienstock & Heuer, 2022; Kobayashi, 2005; Mallam et al., 2019; Sellberg, 2017).

Today, simulation technologies continue to evolve, increasingly integrating artificial intelligence, big data, and digital-twin systems. As Luan (2024) argues, simulation is increasingly integrated with artificial intelligence, big data, digital twins, and related technologies, and is evolving towards more intelligent, real-time, and immersive forms. However, significant challenges remain in realising the expected benefits of digital-twin-supported training and operational preparation. Very few studies include robust validation or quantify improvements over existing processes, and many digital twins remain underused due to limited user involvement and weak alignment with real work practices (Jones et al., 2020; Cameron et al., 2018).

Recent industrial experience demonstrates the consequences of insufficient user involvement in simulator and digital-twin development. For example, an organisation in the Norwegian oil and gas sector used advanced, costly digital twin systems to support operational preparation, training, and scenario simulation. However, these systems remain underused because the design process did not sufficiently involve operators. A lack of understanding of user needs led to solutions that did not align with actual work practices. This illustrates a recurring Human Factors challenge across high-risk domains: technological solutions fail when divorced from a deep understanding of 'work-as-done'. Interviews with control-room operators offshore support this, indicating that digital twins are not used in practice.

15.2.1 Key Developments in Simulator-Based Training

Major developments in simulator-based training can be observed in Figure 15.1. They are as follows.

- **18th–19th centuries**: The emergence of formalised war-gaming practices, that is, most notably the Prussian *Kriegsspiel*, provided structured, risk-free environments for rehearsing tactical and operational decision-making.

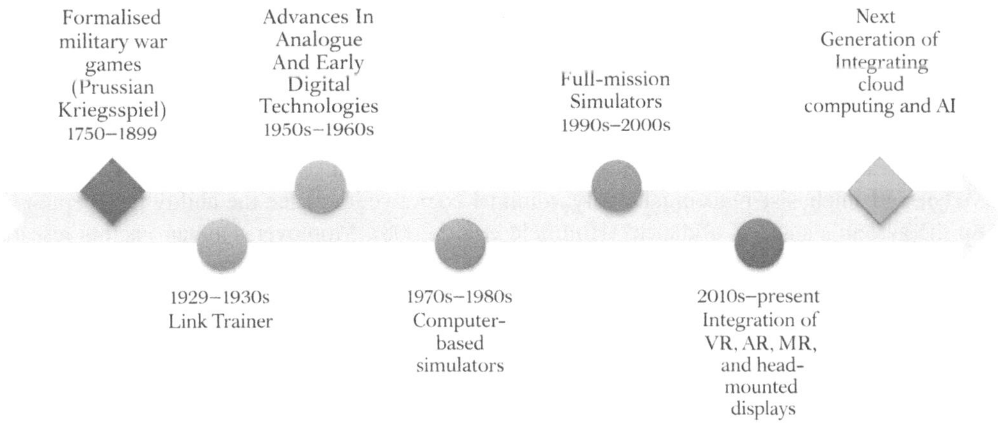

FIGURE 15.1 Timeline of key developments in simulator-based training.

- **1929–1930s**: The introduction of the Link Trainer marked the beginning of large-scale electromechanical flight simulation. It established a new standard for pilot training and accelerated the use of simulators in aviation.
- **1950s–1960s**: Advances in analogue and early digital technologies enabled the development of more sophisticated simulators capable of replicating realistic system behaviour across military and industrial domains.
- **1970s–1980s**: Widespread adoption of computer-based simulators in aviation, nuclear operations, and maritime sectors supported increasingly complex procedural and technical training programmes.
- **1990s–2000s**: The proliferation of full-mission simulators, such as bridge, engine room, and control-room systems, coincided with a growing focus on team coordination and non-technical skills, reinforced by regulatory frameworks, including the STCW Convention.
- **2010s–present**: The integration of virtual reality (VR), augmented reality (AR), mixed reality (MR), and head-mounted displays has enabled more portable, immersive, and adaptive simulation-based training across maritime, healthcare, industrial, and emergency response fields.

Taken together, these developments show that simulation-based training has continuously adapted in response to advances in technology, operational needs, and risk awareness. As systems become more automated and complex, the pedagogical and Human Factors foundations of simulation must also evolve.

15.2.2 Training and Assessment in the Maritime Domain

The maritime sector has adopted simulation across bridge, navigation, and engine-room operations. Full-mission bridge simulators incorporate wraparound visual systems, ship-handling controls, and integrated navigation instruments, allowing officers to practise collision avoidance, port manoeuvres, and crisis management under varying environmental and traffic conditions (Kim et al., 2021). International Maritime Organization (IMO) standards, particularly the STCW Convention, stipulate performance criteria for simulators and permit simulator training to replace portions of mandatory sea time (Drown, 2021).

Engine-room simulators replicate the dynamics of propulsion plants and auxiliary systems, enabling trainee engineers to manage alarms, diagnose machinery faults, and coordinate emergency responses without risking operational assets. Maritime Crew Resource Management (CRM), often referred to as Bridge Resource Management, adapts aviation CRM frameworks to train teams in communication, situational awareness, and shared decision-making during complex operations. Lessons from real-world incidents highlight that training transfer depends critically on system usability. If interface displays are cluttered, cognitively demanding, or inconsistent between training simulators and real operational systems, trainees risk learning behaviours that will not transfer reliably to real-world operations. Best-practice guidelines in industrial Human–Machine Interface (HMI) design emphasise perceptual clarity, minimal cognitive load, and the ability for operators to grasp the system's state 'at a glance' (Hollifield et al., 2008). Moreover, Human Factors research suggests that ease of use, perceptual clarity, and stable interaction patterns are foundational prerequisites for effective training transfer and robust situational awareness.

When poorly designed interfaces overload operators, their capacity to perceive, comprehend, and anticipate critical elements may degrade, impairing decision-making, especially under abnormal or high-stress conditions (Vicente, 1999; Endsley et al., 2025). This reinforces the principle that system design and training design must evolve together. Investigations in offshore and industrial settings, have demonstrated that mismatches between simulator configurations and real operational interfaces can limit training effectiveness and contribute to operational risk.

Overall, maritime simulators offer safe platforms for practising demanding navigational and engineering tasks, enabling crews to rehearse emergency and contingency procedures under conditions that would be hazardous to replicate at sea. Maritime simulation illustrates how sector-specific risks, international regulation, and teamwork demands shape both training scenarios and assessment criteria. Full-mission bridge and engine-room simulators, together with newer VR solutions, provide a rich test bed for linking Human Factors analysis to training design and for experimenting with more transparent performance assessment. At the same time, variation in how these tools are used underlines the central message of this chapter: simulator-based training and evaluation must be deliberately designed from a Human Factors perspective if they are to move beyond regulatory compliance and genuinely support safe, resilient operations.

15.2.3 Simulator Use and Pedagogical Alignment

While simulator technologies continue to advance, their educational value depends on how effectively they are embedded in a coherent instructional system grounded in Human Factors principles. A recurring theme across research is that simulation technologies alone do not guarantee learning; what matters most is how they are embedded within a coherent pedagogical framework founded on task understanding, human factors principles and organisational conditions. Giskeødegård and Wahl (2025) traced the shift from traditional apprenticeship-based learning towards simulation-rich approaches in Maritime Education and Training (MET). They emphasised that choices regarding simulator fidelity, modality, and complexity must be pedagogically based on realism rather than treated merely as regulatory or technical obligations. Their analysis demonstrates that simulator deployment is most effective when aligned with clear learning objectives and instructional design principles. Table 15.1 presents an overview of the main simulator types used across the maritime, aviation, healthcare, and industrial process domains, including their fidelity, strengths, limitations, and pedagogical suitability or best use case.

A key lesson across high-risk industries is that effective training is based on systematic task analysis. This aligns with decades of cognitive engineering research showing that training tailored to actual task demands, rather than just procedure lists, leads to better transfer, quicker decision-making, and enhanced anomaly management (Vicente, 1999; Mumaw et al., 2000). Methods such as Goal-Directed Task Analysis (GDTA) and/or Cognitive Task Analysis (CTA) identify perceptual cues, SA, decisions, workload demands, and potential error traps that characterise real operations.

Sharma et al. (2019) conducted a GDTA of maritime navigation. They identified the essential situation-awareness information requirements for navigators, providing empirically based guidance on how HMIs should support real operations. Training and on-the-job learning aligned with the task analysis enhance knowledge-based understanding, encourage transfer of learning, strengthen

TABLE 15.1
Comparison of Simulator Types in Safety-Critical Industries

Simulator Type	Fidelity Level	Strengths	Limitations	Best Use Cases	References
Full-Mission Simulator (FMS)	Very high (physical controls, 360° visuals, real system dynamics)	✓ Highest realism and immersion ✓ Supports team training & CRM ✓ Enables rehearsing rare, high-risk events ✓ Strong transfer to real operations	✓ Extremely costly ✓ Requires specialised facilities ✓ Limited availability ✓ Difficult to modify scenarios quickly	✓ Bridge, engine-room, nuclear control room ✓ Emergency procedures ✓ Team coordination & NTS	Kim et al. (2021); Mallam et al. (2019); Sellberg (2025)
Desktop/PC-Based Simulator	Low–Medium	✓ Low cost ✓ Easy to deploy remotely ✓ Good for initial familiarisation ✓ Flexible for self-paced learning	✓ Limited realism ✓ Weak physical interaction feedback ✓ Not ideal for complex team scenarios	✓ Basic navigation ✓ Introductory system familiarisation ✓ Procedure walk-throughs	Mallam et al. (2019); Tusher et al. (2023)
Virtual Reality (VR) Simulator	Medium–High (immersive audiovisual experience)	✓ High immersion at lower cost ✓ Excellent for spatial awareness & procedural training ✓ Highly scalable ✓ Supports remote training	✓ Motion sickness for some users ✓ Not always aligned with regulatory needs ✓ Limited tactile feedback ✓ Assessment tools still evolving	✓ Spatial tasks ✓ Procedural drills ✓ Remote-operation familiarisation	Renganayagalu et al. (2021); Mallam et al. (2019); Sellberg (2025)
Augmented/Mixed Reality (AR/MR)	Medium	✓ Blends real environment + virtual overlays ✓ Ideal for maintenance, inspection, & step-by-step guidance	✓ Requires hardware calibration ✓ Visual clutter possible ✓ Limited team-training value	✓ Equipment maintenance ✓ Emergency equipment familiarisation	Mallam et al. (2019); Elendu et al (2024)
Digital Twins (Operational Replicas)	Varies (medium–high depending on data quality)	✓ Real-time data from live systems ✓ Excellent for decision support & scenario testing ✓ Supports remote operations	✓ Expensive to build & maintain ✓ Empirical validation often missing ✓ Not widely used by operators due to usability gaps	✓ Design validation ✓ Operator preparation ✓ Predictive maintenance ✓ Remote operations	Cameron et al. (2018); Jones et al. (2020)

non-technical skills, and reduce the likelihood of errors in novel situations caused by poor procedure design (Mumaw et al., 2000; Sharma et al., 2019). Without such analysis, training risks focusing on superficial procedures rather than the deeper cognitive strategies necessary for safe performance. An example of how task analysis can support learning design is a common but mentally challenging maritime task: ship docking. Initially, docking may seem to be just a single learning goal in a course outline. However, a detailed task analysis reveals that it involves multiple interconnected perceptual, cognitive, and coordination skills. A GDTA or CTA would usually identify subtasks such as interpreting visual and radar cues for distance and speed, coordinating intentions with the bridge team and port authorities, adjusting thruster and engine settings, monitoring wind and current effects, and preparing contingency plans if alignment or control is lost. Breaking down a complex operation into observable parts aligns with how performance measures for pilotage and berthing are structured in assessment tools like CAPA (Ernstsen and Nazir, 2020).

Based on such an analysis, these elements can be translated into clear and quantifiable learning objectives. For instance, the training design might specify that the trainee shall:

- Maintain a safe approach speed below a defined limit within a given distance of the berth,
- Integrate visual, radar and environmental cues to keep the vessel within a narrow lateral tolerance band, and
- Conduct the required bridge-team communication steps at predefined decision points.

From these objectives, the training programme can then derive performance indicators and, where appropriate, key performance indicators that support transparent assessment, such as:

- Approach speed over time (logged automatically by the simulator),
- Deviation from planned track or lateral alignment (time series during the docking phase),
- Completeness and timing of communications (coded from audio or observer notes), and
- Use of thrusters, rudder, and engine commands (compared with patterns observed in expert performance).

These indicators not only support trainee assessment but also serve as a basis for evaluating whether the fundamental system design genuinely promotes safe and efficient docking. If repeated simulator sessions show that trainees consistently miss certain cues, struggle to detect drift, or compensate for automation behaviour in unexpected ways, this alerts designers to issues with information presentation, HMI layout, and control logic. In that sense, the docking example demonstrates a logical sequence: task analysis informs learning objectives, learning objectives shape performance indicators, and these indicators can be utilised both for assessing individuals and for validating and improving system design. Harrington (2025) offers a structured design framework derived from the 4C/ID (Four-Component Instructional Design) model, emphasising the need for coherence among whole-task practice, scaffolding strategies, and the briefing–scenario–debriefing cycle.

Sellberg and Erntsson (2025) analysed how briefing practices shape expectations, roles, and learning goals before an exercise, while Sharma (2025) examined debriefing as the main area where performance is interpreted, linked to theory, and transformed into lasting competence. Together, their work and the present chapter emphasise that learner development depends as much on preparatory and reflective activities as on performance during the scenario itself. Harrington (2025) further illustrates the complex instructional decisions facilitators must navigate in real time during simulations, reinforcing the importance of instructor expertise.

15.2.4 Aviation

Modern flight simulators play a vital role in pilot training, licensing, and ongoing assessments in both civil and military aviation. High-fidelity Full Flight Simulators (FFSs) replicate aircraft cockpits, flight dynamics, instruments, and motion cues, enabling trainees to practise high-risk, rare scenarios in a controlled, risk-free environment. Regulatory authorities allow a certain proportion of required flight hours to be obtained using certified simulators, recognising their capacity to support both technical and non-technical skill development (Cross & Ryley, 2025).

Extended reality (XR) training systems are increasingly supplementing traditional simulators by providing cost-effective, versatile, and scalable training options (Ross, 2022). A key benefit of simulator-based training is the ability to practise abnormal or emergency procedures, such as engine failures, severe weather, and avionics faults, without risking personnel or equipment. These systems also aid in familiarisation with cockpit automation, navigation technologies, and checklist procedures, especially as periodic software updates make real-aircraft training less feasible. Simulators are central to CRM training, enabling flight crews to practise teamwork, communication, workload management, and decision-making under realistic stress conditions (Crichton, 2017). Recurrent simulator checks ensure continued proficiency and are used by airlines and regulators to validate new cockpit procedures before operational deployment (Tuhkala et al., 2024).

Taken together, the aviation examples demonstrate what a mature simulator ecosystem can look like when Human Factors, training design, and assessment are closely integrated. The widespread use of simulators, ongoing emphasis on training, and strong focus on safety through procedures, standards, technology, and design have all helped make aviation operations significantly safer than they were 30 years ago. This experience from aviation shows how mature simulator ecosystems depend heavily on human-factors-informed design, especially in areas such as workload management, communication protocols, and assessment standardisation. These insights directly inform how process-industry simulators could better align training objectives, task demands, and performance evaluation.

15.2.5 Healthcare

Simulation in healthcare covers a wide range of methods, from low-fidelity task trainers to high-fidelity human-patient simulators and VR-based procedural systems. These tools allow healthcare professionals to practise interventions and handle acute clinical events such as cardiac arrest, trauma, or obstetric emergencies without risking patient safety.

Basic task trainers offer anatomically accurate platforms for practising procedures such as venipuncture or intubation. High-fidelity mannequins feature programmable physiology, including pulse, breath sounds, and drug responses, enabling immersive scenario-based training. VR and laparoscopic simulators provide force feedback and three-dimensional visualisation, supporting surgical skill development and reducing learning curves (Elendu et al., 2024; Pandey & Wolfe, 2012). Simulation-based education strengthens both technical skills and non-technical abilities such as communication, leadership, and teamwork. Standardised patient programmes enable learners to practise interpersonal skills and clinical reasoning, while hybrid simulation combines actors with mannequins to facilitate complex team-based exercises. Evidence from systematic reviews shows that high-fidelity simulation improves skill retention, performance, and learner confidence (Issenberg et al., 2005; McGaghie et al., 2010). Residents trained through simulation perform more effectively in real clinical environments, thereby improving patient safety (Nishisaki et al., 2007; Marzouki and Benyass, 2025). Several review articles have demonstrated the impact of simulator training in healthcare (Elendu et al., 2024; Murphy et al., 2016; Saleem & Khan, 2023).

Accreditation bodies increasingly promote standards for simulation centres and instructor development, reflecting the central role of simulation in reducing medical error and enhancing clinical outcomes (Elendu et al., 2024). The healthcare cases demonstrate that simulation only delivers its potential when it is closely coupled with real clinical workflows, cognitive demands, and team coordination challenges. Simulation centres that focus on crisis resource management, handovers and interdisciplinary teamwork illustrate how Human Factors concerns can be translated into scenarios, debriefings, and assessment criteria. This reinforces the broader argument that effective simulator design, regardless of sector, depends on an understanding of work-as-done and on using assessment to support learning rather than merely to certify.

15.2.6 Industrial Process and Nuclear Industries

Simulator-based training plays a crucial role in high-risk process industries, especially nuclear power, where operators must manage complex systems and respond effectively to abnormal events. After the Three Mile Island accident in 1979, U.S. regulators mandated simulator-based training for operator licensing, setting an international example (World Nuclear Association, 2022). Operator Training Simulators (OTS) support routine operations as well as low-probability, high-consequence incidents. They also serve as vital platforms for human reliability analysis (Boring, 2009). Research has shown measurable physiological stress responses during simulated emergencies, such as changes in heart-rate variability, illustrating how stress affects operator decision-making and performance (Pakarinen et al., 2018).

In the process industries, the roles of field operators and control-room operators differ considerably in terms of task demands, perceptual cues, and operational responsibilities. However, both groups share the common goal of running the plant efficiently while ensuring the highest safety standards. The dynamic interaction between these roles offers a valuable context for examining key Human Factors concepts such as distributed situation awareness, mental workload, communication quality, and team coordination. Given its extensive and well-established history of simulator-based training, the process-industry sector offers valuable insights into how complex socio-technical systems can effectively design, implement, and evaluate training to support high-reliability performance. Recent advances incorporate automation research, eye-tracking metrics, machine learning, and AI-driven analytics to examine operator fatigue, trust calibration, and human–machine interaction. For example, predictive models based on blink rate, percentage of eyelid closure (PERCLOS), and mouse-movement patterns have shown high accuracy in detecting fatigue (Dai et al., 2023). AI-powered simulations further aid optimisation of control-room interfaces and maintenance strategies (Khamaj et al., 2024).

In nuclear and other high-reliability process industries, operator training simulators demonstrate how realistic, model-based environments can support the safe handling of complex, tightly coupled systems and low-frequency, high-consequence disturbances. When scenario design is based on task analysis and performance indicators are systematically utilised, simulator exercises serve as a way of testing both operator competence and the adequacy of procedures, automation, and control-room concepts. These practices emphasise the opportunity for broader process-industry training to adopt similarly structured, evidence-based approaches to simulator-based training and performance assessment.

15.3 VR SIMULATORS

Virtual reality (VR) broadens the simulation spectrum by providing immersive, interactive, and repeatable environments for safe practice. In MET, VR is increasingly regarded as a complementary modality to full-mission simulators, especially for procedural tasks, spatial orientation, and remote-operation scenarios (Mallam et al., 2019; Renganayagalu et al., 2021).

Recent empirical research strengthens this perspective. Tusher et al. (2023), using a hybrid multi-criteria decision-making (MCDM) approach, demonstrated that while VR offers clear advantages in flexibility and cost, its pedagogical value depends on careful alignment with learning objectives and regulatory requirements. Their findings demonstrated that:

- Full-mission simulators remain the preferred modality.
- VR ranks second, valued for flexibility and cost-effectiveness.
- Regulatory compliance, relevance, and instructional control outweigh novelty.

In MET, VR and related XR technologies have been proposed as complements, not simple replacements, for established full-mission bridge and engine-room simulators, particularly for procedural practice, spatial orientation, and distributed or remote operation scenarios (Mallam, Nazir, & Renganayagalu, 2019; Renganayagalu et al., 2021). However, the promise of VR cannot be reduced to greater *realism or more technology*; from a Human Factors perspective, its effectiveness depends on how well the virtual environment aligns with specific learning objectives, user needs, organisational constraints, and assessment strategies.

Tusher et al. (2024a) show, using empirical data, that VR and other simulators must be carefully positioned rather than adopted blindly. In a hybrid MCDM study, the authors compared maritime simulator modalities from technical, instructional, and organisational perspectives. Their results indicate that instructors and experts consistently rated full-mission simulators as the most preferred option overall, with VR-based simulators ranked second, valued for flexibility and cost but not seen as a universal substitute. Crucially, the study showed that regulatory compliance, learning relevance, and instructional control were weighted more heavily than the technology's cost or novelty. This suggests that VR should only be adopted when it clearly meets regulatory and pedagogical needs, not merely because it is accessible or trendy.

In parallel, Tusher et al.'s (2023) study of assessment practices in maritime simulation-based training shows that introducing VR requires rethinking how performance is evaluated. Examining current methods, the authors state that assessment remains mainly based on traditional rating scales and instructor judgements, even though newer tools like VR offer richer data collection (behaviour traces, patterns over time, interaction logs) that could support more realistic and outcome-oriented assessments. It underscores the importance of aligning scenario design, performance indicators, and feedback mechanisms clearly, and advocates for assessment methods that include both technical and non-technical skills in realistic team settings. Overall, this body of work indicates that VR can be a valuable addition to maritime simulator environments, but only when integrated into a purposeful training and assessment framework, where technology choices are driven by a careful analysis of learning objectives, user profiles, and organisational commitments, rather than vice versa.

15.4 ASSESSMENT

Assessment remains one of the most challenging and arguably least developed areas of training (Timmis et al., 2016). Although simulators are widely used, assessment practices tend to be informal, varied, and poorly connected to learning outcomes (Sellberg, 2017). Assessment is essential to ensure that simulation-based training results in demonstrable competence. A significant challenge is that assessment must be aligned with the same task analysis that informs scenario design. Without clear performance models detailing the expected cognitive, perceptual, and coordination demands, assessment risks becoming subjective or inconsistent.

Despite the importance of assessment for competence assurance, many simulation programmes still lack validated, transparent, and reliable evaluation tools. Traditional performance evaluation depends on expert observers; however, studies in both maritime and aviation fields highlight ongoing inter-rater variability, especially for non-technical skills. Robust assessment, therefore, necessitates structured rubrics, standardised scoring models, and systematic debriefing practices. This reflects issues seen across simulation-based education more broadly (Chernikova et al., 2020).

Tusher et al. (2024b) demonstrate that VR offers new possibilities for capturing performance data, behavioural traces, interaction logs, and temporal patterns, yet assessment in MET still heavily depends on subjective ratings. Overall, VR can enhance maritime training, but only when pedagogically integrated and supported by coherent assessment strategies. VR presents new opportunities but also introduces challenges. Systematic reviews highlight inconsistent study quality and limited objective measures (Renganayagalu et al., 2021; Tusher et al., 2024b). Emerging research examines ethical considerations (Tusher et al., 2022) and suggests AI-based classifiers for VR performance assessment. Although in early stages, these efforts indicate the potential for richer, more objective evaluation but they also emphasise the importance of transparency and human oversight. Ernstsen and Nazir (2018) promote deriving performance indicators from task analyses and making criteria explicit. Their subsequent Computer-Aided Performance Assessment (CAPA) tool, which combines Analytic Hierarchy Process (AHP) and Bayesian networks, significantly enhances reliability for technical skills compared with instructor Judgment. However, assessing teamwork remains challenging. A strong assessment and learning culture also depends on psychological safety. Learners must feel able to ask questions, reveal uncertainty, discuss shortcuts and work-as-done, and make mistakes without fear of punishment. Without such conditions, authentic performance remains hidden, reporting culture declines, and meaningful learning becomes unlikely. Establishing a Just Culture, where behaviours are judged fairly, and errors are examined for systemic learning rather than blame, is vital for fostering constructive dialogue, accurate assessment, and ongoing safety improvement. This cultural foundation is just as important as the technical tools used for measurement.

In this chapter, CAPA refers to the use of structured performance indicators and computational models to generate transparent assessment scores from simulator exercises. In the context of maritime pilotage operations, Ernstsen and Nazir (2020) developed a CAPA tool for full-mission simulators by combining the AHP to weight performance indicators with a Bayesian network to represent probabilistic relationships between these indicators and overall competence (see also Ernstsen & Nazir, 2018). The AHP is a structured decision-making framework that decomposes a complex problem into a hierarchy of goals and criteria and uses pairwise comparisons to translate expert judgements into quantitative weights (Saaty, 2008). A Bayesian network is a probabilistic graphical model that depicts variables and their conditional dependencies in a directed acyclic graph, allowing uncertain evidence from multiple indicators to be combined into a coherent estimate of competence (Jensen & Nielsen, 2007; Fenton & Neil, 2013).

In simulator-based training, combining CAPA, AHP, and Bayesian networks enables more consistent scoring, structured use of expert judgement, and explicit modelling of uncertainty. This provides a robust complement to traditional instructor assessments and can help reduce issues such as subjective bias and inter-rater variability (Ernstsen & Nazir, 2018, 2020). A related concern is the gap between formal reporting systems and what is actually documented in practice. Recent investigative reports in one of the Scandinavian petroleum sectors have described cases in which written reports on hazardous work and potential gas exposure by contracted workers did not enter the operator's formal reporting systems, despite being submitted locally (Folkvord, 2025). Such accounts, while journalistic rather than academic, illustrate how structural and cultural factors can filter or suppress weak signals from the sharp end, thereby further constraining the information base available for improving training, assessment, and system design. A conceptual Bayesian model for navigation (Ernstsen and Nazir, 2018) and work on GDTA (Sharma et al., 2019) further demonstrate the value of anchoring assessment in explicit competence models. Systematic reviews (Tusher et al., 2023) show that assessment methods range from subjective ratings to physiological measures such as eye-tracking and electroencephalography (EEG), each with context-dependent strengths. No single method is universally appropriate; assessment must align with the targeted competencies and the training context.

Sellberg (2025) highlights ongoing tensions in simulation-based assessment, especially the challenge of balancing validity, reliability, fairness, and the inherently situated nature of knowledge-in-action. Sellberg (2025) emphasises that clearer, research-informed assessment models and better assessor training are necessary to reduce inter-rater variability and improve the defensibility of evaluative judgements. In her broader synthesis, Sellberg (2025) describes simulation pedagogy as a mature yet evolving field requiring continuous knowledge development. Lundin et al. (2025) build on this argument by conceptualising instructors as co-researchers who collaboratively refine training and assessment practices. Their work emphasises simulation as a space for practitioner inquiry and organisational learning rather than a static instructional tool. Collectively, these contributions demonstrate that contemporary MET simulation is concerned not only with technology and fidelity but also with pedagogical design, briefing and debriefing processes, assessment methodology, and instructor professional development.

Taken together, these assessment approaches highlight a central pedagogical message: practical assessment requires explicitly defined performance models, standardised scoring methods, and a learning culture that supports open reflection rather than blame. Without these foundations, even high-fidelity simulators cannot reliably produce competence or support safe operations.

15.5 DIGITAL AND DATA-DRIVEN EVALUATION

Modern simulators automatically record trainee actions, system states, and communication patterns, creating digital traces that support objective measurement and replay-based feedback. CAPA tools can produce consistent evaluations for technical tasks, although teamwork and communication remain challenging to quantify (Ernstsen and Nazir, 2018).

Advances in AI and learning analytics enable personalised feedback, predictive modelling of performance trends, and adaptive scenario design. Instructor dashboards and performance visualisations support targeted remediation and more precise alignment of training with learner needs. Aggregated assessment data provide insight into systemic skill gaps, recurring procedural errors, or communication challenges, informing curriculum development and operational improvement. In this way, simulation-based assessment not only certifies individual competence but also contributes to continuous organisational learning and safety enhancement. The capabilities that Artificial Intelligence offers for interpreting and extrapolating data efficiently, effectively, and robustly create significant opportunities for advancing digital evaluation. Many of these benefits have already been documented in the emerging literature.

However, as the author notes, there are significant risks associated with AI-driven assessment that must not be overlooked and require careful consideration and safeguards. Evaluation and assessment aimed at making improvements, whether in behaviours, skills, or knowledge, are not solely technical processes. They also involve feedback, mentorship, and the relational dynamics that develop between trainees and instructors. As the pace of digitalisation accelerates, longitudinal research will be essential to address these issues and to ensure that digital evaluation solutions support learning responsibly and sustainably. Simulator-based training and performance assessment are not merely educational tools; they also serve as means for verifying whether system design, interfaces, and procedures support human performance. When employed systematically, simulators help confirm whether system design aligns with human capabilities, uncover mismatches between work-as-imagined and work-as-done, and prepare operators for situations that cannot be safely rehearsed elsewhere. In this regard, training becomes a vital element of Safety by Design, as it highlights mismatches between work-as-imagined and work-as-done, identifies usability barriers, and reveals where automation or procedures fail to support operators during unexpected events.

When supported by solid pedagogy and dependable assessment, simulators foster adaptive expertise, enhance non-technical skills, and prepare operators for rare, high-risk events. Training must also equip operators to handle the unexpected. Resilience engineering research shows that rare events do not follow scripted procedures and require adaptive expertise, cue recognition, hypothesis testing, and improvisation under uncertainty (Hollnagel, 2017). These findings are supported by aviation research, where exposure to unexpected events in simulators notably improves pilot resilience and decision-making (McGaghie et al., 2010). Simulator training that includes unexpected events, ambiguous cues, degraded data, and conflicting goals has been shown to enhance operators' ability to manage unforeseen disturbances in nuclear, maritime, and aviation environments (McGaghie et al., 2010; Pakarinen et al., 2018). Creating such scenarios requires careful task analysis and well-defined learning objectives. Without this support, even advanced technology risks becoming an expensive but ineffective practice room. Effective training must be based on the cognitive mechanisms underlying skilled performance. Research consistently demonstrates that whole-task practice, scenario variability, guided reflection during debriefings, and explicit modelling of perceptual cues all improve transfer to real-world operations (Issenberg et al., 2005; Sharma, 2025).

Looking ahead, there is no crystal ball to predict how training and assessment will evolve over the coming decades. The past five years have already expanded the possibilities for using technology in learning, particularly through VR, digital twins, and AI-based tools. At the same time, recent research and practical experience remind us that technology alone can introduce new forms of complexity, distraction, and inequity if it is not grounded in Human Factors principles and an understanding of work-as-done. A key challenge for the coming years will therefore be to balance the social and collaborative aspects of learning (e.g., team-based simulator training and debriefing) with more individual, technology-mediated learning using tools such as VR.

One plausible development is a closer integration of AI and VR, where intelligent tutoring systems serve as a kind of critical friend inside or alongside immersive scenarios, guiding, questioning, and supporting learners in real time, with feedback adapted to individual needs. Such ideas are promising, but their long-term effects on competence, professional evaluation, and team skills remain uncertain and require careful evaluation rather than technological enthusiasm. Across sectors, there is also significant potential for cross-fertilisation: although cognitive demands and regulatory frameworks differ between aviation, maritime operations, healthcare, and the process industries, the ways people learn, coordinate, and interact with technology show essential similarities. Systematic sharing of lessons across these fields may therefore be as vital for the future of simulator training and performance assessment as any specific new technology.

15.6 CONCLUSIONS AND FUTURE DIRECTIONS

Training that only concentrates on procedures without helping operators to recognise patterns, handle uncertainty, and adapt to new disturbances is unlikely to prepare them for the realities of modern, highly automated work. A strong assessment culture also relies on psychological safety and an open reporting attitude. Learners need to feel able to ask questions, admit uncertainty, discuss shortcuts and work as done, and make mistakes without fear of punishment. Without these conditions, genuine performance cannot be observed, and meaningful learning is unlikely. In MET, progress is clear in scenario design, instructor practise, assessment frameworks, and VR integration. However, ongoing challenges include underspecified criteria, inconsistent rating practices, limited longitudinal research, and premature enthusiasm for emerging technologies such as VR and AI.

On this basis, several priorities for future work can be identified:

- More longitudinal and transfer-focused studies are needed to connect simulator performance to behaviour and outcomes in real operations, and to understand retention and decay of competencies.
- Existing tools should be refined, shared, and validated across disciplines, with particular attention to non-technical skills, distributed situation awareness, and team-level performance.
- VR and related technologies should be integrated as complementary modalities, selected for clearly defined learning goals and assessment strategies, rather than as generic replacements for full-mission simulators.
- Rich simulator and VR data can support more nuanced feedback and early risk detection but must be governed by transparent criteria, ethical safeguards, and strong instructor involvement.
- Comparative work across maritime, aviation, healthcare and process industries, and collaborative research with instructors as co-developers, will be crucial for building robust, context-sensitive simulation pedagogies.

We are also experiencing a particularly exciting period for simulator training and performance assessment. Digitalisation and the swift availability of AI-related tools have broadened the range of technical possibilities, with new combinations of simulators, VR, digital twins, and intelligent support systems emerging almost monthly. At the same time, we must be cautious not to be swept away by technological advances and overlook the human aspects of learning: interaction, skill acquisition, team processes, and professional judgement. I hope this chapter will inspire colleagues and practitioners to reflect on these issues, consider which parameters truly matter in their context, and continue developing both the use of simulators and the methods for evaluating training outcomes.

ACKNOWLEDGEMENT

ChatGPT was used to assist with language editing for clarity, flow, readability, and consistency.

The author acknowledges financial and organisational support from the Centre of Excellence in Maritime Simulator Training and Assessment (COAST), hosted at the University of South-Eastern Norway.

REFERENCES

Aguinis, H., & Kraiger, K. (2009). Benefits of training and development for individuals and teams, organizations, and society. *Annual Review of Psychology*, 60(1), 451–474.

ASTD. (2010). *State of the Industry Report*. Alexandria, VA: American Society for Training & Development.

Badu, J., Andreassen, N., Elvegard, R., & Nazir, S. (2025). From gaps to gains: Exploring how professional diversity influences situational awareness in collaborative environments. In: Salman Nazir (ed.), *AHFE International Conference on Training, Education, and Learning Sciences*, AHFE Open Access (vol. 193). AHFE International. http://doi.org/10.54941/ahfe1006651

Bienstock, J., & Heuer, A. (2022). A review on the evolution of simulation-based training to help build a safer future. *Medicine*, 101(25), e29503.

Boring, R. L. (2009, April). Using nuclear power plant training simulators for operator performance and human reliability research. In *Sixth American Nuclear International Topical Meeting on Nuclear Plant Instrumentation, Control, and Human-Machine Interface Technologies* (pp. 5–9).

Cameron, D.G., Waaler, A., & Komulainen, T.M. (2018). Oil and Gas digital twins after twenty years. How can they be made sustainable, maintainable and useful? In *Proceedings of the 59th Conference on Simulation and Modelling (SIMS'59)*, 26–28 September, Oslo Metropolitan University, Norway.

Cedefop. (2019). *Continuing Vocational Training in EU Enterprises: Developments and Challenges Ahead* (Cedefop research paper No. 73). Publications Office. https://doi.org/10.2801/704583

Chernikova, O., Heitzmann, N., Stadler, M., Holzberger, D., Seidel, T., & Fischer, F. (2020). Simulation-based learning in higher education: A meta-analysis. *Review of Educational Research*, 90(4), 499–541

Crichton, M. T. (2017). From cockpit to operating theatre to drilling rig floor: Five principles for improving safety using simulator-based exercises to enhance team cognition. *Cognition, Technology & Work*, 19, 73–84.

Cross, J. I., & Ryley, T. (2025). Assessing evidence-based training in a collaborative virtual reality flightsimulator. *The Aeronautical Journal*, 129(1332), 261–281.

Dai, L., Li, Y., & Zhang, M. (2023). Detection of operator fatigue in the main control room of a nuclear power plant based on eye blink rate, PERCLOS and mouse velocity. *Applied Sciences*, 13(4), 2718. https://doi.org/10.3390/app13042718

Dewan, M. H., Godina, R., Chowdhury, M. R. K., Noor, C. W. M., Wan Nik, W. M. N., & Man, M. (2023). Immersive and non-immersive simulators for the education and training in Maritime Domain—A review. *Journal of Marine Science and Engineering*, 11(1), 147. https://doi.org/10.3390/jmse11010147

Dimitrakieva, S., Marinova-Stoyanova, M., & Gramchev, B. (2025). Integrating artificial intelligence into the management decision-making process in the new era of Industry 4.0. In *Proceedings of the International Scientific and Practical Conference on Environment, Technology and Resources* (vol. 4, pp. 91–98). https://doi.org/10.17770/etr2025vol4.8451

Drown, D. F. (2021). Simulation and the revised 1978 STCW convention. In M. S. Chislett (Ed.), *Marine Simulation and Ship Manoeuvrability* (pp. 245–255). London: Routledge.

Elendu, C., Amaechi, D. C., Okatta, A. U., Amaechi, E. C., Elendu, T. C., Ezeh, C. P., & Elendu, I. D. (2024). The impact of simulation-based training in medical education: A review. *Medicine*, 103(27), e38813.

Endsley, M. R., Dixon, J., Endsley, T., Jamrog, D., Smith-Velazquez, L., & Pfeffer, A. (2025). Divergence in situation awareness and workload. *Ergonomics*, 68(10), 1618–1634.

Ernstsen, J., & Nazir, S. (2018). Consistency in the development of performance assessment methods in the maritime domain. *WMU Journal of Maritime Affairs*, 17(1), 71–90.

Ernstsen, J., & Nazir, S. (2020). Performance assessment in full-scale simulators–A case of maritime pilotage operations. *Safety Science*, 129, 104775.

Fenton, N., & Neil, M. (2013). *Risk Assessment and Decision Analysis with Bayesian Networks*. Boca Raton, FL: CRC Press.

Folkvord, M. S. (2025). *Melkøya: Varslene som forsvinner*. VG. Retrieved from https://www.vg.no/nyheter/i/8plG3Q/vg-avsloerer-varslene-som-forsvinner-paa-melkoeya

Freifeld, L. (2023). 2023 training industry report. *Training Magazine*. https://trainingmag.com/2023-training-industry-report

Giskeødegård, M. F., & Wahl, A. M. (2025). Signature pedagogies of Norwegian simulator-based nautical training. *Discover Education*, 4(1), 148.

Green, M., & McGill, E. (2011, November). The 2011 state of the industry: Increased commitment to workplace learning. *Training + Development (T+D)*, 65, 44–50.

Grensing-Pophal, L. (2024, January 3). Insights from the 2023 training industry report. *Edloomio/Training Industry Magazine*. https://edloomio.com/insights-from-the-2023-training-industry-report

Harrington, S. (2025). Designing simulation for learning. In C. Sellberg (Ed.), *Simulations in Maritime Education and Training* (pp. 17–32). Cham: Springer.

Hollifield, B., Oliver, D., Nimmo, I., & Habibi, E. (2008). *The High-Performance HMI Handbook: A Comprehensive Guide to Designing, Implementing and Maintaining Effective HMIs for Industrial Plant Operations*. Houston, TX: PAS Publishing.

Hollnagel, E. (2017). Resilience engineering and the future of safety management. In N. Möller, S. O. Hansson, J.-E. Holmberg, & C. Rollenhagen (Eds.), *Handbook of Safety Principles* (pp. 23–41). Hoboken, NJ: John Wiley & Sons.

Issenberg, S. B., McGaghie, W. C., Petrusa, E. R., Lee Gordon, D., & Scalese, R. J. (2005). Features and uses of high-fidelity medical simulations that lead to effective learning: A BEME systematic review. *Medical Teacher*, 27(1), 10–28.

Jensen, F. V., & Nielsen, T. D. (2007). *Bayesian Networks and Decision Graphs* (2nd ed.). New York: Springer.

Jeon, C. (2015). The virtual flier: The link trainer, flight simulation, and pilot identity. *Technology and Culture*, 56(1), 28–53.

Jones, D., Snider, C., Nassehi, A., Yon, J., & Hicks, B. (2020). Characterising the digital twin: A systematic literature review. *CIRP Journal of Manufacturing Science and Technology*, 29, 36–52.

Karahalil, M., Lützhöft, M., & Scanlan, J. (2024). Factors impacting curricula in maritime simulator-based education. *WMU Journal of Maritime Affairs*, 1–29.

Khamaj, A., Ali, A. M., & Saminathan, R. (2024). Human factors engineering simulated analysis in administrative, operational and maintenance loops of nuclear reactor control unit using artificial intelligence and machine learning techniques. *Heliyon*, 10(10), e30866. https://doi.org/10.1016/j.heliyon.2024.e30866

Kim, T. E., Sharma, A., Bustgaard, M., Gyldensten, W. C., Nymoen, O. K., Tusher, H. M., & Nazir, S. (2021). The continuum of simulator-based maritime training and education. *WMU Journal of Maritime Affairs*, 20(2), 135–150.

Kobayashi, H. (2005). Use of simulators in assessment, learning and teaching of mariners. *WMU Journal of Maritime Affairs*, 4(1), 57–75.

LearnExperts. (2024). *How Much Do Companies Spend on Training Per Employee?* LearnExperts.ai. Retrieved 5 November 2025, from https://learnexperts.ai/blog/how-much-do-companies-spend-on-training-per-employee

Luan, T. (2024). A comprehensive review of simulation technology: Development, methods, applications, challenges and future trends. *International Journal of Emerging Technologies and Advanced Applications*, 1(5), 9–14. https://doi.org/10.62677/IJETAA.2405119

Lundin, M., Lantz-Andersson, A., Skantz, E., & Williams, P. (2025). Instructors as co-researchers: Strengthening professional practice through collaborative research. In C. Sellberg (Ed.), *Simulations in Maritime Education and Training: A Companion to Simulator Instructors* (pp. 95–108). Cham, Switzerland: Springer Nature.

Mallam, S. C., Nazir, S., & Renganayagalu, S. K. (2019). Rethinking maritime education, training, and operations in the digital era: Applications for emerging immersive technologies. *Journal of Marine Science and Engineering*, 7(12), 428. https://doi.org/10.3390/jmse7120428

Marzouki, Kamal & Benyass, Aatif. (2025). Can simulation-based training address the challenges of residents' training in cardiology? A combination of a survey and a literature review. *SAS Journal of Medicine*, 12, 764–775.

McGaghie, W. C., Issenberg, S. B., Petrusa, E. R., & Scalese, R. J. (2010). A critical review of simulation-based medical education research: 2003–2009. *Medical Education*, 44(1), 50–63.

Mumaw, R. J., Roth, E. M., Vicente, K. J., & Burns, C. M. (2000). There is more to monitoring a nuclear power plant than meets the eye. *Human Factors*, 42(1), 36–55.

Murphy, M., Curtis, K., & McCloughen, A. (2016). What is the impact of multidisciplinary team simulation training on team performance and efficiency of patient care? An integrative review. *Australasian Emergency Nursing Journal*, 19(1), 44–53.

Myers III, P. L., Starr, A. W., & Mullins, K. (2018). Flight simulator fidelity, training transfer, and the role of instructors in optimizing learning. *International Journal of Aviation, Aeronautics, and Aerospace*, 5(1), 6.

Nazir, S., Øvergård, K. I., & Yang, Z. (2015). Towards effective training for process and maritime industries. *Procedia Manufacturing*, 3, 1519–1526.

Nishisaki, A., Keren, R., & Nadkarni, V. (2007). Does simulation improve patient safety? *Anesthesiology Clinics*, 25(2), 225–236.

Pakarinen, S., Korpela, J., Torniainen, J., Laarni, J., & Karvonen, H. (2018). Cardiac measures of nuclear power plant operator stress during simulated incident and accident scenarios. *Psychophysiology*, 55(7), e13071.

Pandey, V. A., & Wolfe, J. H. (2012). Expanding the use of simulation in open vascular surgical training. *Journal of Vascular Surgery*, 56(3), 847–852.

Reason, J. (1997). *Managing the Risks of Organizational Accidents*. Aldershot, UK: Routledge.

Renganayagalu, Sathiya Kumar, Mallam, Steven, & Nazir, Salman. (2021). Effectiveness of VR head mounted displays in professional training: A systematic review. *Technology, Knowledge and Learning*, 26, 1–43.

Ross, G. A. (2022). Extended reality (xR) flight simulators as an adjunct to traditional flight training methods: a thesis presented in partial fulfilment of the requirements for the degree of Master of Aviation at Massey University, New Zealand (Master thesis), Massey University.

Saaty, T. L. (2008). Decision making with the analytic hierarchy process. *International Journal of Services Sciences*, 1(1), 83–98.

Saleem, M., & Khan, Z. (2023). Healthcare simulation: An effective way of learning in health care. *Pakistan Journal of Medical Sciences*, 39(4), 1185.

Sellberg, C. (2017). Simulators in professional maritime education: Understanding the instructor role. *WMU Journal of Maritime Affairs*, 16, 1–19.

Sellberg, C. (2025). Advancing simulation pedagogies in MET. In C. Sellberg (Ed.), *Simulations in Maritime Education and Training* (pp. 109–120). Cham, Switzerland: Springer.

Sellberg, C., & Erntsson, D. (2025). Preparing for simulation through briefing. In C. Sellberg (Ed.), *Simulations in Maritime Education and Training* (pp. 33–48). Cham, Switzerland: Springer.

Sharma, A. (2025). Facilitating debriefings for feedback, discussion, and reflection. In C. Sellberg (Ed.), *Simulations in Maritime Education and Training* (pp. 63–78). Cham, Switzerland: Springer.

Sharma, A., Nazir, S., & Ernstsen, J. (2019). Situation awareness information requirements for maritime navigation: A goal directed task analysis. *Safety Science*, 120, 745–752.

Timmis, S., Broadfoot, P., Sutherland, R., & Oldfield, A. (2016). Rethinking assessment in a digital age: Opportunities, challenges and risks. *British Educational Research Journal*, 42(3), 454–476.

Tuhkala, A., Heilala, V., Lämsä, J., Helovuo, A., Tynkkynen, I., Lampi, E., ··· & Kärkkäinen, T. (2024). The interconnection between evaluated and self-assessed performance in full flight simulator training. *Vocations and Learning*, 17, 253–276. https://doi.org/10.1007/s12186-023-09339-6

Tusher, H. M., Munim, Z. H., & Nazir, S. (2024a). An evaluation of maritime simulators from technical, instructional, and organizational perspectives: A hybrid multi-criteria decision-making approach. *WMU Journal of Maritime Affairs*, 23(2), 165–194.

Tusher, H. M., Mallam, S., & Nazir, S. (2024b). A systematic review of virtual reality features for skill training. *Technology, Knowledge and Learning*, 29(2), 843–878.

Tusher, H. M., Nazir, S., Ghosh, S., & Rusli, R. (2023). Best practices of assessment in maritime simulator training. *TransNav*, 17(1), 105–114.

Tusher, H. M., Nazir, S., Mallam, S., Rusli, R., & Botnmark, A. K. (2022). Learning from accidents: Nontechnical skills deficiency in the European process industry. *Process Safety Progress*, 41(S1), S4–S9. https://doi.org/10.1002/prs.12344

van de Merwe, K. (2024). Agent transparency and human performance in supervisory control (Doctoral dissertation), University of South-Eastern Norway. USN Open Archive. https://openarchive.usn.no/usn-xmlui/handle/11250/3145497

Vicente, K. J. (1999). *Cognitive Work Analysis: Toward Safe, Productive, and Healthy Computer-based Work*. Mahwah, NJ: Lawrence Erlbaum Associates.

White, K. P., & Ingalls, R. G. (2015). Introduction to simulation. In *Winter Simulation Conference* (pp. 1741–1755). IEEE.

World Nuclear Association. (2022). *Three Mile Island Accident*. https://world-nuclear.org/Information-Library/Safety-and-Security/Safety-of-plants/Three-Mile-Island-accident

Part 5

Learning from Practice and Accidents

16 Learning from Normal Work

Piotr Cichowicz and Marcin Nazaruk

16.1 BACKGROUND AND RATIONALE FOR LEARNING FROM NORMAL WORK

16.1.1 Learning from Normal Work: Why Do We Need It?

Over the years, many industries have dramatically improved their safety performance, significantly reducing the number of accidents (Brown, 2020). This is undoubtedly excellent news. However, that success also brings a new set of challenges. Historically, accidents, disasters, and catastrophes have been among the primary drivers for safety improvements. A notable example is the Piper Alpha catastrophe, following which Lord Cullen's recommendations were implemented by both industry and government, leading to significant safety advancements (Broadribb, 2015). Yet, this presents the first modern challenge: a reduced number of accidents can limit our ability to learn proactively. If failures no longer occur frequently, one might assume there's nothing more to learn – after all, everything seems to be functioning well. Unfortunately, this assumption is not always correct. In the shadow of such success, three new challenges stand out – let's look at each in turn. Together, these three challenges highlight the limits of traditional safety approaches and the need for new ways of learning from the everyday reality of work.

16.1.1.1 Challenge 1 – The Absence of Accidents Is Not a Reliable Indicator of Safety

When accidents become rare, it is easy to assume that the workplace is fully under control. The Macondo disaster shows how misleading that belief can be. On 20 April 2010, the Deepwater Horizon crew was celebrating seven years without a lost-time incident when one of the most significant industrial catastrophes of the modern era unfolded (Sutton, 2013).

16.1.1.2 Challenge 2 – Small Numbers Distort Traditional Indicators

Additionally, a low frequency of accidents generates lagging indicators that are increasingly less effective (Cadieux, Roy, and Desmarais, 2006). I remember an internal Health and Safety Executive (HSE) Day at one of the construction companies I worked with. I was scheduled to present later in the session, and just before my turn, the HSE Director shared the company's safety performance results. She highlighted a change in the Total Recordable Incident Rate (TRIR) – from 0.27 to 0.4 – based on the most recent data.

As she walked through the figures, one of the operations leaders raised his hand and asked, 'So does that mean we're 50% less safe than last year?'

To put it in perspective: imagine a company with 7,500 employees. Excluding weekends, days off, and public holidays, each worker might be active for roughly 250 working days per year, resulting in about 15 million person-hours annually. If there are 20 accidents in one year, the TRIR (Total Recordable Incident Rate) is 0.2667. In the following year, if the number of accidents rises to 30, the TRIR becomes 0.4.

$$\text{TRIR} = \frac{(\text{number of recordable cases} \times 200{,}000)}{\text{total hours worked}}$$

That question highlights the limitations of traditional lagging indicators.

DOI: 10.1201/9781003741824-21

16.1.1.3 Challenge 3 – Repeated Accidents and the Safety Plateau

Despite complex safety management systems, safe work procedures, campaigns, awareness sessions, and safety days, many companies still encounter a plateau in safety performance, with frustratingly repeated accidents. I remember working at one of the largest oil and gas companies. Despite significant effort and numerous excellent initiatives focused specifically on hand safety, these types of accidents still happen every single year. The teams involved were highly competent, and strong initiatives were implemented, yet the incidents kept recurring. That's when the third critical problem became clear to me: why do repeated accidents still happen, and why does safety performance stall despite all these efforts?

To address these challenges, I knew we needed a framework that would allow us to draw from everything that had worked in the past, while also helping us see things from a different angle – a framework that lets us look beyond the 20 accidents counted in our TRIR and instead examine the 15 million person-hours that didn't result in injuries. More importantly, it had to be practical and straightforward – something we could actually apply in real-world situations, not just debate in academic discussions.

Learning from Normal Work (LFNW) (Nazaruk, 2023) is a practical framework that focuses on understanding how people adapt to changing conditions and everyday challenges in their work. These adaptations – adjusting, compensating, working around gaps – are not exceptions; they are part of what makes the job possible. That is what we mean by 'normal work'.

16.1.2 Performance Variability

It's tempting to assume that a task without an incident is a 'success' and that, when an incident does occur, someone or something must have failed. But the truth is not that simple.

At one of the companies I worked for, we decided to take a closer look at how issues actually surface in day-to-day operations. We reviewed two sets of documents side by side: last year's accident investigation reports and a selection of routine records, including safety observations, toolbox talks, stop-work cards, and near-miss reports. What we found was striking: the same themes appeared in both.

The contributing factors that showed up in incident investigations were also present in the background of daily operations, just without the incident. The following are some typical issues:

- Limited time
- Missing tools
- Outdated or incomplete procedures
- Risk assessments that overlooked key steps
- Equipment designs that confused operators
- Drawings that didn't match field conditions
- A five-person task attempted by four
- Someone working in the line of fire

These were not isolated findings or signs of failure but reflections of something much more common: the normal challenges of real work. Even in repetitive, standardized tasks, some level of performance variability is inevitable – differences from moment to moment, between people, or across changing conditions.

Workers respond to these constraints through adaptation (Bieder & Bourrier, 2013): they adjust, improvise, and work around what's missing or misaligned (Hale & Borys, 2012). In fact, without these everyday adaptations, much of our work simply wouldn't get done. These adjustments aren't inherently good or bad – they are just part of how work happens (EuroControl, 2014). But that doesn't mean we should overlook them. Understanding how and why people adapt can reveal powerful insights, and that's where real learning begins.

To illustrate how this variability plays out in practice, consider two contrasting examples:

16.1.2.1 Scenario 1 – When Adaptation Increases Risk

A lifting and rigging technician attached slings to lift points positioned three metres high. The procedure required the use of a mobile elevating work platform (MEWP). However, on that day, the MEWP had been redirected to repair an electrical cable powering the entire workshop bay. The truck driver and the client waited, and the plan indicated that the lift had to proceed immediately. The technician faced a dilemma: stopping the job would lead to delays, frustration from the manager, and potential dissatisfaction from the client and the truck driver. On the other hand, strictly following the procedure would avoid increasing the risk of an accident, even though it meant accepting the delay.

The technician adapted to avoid schedule disruptions. Spotting a nearby ladder, they asked a colleague to secure it and then used the ladder. The job moved forward, but the overall risk of an accident increased.

16.1.2.2 Scenario 2 – When Adaptation Reduces Risk

At another site, workers faced recurring errors during shift handovers. The official instructions were 40 pages long and rarely followed, so the team created its own half-page checklist. The shorter document was quick, efficient, and easy to use – and it actually lowered the chance of missing critical information.

The goal of learning from 'Normal work' isn't to label adaptations as 'good' or 'bad' but to study them, reinforce the ones that reduce risk, and redesign the conditions behind those that raise it.

16.1.3 Work as Imagined versus Work as Done

This leads us to a critical concept that exists at different levels in every organization: the gap between work as imagined and work as done. A concept originated from the French Ergonomic tradition in the in the 1950s, highlighted by Hollnagel (2014), was also used in the investigation of the Macondo disaster (CSB, 2016).

Planners, leaders, engineers, and directors who influence how work is shaped by creating plans, rules, procedures, and schedules – are relying on their experience, education, knowledge, skills, past decisions, and so on. However, the model that emerges from this process is shaped by those influences and is ultimately a projection – a view of reality based on their own experiences, assumptions, and beliefs. That's what we refer to as 'Work as Imagined' (IOGP, 2022).

Meanwhile, the actual work happens elsewhere – on construction sites, factories, rig floors, and workshops. It unfolds in real time, under real-world conditions, shaped by resource limitations, last-minute changes, and unforeseen constraints. This is 'Work as Done' – the reality of how tasks are actually carried out in the field (Hollnagel and Clay-Williams, 2022).

It's essential to recognize that there will always be a gap between work as imagined and work as done. But that gap presents an enormous opportunity. If we have methods or tools to explore that space – to investigate and learn from it – we can significantly reduce risk in the workplace. That can also lead to improvements not just in safety but also in quality and efficiency.

Now, let's take that and apply it to an industrial setting, going through a case.

'Every time the machine's cover is opened, the full LOTO process must be followed'. That's what the procedure says.

LOTO stands for Lockout-Tagout. It's a safety method used in industrial settings to make sure that machines are fully shut down and cannot be started again while someone is working on them. This helps prevent injuries from unexpected startup, electrical shock, or release of stored energy – like pressure or tension in moving parts.

The process usually involves turning off the equipment, isolating it from all energy sources, placing physical locks to prevent reactivation, tagging the locks with the worker's name and reason, verifying that all energy is gone, and often notifying a supervisor or filling out documentation. It is detailed and important, especially for higher-risk maintenance tasks. But in this case, the machine cover needs to be opened about ten times a day – for quick checks or small adjustments. Each time, the full LOTO process takes more than half an hour. Doing that repeatedly would take up a large portion of the day and disrupt the work. So, workers adapt. They don't ignore the risks – they've created their own way of doing it that reduces delays while still trying to keep the risk of injury low. For example, they might cut the power and check that everything has stopped moving, even if they don't follow all the formal steps every time.

Here, the gap between work as imagined and work as done is obvious – and it exists for a reason. The written procedure assumes that every instance should follow the full LOTO process. But the way the task actually unfolds in practice makes that difficult to apply. These adaptations show how people respond when the official process doesn't match the way the task actually needs to be done.

We need a way to explore that gap, learn from it, and improve our systems.

However, these gaps are not limited to procedures alone. They can also appear in areas like training, resource management, and broader organizational assumptions.

Work as imagined does not refer only to procedures and plans; it can also reflect broader organizational assumptions. For example, I once asked an HSE manager what skills were required for safe work at height. His answer: 'We have a dedicated training program. Every employee goes through it. A well-known company provides it. So, we're covered'. But no one had asked the workers how effective the training actually was. As it turned out, the course was great, but it focused on a different type of fall protection system than the one used at the site. So no, they weren't covered at all.

Ironically, the best planners know that no plan is ever perfect. Recognizing the gap between how work is imagined and how it is actually done gives us the chance to bring those two worlds closer together.

16.1.4 Context-Driven Behaviour and Interpreting Workarounds and Non-Compliance

Another factor that plays a significant role in Learning from Normal Work is context, and how context drives behaviour in the workplace. This is particularly important because it influences how we view and interpret certain situations. The way we interpret behaviour can either limit our learning or enhance it.

Imagine a person using an overhead crane to move a large metal element. You notice them holding the load with one hand and operating the crane remote control with the other. At first glance, this might appear to be a textbook case of unsafe behaviour. On this site, everyone knew that touching the load was against the rules and could pose a serious danger.

If we stop our curiosity there, our response options are limited. We might instruct the person, retrain them, or issue a warning. But what seems like unsafe behaviour might, in fact, be something very different.

If we start asking open, honest questions that help the person feel comfortable enough to explain, we might learn that the lifting points were positioned so that the load becomes unstable when lifted, causing it to swing. In response, the person instinctively holds the load to stabilize it, not out of carelessness but as a practical compensation for a design flaw. Or we might find out that the crane remote is wired, requiring the operator to walk alongside the load, limiting their ability to maintain a safe distance. Or perhaps the workshop is so crowded that the path the load moves along isn't separated from pedestrians or operating machinery. Hence, the person stays close, not to defy safety rules but to visually monitor the route and warn others in real time.

How we choose to interpret the behaviour directly shapes the learning we gain from it. If it's treated simply as unsafe behaviour, and we don't take the time to explore what might be driving it, the chance to learn remains limited (Sherratt & Ivory, 2019).

The action may be labelled as wrong, but without a deeper understanding, the same thing will likely happen again the next day. However, seeing the behaviour as an adaptation to cope with workplace constraints opens the door to new insights. What looked like a rule violation now points us to issues in the setup – the design of the task, the tools available, or the conditions people are working in. These mismatches are called error traps – factors that make mistakes or workarounds more likely and make the job harder overall, and once we recognize them, we can act. We might involve engineering to improve the stability of the load, replace wired crane control with the wireless model, or redesign the work floor layout to keep people out of the load's path.

This simple example shows how treating workarounds or deviations as signals of system-level challenges, rather than individual failings, can reveal meaningful opportunities for improvement.

16.2 PRACTICAL STEPS TO UPDATE SAFETY PROCESSES WITH LEARNING FROM NORMAL WORK TOOLS

Translating these insights into practical organizational learning requires tools that help reveal how work is done.

16.2.1 The Spectrum of Tools Supporting Implementation

Turning the idea of LFNW into something operational requires more than just awareness – it needs structure, tools, and clarity about where to start. I found that organizations benefit most when they have clear, practical steps without disrupting the safety systems they've already built.

To support clear and practical work, a spectrum of tools is designed to help teams integrate LFNW into their existing processes. Some are easy to introduce and use with minimal preparation, while others involve more planning, deeper facilitation, or long-term investment. The key is to find a starting point that fits the current organization's context and capacity.

16.2.1.1 Tool 1: PATH – Learning Conversation

Safety Conversations are a simple tool used by many of the companies we've worked with over the years but to support leaders who facilitate these conversations and increase the depth of findings, we developed a tool called PATH – Learning Conversation.

PATH Learning Conversation offers a simple but carefully designed structure: Present, Ask, Think deeper, Help.

The strength of PATH lies not just in the questions themselves but in their sequence and focus. Each question is crafted to explore the task rather than the person's characteristics, shifting the tone from judgement to curiosity about the task carried out. The goal isn't to test knowledge or catch someone doing something wrong – it's to understand how the job is really done.

For example, one of the PATH questions asks:

'If you were teaching an apprentice how to do this job, what would you warn them about?'

It's a simple, practical question that often leads to insights rarely heard in standard safety conversations.

I remember a senior HSE leader in Malaysia admitting that, despite years of running safety conversations, getting much real value out of them was hard. However, after using the PATH approach, they uncovered a whole range of error traps in just 20 minutes. What stood out was not just the quality of insight but how the conversation felt:

'It didn't feel like an interrogation or a check-up. People opened up. It felt like a real conversation – like we were genuinely interested in their work'.

This impact comes from a unique combination: questioning techniques designed for learning, a mindset of partnership rather than inspection, a carefully considered order, and a focus on external factors that shape the task. In this case, PATH enabled insights that had not surfaced during years of more traditional, unstructured conversations.

16.2.1.2 Tool 2: Pre- and Post-Job Briefings

From my experience on oil and gas drilling rigs, a typical pre-job briefing or toolbox talk often went something like this:

> *Is everyone familiar with the hazards? There's pressure involved, so watch out for the pipes. Equipment is ready, right? Do we have all the connections we need? The permit is signed, so we can start – watch out for trips and falls.*

It was routine, and while the intent was good, I always felt it could be better – more engaging, more focused on the realities of the work. I addressed that in the Guidance on Human Factors in Task-Based Risk Assessment, published by the Energy Institute, co-authored with M. Nazaruk (Energy Institute, 2025). Still, here, I want to highlight a few practical ideas that can help improve the quality and depth of these meetings.

16.2.1.2.1 Pre-Job Briefings

One useful approach is to use the Energy and Error Wheel as a discussion aid and guide workers beyond the standard checklist. For example, instead of stopping at 'Is everyone ready?', you can take them through a structured reflection:

- Ask workers to describe:
  - o Where will they be working today?
  - o What they will be doing, step by step.
  - o The key phases of the task.
  - o Which steps are critical?
- Then, using the Energy and Error Wheel:
  - o Identify the hazards present in the critical phases.
  - o Discuss controls for each of those hazards.
  - o Identify potential error-producing conditions.
  - o Discuss controls for those conditions as well.

This structure doesn't take much more time, but it encourages workers to think more deeply about the task ahead and surfaces issues that standard briefings often miss.

16.2.1.2.2 Post-Job Briefings

In most places I've worked, post-job discussions usually happened informally. People would talk about the job, but there wasn't much structure, and the focus varied depending on the supervisor. Sometimes we talked about documentation, and other times we talked about equipment or planning. Valid points came up, but it wasn't always clear how to turn those conversations into learning.

That's why we started using a more structured set of questions – something that could be repeated, shared, and lead to clearer improvements. Here's a list that's worked well in practice to uncover constraints, issues, and learning opportunities:

- Was there anything you wish you had known or planned for differently before starting the job?
- Were there any surprises?
- What documents could we improve?
- What skills do we need next time?
- Can you think of an error you made and what we could learn from it?
- Can you think of a situation that made you improvise or use a workaround?

The Energy Institute guidance (2025) has more examples, but even this short list can uncover a wide range of constraints and practical issues. It also helps shift the focus from one-off problems to more systemic-level reviews, which are still short and useful on the frontline.

16.2.1.3 Tool 3: Walk-Through-Talk-Through (WTTT)

The idea behind Walk-Through–Talk-Through (WTTT) is simple: break the task into steps, figure out what can go wrong, what is challenging or increases the chances of making a mistake at each step, and what can be improved (Kirwan and Ainsworth, 1992). It's a structured conversation between the facilitator and the worker, carried out at the actual place where the work happens.

I remember using this approach during a task review in a pressure testing laboratory. The job involved inspecting and controlling a nitrogen station located outside the building. Initially, we used a standard review format.

The conversation went along familiar lines:

- What are the main hazards here?
- What equipment do you inspect?
- How often is the inspection done?
- Where do you report the results?
- Do you see any problems with this task?

The technician responded as expected. The main hazards were pressure and low temperatures, since the expanding gas was cold. A leak was mentioned, but since the station was outside and used nitrogen, it wasn't seen as harmful. The inspection focused on pressure gauges and valves; the task was carried out twice a day, and results were entered into an Excel file. At the end of the review, the conversation turned to behavioural advice – watch for leaks, wear gloves, and concentrate when entering data.

Some points were raised, but nothing particularly new or actionable.

A few days later, we revisited the task using a Walk-Through–Talk-Through. This time, I asked the technician to walk me through the inspection step by step. For each step, I asked what could go wrong, what makes mistakes more likely, and what could be done differently.

The quality of the insights changed completely. We found that the inspection involved many detailed elements, and it was easy to overlook a step, particularly in the absence of a checklist. Access to the equipment was also an issue: the nitrogen station was fenced off, and the lab manager kept the key to the gate. Technicians sometimes took readings from outside the fence if he wasn't available. The gauges were small and not easy to read from a distance.

Then there was the Excel register. Its layout made it easy to confuse values, especially since the fluid level was recorded in inches, while the pressure was measured in bars. The numbers often fell within a similar range (100–150), so it was easy to mix them up. There was no double verification or clear formatting to prevent this. A mistake here could easily affect the outcome of a pressure test.

Compared to the standard review, the WTTT uncovered a far more detailed picture of the task and its challenges. We identified specific constraints and improvement opportunities: introducing a checklist, improving access to the fenced area, resizing or relocating gauges for better visibility, and redesigning the data entry form to reduce the risk of mix-ups

That experience reinforced for me how powerful this tool can be. I've used WTTT across industries – from labs to heavy industry to field operations – and it consistently brings out insights that would never come up in a typical review or checklist discussion.

16.2.1.4 *Tool 4: Learning Teams*

The next tool in the LFNW spectrum is Learning Teams. This method brings together people from different parts of the organization to explore how work is done in practice, identify constraints, and develop practical improvements (Conklin, 2016).

A Learning Team typically moves through three phases: preparation, the learning and improvement session, and follow-up actions.

16.2.1.4.1 Preparation

This phase is about setting the scope and organizing the session. A few key questions need to be answered:

- What task or activity are we going to explore?
- Who should participate – and who brings the right mix of perspectives?

- How can we secure leadership support?
- Who will facilitate the session?
- When and where will it take place, and what arrangements are needed?

Learning Teams work best for tasks that involve human interaction – people working with equipment, navigating processes, and coordinating with others. They are less useful for isolated equipment failures like a broken pump but highly effective in areas where people interact with systems and each other.

16.2.1.4.2 Learning and Improvement Session

This is the core of the process and includes two parts.

The first part focuses on learning – understanding how the task is actually carried out, what makes it difficult, and what increases the chance of error. The goal is to identify real-world constraints and challenges based on the experience of those who do the work.

The second part is about improvement. The group reflects on what has been shared and starts developing ideas for changes to support the task, making it safer, more reliable, or easier to do.

16.2.1.4.3 Follow-Up and Communication

The final phase involves turning insights into action. That includes prioritizing what needs to be addressed, preparing recommendations, and ensuring the right people hear it. Leadership involvement here is key – both to support action and to ensure that the effort leads to real change.

One of the most valuable Learning Teams I was involved in took place during the early stages of the COVID-19 pandemic. The goal was to prepare a large site for reopening. The team included frontline workers, supervisors, site leaders, a maintenance manager, a security representative, and a cleaning services provider.

This mix of roles brought different perspectives into the discussion, and that made a real difference.

We explored what challenges could arise when people return to the site. Without this range of viewpoints, we would have missed a lot: things like how couriers and delivery drivers move differently than regular staff, the flow of people in the canteen, overlooked interactions – like workers needing to visit security for access rights – or technicians performing inspections across different buildings, interacting with multiple teams.

Those conversations helped us identify and address issues early. Without this tool, many of these challenges would likely have gone unnoticed, and the site would have been less prepared for safe reopening.

16.2.2 Implementation Strategy

Now that we have a clear view of the toolbox available, the next question naturally arises: how do we implement Learning from Normal Work?

Despite the complexity of modern companies and the variety of their operations and safety systems, the implementation path is surprisingly straightforward. Successful implementation always revolves around three core elements:

- The role of leadership
- Integration with existing processes
- Building organizational capability

16.2.2.1 Leadership Role

The most critical factor in LFNW is the leadership mindset, which can be precisely described.

First, leaders must understand the modern view of accident causation (Grant et al., 2018; Dekker, 2017). If they believe that accidents are primarily caused by human error or individual shortcomings, we will not move forward. Errors are rarely isolated failures; they are usually shaped by conditions and the error traps present in the system.

The next key shift is how leaders interpret behaviours and workarounds. Behaviours and workarounds are not the root problem. They are symptoms of weaknesses in the system (Wickens et al., 2003). Fixing the system, not blaming the worker, must become the focus.

Finally, leaders must reconsider how they respond to failure and interact with workers. If failures are met with punishment – even subtle forms of blame – learning will not happen. The same applies to the questions they ask:

- 'Did you follow the procedure?' or 'Why did you make this mistake?' – These questions carry the hidden assumption that the worker is at fault.
- In contrast, 'Help me understand your task', or 'What makes following these steps difficult?' are based on the assumption that workers know valuable things and that learning is possible.

In short, how leaders think, speak, and act either opens or closes the door to learning (Table 16.1).

I once facilitated a training session for oil and gas company leaders. We discussed the mindset shift required to move beyond seeing accidents resulting from individual shortcomings. During the session, one of the leaders shared a perspective I've heard many times before. He said:

> I have years of experience, and accidents happen because people act irresponsibly. The only solution that works is to replace them with people who understand and behave responsibly.

In reply, I asked him a simple question: 'If that approach is effective, why do you still see so many accidents?'

There was a pause – a moment of reflection. We followed up after the session, and he agreed to try something different – to shift focus from fixing people to improving the environment and context in which they work.

I spoke with him again last year. He told me that he's still working with the same team but that things have changed. The team is actively involved in identifying error traps and working with him to improve the conditions around the work. Engagement is high, and safety performance has improved – not because the people changed but because the approach did.

TABLE 16.1
Leadership Mindset Transformation

From	To
Safety is the lack of accidents.	Safety is the capacity to succeed under varying conditions.
Non-compliance with procedures is a cause of accidents.	Non-compliance may be necessary if procedures don't match reality.
Punishment improves behaviour.	Punishment may be detrimental to safety.
Behaviour is driven by personal motivation, independent of environment.	Behaviour is shaped by context and local constraints.
Leaders should catch and punish non-compliance.	Leaders should listen and address what makes work difficult.

The takeaway from that experience is one I've seen repeated across industries: you can apply the best tools and build the most well-designed strategy, but without a shift in mindset, the results will always fall short.

16.2.2.2 Integration with Existing Safety Processes

The tools described earlier are not standalone methods – they serve as vehicles for integration. They translate the mindset and theoretical approach of LFNW into daily practices. But for this shift to be consistent and sustainable, it's not enough to introduce tools on their own. Existing safety protocols must also be updated to align with this thinking. Without that alignment, the organization risks sending mixed signals – promoting learning in some areas while maintaining rigid, outdated approaches in others.

The smart move is not to abandon existing safety protocols but to build on them, drawing as much value as possible while making targeted updates where needed to modernize the approach and embed LFNW into existing systems.

This way of working can be integrated into many areas. At a minimum, focusing on risk assessments and a system approach to Behavior-Based Safety (BBS) programmes makes sense. Traditional methods often fail to capture the full context of work in these areas, and the tools described earlier can help uncover practical challenges and support better decisions.

Depending on the process, integration might involve small but fundamental tweaks, as in the case of risk assessments, or broader shifts in how conversations are framed and actions are developed (Energy Institute, 2025). In practice, modifying and integrating are far easier – and often more sustainable – than reinventing.

In the following sections, we'll look at these processes in more depth and explore how LFNW can be practically applied.

16.2.2.2.1 Risk Assessment

According to the HSE in the United Kingdom, a typical task-based risk assessment consists of five major steps: identifying hazards, assessing the risk, controlling the risk, recording findings, and reviewing effectiveness (Health and Safety Executive, 2025).

Within this structure, the process of assessing risk relies heavily on identifying hazards. However, in practice, identifying hazards alone often fails to capture the full picture of risk.

One example observed on a site involved technicians monitoring pressure gauges during a pressure test to verify the integrity of equipment. The setup included two pressure lines, each fitted with a manometre that looked identical: same indicators, same layout, same font (Figure 16.1).

FIGURE 16.1 Identical pressure gauges with differing units highlight how error traps can lead to critical workplace mistakes.

The only difference was in the unit of measurement. One gauge displayed pressure in bars, the other in megapascals (MPa).

It's easy to misread the units during operations, especially under time pressure. A person may glance at one gauge, assume the second uses the same scale, and report the values accordingly. If those readings are passed on to an engineer without clarification, it could result in a decision to increase pressure, believing it's still within safe limits. The difference in units is critical: 1 MPa equals 10 bars. A misreading in this context could lead to over-pressurization, with potentially severe consequences.

Traditional risk assessments are focused on hazards – things that can directly cause harm – so in the risk assessment, you will see entries like pressure, explosion, or projectiles (First, 1983).

However, as is evident from the above, the overall risk is influenced not only by hazards. Visual similarity between pressure gauges could contribute to risk in a dramatic manner – yet we won't find it in a typical risk assessment.

So, what can we do to make it better? Improving the risk assessment template by adding error traps and corresponding control measures is a good start, but it won't be sufficient on its own.

We also need to improve the quality of the questions asked during the assessment to identify potential error traps better. Employees involved in the risk assessment process must be educated about LFNW and error traps.

Finally, we need to update our existing risk assessments and verify them on the job to ensure they reflect the real conditions and challenges workers face. For example, during a verification walk, workers might highlight that two types of safety valves look almost identical but operate at different pressures – something not captured in the original assessment but critical for safe operation.

16.2.2.3 Building Organizational Capability

Building internal capability is the final piece of an effective strategy for implementing LFNW.

This involves three essential components:

- Champions – individuals who will support implementation, guide teams, coach others, and ensure consistency in using tools and the thinking behind them. (This is in line with the 'Leading Change' approach from Kotter.)
- Leader training – to equip supervisors and managers to lead conversations, recognize systemic factors, and ask better questions.
- Workforce awareness – to ensure that employees understand the basics, such as the difference between hazards and error traps, and how the tools support safer, more reliable work.

Let's focus on the first point – champions – because their role is key.

These people facilitate the use of tools, support teams in conversations, and help ensure that what's being done reflects the intent behind LFNW, not just the appearance. They provide guidance when things are unclear, answer complex (or difficult) questions, and help maintain direction. Without them, implementation often becomes inconsistent and shallow.

I've seen many organizations attempt to move forward with tools, updated safety protocols, and leader training, but skip the step of building this internal capability. In some cases, they rolled out the tools and changed procedures but didn't invest in people who could support the work over time.

The results were predictable. I saw tools being used, but the quality of output dropped. People confused error traps with hazards. Blame-oriented phrases reappeared. Teams focused on individual behaviours instead of systemic issues. In some cases, Walk-Through Talk-Through was even merged with risk assessments, which completely missed the point of the tool.

I've also seen what happens when leaders start asking important questions – about causation, accountability, learning, or how to respond to certain situations – and no one internally can provide clarity, offer perspective, or explain the thinking behind the approach. Without someone to handle those questions, misunderstandings take hold. And once the wrong interpretation spreads, bringing the conversation back on track is hard.

We have now covered the spectrum of tools; the implementation strategy is built around three central pillars and touched on the processes that can be improved.

As mentioned at the beginning, theory is one thing, but it must hold up against the realities of day-to-day work and be truly implementable.

With that in mind, let's take a closer look at how we can upgrade some of the key safety processes, like risk assessment and BBS programmes, from the practical perspective of the people doing the work.

16.2.3 Behaviour-Based Safety Programmes (BBS)

Across industries, many organizations are beginning to question the value of their BBS programmes in reducing risk (Krause, Seymour, and Sloat, 1999; Hopkins, 2006; Oswald et al., 2018). In many cases, the programmes no longer contribute to meaningful improvement. Observation data becomes repetitive and adds little insight, and the entire process risks becoming a tick-box exercise.

This often happens when behaviour is treated as the primary cause of accidents rather than a symptom of deeper system conditions. With that mindset, the focus shifts towards correcting individuals, reminding them how they should act, often by pointing to procedures or expected behaviours described in risk assessments. The assumption is that systems are well designed, and any deviation is a failure to comply.

But real work rarely follows a script. People adapt to missing information, conflicting goals, and tools, time, or space limitations. These adaptations are often interpreted as violations, which leads to more control and less curiosity. Opportunities to understand how work is really done are lost.

Another issue is that observations capture only a narrow moment in time. They rarely reveal the deeper context: the constraints, pressures, or trade-offs involved in getting the job done (DeJoy, 2005). Observers typically don't see workers' decisions in response to unclear procedures, gaps in planning, or the limits of available equipment. What appears as a behaviour is often just the visible end of a much more complex situation.

To move forward, BBS programmes should shift focus from observation to conversation. This means giving workers space to describe what they're doing and why and treating those insights as critical input for learning. Making this shift requires psychological safety, strong listening and questioning skills, and performance indicators that support reflection and learning rather than reinforcing thoughtless compliance. The entire BBS ecosystem – observation forms, training materials, expectations, feedback loops, and data interpretation – needs to be realigned to support this change.

The goal is to create a system where behaviours are not judged in isolation but understood in context, where observers are not there to catch unsafe behaviours but to uncover the conditions that shape performance. In addition to human-facilitated tools, emerging technologies now provide new support mechanisms for learning from normal work.

16.3 LEVERAGING ARTIFICIAL INTELLIGENCE TO ENHANCE LEARNING FROM NORMAL WORK IMPLEMENTATION

Artificial intelligence (AI) is already helping improve safety through technologies like predictive maintenance and real-time monitoring (Shin, Han, and Rhee, 2021). However, these applications mainly focus on systems and equipment.

Large language models (LLMs) offer a different kind of value. They work with language, which makes them especially useful for supporting the human side of learning (Dai et al., 2023), such as practising questioning techniques or reviewing written procedures to identify potential error traps.

Unlike traditional AI tools, LLMs are designed to engage with how people think, speak, and reason about work. This makes them well-suited to supporting LFNW, where insights often come from how tasks are understood and discussed. Because LLMs can produce plausible but incorrect outputs, their use in LFNW should remain advisory and should always be subject to human review by competent practitioners.

Another key advantage is accessibility. These tools can be used in almost any language, making scaling capability across teams, sites, and countries easier.

They can be used to support internal capability in a very practical way – for example, by helping champions and leaders practise the right types of questions, check the clarity of procedures, or reflect on how real work is described. They can do this quickly, consistently, and across many different settings.

16.3.1 Practising Questioning Skills – FieldTalk

One of the most critical areas is developing strong questioning skills. Mastering the ability to ask effective questions is a key part of the skill set required for successful LFNW practitioners, and, like any skill, it demands consistent practice to improve.

However, several real-world challenges make practising questioning skills difficult:

- Workers often don't have easy opportunities to practise, due to language barriers, limited access to the frontline, or operational constraints.
- Language differences make questioning harder to apply. Techniques like TEDS (Tell, Explain, Describe, Show) and open-ended questions don't always translate well across sentence structure, tone, or meaning, especially on multicultural worksites.
- People with auditing or compliance backgrounds often fall back on closed or judgemental questions unless they've been actively retrained.
- Emotional tone matters. The way a question is asked and the purpose behind the conversation can affect trust. Practising on real people without skill can do more harm than good.

This is precisely how AI, specifically a custom-trained conversational agent, can offer a powerful solution.

At Psychology Applied, we have developed an AI-based training tool – FieldTalk that allows practitioners to practise questioning techniques in a realistic but low-risk environment. The AI agent plays the role of a frontline worker, responding to different types, tones, and structures of questions with technical answers and simulated emotional reactions. This provides immediate feedback on the conversation's content and its emotional impact.

The tool creates a safe environment where users can develop better questioning habits, refine follow-up strategies, and adapt to subtle conversational signals – all in their native language and at their own pace.

Figure 16.2 is a screenshot showing the tool in action.

16.3.2 Identifying Procedural Uses – FlowRight

Procedures play a central role in most safety management systems. While they are meant to support task execution and reduce risk, under certain conditions they can become a source of problems – and act as error traps themselves.

In practice, procedures are often not followed because:

- They are outdated or inaccurate.
- They are challenging to use or overly complex.
- They slow down the task or don't match how work is actually done.
- Key information is hard to find.

LFNW Coach – Introductory Instructions
My response will consist of two parts:

1. I will first respond as a worker. I was trained to modify my response depending on your question. If the question is good, I'll share information about my situation. However, if the question is poor, I will tell you little or even stop the chat.
2. Then, I will respond as a coach. I'll share with you hints on how to improve your questions.

Try to apply my guidance so that you get better with every question. I will score your questions. You will get:

- 10 points for TEDS question focused on error traps
- 8 points for a good use of open-ended questions focused on error traps
- 3 points for a closed or a leading question focused on hazards, compliance, early solutions, danger
- 0 points for blame-implying or aggressive question

Are you ready to start? Yes / No

User Response:yes

Scenario: You entered the yard nearby a large warehouse and observed a forklift driver offloading long pipes without the pipe clamp installed. You know the site rules require all forklift drivers to use the pipe clamp to hold and stabilize the pipes. You see how the long pipes wobble on the forks, and they could roll and fall. You see the truck driver near the forklift. If the pipes fall, they could hit the truck driver. Your task is to talk to the forklift driver about the situation. You should put them at ease and ask questions to learn about the error traps. Ask your first question.

User Question: why did you do this?

Answer as a Worker: Mate, what the hell? I'm just trying to get the job done, alright? We're under pressure and you show up asking "why" like I'm some kinda idiot. Maybe talk to my boss if you've got a problem.

Coach's Feedback

- **The type of question asked:** Blame-implying "why" question
- **The worker's emotion experienced in response to the question:** Defensiveness and irritation
- **The focus of the question:** On the person – implied blame
- **The score:** 0 points

Note: "Why" questions almost always sound like blame, even if you don't mean it that way. They trigger defensiveness and shut down conversation.
A better focus would be on what made the situation difficult, or what information or equipment the worker might have lacked. You want to show curiosity, not judgment.
Please try again with a question that starts with *What / When / Who / Where*, or use a TEDS prompt (*Tell, Explain, Describe, Show*).

FIGURE 16.2 Screenshot from the FieldTalk – showcasing practice session.

As highlighted in the *HPOG Guide to Best Practices in Procedure Formatting (HPOG Steering Committee, 2021)*, procedures should be written with real people in mind, which means structure, clarity, and usability are critical. Reviewing or writing procedures with these factors in mind is not easy, especially when time and resources are limited.

At Psychology Applied, we've developed a practical AI tool called FlowRight to support this process. It can be used to:

- Improve the quality of new procedures during drafting.
- Review existing procedures and identify potential error traps or weak points.

When a procedure is uploaded, FlowRight analyses the content for issues related to structure, vocabulary, step formatting, layout, and references. It provides a simple summary with specific, actionable suggestions – each one supported by clear reasoning.

It can also act as a support tool for HSE professionals, engineers, or anyone writing or reviewing procedures, helping ensure they reflect how tasks are done, not just how they're imagined on paper.

This approach helps make procedures clearer, easier to use, and less likely to introduce hidden risks into day-to-day work.

Figure 16.3 is a screenshot showing Procedure Coach in action.

16.3.3 Using AI to Review Accident Investigation Reports – DeepFind

It's easy to fall into the full range of cognitive biases and other common pitfalls during incident investigation – things that can distort our findings and seriously limit learning. Our AI-based DeepFind agent was built to analyse reports through the lens of these typical issues, and even coach investigators on *why* certain elements should be improved in the first place.

The tool helps investigators identify gaps in several key areas:

- Cognitive biases: It draws on well-known patterns like hindsight bias, what-you-look-for-is-what-you-find bias, and fundamental attribution error. For example, if a report frames the outcome as obvious in retrospect or quickly concludes that non-compliance was the root cause, the tool will challenge this thinking and prompt the investigator to revisit the reasoning. It encourages a more accurate and fair representation of how the situation looked at the time.
- Behavioural causes: When a report includes phrases like 'the worker skipped a step in the procedure', the tool prompts the investigator to dig deeper: *What was the context? Why might that behaviour have made sense in the moment?* It supports a shift away from blaming individuals towards understanding the system conditions that shaped their actions.
- Language matters: The tool detects emotionally loaded or judgemental language – terms like 'rule violation', 'complacency', or 'carelessness' – and proposes more neutral, precise alternatives. It explains how such words can bias the reader, reduce curiosity, and block meaningful organizational learning.
- Confusing error traps with behaviours or internal states: It flags when the report mixes up system-induced traps (like poor interface design) with observed behaviours or cognitive states. Statements such as 'the operator didn't know about the risk assessment' or 'pressed the wrong button' are explored more deeply – was this a knowledge gap, a communication failure, or a design flaw?
- Mismatch between findings and corrective actions: The tool also checks whether the recommended actions logically follow from the findings. If the report concludes with behavioural advice like 'pay more attention' or 'keep your mind on the task', the AI will prompt the investigator to consider more systemic solutions, like improving equipment design, addressing unclear procedures, or updating training requirements for specific roles.

By using AI as a second set of eyes, specifically trained to detect common pitfalls, we can produce investigation reports that are not only more accurate but also far richer in learning value.

Instead of reinforcing simplified narratives about individual mistakes, AI-supported analysis can help organizations build a deeper, more systemic understanding of how accidents emerge and how they can be better prevented in the future.

Procedure Review Agent - Uploaded Document: *Caustic Unload SOP.docx*

User Request:
Please run the analysis of this SOP

Thank you for providing the SOP titled **"UNLOADING 50% CAUSTIC FROM RAILCAR"** (Procedure ID: 402-089). Below is a structured review based on Sections **8.1 to 8.10 of the HPOG Procedure Best Practice in Procedure Formatting (Rev. 1, August 2021)**. The aim is to enhance usability, readability, and reduce the likelihood of human error.

☑ **Format Review – Aligned with HPOG Sections 8.1 to 8.10**

8.1 Structure of Procedure Steps
Summary:
Many steps start with action verbs (e.g., "Connect," "Open," "Verify"), which is a strength. However, there are several areas needing improvement.
Issues Identified:

- **Multi-action steps** reduce clarity and create risk of skipped actions.
    - *Example:* "Walk unloading line down from railcar to unloading header…" combines walking, lifting hose, and optional air blow-off.
- **Vague equipment references** are used.
    - *Example:* "Open valves from railcar to unloading pump" – which valves?
- Some sentences are overly long and not broken into single actions.

Recommendations:

- Break compound steps into discrete sub-steps.
- Include equipment ID/tag for all manipulable items.
- Use parallel grammatical structure for similar tasks.

Would you like to collaborate on rephrasing a few of these steps to improve clarity and action focus?

8.2 Step Numbering and Indentation
Summary:
Step numbering is entirely missing, which impairs usability and makes referencing difficult.
Issues Identified:

- Safety statements and procedural steps are intermingled.
- Indentation and hierarchy (main vs substeps) are not applied.

Recommendations:

- Implement consistent step numbering (e.g., 1., 1.1, 1.1.1).
- Separate preparatory checks, actions, and follow-ups visually and numerically.
- Use bullets only for unordered lists and numbered steps for ordered procedures.

Would you like help developing a step-numbering template for this SOP?

8.3 Vocabulary
Summary:
Language is mostly functional, but there are opportunities to make it clearer and more standardized.
Issues Identified:

- Inconsistent verbs (e.g., "Insure" should be "Ensure"; "Use caution" is vague).
- Ambiguous adjectives/adverbs ("slightly," "ca…fully," "good idea") create uncertainty.
- Technical phrasing sometimes lacks clarity.

FIGURE 16.3 Screenshot from the FlowRight – real-world example of procedure analysis.

16.4 SUMMARY

While traditional approaches to safety have helped reduce accidents, many industries now face a plateau: risks remain, even if incident numbers are lower.

People work in complex environments shaped by context – tools, time pressure, procedures, equipment – not in isolation. Plans can never cover everything, and a gap always exists between how work is imagined and how it is actually done.

If we treat workarounds or non-compliance as discipline problems, we shut down learning. If we see them as adaptations to system weaknesses, they become opportunities for improvement.

LFNW builds on existing systems. It asks us to stay curious, connect with workers, and explore why actions made sense at the time. It reveals error traps, hidden system weaknesses, and the real-world dependencies that shape risk.

LFNW offers practical tools like PATH-Safety Conversations, pre- and post-job briefings, and walk–through–talk–through methods. Success depends on a leadership mindset, smart integration into existing processes, and building real organizational capability.

Finally, AI provides new ways to practise questioning skills, review procedures, and detect investigation biases, helping reinforce systemic thinking over blame.

I remember the first time I came across this approach. For the first time, I thought – 'Finally, here's something practical. Something grounded in the field, that values the knowledge and experience of the workforce and doesn't add another administrative burden for already busy supervisors and managers'. Years later, after implementing it in practice across many companies, I still feel the same. It works because it respects how work is really done and helps us build from there.

REFERENCES

Bieder, C., & Bourrier, M. (2013). *Trapping safety into rules: How desirable or avoidable is proceduralization?* CRC Press.

Broadribb, M. P. (2015). What have we really learned? Twenty five years after Piper Alpha. *Process Safety Progress*, *34*(1), 16–23. https://doi.org/10.1002/prs.11691

Brown, J. (2020). Nearly 50 years of occupational safety and health data. *Beyond the Numbers: Workplace Injuries*, 9(9). U.S. Bureau of Labor Statistics.

Cadieux, J., Roy, M., & Desmarais, L. (2006). A preliminary validation of a new measure of occupational health and safety. *Journal of Safety Research*, *37*(4), 413–419. https://doi.org/10.1016/j.jsr.2006.04.008

Conklin, T. (2016). *Pre-accident investigations: Better questions – An applied approach to operational learning*. CRC Press.

CSB. (2016). U.S. Chemical Safety and Hazard Investigation Board investigation report executive summary (Report No. 2010-10-I-OS). Drilling rig explosion and fire at the Macondo Well.

Dai, W., Lin, J., Jin, H., Li, T., Tsai, Y.-S., Gašević, D., & Chen, G. (2023). Can large language models provide feedback to students? A case study on ChatGPT. In *IEEE International Conference on Advanced Learning Technologies (ICALT)* (pp. 323–325). IEEE. https://doi.org/10.1109/ICALT58122.2023.00100

DeJoy, D. M. (2005). Behavior change versus culture change: Divergent approaches to managing workplace safety. *Safety Science*, *43*(2), 105–129. https://doi.org/10.1016/j.ssci.2005.02.001

Dekker, S. (2017). *The field guide to understanding "human error"*. CRC Press.

Energy Institute. (2025). *Guidance on human factors in task-based risk assessment* (1st ed., EI 3579). Energy Institute.

EuroControl. (2014). Systems thinking for safety: Ten Principles, A white paper. Moving towards Safety-II. Eurocontrol.

First, M. W. (1983). Engineering control of occupational health hazards. *American Industrial Hygiene Association Journal*, *44*(9), 621–626. https://doi.org/10.1080/15298668391405445

Grant, E., Salmon, P. M., Stevens, N. J., Goode, N., & Read, G. J. (2018). Back to the future: What do accident causation models tell us about accident prediction? *Safety Science*, *104*, 99–109. https://doi.org/10.1016/j.ssci.2017.12.018

Hale, A., & Borys, D. (2012). Working to rule, or working safely? Part 1: A state of the art review. *Safety Science*, *55*, 207–221. https://doi.org/10.1016/j.ssci.2012.05.011

Health and Safety Executive. (2025). *Steps Needed to Manage Risk*. https://www.hse.gov.uk/simple-health-safety/risk/steps-needed-to-manage-risk.htm

Hollnagel, E. (2014). *Safety-I and Safety-II: The past and future of safety management*. Ashgate.

Hollnagel, E., & Clay-Williams, R. (2022). Work-as-imagined and work-as-done. In *Implementation science* (pp. 175–177). Routledge.

Hopkins, A. (2006). What are we to make of safe behaviour programs?. *Safety science*, *44*(7), 583–597.

HPOG Steering Committee. (2021). *Best practice in procedure formatting* (Rev. 1). https://www.hpog.org
IOGP. (2022). *Learning from normal work*. International Association of Oil & Gas Producers (IOGP). Retrieved from https://www.iogp.org/bookstore/product/learning-from-normal-work/
Kirwan, B., & Ainsworth, L. K. (1992). *A guide to task analysis*. CRC Press.
Krause, T. R., Seymour, K. J., & Sloat, K. C. M. (1999). Long-term evaluation of a behavior-based method for improving safety performance: a meta-analysis of 73 interrupted time-series replications. *Safety science*, *32*(1), 1–18.
Nazaruk, M. (2023). Learning from normal work: How to proactively reduce risk when nothing goes wrong. *Professional Safety*, *68*(11), 14–21.
Oswald, D., Sherratt, F., & Smith, S. (2018). Problems with safety observation reporting: A construction industry case study. *Safety science*, *107*, 35–45.
Sherratt, F., & Ivory, C. (2019). Managing 'a little bit unsafe': Complexity, construction safety and situational self-organizing. *Engineering, Construction and Architectural Management*, *26*(11), 2519–2534. https://doi.org/10.1108/ECAM-09-2018-0376
Shin, W., Han, J., & Rhee, W. (2021). AI-assistance for predictive maintenance of renewable energy systems. *Energy*, *221*, 119775. https://doi.org/10.1016/j.energy.2021.119775
Sutton, I. S. (2013, May). Summarizing the deepwater horizon/macondo reports. In *Offshore technology conference*. (pp. OTC-24027). OTC.
Wickens, C. D., Lee, J. D., Liu, Y., & Gordon-Becker, S. E. (2003). *An introduction to human factors engineering* (2nd ed.). Pearson.

17 A Structured Approach to Identifying Human Factors in Accident Investigations

Stig Winge, Stine Skaufel Kilskar, and Stig Ole Johnsen

17.1 INTRODUCTION

17.1.1 Motivation and Background

Although physical ergonomics has been recognized in the Norwegian offshore oil and gas industry, the broader field of human factors (HF) has not been sufficiently prioritized by safety authorities, management, or engineering (Johnsen et al., 2017). This view aligns with recent findings from Havtil (Norwegian Ocean Industry Authority) (Bergh et al., 2024), which conclude that both development projects and operational activities still lack adequate attention to HF. Havtil has identified a need to increase the use and knowledge of HF, as well as a need for methods and tools to identify and analyze HF in accident investigations. Havtil hired SINTEF Digital to develop an HF methodology for use in accident investigations and a supporting training curriculum.

The Safety Forum's report on learning from incidents (Sikkerhetsforum, 2019) highlights the need for both authorities and companies to improve their investigations. Among the recommendations was that investigation teams should possess expertise in HF, organizational factors, and operational management at the same level as technical competence. They also recommended that companies and regulatory authorities adopt investigative approaches that ask, "Why did it make sense to act the way they did?" rather than "What did they do wrong?" These recommendations emphasize, among other things, the need to establish concrete HF tools to more effectively and thoroughly examine human and organizational aspects of serious incidents in the Norwegian petroleum industry.

This chapter reflects the Safety by Design principle of learning through systemic understanding. By integrating HF analysis directly into accident investigations, the approach moves beyond identifying individual errors to uncovering design, organizational, and contextual factors that shape human performance. In this way, investigation becomes an instrument for redesign – transforming insights about *why it made sense* into safer systems and practices.

We use the IEA (2000) definition of Human Factors:

> Human Factors is the scientific discipline concerned with the understanding of interactions among humans and other elements of a system, and the profession that applies theory, principles, data, and other methods to design in order to optimize human well-being and overall system performance.

A guide for identifying HF in accident investigations was developed. The guide is based on relevant approaches, best practices, methods, and HF literature. The guide is first and foremost based on the theory of Situational Awareness (Endsley, 1995), the "Human Factors Analysis and Classification System" (HFACS) (Shappell & Wiegmann, 2000), "Demystifying Human Factors" (IOGP, 2018),

DOI: 10.1201/9781003741824-22

"Learning from adverse events" (CIEHF, 2020), "Introduction to Human Factors in Accident Investigation" (Bridger, 2021), "Human Factors in the NSIAs safety investigations" (NSIA, 2022), and "Human Factors Investigation Toolkit" (HPOG, 2023).

The guide employs important theoretical approaches and tools, including categories of decisions and actions, theories of situation awareness and distributed situation awareness (DSA), and PSFs.

The aims of using HF in general and in accident investigations are to (1) improve safety (i.e., reducing the risk of injury and death); (2) improve performance in safety-critical situations (i.e., increasing quality, productivity, and efficiency); and (3) support satisfaction and usability (i.e., increasing acceptance, comfort, and well-being).

A goal was that the guide should be relatively simple to use, as many of the potential users are not HF experts. Some of the references to theory and literature are therefore simplified. Using the guide still requires basic knowledge about HF and accident investigation methods.

This chapter builds upon and expands a paper that describes the guide (Winge et al., 2023). It adds further detail from the guide and clarifies the approach based on an interview with Havtil in May 2025, together with our subsequent experience and reflections.

This chapter aims to document our approach and provide inspiration for others who are developing or using similar approaches. More specifically, the aim is to (1) describe how different approaches have influenced this guide, (2) describe the rationale for the selection of theoretical inspirations and choices, (3) describe the approach itself, and (4) discuss future use and development of the guide.

17.1.2 Theoretical Background

Interviews and literature consistently emphasized that HF must be viewed from a systems perspective (see, e.g., Bridger, 2021; CIEHF, 2020; Reason, 2016). In the past, HF/ergonomics was commonly understood through a micro-perspective, focusing on the individual and their immediate context. However, this view has evolved. A system can be defined as an organization of people and the machines they operate and maintain, performing tasks necessary to achieve the system's objectives (ISASI, 2016). A systems approach recognizes that all elements within the system may contribute to an incident (ibid.). A systems perspective means that accidents are seen not as the result of isolated errors but as emergent outcomes of complex interactions between people, technology, and organizations. This understanding shifts the focus of investigation from attributing blame to identifying systemic design conditions – the very foundation of *Safety by Design*.

Simply put, a systems perspective in the context of HF means viewing human behavior in relation to the broader system of which people are a part. This perspective acknowledges that complex systems must be understood as socio-technical systems and that the causes of high-consequence incidents are systemic in nature (CIEHF, 2016, 2020). The term "socio-technical" reflects the understanding that the social, cultural, and technical contexts influence how the system functions. Most serious incidents occur through interactions among the many elements that comprise such socio-technical systems; in other words, they are *systemic* and must be understood in terms of the interactions, communication, interdependencies, and control mechanisms across different levels of the system (CIEHF, 2016).

The systems perspective is closely linked to the MTO (Man–Technology–Organization) methodology. A known limitation of the MTO approach is the tendency to analyze Man, Technology, and Organization as separate categories, which can obscure the interactions between them. The guide therefore emphasizes the links and interdependencies across these elements.

The systems perspective on accidents and investigations can be illustrated through Reason's (2016) description of the "anatomy" of an organizational accident. The incident "begins" with the negative consequences of organizational processes, such as planning, design, management, communication, and culture. Some of these issues are transferred to local workplaces in the form of conditions that contribute to human errors and rule violations, for example, high workload, time pressure, lack of knowledge and experience, and inadequate equipment. At the "sharp end," the combination of workplace conditions and the potential for human error and rule violations can lead to *unsafe acts*.

In complex socio-technical systems, many interacting factors can influence one another and contribute to an incident. Viewing HF from a systems perspective has important implications for investigations. It means that the HF perspective should be a central component of the investigation, not merely a tool for examining individual human behavior. Therefore, it is essential to use models and tools that go beyond simple cause-and-effect and that support the identification of interactions among contributing factors.

The first step in this HF guide is to identify key decisions and actions that contributed to the incident. The concept of "human error" is a subject of debate. It is necessary to clarify what is meant by "errors" and why it is essential to identify and analyze decision-making actions as key factors during events leading up to an accident.

People frequently intervene during events, preventing incidents and reducing their consequences. At the same time, "human errors" have played a critical role in many major and well-known accidents (Stanton et al., 2017a). The focus of an investigation should be on what can be learned, not on who is to blame. Human behavior dominates risk in modern technological systems (Reason, 1995). When people at the "sharp end" are part of the barrier system, accidents must be expected – due to the "inherent variability in human performance" (Kjellén & Albrechtsen, 2017, p. 142). In many safety-critical systems, humans represent the final line of defense (Dekker, 2014).

It is debatable whether the term "human error" is appropriate, as the systems influence the people they are part of, and because the action in question may have made perfect sense to those involved at the time. For this reason, the term is often placed in quotation marks. Human error is a complex construct (Stanton et al., 2017a, Chapter 6). As Shorrock (2014) points out, even after decades of research, there is little consensus on what "human error" means. What these definitions have in common is that they refer to actions that had unintended negative consequences or failed to achieve the desired result (Dekker et al., 2010).

As many scholars have noted, errors are consequences, not causes (e.g., Reason, 2016). Most safety researchers today agree that accidents in complex systems are never caused by a single factor, such as human action, but rather by a combination of contributing factors. Identifying an error is the beginning – not the end – of an investigation. People rarely make errors intentionally. More often, errors occur due to so-called error traps (Federation of Norwegian Industries, 2025). While basic human abilities cannot be changed, the conditions under which people work can be (Reason, 2016).

The role of HF in investigations is to understand the behavior of individuals and groups, as well as how they interact with the broader socio-technical system, including variability in human performance. This approach opens the door to improved design and overall system performance (CIEHF, 2020). Regardless of whether the term "human error" or another concept is used, it is helpful to apply taxonomies to assess the type of error. The purpose of categorizing actions in this guide is to help identify the performance-influencing factors that affect situation awareness and decision-making. Classifying actions by "error" type provides a basis for analyzing their causes and developing preventive measures.

17.1.3 Experiences with the Guide from Havtil

Two years after producing the guide, a semi-structured interview was conducted with two representatives from Havtil about their experience with implementing and applying the guide to accident investigations. Overall, the interviewees expressed a positive view of the HF guide. They have experienced that traditional investigation methods do not sufficiently focus on HF, and therefore, the guide is an essential contribution to identifying HF in accident investigations. Furthermore, their positive feedback was that (1) the structure is logical and well-developed, (2) it offers valuable prompts for considering PSF that might otherwise be overlooked, (3) the questions help structure investigations, and (4) the terminology is suitable for those familiar with HF principles.

Some challenges have also been identified. While the terminology and model used in the guide are generally appropriate for HF professionals, the model and specific terms are not widely understood by non-specialists. The key factors, steps, and the original model (Figure 17.2) are difficult for some users to understand. One key issue is that the investigation starts at the bottom right of the model, while the causes originate at the top. Like many accident models (e.g., Reason, 2016), the causal direction flows in one direction, while the investigative steps proceed in the opposite direction, moving backwards. To clarify this, we have made some changes to the model (Figure 17.1), which describes the HF questions to be asked at different steps of an investigation. We have numbered the questions in sequence and added the numbers to Figure 17.2.

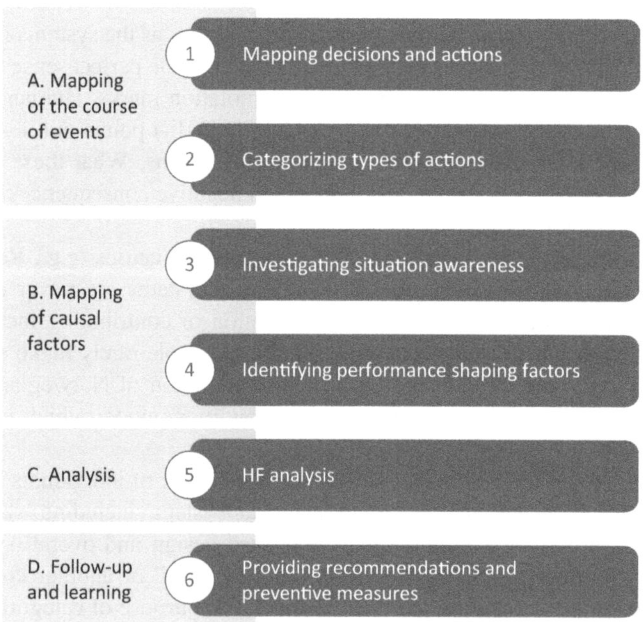

FIGURE 17.1 Overall framework for HF steps (1–6) related to main steps in a generic investigation approach (A–D).

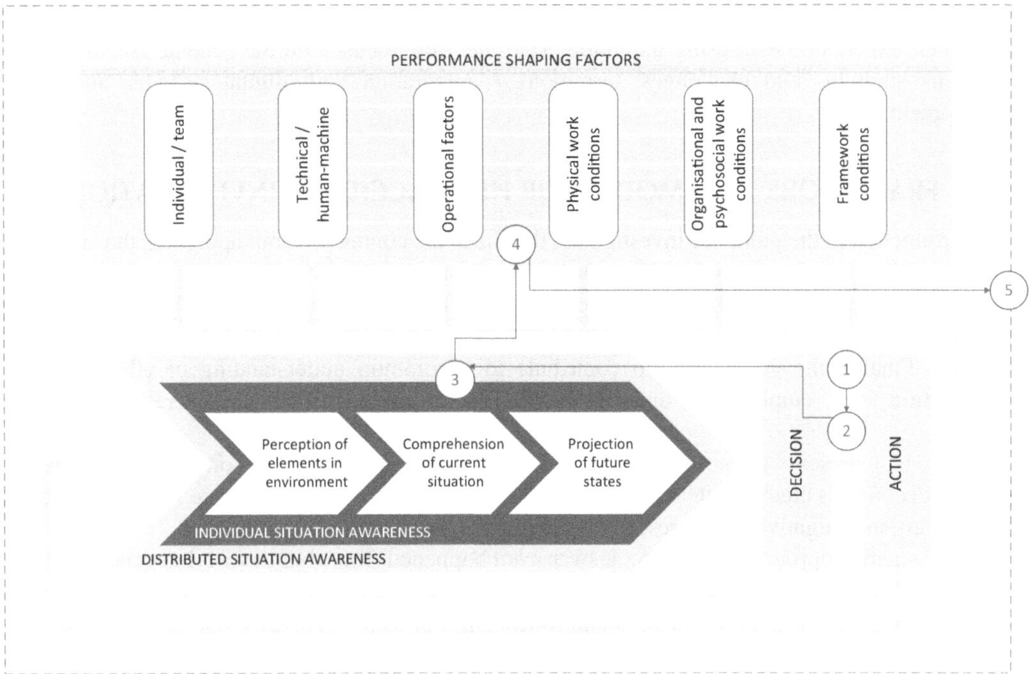

FIGURE 17.2 Model for examining HF in accident investigation (based on Endsley, 1995, and NSIA, 2022). (1–5 is from Figure 17.1: 1, Mapping decisions and actions. 2, Categorizing types of actions. 3, Investigating situation awareness. 4, Identifying performance-shaping factors. 5, HF analysis.)

17.2 MATERIALS AND METHODS

Because the guide is intended for interdisciplinary investigation teams, clarity and usability were prioritized. The method therefore emphasizes accessible concepts and practical steps, supported by but not dependent on deep HF expertise. The development of the guide was based on two literature reviews, interviews, and workshops with Havtil.

The scope of the first literature review was to identify relevant publications on HF and accident investigations. The review was based on keyword searches for publications in Web of Science, Scopus, and Google Scholar for the years 2000–2022. Relevant publications were analyzed, and further relevant papers were selected based on snowballing. The keywords used were "Human Factors," "accident/incident," and "investigation/method/tool/classification."

In the second review, literature on situation awareness was obtained through searches on article titles in Scopus, aiming to identify articles discussing situation awareness in the context of accident investigations as well as articles on DSA.

Individual interviews and group interviews were conducted with experts on HF and accident investigations, as well as representatives of authorities with experience in HF and accident investigations. In addition to the group interviews with Havtil, 11 experts were interviewed. The workshops with Havtil prioritized the overall scope of the project, selecting essential approaches and tools to include, and identifying and categorizing the most relevant PSFs to include in the guide.

A semi-structured interview was conducted with two representatives from Havtil in May 2025 about their experience with using the guide. The interview focused on the general usability of the guide, its strengths and challenges, terminology, integration with digital systems, and future development needs.

17.3 FRAMEWORK AND MODEL FOR HF IN ACCIDENT INVESTIGATIONS

The introduction to the guide for investigators describes the context, overall approach, the aim of the guide, and the system perspective. The guide aims to support the investigation team in identifying why people involved acted as they did and to identify PSFs. The aim is not to identify scapegoats but to guide investigated enterprises in improving key HF issues such as safety, efficiency, and usability. Finally, the guide aims to contribute to a common understanding of HF among the investigation team, emphasizing that HF must be an integrated part of the overall investigation approach.

Figure 17.1 presents the six HF steps (A–D) and shows how they map onto a generic investigation process. This creates a clear analytical workflow and helps teams integrate HF considerations into existing investigative structures.

In the generic approach, A seeks to answer *what* happened, B is about *why* it happened, and C is the *analysis* of causal factors and the connections between them. In contrast, D is about developing and implementing *prevention measures* and follow-up. For step D, Havtil's responsibilities include providing guidance, addressing deviations, promoting improvements, and issuing formal orders to companies. The six HF steps in accident investigations are described in detail in Section 17.4.

Figure 17.2 provides the core analytical model used in the guide. It brings together decisions and actions, situation awareness, and performance-shaping factors into a coherent structure that investigators can apply repeatedly for each key action. The model in this chapter has been slightly modified based on feedback from Havtil (see Section 17.1.3). The overall context of the model is to ensure that HF is considered in a systemic perspective, that is, including MTO (see Kjellén & Albrechtsen, 2017). The model consists of three main elements: (I) decisions and actions, (II) situation awareness – individual and distributed, and (III) performance-shaping factors.

The large arrows in the model illustrate the links between elements in a dynamic decision-making process. In contrast, the thin arrows show the sequence of investigating HF elements with respect to steps 1–5 from Figure 17.1.

The model was developed based on the literature review, similar approaches, interviews, and discussions with Havtil. This model is inspired by a model used by the Norwegian Safety Investigation Authority (NSIA) (2022), where Endsley's (1995) theory of situation awareness is central. Other alternatives to this approach were also evaluated, foremost among them HFACS (Shappell & Wiegmann, 2000). HFACS has been applied to multiple domains and provides a consistent structure for accident analysis; however, it has been criticized for poor levels of inter-rater reliability (Stanton et al., 2017a). Based on discussions with Havtil, our assessment was, however, that HFACS could be too complicated for the multi-disciplinary investigation teams in Havtil. We judged that the suggested approach (Figure 17.2) has greater potential for producing a common understanding of the approach to HF and accidents among the investigators than HFACS. However, we encourage Havtil to utilize HFACS in combination with this tool. HFACS offers a structured, non-blaming approach to analyzing accidents by linking human actions to organizational conditions. Yet its starting point in "unsafe acts" can inadvertently shift attention toward individuals, unless it is complemented by analysis of the broader design and contextual influences on behavior.

The model is used to analyze actions (or non-actions) that directly or indirectly influenced the course of events. The method examines one decision or action at a time. Often, there are several contributory actions to the event, and the model can be used multiple times. Since actions and the actors are key elements in building an understanding of the incident, we recommend using Sequentially Timed Events Plotting (STEP) (Hendrick & Benner, 1986) to document the course of events.

Understanding a situation is crucial for making informed decisions and determining appropriate actions, whether or not they are taken. Once a key action has been identified and described, it is important to understand the situation awareness of the people involved before and during the event. Situation awareness not only affects the actual decision but also influences the decision-making process itself. The next step in the model is to identify PSFs that affect situation awareness, as well as decision-making and the performance of actions. The lines between the categories of PSFs in Figure 17.2 illustrate that there are often relations between PSFs in the different categories.

An actor deals with the situation by making decisions and acting. Like Endsley (1995), the model incorporates a feedback loop from the performed action to the state of the environment in which the situation unfolds. An action based on current situational awareness will change the condition of elements in the environment, which in turn affects the information perceived by the person(s). These new perceptions influence the comprehension of the situation and the projection of future states, that is, situation awareness.

The following example is provided in the guide: the members of a ship crew become aware of another vessel with a given direction and velocity (states of the element) and realize that it is on a collision course. They contact the crew of the approaching vessel, inform them about the situation, and request that they change their course (affecting the state of the elements). This information exchange changes the understanding of the situation (as less critical).

17.4 KEY HF STEPS IN ACCIDENT INVESTIGATIONS

17.4.1 Mapping Decisions and Actions

Mapping decisions and actions is the starting point of the HF analysis. Rather than asking who failed, investigators should examine how each action emerged from the actor's goals, perceptions, and constraints. When used in this context, the term is written with quotation marks because it is often associated with blaming individuals. The guide emphasizes that investigators should view actions as consequences, not causes (Reason, 2016), that "errors" are often the result of error traps, and that identifying an act that contributed to an accident is the beginning, not the end, of an investigation. Identifying "human error" is a step to determine conditions, such as the PSFs, that can be changed.

The guide recommends that the investigation team use STEP (Hendrick & Benner, 1986) to identify the course of events, relevant actors, decisions, and actions that contributed to the incident. The guide describes four questions to help the investigation team: (1) Which actions contributed to the incident? (2) How did the actions contribute to the incident? (3) If a similar situation should occur, will a person in a similar situation act differently? (4) What type of action was it?

17.4.2 Categorizing Types of Actions

Five types of "unsafe acts" are described in the guide based on HFACS (Shappell & Wiegmann, 2000), which is based on Reason (1990). The first three are *unintentional "errors,"* and the last two are *intentional violations*:

1. *Decision "errors"* occur when a behavior or action proceeds as intended, but the chosen plan proves problematic in achieving the desired state.
2. *Skill-based "errors"* occur in the execution of a routine, highly practiced task with little or no conscious thought.
3. *Perceptual "errors"* arise when degraded or ambiguous sensory input – darkness, noise, poor visibility – leads to mistaken inferences about the situation.

4. *Routine violations* are departures from rules, often in a context where such behavior is perceived to be tolerated or justified by an overriding objective.
5. *Exceptional violations* are isolated departures from authority, not necessarily indicative of an individual's typical behavior pattern nor condoned by management.

Categorizing unsafe acts related to the five types of unsafe acts provides valuable insight into why individuals behaved the way they did, and it guides subsequent questions related to situation awareness, PSF, and appropriate safety interventions. The aim is not to blame but to identify PSFs (or "error traps") that can be improved.

17.4.3 Investigating Situation Awareness

As highlighted by others before us, the definition of situation awareness has changed over the past few decades. Focusing on the individual, Endsley (1995) defines situation awareness as "the perception of the elements in the environment within a volume of time and space, the comprehension of their meaning, and the projection of their status in the near future (p. 36)." Later research interest in situation awareness has expanded from the individual level to the team level and, more recently, to the level of socio-technical systems (Stanton et al., 2017b). The guide presented in the current chapter includes investigations of both individual and DSA.

17.4.3.1 Individual Situation Awareness

The most cited paper on situation awareness is Endsley's (1995) "*Toward a Theory of Situation Awareness in Dynamic Systems*," in which she presents a theoretical model of situation awareness based on its role in dynamic human decision-making. Endsley's three-level model describes situation awareness as the perception of relevant cues (Level 1), comprehension of their significance (Level 2), and projection of future states (Level 3).

That is, individual situation awareness can be seen as a chain of information processing. Information in this context is everything that can be perceived through one's senses, that is, what is seen, heard, smelled, tasted, or felt. Elements in the environment can be other actors, technical systems, infrastructure, weather conditions, and other local conditions at the particular point in time.

To investigate the individual situation awareness held by actors who were involved, investigators ought to seek answers to questions such as the following:

- What information was perceived by the person before (and/or during) the event?
- How did the person use this information to interpret/comprehend the situation?
- In what way did the person use their comprehension to foresee what was likely to happen?
- What decisions did the person make based on his/her comprehension of the situation and prediction of future states?

If more than one person was central to the decision and action, the questions should be investigated concerning all the people involved.

17.4.3.2 Distributed Situation Awareness (DSA)

"When loss of situation awareness seems to have played a role in an adverse event, accident investigators need to examine why the system's DSA was degraded, not who lost awareness" (Salmon, Walker, and Stanton, 2016). Hence, investigators also need to assess the DSA.

Whereas Endsley (1995) mentions that situation awareness can also be understood on a team level, Garbis and Artman (1998) were the first to argue that situation awareness is distributed throughout the artifacts that team members use. The idea that socio-technical systems hold situation awareness is often referred to as DSA. Stanton et al. (2006) define DSA as "activated knowledge for a specific task within a system" (p. 5). From this perspective, situation awareness is a dynamic and collaborative process that can be understood through the analysis of interactions between agents within the system. Agents can be both human (e.g., colleagues within the same company or people from other companies) and non-human (i.e., technological artifacts). Non-human agents have some level of situation awareness, at least in the sense that they hold contextually relevant information (Stanton et al., 2006).

Stanton and Harvey (2017) state that "most, if not all, accidents and near misses are caused, at least in part, by the failure to communicate information between agents and tasks" (p. 231). In a paper analyzing the Air France 447 accident, Salmon, Walker, and Stanton (2016) explain how transactions between agents are what enable a system to maintain DSA throughout the course of a task. Transactions represent an exchange of situation awareness between agents, thus referring to more than just the communication of information. The authors distinguish between four forms of transaction failures: (I) absent transaction: a transaction in awareness is required but is not initiated; (II) inappropriate transaction: a transaction in awareness is initiated, but the content is incorrect; (III) incomplete transaction: an appropriate transaction is initiated, but the content is incomplete; and (IV) misunderstood transaction: the receiver misunderstands what is being transacted (ibid. p. 74). In the guide, these categories are, however, simplified, emphasizing that DSA must be understood in terms of the relevant, correct, and timely exchange of information.

The current guide includes several questions to help the investigation team in examining transaction failures:

- What (human and non-human) actors were relevant for the course of events?
- What information did these actors need to perform their tasks?
- What critical information did each of the actors have?
- What information needs to be exchanged with other actors?
- What necessary exchange of information was either not initiated, initiated with incorrect or incomplete content, or misunderstood?
- What are the probable links between identified failures in information exchange, actions, and the course of events?

Salas et al. (2005) systematize what research knows about teamwork into five main components and describe how these are based on three central coordinating mechanisms: *(I)* **shared mental models**, that is, that team members have sufficiently similar and compatible mental models (e.g., of available resources, the members' competence, likely scenarios, etc.) that guide them toward the same objectives; *(II)* **closed-loop communication**, that is, ensuring that sent communications are heard and accurately understood; *(III)* **mutual trust** to ensure that all team members exchange critical information without hesitation.

To analyze the DSA with respect to coordination mechanisms, accident investigators can look for answers to the questions such as the following:

- Did the team members have clear and sufficiently similar understandings of each individual's role, the available resources, shared information, and each member's competence?
- Was there a perception among the people involved that communication within the team was effective, especially with respect to open and clear communication in critical operations?
- Does the person describe mutual trust to their team members?
- Does the person describe mutual trust to other actors who were involved in the event?
- Was all critical information exchanged on time with all relevant actors?

The last question is closely related to the questions regarding transaction failures that were discussed earlier. It is necessary to emphasize that the questions listed here and in Section 17.4.2 are directed at the investigator(s) and are not suitable as interview questions during the investigation. The guide also includes examples of interview questions that address these topics but using simpler and less academic language.

17.4.4 Identifying Performance-Shaping Factors

HF can be described as "··· simply those things that can influence what people do" (IOGP, 2018, p. 9). These "Performance-Shaping Factors" may include factors related to the *job* people do (e.g., time available or control panel design), *personnel factors* (e.g., fatigue, capability), and *organizational factors* (such as roles and manning levels).

To determine which PSFs should be included in the guide and in which category they should be placed, a workshop was held with the researchers and Havtil. Beforehand, the researchers had presented lists of different often-used PSFs and categories of PSFs, based on literature and other approaches (e.g., Bridger, 2021; NSIA, 2022; Shappell & Wiegmann, 2000). During the workshop, the PSFs and their categorization were adjusted to meet the needs of Havtil, considering their overall approach to accident investigations, roles, tasks, and the regulations they oversee. This adjustment resulted in 82 individual PSFs organized in categories. The categories and examples of PSFs are described in the following sections. For each category, the investigator should ask whether there were factors that affected the person's situational awareness and contributed to the incident.

17.4.4.1 Individual/Team

This category consists of two subcategories. The first concerns the individual and includes factors such as vision and hearing; aids for impaired vision or hearing; the function of such aids; medication or drugs; illness; working hours; perceived time pressure, task pressure, and stress; cognitive overload; many tasks; tiredness, fatigue, exhaustion; training; practice; experience; motivation; physical performance; concerns; and compliance with instructions, rules, and procedures. The second subcategory concerns the team and personnel, including factors such as communication and information flow, affirmative communication, collaboration, coordination, trust among actors, access to the same information, shared mental models, and the availability of personnel.

17.4.4.2 Technical/Human–Machine Interface

This category is also divided into two. The first subcategory deals with machines and equipment, while the second concerns the interaction between humans and machines. Factors included in the first category are technical failure of equipment/machine, functionality of equipment/ machine, condition of equipment/machine, availability of equipment/machine, and technical barrier elements. Factors included in the second category are appropriate design, user-friendliness, technology adapted to tasks, user-centered design, a human–machine interface designed to support good situation awareness, automation, mental workload, alarms, digital decision support, and search-ability in systems.

17.4.4.3 Operational Factors

This category includes factors such as the nature of the task, operational goals/expectations for the task, many tasks, many simultaneous tasks, conflicting tasks, monotonous tasks, task distribution, distractions during work, interruptions during work, time pressure, work pressure, work demands and expectations relative to resources, as well as differences between planned and actual work execution.

17.4.4.4 Physical Work Environment

Workplace design/layout; distance to machines, equipment, and aids; access to work areas; noise; temperature; air quality; lighting; chemicals; weather conditions; and ergonomics are factors included in this category.

17.4.4.5 Organizational and Psychosocial Work Environment

This category addresses factors such as responsibility, relationships, role clarity, work planning, leader support and follow-up, colleague support, risk assessments, distribution of tasks, guidance, control and follow-up, routines and procedures, safety culture, "blaming culture," organizational factors in the design phase, and knowledge sharing and learning.

17.4.4.6 Framework Conditions, Use of PSFs, and Further Analysis

Framework conditions include, for instance, staffing, collaboration models and contracts, hiring practices, and regulatory principles and supervision.

Investigators are also encouraged to use other tools and approaches when a more in-depth analysis is needed, particularly HFACS and CRIOP. HFACS is especially useful for analyzing management and leadership factors. CRIOP (Johnsen et al., 2025) is a methodology used to verify and validate the ability of control centers and similar entities to handle all modes of operation safely and effectively.

CRIOP consists of three elements. First, it includes a set of best-practice HF requirements describing 'work as imagined.' Second, it includes a CRIOP scenario analysis in which safety-critical tasks are explored together with users in order to examine 'work as done.' Third, it is based on a structured discussion of challenges and mitigating actions involving the workforce, management, and HF experts. This helps ensure that proposed changes are supported and implemented and can also promote double-loop learning. CRIOP is a useful methodology for ensuring that HF is incorporated into both the early concept and design phases and into operations, thereby supporting the application of best practices. Adapted versions of HFACS and CRIOP are included in a toolbox alongside other tools for analyzing HF in greater depth.

17.4.5 HF Analysis

It can be argued that analysis is a part of all the investigation steps. The aim of the analysis at this stage is to gain an overview of the accident from a systemic perspective (see Section 17.1.2) and to establish causal links between decisions and actions, types of actions, situation awareness, and PSFs. In the analysis, it can be useful to apply the models (Figures 17.1 and 17.2).

Factors identified in investigations are influenced by the accident models and checklists used, described as the WYLFIWYF principle (What-You-Look-For-Is-What-You-Find) (Lundberg et al., 2009). The models and checklists do not necessarily point to all relevant factors and combinations of factors, and they can lead to linear causal thinking. Most accidents result from a combination of factors that are not necessarily linear. Therefore, it is especially important to consider if essential factors have been omitted. It will also be helpful to use HFACS and CRIOP to explore leadership, management, and system design in more depth.

17.4.6 Providing Recommendations and Preventive Measures

In the analysis, the model (Figure 17.2) should be used to map connections between actions, situation awareness, and PSFs. The results of the HF analysis can be very useful in the follow-up of the enterprises investigated, providing insights into deviations, areas for improvement, recommendations, and directives. The follow-up should prioritize preventing individuals from coming into similar situations, removing error traps, and hence reducing the probability of "human errors."

Recommendations from an HF perspective should be directed at the level in the system that is most relevant, for example, equipment design, interfaces, tasks, working environment, and organizational factors. Recommendations targeted at the system level have a greater potential impact than those directed at the frontline and sharp-end workers.

HF investigations are valuable only when findings inform system design improvements. Recommendations should prioritize design-level interventions – such as redesigning equipment, interfaces, or work processes – before implementing procedural or training measures. This follows the hierarchy of controls principle: eliminating or redesigning hazards before relying on human vigilance or compliance.

In the follow-up of the enterprise, a version of the hierarchy of controls adapted to HF (see HPOG) can be utilized. In a comprehensive literature review of safety interventions for preventing work accidents, Dyreborg et al. (2022) found that the relative effectiveness of workplace safety interventions aligns with the Public Health Hierarchy of Hazard Control. In particular, the most effective interventions are those that eliminate risk at its source through engineering controls or by physically separating workers from hazards. Based on the hierarchy of controls, the questions that should be asked in developing recommendations and measures are as follows: (1) Can the hazard be removed? (2) Can the role of humans in safety-critical situations be reduced or removed, for example, by automation? (3) Can consequences of "human failure" be prevented, for example, by additional barriers? (4) Can human performance be assured by using an interlock or other engineered means? (5) Can the factors that shape and motivate behavior be optimized? The hierarchy of controls is designed for occupational health and safety. Principles for barrier management, as described in the Barrier Memorandum (PSA, 2017), are like the hierarchy of controls and should also be applied in developing preventive measures.

Key takeaways:

- Focus on *why* actions made sense.
- Map situation awareness and PSFs to uncover systemic conditions.
- Combine this method with tools like HFACS and/or CRIOP for a more comprehensive learning experience.
- Use results to redesign interfaces, procedures, or organizational structures.
- Investigations are opportunities for learning and design improvement.

17.5 DISCUSSION

This chapter aimed to (1) describe how different approaches have inspired this guide, (2) describe judgments behind the selection of theoretical inspirations and choices, (3) describe the approach itself, and (4) discuss future use and development of the guide.

This guide is a relatively simple approach based on a taxonomy of decisions and actions, situation awareness (Endsley, 1995), and PSFs. The model was developed based on a literature review, similar approaches, interviews, and discussions with Havtil. Other alternatives to this approach were also considered, foremost among them HFACS (Shappell & Wiegmann, 2000) and HFACS for the oil and gas industry (Theophilus et al., 2017). HFACS has been applied to multiple domains and provides a consistent structure for accident analysis; however, it has been criticized for poor levels of inter-rater reliability (Stanton et al., 2017a). Based on discussions with Havtil, we assessed that the approach presented here (Figure 17.1) has greater potential for producing a common understanding of the HF and accident investigation approach among investigators at Havtil than HFACS. We believe that HFACS is more effective at identifying and analyzing leadership and management factors, as well as analyzing accident samples. Depending on the purpose, HFACS can therefore be a suitable choice. We recommend, however, that Havtil combine the two approaches and utilize HFACS in accidents where a more thorough analysis of leadership and management factors is necessary.

One possible risk associated with a highly specific HF approach in investigating accidents is that it may not be integrated into the broader investigation strategy. The guide therefore emphasizes that the aim is to contribute to a common understanding of HF within the investigation team, and that the investigation of HF must be an integral part of the overall investigation approach to ensure a systemic perspective.

One possible limitation of the guide is that it is relatively simple, and that some of the literature described and methods used are simplified. This is because the goal was for the guide to be easy to use, as many potential users are not HF experts. To investigate HF more thoroughly, we recommend that a smaller group of Havtil investigators receive more education in HF, and that another guide for conducting more in-depth HF investigations be produced.

Another limitation of using frameworks with fixed factors in accident investigations is that they often reveal what the investigator is looking for (Lundberg et al., 2009). Therefore, the guide emphasizes that investigators must also look for factors not included in the guide and must be trained in identifying HF. Combining this approach with other methods, such as HFACS, is encouraged by the guide.

A recent interview with representatives from Havtil showed positive experiences with the HF guide. People at Havtil experience that traditional investigation methods often do not adequately address HF, making the guide a valuable resource for identifying HF elements in accident investigations. Feedback emphasized that the guide is logically structured, well-developed, and that the questions included help organize and direct the investigative process. However, a key challenge is that the core concepts, investigative steps, and the original model (Figure 17.2) are not widely understood by those without HF expertise. Specialized HF knowledge is essential for applying the six HF steps described in this chapter and for developing effective preventive safety recommendations and measures. This is consistent with recommendations from the Safety Forum (Sikkerhetsforum, 2019), which emphasized that investigation teams should have expertise in HF, organizational factors, and operational management at a level equivalent to their technical expertise.

REFERENCES

Bergh, L. I. V., Teigen, K. S., & Dørum, F. (2024). Human performance and automated operations: a regulatory perspective. *Ergonomics*, *67*(6), 744–758. https://doi.org/10.1080/00140139.2024.2321457

Bridger, R. S. (2021). *Introduction to human factors in accident investigation*. RS Bridger.

CIEHF (Chartered Institute of Ergonomics & Human Factors) (2016). *Human factors in barrier management*. Chartered Institute of Ergonomics & Human Factors.

CIEHF (Chartered Institute of Ergonomics & Human Factors) (2020). Learning from adverse events. White Paper.

Dekker, S. (2014). *The field guide to understanding 'Human Error'*. CRC Press.

Dekker, S., Cook, D. R. I., Johannesen, D. L., Sarter, D. N. & Woods, P. D. D. (2010). *Behind human error*. Farnham: Ashgate Publishing, Ltd.

Dyreborg, J., Lipscomb, H. J., Nielsen, K., Törner, M., Rasmussen, K., Frydendall, K. B., ··· & Kines, P. (2022). Safety interventions for the prevention of accidents at work: A systematic review. *Campbell Systematic Reviews*, *18*(2), e1234.

Endsley, M. R. (1995). Toward a theory of situation awareness in dynamic systems. *Human Factors 37*(1), 32–64.

Federation of Norwegian Industries (2025). *Safety, leadership, and learning – A practical guide to HOP*.

Garbis, C., & Artman, H. (1998). Team communication and coordination as distributed cognition. In T. Green, L. Bannon, & C. Warren (Eds.), Proceedings of 9th Conference of Cognitive Ergonomics: Cognition and Cooperation, pp. 151–156.

Hendrick, K., & Benner, L. (1986). *Investigating accidents with STEP* (Vol. 13). CRC Press.

HPOG (Human Performance Oil and Gas) (2023). *Human factors investigation toolkit*. Retrieved from www.hpog.org, April 1, 2023.

IEA (International Ergonomics Association) (2000). *Definition and domains of ergonomics*. Retrieved from https://iea.cc/about/what-is-ergonomics

IOGP (International Association of Oil & Gas Producers) (2018). Demystifying human factors: Building confidence in human factors investigation. Report 621.

ISASI (International Society of Air Safety Investigators) (2016). *Guidelines for the investigation of human factors in accidents and incidents*.

Johnsen, S. O., Kilskar, S. S., & Fossum, K. R. (2017). Missing focus on Human Factors–organizational and cognitive ergonomics–in the safety management for the petroleum industry. *Proceedings of the Institution of Mechanical Engineers, Part O: Journal of Risk and Reliability*, *231*(4), 400–410.

Johnsen, S. O., Thieme, C., Aminoff, H., Park J., Fernander, M., Boren, M., Kjøl, Å. A., & Kirwan, B. (2025). *CRIOP A scenario method for crisis intervention and operability analysis*. Norway: SINTEF Technology and Society.

Kjellén, U., & Albrechtsen, E. (2017). *Prevention of accidents and unwanted occurrences: Theory, methods, and tools in safety management*. CRC Press.

Lundberg, J., Rollenhagen, C., & Hollnagel, E. (2009). What-You-Look-For-Is-What-You-Find–The consequences of underlying accident models in eight accident investigation manuals. *Safety Science*, *47*(10), 1297–1311.

Norwegian Safety Investigation Authority (NSIA) (2022). *Human factors in the NSIA's safety investigations* [Menneskelige faktorer i Havari-kommisjonens sikkerhetsundersøkelser] (1st ed.). Lillestrøm: Norwegian Safety Investigation Authority.

PSA (Petroleum Safety Authority Norway) (2017). *Principles for barrier management in the petroleum industry* [Barrierenotat 2017, in Norwegian]. Stavanger, Norway: PSA.

Reason, J. (1990). *Human error*. Cambridge University Press.

Reason, J. (1995). Understanding adverse events: Human Factors. *BMJ Quality & Safety*, *4*(2), 80–89.

Reason, J. (2016). *Managing the risks of organizational accidents*. Routledge.

Salas, E., Sims, D. E., & Burke, C. S. (2005). Is there a "Big Five" in Teamwork? *Small Group Research*, *36*(5), 555–599.

Salmon, P. M., Walker, G. H., & Stanton, N. A. (2016). Pilot error versus socio-technical systems failure: a distributed situation awareness analysis of Air France 447. *Theoretical Issues in Ergonomics Science*, *17*(1), 64–79.

Shappell, S. A. & Wiegmann, D. A. (2000). *The human factors analysis and classification system-HFACS*.

Shorrock, S. (2014). *'Human error': Still undefined after all these years*. Humanistic Systems.

Sikkerhetsforum. (2019). *Learning from incidents: Report from the Safety Forum 2019* [In Norwegian]. Norway: Stavanger.

Stanton, N. A., & Harvey, C. (2017). Beyond human error taxonomies in assessment of risk in socio-technical systems: a new paradigm with the EAST 'broken-links' approach. *Ergonomics*, *60*(2), 221–233.

Stanton, N. A., Salmon, P. M., Rafferty, L. A., Walker, G. H., Baber, C. & Jenkins, D. P. (2017a). *Human factors methods: a practical guide for engineering and design*. CRC Press.

Stanton, N. A., Salmon, P. M., Walker, G. H., Salas, E., & Hancock, P. A. (2017b). State-of-science: Situation awareness in individuals, teams, and systems. *Ergonomics*, *60*(4), 449–466.

Stanton, N. A., Stewart, R., Harris, D., Houghton, R. J., Baber, C., McMaster, R., Salmon, P., Hoyle, G., Walker, G., Young, M. S., Linsel, M., Dymott, R., & Green, D. (2006). Distributed situation awareness in dynamic systems: theoretical development and application of an ergonomics methodology. *Ergonomics*, *49*(12–13), 1288–1311.

Theophilus, S. C., Esenowo, V. N., Arewa, A. O., Ifelebuegu, A. O., Nnadi, E. O., & Mbanaso, F. U. (2017). Human factors analysis and classification system for the oil and gas industry (HFACS-OGI). *Reliability Engineering & System Safety*, *167*, 168–176.

Winge, S., Kilskar, S. S., & Johnsen, S. O. (2023). A guide for identifying human factors in accident investigations. *Proceedings of the 33rd European Safety and Reliability Conference*.

18 Evolution of Human Factors Standards

Challenges from Robotics and Automation

Jonathan Earthy

18.1 INTRODUCTION

In robotics and automation, Safety by Design necessitates striking a balance between technical autonomy and meaningful human oversight. ISO 9241-810 and related human-system interaction standards emphasise designing transparent interactions, shared situational awareness, and adaptive interfaces that maintain operator engagement and trust even as systems become more intelligent and self-directed.

The goal of this chapter is to get you started in using these ergonomics standards to address challenges related to robotic, intelligent, and autonomous systems.

18.1.1 Why Standards?

Many people do not use standards, citing cost and lack of awareness. However, Dresner (2002) observes,

> *I'm not sure why standards get such short shrift. After all, they are a distillation of best practice by the peers of those who should be glad of their support. The need for standards to apply to so many 'unique' situations means that some of their advice will be honed down to be generic and inoffensive – that makes for polished, pure elements that provide a strong foundation to be built on.*

In the context of Safety by Design, standards represent codified learning from decades of engineering and Human Factors (HF) experience. They provide structured ways to embed safety, usability, and trustworthiness into systems long before they reach operation.

The chapter will take you from what standards are, through the history and work of ISO TC159/SC4 (the ISO committee for human-system interaction), to a study of how ergonomics standards need to evolve to address Robotic, Intelligent and Autonomous (RIA) systems. It then focuses on the issues of control of automation and robotics. It discusses how standards developed by SC4 and related committees can be applied and are being developed for this application. An example is presented that addresses these issues in the maritime sector by applying the introduced standards. Compliance, regulation, lifecycle, evaluation, and other issues are discussed, and the chapter concludes with some recommendations for increasing the use of standards.

18.1.2 What Is a Standard?

The primary purpose of standards used to be stated as 'the facilitation of world trade'.

DOI: 10.1201/9781003741824-23

Nowadays (ISO 2024), the scope is set more widely:

> *Think of them as a formula that describes the best way of doing something.*
> *It could be about creating a product, managing a process, delivering a service or supplying materials – standards cover a vast range of activities.*
> *Standards are the distilled wisdom of people with expertise in their subject matter and who know the needs of the organizations they represent – people such as manufacturers, sellers, buyers, customers, trade associations, users, scientists and regulators.*

In simpler terms, a standard represents a consensus on properties, practices, and shared knowledge, and, increasingly, shared values. Understanding not only what standards are but also what they do is crucial for applying them effectively in design and certification of emerging robotic and autonomous technologies.

18.1.3 What Are They for?

The process of conformity assessment demonstrates whether a product, service, process, claim, system, or person meets the relevant requirements. Such requirements are stated in standards, regulations, contracts, programmes, or other normative documents. Examples of conformity assessment include the following:

- Certification – provision of written assurance by an independent body
- Testing – helps to determine one or more characteristics of an object or product
- Inspection – regular checking of a product to make sure that it meets specified criteria
- Mutual recognition – increases confidence in conformity assessment between countries (or other entities)
- Trade – standards harmonise conformity assessment activities. This reduces the need for duplicate testing when acquiring or supplying goods. It is a key factor in increasing the efficiency and scale of trade.
- Safety improvements and avoiding legal consequences – suppliers have been found guilty in court due to not following best practices as described in standards.

Put simply, standards are instruments for building trust. In a Safety by Design perspective, they enable designers, suppliers, and users to align expectations and operationalise trustworthiness throughout a system's lifecycle.

18.2 TYPES OF STANDARDS

As indicated in the preceding section, standards are evolving. Figures 18.1 and 18.2 present Steedman's (2012) model of evolution of standards. This provides a useful knowledge structure and the basis of a framework for coherence. This model is gaining more relevance with time.

Steedman's model illustrates how standards evolve with societal and technological change: from specifying technical requirements, to defining processes, and finally to expressing values and principles that guide ethical and behavioural expectations.

Figure 18.1 introduces the three generations in the evolution of standards.

In terms of ergonomics: The first generation focused on matching machines to human capabilities and limitations. The second generation was integrating human and organisational performance, and now, the third generation is about human-system values, addressing the right problems when designing for human-system harmony. Thus, in using standards, a key issue is to identify and explore suggested processes and outcomes and highlight the benefits/takeaways.

First generation Product	Technical specifications Better products
Second generation Process	Management systems Better processes
Third generation Potential	Values Principles Better behaviours

FIGURE 18.1 Types of standards – the standards landscape.

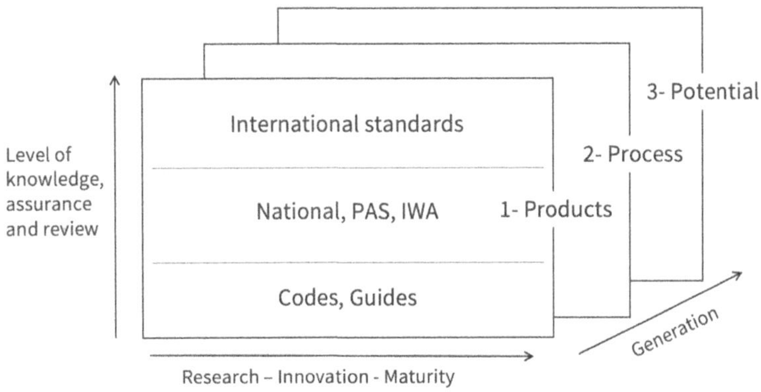

FIGURE 18.2 Types of standards – framework for coherence.

Figure 18.2 presents Steedman's framework for coherence. It clarifies the relationship between international standards (e.g., ISO, IEC) and less widely agreed or validated standards (e.g., de facto, sector, industrial) as evolutionary but not inimical to each other. Abbreviations are (PAS) Publicly Available Specification and (IWA) International Workshop Agreement.

In this chapter, the Steedman model is used as part of the description of the standards that are introduced. The following section introduces the committee that develops them. In this chapter, standards and the Technical Committees (TCs) that manage them are referred to by their reference numbers. The full titles of all standards are listed in the References section. Standards are regularly reviewed and updated. Use of the latest version is advised. For more information on any standard, use the standard's number in the search function on the ISO (www.iso.org) or IEC (www.iec.ch/) website. For ISO/IEC standards, use the ISO website. Titles of, and information on, TCs are available from the same sites.

18.3 ISO TC159/SC4 EVOLUTION, SCOPE, AND CONCEPTS

Sub-Committee Four (SC4) of ISO TC 159 *Ergonomics* was formed 40 years ago to address ergonomics requirements for office work with Visual Display Terminals (VDTs). As Information Technology (IT) use became more widespread, we added control centres and renamed the Committee to *Ergonomics of human-system interaction*. SC4's scope has progressively expanded from office ergonomics to encompass the full range of human-system interaction, including emerging areas such as AI, robotics, and immersive systems. Today, SC4's scope reflects this expansion and covers the full breadth of human-system interaction:

> *Standardization in the field of HFE (Human Factors Engineering), the interaction between and within systems (often computer-based) and the people (who use, operate, and are affected by them as well as those who develop, manufacture, evaluate, install and maintain them). 85 active standards covering:*
>
> *hardware ergonomics (including input, displays, and interactive devices),*
> *software ergonomics (including interaction and interface design),*
> *ergonomics of the context of use (including tasks, environments, workplaces, control centres), and*
> *human-centred design processes and methods (including usability engineering, accessible design, and participative design methods).*

Together, these standards represent a comprehensive framework for designing usable, dependable, and safe interactive systems, and provide a foundation for applying Safety by Design in practice.

The core and founding concepts of SC4 are as follows:

- Focus on ergonomics of the interface, clarifying that SC4 addresses interaction quality rather than system engineering or user competence,
- Human-centred approach (originally called people-centred ergonomics) as part of lifecycle management,
- Context of use and evaluation in context.
- User-centred evaluation (by users where possible),
- Usability as a measure (task-based evaluation).

These core concepts provide a structured basis for specifying, evaluating, and measuring ergonomics issues in interaction with increasingly complex software-intensive systems.

18.4 THE HUMAN-CENTRED APPROACH

Paul Branton was the founder and first Chair of SC4. He established the human-centred approach (originally called 'person-centred ergonomics' as a counter to Taylorism), which is the basis of standards from SC4. The 'Brantonian view' gives a precise rationale and definition of value, not the vague platitudes about rights and empathy, which tend in the direction of 'human-centric', with all the associated political and environmental implications of that ideology.

The following is an updated summary of the introduction to *'Person-centred ergonomics' – a Brantonian view of Human Factors* (Oborne *et al.*, 1993).

Why do we need people in systems and environments, especially those shaped by automation? The human-centred approach holds that people bring purpose and agency to a system – they act with goals, values, and the capacity to interpret and adapt. When the purpose (the goal) is understood, it is possible to begin to design a system to facilitate it. Only by understanding the person/worker will we begin to understand a man-made system, and when we understand the person's variabilities, we will be in a better position to adapt the system to accommodate and even to make best use of them. And when we can do this, we should be well placed to design the system for maximum efficiency, effectiveness, user satisfaction, and safety.

The human-centred approach takes as its central point the need to accommodate the human attributes which the person brings to the system. The emphasis is on accentuating the positive and reducing the negative effects of a person's interactions.

To take a simple example, the design of controls for safe and efficient operation. Traditional design works from the viewpoint of the system's requirements as they impinge on the operator's wishes and abilities. The operator's abilities are recognised as important, but from the traditional viewpoint it is the system which defines the interaction and thus the design. The human-centred approach is one in which the operator and their abilities define the working system. From a human-centred perspective, design must begin with the operator, supporting their goals, variabilities, and decision-making capabilities. They are the component which is designed to activate the system and to maintain its efficient running. The goal is to create supportive dynamic environments which enable people to work at their most safe and effective levels, not just to design the environment to 'fit' the person in some static sense.

In human-centred design, the user is seen as self-controlled, necessarily possessing value standards and interests in social relations, perpetually seeking, and evaluating information from the surroundings. The search varies in intensity, depending on the level of motivation. Design considerations include the form of mental operations, their contents being material either taken immediately from the surrounding world or from stored, primarily emotive experiences. The explanation is purposive, rather than causal, as it is argued that the thoughts which determine behaviour are forecasts of future states of affairs and their consequences for the person, rather than past experiences that are in themselves of speculative origin.

The human-centred approach represents a major shift in the way in which designers view the role of the operator within the working situation. It supports people's ability to learn and compensate for their biological (and other) 'weaknesses'. The task is to design a supportive enough environment to facilitate such compensating behaviour. The human being is at the centre of the working situation, and we must understand the abilities, responsibilities, and requirements which people bring to the situation (not just their shortcomings) in order to be able to deal adequately with the system:

1. Human-centred design is from the viewpoint of the human being within the system rather than the system's requirements.
2. In doing so, the whole person must be considered. This includes a person's view of the system, purpose, and responsibility within the system, and the dynamic nature of the interaction. In particular, the thought processes of people need to be studied as well as the traditionally observable behaviours, and full cognisance must be taken of the values of people within the system.
3. The values that humans bring to a working system are that they are purposive, information-seeking, uncertainty-reducing, and responsible.

As Branton developed this philosophy, human interaction with systems evolved from industrial work with machinery to include control and monitoring of complex computer-based systems. Oborne *et al.* (1993) end with Branton's discussion of Visual Display Unit stress with particular reference to the effect of complex, computer-based automation and monitoring or what we would now call 'meaningful human oversight and control'. Although Information and Communication Technology (ICT) has evolved and its application has expanded, the Brantonian view continues to apply to control. From a design perspective, the values humans bring to the interaction remain unchanged. Interaction with robots may or may not present a new challenge, but if meaningful human oversight and control of technology is what regulation requires (and at least in Europe it is (EU 2023, 2024)), an alternative to the human-centred approach is not obvious. With respect to the need to oversee and control complex computer-based systems, this is no longer a rare task. Such systems are now within most technology that humans are both required to and choose to use.

In essence, Branton anticipated today's demand for meaningful human oversight: systems must support human purpose and judgement rather than merely constrain behaviour.

18.5 SPREADING THE WORD, TC 159/SC4 COLLABORATIONS

Because SC4 focuses on interactive systems and their performance in context of use, SC4 has always had a strong theme of application in its work, and as a result has found mutual benefit from collaborations with a wide range of standards bodies and other organisations. The following sections present examples of these collaborations and the resulting standards. For each section, the Steedman framework generation is indicated. Later sections in this chapter will address the use of many of these standards for automation and robotics.

18.5.1 IT Hardware and Peripherals

Partnering with the whole industry, 1983: Many ergonomics issues need to be addressed with new devices and a new type of work. The results were first-generation standards used industry-wide for displays, keyboards, and user input devices; test measures were based on user's purpose/goals, specification of the basis for suitable tests.

18.5.2 Furniture and Office Environment

Partnering with health and safety regulators, 1983: The ergonomics problem(s) to be addressed were seating, lighting, keyboards and screens, and hours of use. The result at the time was a first-generation standard: ISO 9241-5 *Workstation layout and postural requirements*. As work moves out of offices, these are joined by second- and third-generation standards: ISO 9241-500 *Ergonomic principles for the design and evaluation of environments of interactive systems*, ISO 9241-520 *Ergonomics aspects of Activity Based Working*.

18.5.3 Systems and Software Engineering

Partnering with ISO/IEC SC7: The initial need was first-generation standards to specify quality of software and usability of user documentation. This was later extended to the software and systems lifecycles, then detail of processes and process capability assessment. The Ergonomics problem(s) to be addressed were (and are) user documentation, inclusion of human-centred design, and usability in quality and process models. Results include the concept of quality in use ISO/IEC 25019 *Quality-in-use model*; second-generation system and software quality models; Human-centred design in ISO/IEC/IEEE 15288 *System life cycle processes*; usability in ISO/IEC/IEEE 12207 *Software life cycle processes*; ISO/TR 25060 *General framework for Common Industry Format (CIF) for usability-related information*; and ISO-compliant process capability assessment in SC4 using ISO/IEC 33000 *Process assessment* standards.

18.5.4 Ergonomics Principles, Concepts, and Process

Partnering with ISO TC159/SC1: TC159 needed to revise ISO 6385 *Ergonomic principles in the design of work systems*. The Ergonomics problem(s) to be addressed were the evolution of ergonomics thinking, new areas of application. Results include the third-generation standards ISO 26800 *General approach, principles and concepts*; ISO 27500 *The human-centred organisation – Guidance for executives*; and ISO 27501 *Guidance for managers* – principles including human-centred, usability, and unified ergonomics-oriented design process. IEA/ILO (2021) is a good example of a guidance document that integrates and elaborates these standards for organisations.

18.5.5 Medical Devices

Partnering with IEC SC 62A and ISO TC 210: The context was the growing complexity of medical devices and reporting of related safety issues. Results include IEC 63266 *Application of usability engineering to medical devices* and useful new terminology and concepts for use in Ergonomics ('correct use', 'use error', 'normal use'), and the application of usability engineering beyond safety for medical devices.

18.5.6 IT Governance and Management

Partnering with ISO/IEC SC40: There is concern that the new third-generation version of ISO/IEC 38500 *Corporate governance of information technology* is too abstract. A new standard, ISO 9241-250 *Stakeholder engagement in IT Governance during specification, procurement and operation*, is under development. This acts as a bridge between an ISO 38500 principle for 'stakeholder engagement' and human-systems interaction standards.

18.5.7 Robotics

Partnering with ISO TC299: The challenge is the evolving human-system issues as autonomous robots interact more closely with people. The Ergonomics problem(s) to be addressed include the safety of autonomous robots, cobots, care robots, and social interaction with robots. The results so far are the discussion of third-generation issues in ISO/TR 9241-810 *Robotic, intelligent and autonomous systems*, a developing ISO 9241-812 on AI system ergonomics, and support for Ergonomics in Robotics standards.

18.5.8 Dependability

Partnering with IEC TC56: TC56 addresses the general issue of dependability (availability and reliability of systems). The context was a growing understanding that most systems depend on people to maintain operation or recovery from failure (described as 'external mitigation'). The Ergonomics problem to be addressed was the need to address human-machine relationships throughout the system lifecycle. The result was a greater understanding of high-integrity systems development in relation to Human Centred Design (HCD) and usability, and IEC 62508:2010 *Guidance on human aspects of dependability* (addressing 'human-orientated design'). This second-generation standard is under revision to include guidance on human reliability assessment.

18.5.9 Complex Systems Process Capability

Partnering with the critical systems sector(s) on the use of process capability assessment to reduce contract risk in acquiring software-intensive systems: The requirement was to assess/assure human-systems integration (also known as HF integration) to manage project risk. The challenge for Ergonomics was to develop a process capability assessment model (a second-generation standard) for human-centred design. The result was ISO/TS 18152 *Specification for the process assessment of human-system issues*. This aligns with ISO/IEC/IEEE 15288 and defines the role and activities of HF in large/critical projects. The concepts developed were used in ISO 9241-220 *Processes for enabling, executing and assessing human-centred design within organizations*.

18.6 ISO 9241-810 ERGONOMICS OF RIA

ISO/TR 9241-810 *Ergonomics of human-system interaction – Robotic, intelligent and autonomous systems* was published by ISO in 2020. This section and Sections 18.7.4 and 18.7.5 summarise the main concepts and considerations of that standard (Earthy and Downs, 2018). Readers are referred to ISO/TR 9241-810 for more detail and informative annexes that discuss human-RIA system issues, present case studies of ergonomics issues for RIA systems, the necessary developments to ergonomics and changes to ergonomics standards, a report of the method followed in developing the TR and an extensive bibliography of the source references and standards.

18.6.1 Ergonomics Concepts

18.6.1.1 Concern for RIA Systems

There is a range of interacting influences to be considered in the design of work involving RIA systems. Adaptation/evolution of ergonomics and its standards are required to better address RIA system issues/problems. Ergonomics has traditionally concerned itself with physical and cognitive acts in a physical world. It is now addressing interactive systems and the digital world that extends to information, both about the physical world and of itself, including science, business information systems, entertainment, and knowledge. For RIA technology, Ergonomics needs to consider social ergonomics to assess and assist user experience, accessibility, usability, and avoidance of harm. Ergonomics will need to extend its scope to take account of the effect of social interaction with and through RIA systems, including social media, human-machine teaming, and adaptive interfaces and environments. Likewise, Ergonomics needs to explore the impact on non-users of RIA systems who are in the environment in which the RIA systems are operating but are not using the RIA systems. Ergonomics also needs to take account of legal, regulatory, and governance requirements that are being set by other parties, largely or even completely without ergonomics input.

Having outlined why RIA systems demand new ergonomic thinking, the next step is to explore which design paradigms define the human-machine relationship.

18.6.1.2 Design Approaches for RIA Systems

There is a range of approaches that can be taken when designing the relationship between an RIA system and its users, both human and other RIA systems. These are described in Table 18.1. They have ergonomic, sociotechnical, ethical, and possibly political implications. Ergonomics applied to RIA systems should consider the most suitable paradigm for an application and the effect on individuals and society.

Selecting the appropriate paradigm – whether augmentation, teaming, or governance – defines not only technical architecture but also human trust and responsibility.

18.6.1.3 Perceived Autonomy

From a human-centred point of view, apparently autonomous or intelligent behaviour is a human-system issue and should be addressed in design.

TABLE 18.1
Design Approaches for RIA Systems

Design Approach	Description
Augmentation	The system improves human performance (including decision aids, exoskeletons, physical and/or cognitive prosthetics).
Replacement	The system replaces human functions and/or entire human jobs.
Remoting	Allows the user to act on the physical environment at distance.
Teaming	The human and machine work together for a common goal (need to be social).
Symbiosis	The human and the system are closely linked, working together for mutual benefit, such as open, online courses and games in which the users get excitement and knowledge, and the system gets training.
Parasitic	The human is a source of data collected by the system, but with little or no benefit to the human.
Influence	Intelligent systems influencing human behaviour, such as social media chatbots.
Unknown alternatives	Emerging paradigms of organisational or societal relationships with RIA systems that have not yet been defined.
Benevolent governance	Humans/humanity passing governance to AI.

18.6.1.4 Control Loop

The control loop consists of the fundamental control functions, supporting information, affectors and effectors necessary for making adjustments in a process. A user, autonomous agent or other component of a system is said to be in the loop if that user, autonomous agent, or component is in the process flow for making adjustments to the process. With the advent of RIA system agents, finer descriptions of where the user is located in the loop are emerging.

18.6.2 Categories of Human-RIA System Issues

The identified human-RIA system issues fall into six categories. These are summarised in Figure 18.3 and outlined in the following sub-sections.

ISO/TR 9241-810 describes human-system issues that are specific or particularly relevant to RIA systems rather than those that are common to non-RIA systems. It includes examples of ISO standards that can be related to these categories.

18.6.2.1 RIA System – Effects on Humans

The issues in this category are not limited to those experienced by users (those interacting with the RIA system or for whom its effects are intended). Unless an RIA system is totally unobtrusive, it impacts non-users as well, who have no vested interest in what it is doing. With respect to the non-users, an obtrusive RIA system in their environment can alter the way they perceive their environment, prompt an emotional response, and alter their behaviour.

18.6.2.2 Human-RIA System Interaction

This category is where most ergonomics resides today. It addresses the nature of tasks and the design of work to ensure that users can accomplish their intended tasks. The issues in the category specifically cover the consequences and impact of the design of the user interface on user (i.e., individual or team) interactions with the RIA system.

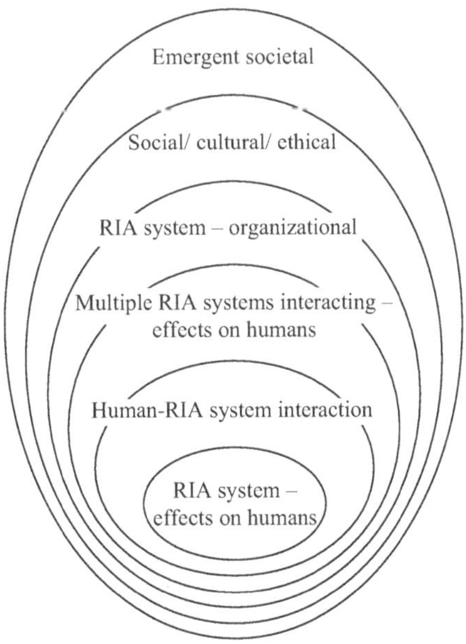

FIGURE 18.3 The six categories of RIA system issues identified.

18.6.2.3 Multiple RIA Systems Interacting – Effects on Humans

Interactions can be obvious to the user, in the case of a physically embodied RIA system, or hidden, as can be the case with software agents, where the user only perceives a single system rather than its constituent elements. Where RIA systems from different suppliers or organisations interact, compatibility and communication issues between one RIA system and another are foreseeable, as well as with the humans present in environments where multiple RIA systems are operating. It is difficult to predict what emergent behaviours will result and their effect on users.

18.6.2.4 RIA System – Organisational

The RIA system is likely to affect the activities that occur in the organisation such as work and organisational processes, and the roles that people in the organisation play. These issues raise the question of how we can optimise the existing organisational structures/working practices, etc., to make the best use of a new RIA system in an existing organisation. Although the issues in this category may not arise for every RIA system, they are likely to arise for any RIA system that is implemented at the level of an organisation.

18.6.2.5 Social/Cultural/Ethical

RIA systems will operate within a social and cultural context. An effective RIA system will seamlessly integrate within that context, as a familiar member of the team or a cooperating entity. Behaviours of an RIA system should be expected to fit social and cultural norms. The issues in this category describe how the design of an RIA system can impact the social interactions, group attitudes and motivation, and collective group behaviours positively or negatively, depending on how seamlessly the system fits within the social and cultural context.

18.6.2.6 Emergent Societal

The widespread or strategic use of RIA systems will redefine humanity's relationship to technology. In the longer-term, intelligent agents may be employed to make important decisions about governance. Assurances about the data and algorithms used, and how they are governed, are likely to be a concern.

18.6.3 Ergonomics and RIA Systems

The categories described in Section 18.6.2 demonstrate that RIA systems influence HF across individual, organisational, and societal levels. Consequently, ergonomics must evolve from designing interfaces to considering the entire system design. The changes in the nature of tasks and the design of work to accommodate the complex, social human-machine interaction of RIA systems are fundamental for ergonomics. These changes will require that ergonomics integrates with the systems approach, adapts its best practices, and expands into areas of psychology and other social sciences that ergonomists do not deal with on a regular basis at present.

RIA systems will be more connected, complex, probabilistic and non-deterministic, social, and augment human capabilities well beyond merely replacing physical work. This will have an impact on human behaviour (Figure 18.4). Interaction with these agents can become a relationship, their interface a personality, and users and agents can form complex human-machine teams, working together towards a shared goal.

Therefore, it is essential that a human-centred approach (human enhancement) is considered at all levels (of a system, organisation and scale of system), where RIA system and overall system design is focused around the decisions and tasks that are retained by the human user (across all potential use cases: optimal, degraded, reversionary, and emergency) with the RIA system supporting these. The human should be kept in the loop to the extent that is needed to achieve the required levels of performance and safety across all system states. This is particularly important when systems transition between states, such as from normal operation into an emergency state. Determining the needed extent of human involvement in the loop is part of system risk assessment.

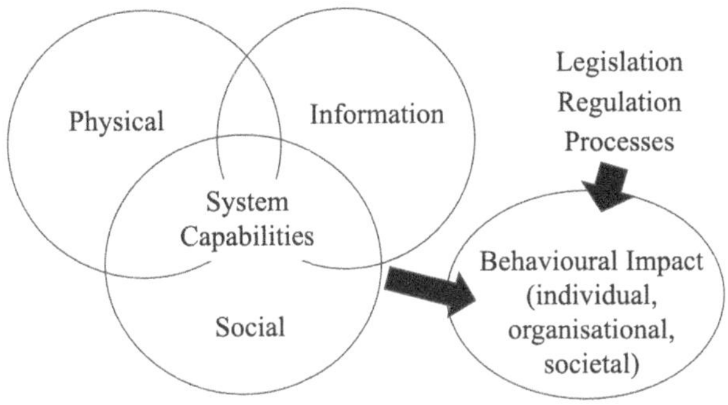

FIGURE 18.4 People's interaction with RIA system.

Similarly, it is critical to seamlessly integrate the RIA system with the social context in which it operates, to ensure that its operation does not disrupt the social and cultural fabric of the environment. Thus, it is important that effective change management procedures are implemented whenever an RIA system is introduced into an environment that includes teams or other groups of people whose social interactions can be affected. Although change management is important during the introduction of any new or modified system, it is especially important for RIA systems to help ensure their acceptance and minimise disruptions of group activities in the environment.

18.7 IMPLICATIONS FOR SAFETY BY DESIGN

Traditional automation design has often aimed to replace human work. A Safety by Design perspective redefines this goal: to enhance human performance, preserve accountability, and maintain system resilience. This section summarises the analysis and recommendations by Earthy and Sherwood Jones (2015).

18.7.1 Problems

The normal approach to automation is 'human replacement automation' (Boorman 2014). This is seen as a technical matter: engineers entering the organisation ask, 'Tell me what you do', and automate where practicable. Automation is seen as organisational lubricating oil to reduce friction and so reduce costs, improve productivity, or remove human error. The 'error reduction fallacy' is widely believed and seems to be persistent. The truth of the matter is that human error is a symptom of trouble deeper inside the system (Dekker, 2002).

The human and organisational problems arising from the human replacement approach have been documented regularly from as early as 1963; as have the adverse social and cultural consequences.

The human role of supervising 'human replacement' automation is problematic. The problems of supervisory control could be summarised as follows:

- Strong silent automation giving the operator an inadequate human window, lack of overview, leading to mode 'errors', and automation 'surprises';
- Disruption of teamwork;
- Additional difficult monitoring tasks, and the need to understand the automation as well as the process under control;
- Additional high-skill planning, customisation, and set-up tasks;
- Clumsy automation that takes care of routine situations (leading to boredom) but cuts out and disrupts the operators during unusual situations.

These problems are felt most acutely when things go wrong, and the operator is faced with the error-producing conditions of time pressure, conflicting goals, and inadequate information.

Introducing automation affects job design (and training needs), team structure, and organisational reporting. Compromises, shortcuts, and effects that have not been anticipated can be considered 'organisational debt' (Blank, 2015), which needs to be addressed, perhaps by formal Management of Change.

Lessons from past automation failures show that safety emerges from the interplay of human competence, system usability, and organisational design – not from removing humans from the system.

18.7.2 Desired Outcomes

The safety of a staffed operational system requires a defined, engineered and maintained balance of system usability and operator competence. To achieve this, the needs of humans must be considered throughout the lifecycle; that is, the design optimises the effectiveness, efficiency, safety, environmental impact, and satisfaction of staff. Standards define and operationalise Human-Centred Quality (HCQ) (ISO 9241-220) and the approach to testing.

18.7.3 Use of HCD to Achieve Outcomes

From a safety-critical systems design perspective, HCD is intended to mitigate the risks resulting from not addressing users' mental and physical characteristics and capabilities. HCD also provides information to mitigate risks caused by deficiencies that are the responsibility of other stakeholders, such as those responsible for safety, performance, engineering, support, and organisational design.

In the human-centred approach,

> *Automation needs to be ingrained in an organization's DNA early on and not deployed later as a replacement measure for existing job functions. It should instead be used to allow people and resources to be more focused; on driving the business forwards, rather than on just keeping the lights on.*
>
> (Boorman, 2014)

This places automation in support of a human-centred organisation (ISO 27500), with operator input to risks and opportunities. From a safety perspective, it recognises the need to prevent and recover from errors, and that people contribute to safe operations as well as make mistakes.

There are four issues in the human aspects of automation that do not seem capable of resolution by traditional design, and for which even a management solution appears difficult. They are:

- Risk compensation, where the performance gains intended for increased safety are used to commercial advantage;
- Automation bias, where people believe the automation when they shouldn't, and frequent errors of omission in manual operation are replaced by major errors of commission in automated operation;
- Moral buffering, where remoteness and/or algorithms lead people to act in an inhumane manner;
- Affect dilemma, where people attribute personality to automation – probably inappropriately, but unavoidably.

These are typically sociotechnical issues and not considered in a Probabilistic Risk Assessment or safety case. A human-centred approach supports consideration of these issues in design and operation. These will increasingly include the man, technology, and organisational (MTO) perspective, considering human readiness, technological, and organisational maturity. Operators will interact with systems that can learn, adapt, and make semi-autonomous decisions, requiring consideration of issues such as explainability and meaningful human oversight and control.

18.7.4 Transparent Interaction and Transparent Users

ISO 9241-810 elaborates on the requirements for Ergonomics to address these challenges as follows.

For humans and RIA systems to work together via transparent interactions, users must be able to understand the system's current state, the functions it performs within the activity, and how it is likely to behave under specific circumstances or conditions. Likewise, the system will need to 'understand' the users' current state, the functions that it performs in the activity being conducted, and predict the behaviour of the user under given circumstances or conditions. This implies that both user state monitoring sensors and models of the task, user goals, and human behaviour have been built into the system. This is likely to be complex due to the wide variation in human characteristics, behaviours, cultures, experience, physiological and cognitive state, personality, etc. Without appropriately accurate models of humans and systems, it is possible that a human will have to adapt to, accommodate and compensate for sub-optimal system behaviour/decisions resulting from inadequacies in the machine models. If the RIA system is capable of learning by modifying its models of humans and systems, careful decisions must be made about whether, and how, that learning is incorporated into the system and applied. Humans are not always free to change how they perform a task when they discover what can be a more efficient or otherwise superior process for performing the task, particularly when functioning in teams or otherwise interoperating with others.

To optimise human-machine teams, it will be important that both the human and the machine elements are working towards a common goal with a shared understanding of the current situation and desired end state. Without a shared understanding of both the goal and the current situation, it is possible that the human-machine team will become inefficient or even have human and machine working in conflict with each other. It may be necessary to develop complementary task models to help facilitate the development of these transparent interactions between human and machine in RIA systems.

18.7.5 Safety Aspects of RIA Systems

ISO 9241-810 discusses the ergonomics of safety of RIA systems as follows. RIA systems can introduce new safety issues, for example, something that is temporarily too fast for humans to address in real time, or something that is otherwise not directly observable, such as online activity. Latency, the time that humans have to take over from an RIA system that detects that it has lost control, is another factor. Or, at a societal level, there is the 'Frankenstein issue', the systemic consequences when the inventor (human or humanity as a whole) does not take timely responsibility for a created entity. Safeguards are necessary, but what form should they take and what part does ergonomics play?

A safety set is a fundamental set of normal/expected human behaviours in context, such as robots at edge states of physical engagement with human bodies. These would probably be phrased as sets of safety requirements ('the system shall do this ... under these circumstances'), or cases. For example, the safe condition for a self-driving car to achieve in the event of loss of control is to park on the hard shoulder with hazard lights on. For online human-systems safety, it can be, for example: 'do what a user does most frequently'; but issues, such as offline actions not accessible to machine learning, can lead to different, dangerous interpretations by an intelligent agent, such as changes in usability related to donning personal protective equipment. In other cases, what is considered ethical can be most appropriate. Thus, the safety set relates to ethics/laws/regulations. It is not easy to decide what is in a universal set.

The user/RIA system relationship approach is likely to influence both safe action and construction of safety/harm. Different safety sets will be required for different cultures, and there is a need to also be aware of the cultural context of the designer and purchaser of a system. A new section describing the cultural construction of safe/safety needs to be added to the context of use statement so that this issue can be considered in design. For example, having selectable safety sets as part of regional customisation, or designing for culture and context awareness so that the RIA is sensitive to the culture of its source of production and the area of use when these entities/systems/end users are not the same.

In either case, there is probably a different safety set for every category (see Section 18.6.2), although there will be interrelated causes and effects between categories (e.g., attitude to robots extends to social/cultural/ethical). To facilitate the consideration of safety, ergonomics standards for the RIA system should include a means of assessment. Consideration should be given to the broadest application of measurement of human-centred quality as a basis of assessment. This includes avoidance of harm, which is a human-centred treatment of safety and other types of harm (both to and from humans). ISO 9241-220, ISO/IEC 25063, and ISO/IEC 25065 contain supporting material on HCQ, context and user requirements, respectively. ISO 9241-220:2019, Annex E.5, contains examples of harm from use.

18.8 EXISTING STANDARDS AND RIA SYSTEM ADAPTATIONS

Standards for ergonomics provide information that is useful in addressing many of the issues and hazards related to RIA systems, particularly those related to user interface, organisational issues, and lifecycle processes. Although these standards may not adequately cover the emerging/new requirements of RIA system characteristics, this section summarises how these standards can already be used to address RIA systems. As autonomy and AI proliferate, these ergonomic standards intersect increasingly with emerging AI governance standards, forming a unified system of human-centred assurance.

18.8.1 Principles of Ergonomics

The following standards define a framework for ergonomics on a strategic level.

ISO 26800 describes the general ergonomics approach and specifies basic ergonomics principles and concepts applicable to the design and evaluation of work systems. For RIA systems, the underlying principles of ergonomics remain the same, although the relative emphasis placed on them will vary.

ISO 27500 presents the rationale and general principles of human-centeredness in a concise form for executive board members and policy makers. It provides a framework for organisational behaviour when RIA systems are considered or implemented, in particular the sociotechnical nature of approaches to automation and the larger categories in Section 18.6.2/Figure 18.3.

18.8.2 Human-Centred Design Process

Design issues: Human-centred design standards such as ISO 9241-210 and ISO 9241-220 provide a structured process for embedding usability and risk management into lifecycle activities that are key for achieving Safety by Design.

ISO 9241-210 provides requirements and recommendations for human-centred design principles and activities throughout the life cycle of interactive systems. These are elaborated in two human-centred design process models. ISO 9241-220 describes the processes that ensure human-centred quality of interactive systems. ISO/TS 18152 describes the processes that address human-system issues in the engineering of systems. At the time of writing, ISO TS 18152 is under revision. This will extend the bridge to systems engineering and emphasise the importance of the HF integration concept for the development of RIA systems.

The processes in these standards go a lot further than the current norm in HCD practice and will apply to RIA systems as much as to other types of systems and any set of human-systems issues. ISO 9241-220:2019, Annex F, provides guidance on risk management and human-centred design.

18.8.3 Interaction and Interface

ISO 9241-110, ISO 9241-112, ISO 9241-130, ISO 9241-129, ISO 9241-154, and ISO 9241-820 set general design principles for the ergonomic design of interactive systems. They provide a framework for applying those principles to the analysis, design, and evaluation of interactive systems. While they are applicable to all types of interactive systems, they do not cover the specifics of every context of use, such as safety-critical systems and collaborative work. The principles are generally independent of any specific design style or application. ISO 9241-129 and 9241-130 specifically address adaptive dialogues. Other parts address novel and multiple modalities, including ISO 9241-820 for immersive environments. Guidance on AI Ergonomics ISO 9241-812 is under development.

18.8.4 Accessibility

ISO 9241-171, ISO/IEC 29136, ISO 9241-129, ISO 9241-971, and ISO/TR 22411 provide guidance on specific aspects of the design of an interactive system that take account of the widest possible range of users. The application of the specific guidance takes place within the application of the human-centred approach to design.

ISO 9241-171 provides ergonomics guidance and specifications for the design of accessible software for use at work, in the home, in education and in public places. Most of the provisions apply to any part of the software that implements or contributes to the software user interface.

ISO 9241-971 provides ergonomic guidance focused on ensuring the accessibility of tactile/haptic interactions with ICT systems. This document deals with modality shifting and accessibility guidance for haptic/tactile modalities such as gestures, vibration, and force feedback.

18.8.5 Workspace and Workload

ISO 6385, ISO 9241-2, and the ISO 10075 series provide ergonomics principles for the design of tasks, work, and work systems. They encourage attention to human, social, and technical requirements in a balanced manner during the design process.

The systems approach in these standards assists in both existing and new situations, such as introduction of an RIA system. Ergonomics evaluations of existing or new work systems will show the need for, and encourage attention to, workload and the role of the worker with/within those systems.

The principles specified in these standards support the design of optimal working conditions with regard to task performance, workload, human well-being, safety, and health. This includes the development of existing skills and the acquisition of new ones, while taking account of technological and economic effectiveness and efficiency.

18.8.6 Context and Environment

ISO 9241-11 provides a framework for understanding the concept of usability and applying it to situations where people experience or use interactive systems (including RIA systems), other types of systems, and products and services. Usability is a scalable, task-based measure of the degree to which users are enabled to achieve goals effectively, efficiently and with satisfaction, taking account of the Context of Use. ISO 9241-11 emphasises that usability is dependent on the context of use (users, goals, tasks, environment, and resources).

Control and control centres present an early and widespread application of RIA systems. The overall strategy for dealing with user requirements in control centres is presented in ISO 11064-1. ISO 11064-2 provides guidance on the design and planning of the control room in relation to its supporting areas. Requirements for the layout of the control room are covered by ISO 11064-3.

Ergonomics requirements, recommendations, and guidelines for the design of workplaces in control centres are established in ISO 11064-4. Displays and controls, human–computer interaction, and the physical working environment are presented in ISO 11064-5 and ISO 11064-6. Evaluation principles are dealt with in ISO 11064-7. At the time of writing, ISO 11064 is under revision. However, the framework seems robust, and the standard will continue to reference 9241 for interaction details.

ISO TC159 sets standards for modelling, measuring, and assessing the impact of properties of the physical and thermal environment. As RIA systems increase their presence in environments, especially when controlling dynamic integrated environments, personal capability, comfort, and safety will increasingly depend on machine application of these standards and correct interpretation of the integrated effect of environmental factors.

18.8.7 Useful AI Standards

Many standards for AI are now available. Those useful for addressing the ergonomics of RIA systems include the following:

ISO/IEC 5338 augments the software life cycle processes in ISO/IEC/IEEE 12207 and the system life cycle processes in ISO/IEC/IEEE 15288 with modifications and additions of AI-specific processes from ISO/IEC 22989 and *ISO/IEC 23053* to support the definition, control, management, execution, and improvement of an AI system in its life cycle stages. These processes can also be used within an organisation or a project acquiring AI systems.

ISO/IEC 22989 establishes terminology and describes AI concepts. This supports communication between the many stakeholders.

ISO/IEC 42005 provides guidance for individuals and societies conducting AI system impact assessments.

ISO/IEC/TR 24028 provides an overview of issues associated with establishing, losing and assessing trust.

ISO/IEC/TR 24027 addresses bias in AI systems, particularly with respect to AI-aided decision-making and *ISO/IEC/TS 12791* addresses treatment of bias in machine learning.

18.9 EXAMPLES FROM THE MARITIME SECTOR

The maritime domain provides a concrete example of Safety by Design in action, where ergonomics standards have been applied to ensure human oversight and trust in autonomous vessel operations.

This section presents examples of the consideration of human-system issues for RIA systems in the maritime sector. The first presents an application of the thinking outlined in Section 18.7.3 to develop a set of safety behaviours for autonomous ships. The second is an example of the application of the thinking outlined in Section 18.8.2 to consider human-system risks beyond strict safety using the ISO 31000 *Risk management* definition 'the effect of uncertainty on objectives' by the EU OCEAN project (Earthy *et al.*, 2025).

18.9.1 Safety Set

The Norwegian HUMANE (Human Maritime Autonomy Enable) project (Lützhöft and Earthy 2024) explored the sociotechnical issues for use of AI in the maritime sector and developed a set of safety behaviours for autonomous ships (which are basically large robots) using the ISO 9241-810 framework, based on existing maritime requirements, the Principles of Human-Centred Artificial Intelligence (HCAI) (*Shneiderman*, 2020), and the HUMANE generalised handover model (Lützhöft and Earthy 2024). These are presented in Table 18.2.

TABLE 18.2
The Maritime Autonomous Ship Safety Set

1a	Be identifiable as under control of an artificial intelligence or
1b	Be identifiable as under remote control
2	Act in a manner that is understandable by seafarers on own and other ships and shore services (e.g., pilot, Vessel Traffic Services, owner)
3	Communicate its status and capability to crew on own and other ships and shore services
4	Be able to explain its intention
5	Be able to explain its course of action
6	Recognise when it requires assistance
7	Request assistance
8	Place itself in a state in which it does not present a hazard to the environment, other shipping, fixed structures, or those rendering assistance
9	Whilst in a safe state, allow authorised personnel to operate the functions necessary to restore, replace or supplement lacking capability (defects, failures and events that are not reasonably foreseeable)
10	Render assistance to other ships and persons in distress
11	Transfer control to authorised personnel on request
12	Protect itself from unauthorised access
13	Be sufficiently dependable to fulfil its intended purpose

These safety behaviours operationalise the principles of ISO 9241-810, embedding transparency, explainability, and resilience into the behaviour of autonomous ships.

HUMANE proposes that all behaviours are to be demonstrated under all circumstances. This means that in many cases or system states (such as failure), the required behaviour will change to address the context and capability of the system in that case or state. Departures from specified behaviours are to be justified. Each behaviour can be elaborated in terms of what it means, how it can be assessed, and the design required to meet it.

This set of behaviours provides the Human Element approval criteria for evidence of risk-based design, or an assurance case for an autonomous ship or system submitted to a regulatory authority.

18.9.2 Integrated Solution

ISO 17894:2005 was the first goal-based standard for maritime programmable systems. When the standard was developed, the only programmable systems on a ship were islands of operational technology (control systems, navigational instruments). Now we have widespread use of software-intensive systems (SIS), increased connectivity and interconnection, and new applications, such as AI. These need both a broader interpretation of risk and the application of risk management thinking with respect to ICT. A proposed revision of ISO 17894 (Earthy and Twomey, 2025) provides general principles for SIS in maritime applications. The purpose is to support seafarers in maintaining meaningful oversight and control of complex, interconnected computer, and communications technology. The International Maritime Organisation has made clear that this responsibility will remain with the seafarer (IMO, 2025). The maritime sector therefore needs people to *responsibly* use and manage SISs, to *trust* that systems are working, will work, and can be used. In other words, be *robust* enough for the job, *usable* enough to support the seafarer in doing the job, and *stable* enough throughout their life to keep on doing that job, whenever required.

Table 18.3 outlines the architecture of the proposed revision to ISO 17894, which supports the principles that mitigate the range of risks associated with the design and use of complex integrated systems in maritime applications. These principles are listed in Table 18.4.

TABLE 18.3
Architecture of ISO 17894

Clause	Title and Contents
1	Scope
2	Normative references
3	Terms and definitions
4	Abbreviations
5	Use of the standard (including means of assessment suitable for the range of modes of use of the document)
6	Intention for marine SIS
7	Product principles for marine SIS (meaningful name, required principle, recommended criteria) ordered to facilitate balanced assessment
8	Lifecycle principles for marine SIS (meaningful name, required principle, recommended criteria) ordered to facilitate balanced assessment
9	Sustainability
10	Conformance
Annex A	Terms and concepts (Software Intensive System; Context of use and the total system; Risk assessment; Systems engineering; Agreement process; System of Interest (SoI) and system integration; System stability – feedback, iteration, and trust)
Annex B	Guidance on the principles for marine SIS (commentary, guidance, source standards)
Annex C	Guidance on the lifecycle of marine SIS (stages, stakeholders, documentation)
Annex D	Application of the principles in the lifecycle (whole lifecycle and its management)

TABLE 18.4
ISO 17894 Principles

Intention for Marine SIS

The SoI shall be demonstrably suitable for the user and the given task in a particular context of use. It shall deliver correct, timely, sufficient, and unambiguous information to its users and other systems. The hardware and software of the SoI shall respond correctly throughout its lifecycle.

This will be achieved if the following product and lifecycle principles are fulfilled by the SoI and its associated elements throughout its life.

Product Principles for Marine SIS

1	Secure – Unauthorised access to the SoI shall be prevented.
2	Useful – The SoI shall provide functions which meet user needs.
3	Performance – The SoI shall maintain specified levels of accuracy, timeliness, and resource utilisation when used under specified operational and environmental conditions.
4	Sharing – Functions shall be appropriately allocated between users and SoI.
5	Expected – The operation of the SoI shall be consistent and shall correspond to user expectations of the underlying process.
6	Controllable – The interaction between the SoI and the user shall be controllable by the user.
7	Usable – The SoI shall be acceptable to the user and support effective and efficient operation under specified conditions.
8	Tolerant – The SoI shall be tolerant of faults and input errors.
9	Maintainable – The SoI shall support proper installation and maintenance, including repair and modification.
10	Non-Hazardous – In the event of failure, the SoI shall remain in or revert to the least hazardous condition.
11	Harmless – The SoI shall be free from unacceptable risk of harm to persons or the environment.

(*Continued*)

TABLE 18.4 (Continued)

Lifecycle Principles for Marine SIS	
The successful realisation and use of a robust and usable marine SIS require a systematic approach throughout the life of the SoI. The key requirements for any approach which aims to meet the product principles given above are as follows.	
12	HCD – Human-centred activities shall be employed throughout the lifecycle.
13	Documented – Suitable documentation shall be produced to ensure that all SoI lifecycle activities can be performed effectively.
14	Planned – All SoI lifecycle activities shall be planned and structured in a systematic manner.
15	QMS – All parties involved in lifecycle activities shall have and use a Quality Management System.
16	Competent – Persons who have responsibilities for any lifecycle activities shall be competent to discharge those responsibilities.
17	Configured – The SoI configuration shall be identified and controlled throughout the lifecycle.
18	V&V – Verification and validation activities shall be employed throughout the lifecycle.
19	Compliant – Existing requirements for marine systems shall be taken into account throughout the lifecycle.
20	Safe – The required level of safety shall be realised by appropriate activities throughout the lifecycle.

18.10 REFLECTIONS AND FUTURE DIRECTIONS

Automation has caused problems for its human operators from the start. The problems are well-known and can largely be mitigated by the use of established HCD standards. It is the author's experience that:

a) these standards are not widely known or used and
b) the risks to safe and effective operation from poor usability are frequently unmanaged.

On the HF side, there is some progress in understanding its shortfalls in dealing with engineers. The need for a human-centred approach will grow with the additional complications of RIA systems.

18.10.1 Assurance

All generations of human-system interaction standards have a purpose. The style of assessment is different for each type of standard. International standards are often third-generation principles to give the widest application and to support innovation in design. These require interpretation for a particular purpose or design. The most suitable means of assessment is a safety case. Second-generation standards specify and use management system certification or process capability assessment. Specific regulations, products, or technologies at a national or industrial sector level need a prescription for the testing and quality assurance of products and systems. However, such an assessment still needs to address risks related to lifecycle and context of evaluation. At a project level, the use of ergonomics standards expands the focus of technology readiness levels.

Ergonomics standards are used in support of regulation. An example is the European Machinery Regulation, which adopts 20 Ergonomics standards, including EN 894 *Ergonomics requirements for design of displays and control actuators* and many AI and robotics standards.

18.10.2 Core Concepts

The design of robotic and automated systems usually concentrates on the technical system and then works out the required competence of the operator. The need to design the human-system interaction is often forgotten or even ignored. The ergonomics concepts that underlie SC4 standards provide tools for the lifecycle of this important part of a complete work system.

Human-Centred Design provides a through-life perspective and should be seen as part of life-cycle management. It assists in identifying the scope of control, supports HCAI and choices about design for human augmentation, teaming and disseminated control (multiple users and other stakeholders in several locations).

For safety-critical interactions, complex, distributed, and embodied systems are increasingly common, and this creates new challenges. These challenges are not confined to the control room or the immediate working environment. To carry out everyday social and financial activities with acceptable risk, people need meaningful oversight and control of complex computer-based systems.

User-centred evaluation, preferably involving actual users wherever possible, is therefore essential. Usability should be assessed through task-based evaluation, with particular emphasis on user-centred measures and evaluation in context. Ideally, the total system should be evaluated in its actual context of use, rather than evaluating subsystems separately in only representative situations. For control-related functions, especially in safety-critical interactions, such evaluation should be scenario-based.

18.10.3 Messages

As automation becomes increasingly pervasive, standards provide a vital bridge between design ambition and operational trustworthiness. They embody collective learning about how to engineer systems that respect and support human purpose, variability, responsibility, and meaningful control. To design for safety effectively, standards should be used not as constraints but as enablers that embed human-centred knowledge throughout every stage of system development, deployment, and evolution.

Standards can also save both time and money, and access to information about them is often free. Extensive public information is available on the websites of ISO and IEC, (ISO.org/home.html, iec.ch/homepage), including descriptions of published standards and standards under development, terms and definitions through the ISO Online Browsing Platform and IEC Electropedia, and details of technical committees and subcommittees, including their work programmes.

This chapter has concentrated on how human-system interaction standards address and are evolving to address issues related to robotic, intelligent, and autonomous systems. If your concerns are not yet addressed in existing standards, consider contributing to their development. Standards evolve through expert participation: iso.org/get-involved.html.

18.10.4 Why Human Factors Standards Matter

HF standards distil decades of collective learning about how people interact with technology. They provide designers, engineers, and regulators with shared principles, definitions, and evaluation methods that ensure systems are usable, safe, and dependable in real operational contexts. By embedding HF standards throughout the lifecycle – from concept development to testing and operation – organisations can reduce design errors, prevent usability-related failures, and strengthen meaningful human oversight. In Safety by Design, HF standards act not as constraints but as enablers: they help teams make informed decisions, avoid repeating past mistakes, and ensure that automation and AI technologies align with human capabilities, values, and responsibilities.

ACKNOWLEDGEMENT

The author would like to thank Lloyd's Register and ISO for permission to publish this chapter, although any opinions expressed are those of the author and do not necessarily reflect those of Lloyd's Register or ISO. International standards work is traditionally anonymous, but a book chapter gives a welcome opportunity to personally thank the experts who made special contributions to the standards and papers reported in this chapter: Jim Carter, Joshua Downs, Daryle Gardner Bonneau, Susan Harker, and Brian Sherwood Jones. The author is also grateful to the editors for their insightful comments and assistance.

REFERENCES

This section is divided into two parts. The first part lists the references cited in the chapter in alphabetical order. The second part lists the standards cited numerically.

Blank, S. (2015). *Organizational Debt Is Like Technical Debt – But Worse*. https://steveblank.com/2015/05/19/organizational-debt-is-like-technical-debt-but-worse

Boorman, C. (2014). *The Need for Human Centric Automation*. http://blog.automic.com/the-need-for-human-centric-automation

Dekker, S. (2002). *The Field Guide to Human Error Investigations*. Aldershot: Ashgate Publishing Ltd. ISBN-13: 978-0754619246.

Dresner, D. (2002). *Guide 275, Desert Island Standards, Guidelines for IT Management*. National Computing Centre. https://ictknowledgebase.org.uk/fileadmin/ICT/pdf/NCC/Desert_Island_Standards.pdf

Earthy, J. V., & Downs, J. L. (2018). *Human-System Interaction Standards for Robotic, Intelligent, Autonomous Systems*. NATO. https://publications.sto.nato.int/publications/STO Meeting Proceedings/STO-MP-HFM-300/MP-HFM-300-18.pdf

Earthy, J., Lee, S., Petersen, E., & Twomey, B. (2025). *Changes to ISO 17894, EU OCEAN, D3.6*, https://ocean-navigation-awareness.eu/wp-content/uploads/2025/09/OCEAN-Changes-to-ISO17894-V1.0c-D3.6.pdf

Earthy, J. V., & Sherwood Jones, B. (2015). Using human-centred design to mitigate the risk of human error in software intensive systems. *Safety Critical Systems Club, Bulletin*, 11–15.

Earthy, J., & Twomey, B. (2025). *Proposed changes to ISO 17894, EU OCEAN D10.6*, https://ocean-navigation-awareness.eu/wp-content/uploads/2025/09/OCEAN-Proposed-changes-to-ISO-17894-V1.0c-D10.6.pdf

European Union (2023). Machinery regulation. Regulation (EU) 2023/1230 of the European Parliament and of the Council of 14 June 2023 on machinery.

European Union (2024). Artificial intelligence act. Regulation (EU) 2024/1689 of the European Parliament and of the Council of 13 June 2024.

IEA/ILO (2021). *Principles and Guidelines for Human Factors/Ergonomics - Design and Management of Work Systems*. International Ergonomics Association/International Labour Organisation. ISBN: 978-92-2-035590-9. https://www.ilo.org/sites/default/files/wcmsp5/groups/public/%40ed_dialogue/%40lab_admin/documents/publication/wcms:826596.pdf

IMO (2025). *Intersessional Working Group on Maritime Autonomous Surface Ships, MSC/ISWG4/MASS*. International Maritime Organization.

ISO (2024). https://iso.org/standards.html

Lützhöft, M., & Earthy, J., eds. (2024). *Human-Centred Autonomous Shipping*. Routledge. ISBN: 9781032554945. https://routledge.com/Human-Centred-Autonomous-Shipping/Ltzhft-Earthy/p/book/9781032549897

Oborne, D. J. *et al.* (1993). *Person-Centred Ergonomics - a Brantonian View of Human Factors*. Taylor and Francis Ltd. ISBN-10.0748400516.

Shneiderman, B. (2020). Human-centered artificial intelligence: Three fresh ideas. AIS Transactions on Human-Computer Interaction, https://aisel.aisnet.org/thci/vol12/iss3/1

Steedman, S. (2012). *Creating a Knowledge Structure for Programme and Portfolio Management*. Presentation given at the Achieving Coherence in Defence Procurement Forum May 2012. BSI Sector Report, https://bsigroup.com/Documents/about-bsi/NSB/BSI-defence-sector-report-UK-EN.pdf

List of standards. Standards are regularly reviewed and updated. Use of the latest version is advised. For more information on any standard, use the standard's number in the search function on the ISO (www.iso.org) or IEC (http://www.iec.ch) website. For ISO/IEC standards, use the ISO website.

EN 894-1 (1997). *Safety of machinery - Ergonomics requirements for design of displays and control actuators - General principles for human interactions with displays and control actuators.*

EN 894-2 (1997). *Safety of machinery - Ergonomics requirements for design of displays and control actuators - Displays.*

EN 894-3 (2000). *Safety of machinery - Ergonomics requirements for design of displays and control actuators – Control actuators.*

EN 894-4 (2010). *Safety of machinery - Ergonomics requirements for design of displays and control actuators - Location and arrangement of displays and control actuators.*

ISO/IEC 5338 (2023). *Information technology – Artificial intelligence – AI system life cycle processes.*
ISO 6385 (2016). *Ergonomics principles in the design of work systems.*
ISO 9241-2 (1992). *Ergonomic requirements for office work with visual display terminals (VDTs) - Part 2: Guidance on task requirements.*
ISO 9241-5 (2024). *Workstation layout and postural requirements.*
ISO 9241-11 (2018). *Ergonomics of human-system interaction - Part 11: Usability: Definitions and concepts.*
ISO 9241-110 (2020). *Ergonomics of human-system interaction - Part 110: Interaction principles.*
ISO 9241-112 (2025). *Ergonomics of human-system interaction - Part 112: Principles for the presentation of information.*
ISO 9241-129 (2010). *Ergonomics of human-system interaction - Part 129: Guidance on software individualization.*
ISO 9241-130 (2026). *Ergonomics of human-system interaction* - Part 130 *User assistance within interactive systems.*
ISO 9241-154 (2013). *Ergonomic requirements for office work with visual display terminals (VDTs) - Part 154: Interactive voice response (IVR) applications.*
ISO 9241-171 (2025). *Ergonomics of human-system interaction - Part 171: Software accessibility.*
ISO 9241-210:2019 *Ergonomics of human-system interaction Part 210: Human-centred design for interactive systems.*
ISO 9241-220 (2019). *Processes for enabling, executing and assessing human-centred design within organizations.*
ISO 9241-250 (under development). Stakeholder engagement in IT Governance during specification, procurement and operation.
ISO 9241-500 (2018). *Ergonomic principles for the design and evaluation of environments of interactive systems.*
ISO 9241-520 (under development) *Ergonomics aspect of Activity Based Working.*
ISO 9241-810 (2020). *Ergonomics of human-system interaction – Part 810 Robotic, intelligent and autonomous systems.*
ISO 9241-812 (under development). AI system ergonomics.
ISO 9241-820 (2024). *Ergonomics of human-system interaction - Part 820: Ergonomic guidance on interactions in immersive environments, including augmented reality and virtual reality.*
ISO 9241-971 (2020). *Ergonomics of human-system interaction - Part 971: Accessibility of tactile/haptic interactive systems.*
ISO 10075 (2006-2024). *Ergonomic principles related to mental workload* (three parts – concepts, design, measurement).
ISO 11064-1 (2000). *Ergonomic design of control centres - Part 1: Principles for the design of control centres.*
ISO 11064-2 (2000). *Ergonomic design of control centres - Part 2: Principles for the arrangement of control centres.*
ISO 11064-3 (1999). *Ergonomic design of control centres - Part 3: Control room layout.*
ISO 11064-4 (2013). *Ergonomic design of control centres - Part 4: Layout and dimension of workstations.*
ISO 11064-5 (2005). *Ergonomic design of control centres - Part 5: Displays and controls.*
ISO 11064-6 (2005). *Ergonomic design of control centres - Part 6: Environmental requirements for control centres.*
ISO 11064-7 (2006). *Ergonomic design of control centres - Part 7: Principles for the evaluation of control centres.*
ISO/IEC/IEEE 12207 (2017). *Systems and software engineering - Software life cycle processes.*
ISO/IEC/TS 12791 (2024). *Information technology – Artificial intelligence – Treatment of unwanted bias in classification and regression machine learning tasks.*
ISO/IEC/IEEE 15288 (2023). *Systems and software engineering - System life cycle processes.*
ISO 17894 (2005). *Ships and marine technology – Computer applications – General principles for the development and use of programmable electronic systems in marine applications.*
ISO/TS 18152 (2010). *Ergonomics of human-system interaction - Specification for the process assessment of human-system issues.*
ISO/TR 22411 (2021). *Ergonomics data for use in the application of ISO/IEC Guide 71:2014.*
ISO/IEC 22989 (2022). *Information technology – Artificial intelligence – Artificial intelligence concepts and terminology.*
ISO/IEC 23053 (2022). *Framework for Artificial Intelligence (AI) Systems Using Machine Learning (ML).*

ISO/IEC/TR 24027 (2021). *Information technology – Artificial intelligence (AI) – Bias in AI systems and AI aided decision making.*
ISO/IEC/TR 24028 (2020). *Information technology – Artificial intelligence – Overview of trustworthiness in artificial intelligence.*
ISO/IEC 25019 (2023). *Systems and software Quality Requirements and Evaluation (SQuaRE) – Quality-in-use model.*
ISO/TR 25060 (2023). *Systems and software engineering - General framework for Common Industry Format (CIF) for usability-related information.*
ISO/IEC 25063 (2014). *Systems and software engineering – Systems and software product Quality Requirements and Evaluation (SQuaRE) – Common Industry Format (CIF) for usability: Context of use description* (under revision).
ISO 25065 (2019). *Systems and software engineering – Software product Quality Requirements and Evaluation (SQuaRE) – Common Industry Format (CIF) for Usability: User requirements specification.*
ISO 26800 (2011). *General approach, principles and concepts.*
ISO 27500 (2016). *The human-centred organisation – Rationale and general principles.*
ISO 27501 (2019). *The human-centred organisation –Guidance for managers.*
ISO/IEC 29136 (2012). *Information technology – User interfaces – Accessibility of personal computer hardware.*
ISO 31000 (2018). *Risk management - Guidelines.*
ISO/IEC 33000 (n/a), *Systems and software engineering - Process assessment* (series).
ISO/IEC 38500 (2024). *Information technology - Governance of IT for the organisation.*
ISO/IEC 42005 (2025). *Information technology – Artificial intelligence (AI) – AI system impact assessment.*
IEC 62366-1 (2015). *Medical devices – Part 1: application of usability engineering to medical devices.*
IEC 62508 (2010). *Guidance on human aspects of dependability.*

Part 6

Concluding Chapter: Safety and Efficiency in Automation, AI, and Remote

19 Safety and Efficiency in Automation, AI, and Remote Operations

Verification, Validation, and Systematic Safety Management through Human-Centred Engineering

Stig Ole Johnsen and Frøy Birte Bjørneseth

19.1 INTRODUCTION

Safety-critical industries are undergoing rapid transitions driven by automation, digitalisation, artificial intelligence (AI), and increasingly remote modes of operation. Autonomous cars and ferries, remote subsea intervention, and AI-supported processes are no longer experimental – they are becoming part of normal operations. As technology matures, it becomes a system shaped by regulations, organisational structures, human capabilities, and competing goals. Following Rasmussen's (1997) risk management framework, safety emerges from the coupled actions of actors at all system levels.

The chapters in this book collectively demonstrate that introducing powerful automation without parallel investment in human-centred design, verification and validation, and organisational learning creates fragile systems. Furthermore, as demonstrated by MiT (2025), deriving value from AI remains challenging; 95% of organisations in that report saw no measurable return on AI investments. Early outcomes from the 5% that do deliver value show that learning-capable systems, when used for specific processes, can provide real benefits. Therefore, essential skills such as concept scoping, analysis, and process improvement are crucial knowledge in safety by design.

Meaningful Human Control (MHC) becomes essential: humans must retain the ability to monitor, understand, and influence automated behaviour, particularly during anomalies or unexpected system states.

The prior chapters indicate that technological capability is evolving more rapidly than safety practice. Complex automation, opaque machine-learning models, and remote supervision are deployed without equivalent investment in human-centred engineering, validation, or training. This creates a gap as systems become technically advanced long before organisations fully understand, govern, or safely integrate them. When this happens, the technology's potential benefits are overshadowed by new forms of brittleness and vulnerability.

The dangers of this gap are familiar from recent history. Advanced automation and incidents such as the Boeing 737 MAX crashes highlighted what can happen when automated logic is introduced without sufficient design, training, or consideration of how pilots will be expected to

DOI: 10.1201/9781003741824-25

respond. In the maritime domain, bridge accidents and near-groundings have revealed that alarm floods, complex displays, and opaque system behaviour can overwhelm crews, especially at night, long before a formal emergency is declared. In the offshore and process industries, control room operators have experienced chronic alarm overload and abnormal conditions. They are often expected to manage these situations with limited time and incomplete information, which can easily overwhelm their capacity to respond.

These cases share two features. First, they demonstrate that the explanation 'human error' is insufficient as a final causal account. Human actions are the visible endpoint of a chain that often begins with project scoping, design decisions, unqualified software or equipment changes, organisational pressures, or undocumented modifications.

Second, they reveal that automation does not eliminate the need for humans; it redistributes tasks and intensifies the need for human oversight and intervention in key situations (Bradshaw et al. 2013; Bainbridge 1983; Hancock 2021). Humans are still responsible for monitoring performance, detecting anomalies, and taking control when automated functions can no longer manage the situation. If systems are not constructed explicitly to support human perception, understanding, and intervention, the operators' opportunity to maintain MHC disappears.

In the United States, the National Institute for Occupational Safety and Health (NIOSH) Prevention through Design (PtD) initiative recognises that "one of the best ways to prevent and control occupational injuries, illnesses, and fatalities is to 'design out' or minimise hazards and risks early in the design process" (Behm et al., 2014). In several accident analyses, poor design has been identified as a key root cause, including the Chernobyl disaster (Insag-7, 1992), the Boeing 737 Max crash (Endsley, 2019), and the USS John S. McCain accident (NTSB, 2019).

The implication is that safety management must become more integrated than before. Safety can no longer be regarded as a separate layer added on top of technology; it must be considered an inherent property of system design and operation. The accumulated perspectives throughout the book demonstrate that Safety by Design should be implemented as a lifecycle approach, from design through operation and organisational learning.

19.2 THEORETICAL APPROACH AND CONCEPTS

This concluding chapter is based on a synthesis of the chapters in this book; thus, it draws on systematic literature reviews, analyses of accident investigations, and interviews with subject experts.

The basic theory is based on Foundations of Safety Science (Dekker, 2019), using both failures and successes as a basis, that is, Safety I and Safety II (Hollnagel, 2017). A central premise is that 'human error' is not a cause but a symptom of deeper systemic conditions. The guiding credo is therefore that learning must take precedence over blame, and that failures and successes alike reveal how socio-technical systems actually function.

19.3 CONTROLLING AND SUSTAINING SAFETY

Building on these theoretical foundations, we now turn to how safety is controlled and sustained in practice. Our research reveals three recurring principles that determine whether organisations successfully adapt to automation, AI, and remote operations:

1. **Open and democratic sharing of information**, which enables systemic learning and early detection of weak signals.
2. **Early, risk-based action through scoping and design**, where hazards are addressed before they become embedded in the system.
3. **Organisational and human readiness**, ensuring that new technologies match user capabilities, operational realities, and contextual demands.

19.3.1 Open Sharing of Information

Norway has achieved the world's best road safety in recent decades. One of the key reasons has been a systematic risk-based approach, open knowledge sharing, and systematic learning (Elvik, 2021). Open democratic processes and the sharing of information among relevant actors are a key foundation for safety, that is, a broad democratic process involving the regulator, organisations, affected people, and the design of technology (Elvik, 2021). If we use the Norwegian fatality rate per km driven as a best-practice measure, we can compare it with other countries, arguing that cars, roads, and people are similar. There are differences in regulations, training, and speed limits that affect risk. In the United States, approximately 40,000 road deaths occur each year. If the Norwegian fatality rate per kilometre driven were applied, this figure would be closer to 10,000. Although illustrative rather than conclusive, this comparison suggests that tens of thousands of lives – potentially up to 30,000 annually – could be saved through systematic, risk-based regulation, speed limits, infrastructure design, and training. This is significant.

These democratic processes are also a vital part of action research aimed at enhancing safety and organisational learning.

Action research, as described by Greenwood and Levin (2006), is a structured, participatory approach to improving safety. It involves cycles of planning, action, observation, and reflection carried out jointly by practitioners, regulators, and researchers. By prioritising actions, implementing mitigating measures, and evaluating outcomes together, organisations create a powerful engine for continuous learning. Action research, therefore, exemplifies a Safety-by-Design mindset: understanding work-as-done (WAD), improving it collaboratively, and strengthening the system.

Figure 19.1 is an example of the impact of action research on safety in offshore supply vessels used in the oil and gas industry (Antonsen et al., 2007). Following a rise in collisions, injuries, and one fatality in 2000, action research was initiated in collaboration with vessel crews, management, and regulatory authorities. A joint working group – referred to as the 'Captains' Forum' – was established to improve safety. This group addressed relevant MTO issues, including procedures and responsibilities, to enhance safety and foster continuous learning. The actions led to significant improvements in accidents (such as collisions) and personal injuries.

A similar improvement in safety has been achieved in offshore helicopter transport in Norway, through open dialogue, collaborative learning, and continuous development – an approach facilitated by the 'Forum for Helicopter Safety', Bye et al. (2018).

These examples highlight a core Safety-by-Design principle: when practitioners, regulators, and designers share operational knowledge openly and analyse WAD collaboratively, they create the organisational preconditions for continuous learning and systemic improvement.

These lessons are similar to those from systematic safety work, focusing on safety measures with the most significant impact on risk. The Hierarchy of Controls (HoC) is an approach supported by regulators and the oil and gas industry in Norway (NIOSH, 2024; Manuele, 2005). HoC provides a structured safety framework, organising interventions by effectiveness – from eliminating hazards at their source to relying on personal protective equipment (PPE) as a last resort.

19.3.2 Eliminating Risks

Recent analyses of accident investigations estimate that a large proportion of adverse events (approximately 40–60%) involve design-related deficiencies in routines and technology (Kinnersley and Roelen, 2007; Moura et al., 2016). This indicates that more focus should be placed on system design. Human Factors (HF) is a key component to ensure MHC, but this is still lagging, as seen in Bergh et al. (2024). Excellent design must be based on identifying the right problem before designing the right solutions. Thus, we start by exploring how to prioritise the right actions through HoC.

The Hierarchy of Controls prioritises interventions that eliminate or reduce hazards at their source. It directs organisations to address risks as early as possible in the design lifecycle, reducing the need

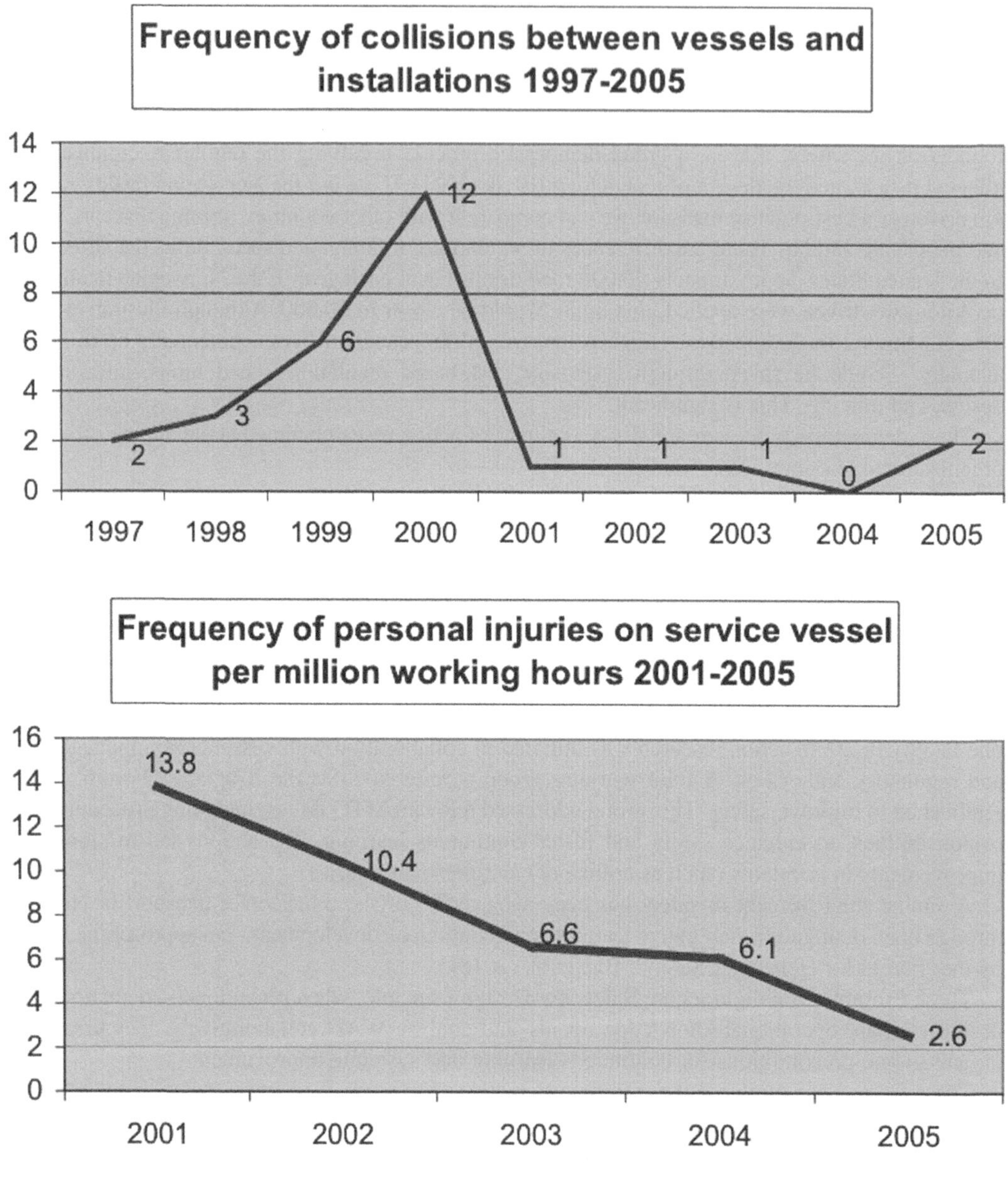

FIGURE 19.1 Effects of action research on offshore supply ship safety (collision and personal injuries).

for compensatory measures later. Although most effective during the design phase, HoC principles should also be applied during operation to reduce error traps and enhance system defences. Several studies support this risk-based design and mitigation approach (Dyreborg et al., 2022).

We have used the hierarchy of controls as defined by NIOSH (2024) and Manuele (2005), listing the most critical issues first (Figure 19.2).

Using the Hierarchy of Controls helps organisations prioritise interventions with the greatest influence on risk, beginning with system-level measures – elimination, substitution, and engineering controls – before resorting to administrative controls or PPE.

19.3.2.1 Elimination

Elimination involves removing hazardous conditions by changing design, equipment, or work processes so that serious incidents cannot occur. Ideally, this should be achieved during the

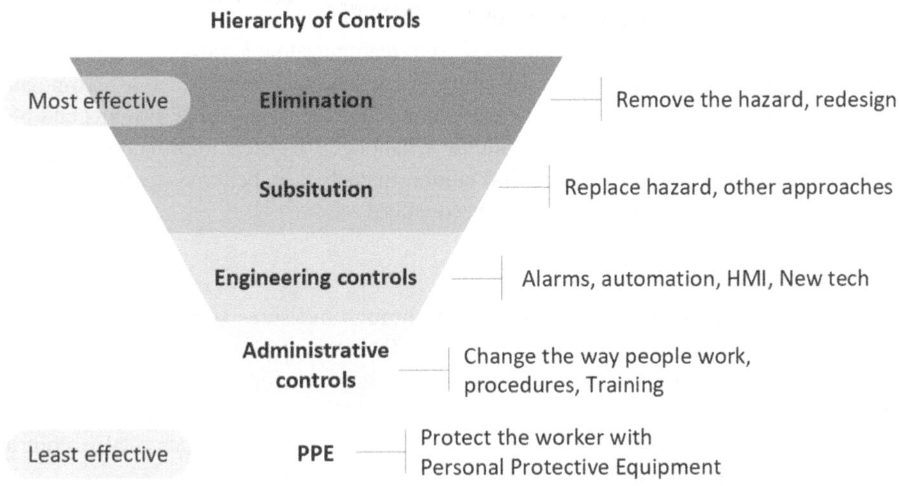

FIGURE 19.2 The hierarchy of controls as a tool to prioritise efforts.

conceptual design phase through user-centred design. Examples include removing personnel from hazardous environments via remote operation. In the Chernobyl disaster, design flaws – such as a counterintuitive reactor design and setup – could have been eliminated by adopting a safer reactor design (Insag-7). The Deepwater Horizon accident occurred in an oil field known to be dangerous due to high pressure and the rock's fragility. Other operators had halted drilling and moved their operations to less risky areas. By halting operations in such a high-risk field, the accident could have been eliminated.

In the Texas City refinery explosion, replacing open blowdown stacks with a closed flare system during design could have removed the underlying hazard.

The Boeing 737 Max crisis illustrates how poor initial design choices, aimed at fitting larger engines to an older airframe, necessitated complex mitigation systems that eventually malfunctioned.

19.3.2.2 Substitution

We replace material, activity, or operation with safer variants, new technology, or systems designed to be error-tolerant. Automation can substitute human operations to reduce risks and avoid dangerous, difficult, or dirty tasks. Examples: The automatic ground collision avoidance system on the F-16 has avoided four operational disasters by substituting human action with automation. In the Bhopal gas tragedy, a less toxic chemical could have been substituted for methyl isocyanate (MIC). In the Fukushima nuclear disaster, they could have reduced the risk of meltdown by substituting older reactor designs with more modern, fail-safe designs.

The use of technology to substitute human operations should use a modern version of Fitts List (Fitts, 1951) to ensure that technology increases safety versus the old way of working.

19.3.2.3 Engineering Controls

Engineering controls (technical controls) involve designing systems that reduce risks independently of human actions (e.g., reducing the likelihood of errors). These include error-tolerant designs, improved alarms, or HMI displays that enable 'situation at a glance' awareness.

Examples include Unified Bridge (UB) design (Bjørneseth, 2021), which reduced cognitive load and helped operators avoid mistakes. Chernobyl example: the control panel was poorly designed, with counterintuitive layouts, a lack of real-time feedback, and misleading alarms.

Deepwater Horizon: poor alarm and HMI issues contributed to the delayed response. Many critical alerts were suppressed, delaying the recognition of the blowout. Three Mile Island: poor alarm and interface design delayed response; operators failed to recognise that a pressure relief valve was stuck open because a misleading indicator on the control panel showed only the valve's command status, not its actual position. Texas City refinery explosion: better pressure relief systems and blast walls might have mitigated the damage. Space Shuttle Challenger disaster: improved O-ring design and testing could have prevented the catastrophic failure. In each case, better system designs could have avoided the accident and/or mitigated the consequences.

19.3.2.4 Administrative Controls

Administrative controls modify how people work through measures such as training, job rotation, communication protocols, and procedural redesigns. These controls often become necessary when technical measures are insufficient. Examples include implementing Crew Resource Management (CRM), developed after the Tenerife Airport disaster to improve cockpit communication and teamwork. In Piper Alpha, inadequate procedures and poor communication also contributed to the tragedy, highlighting the importance of administrative interventions.

19.3.2.5 Personal Protective Equipment (PPE)

PPE serves as the last line of defence when higher-level controls are not feasible. Examples include survival suits, gloves, or respirators. Following the 9/11 attacks, improved respiratory protection for first responders could have mitigated long-term health impacts. Similarly, during the Montara oil spill, proper use of PPE could have reduced workers' exposure to hazardous materials.

The Hierarchy of Controls aligns closely with safety-science thinking. Elimination and substitution act on the system's structure, while engineering controls shape the boundary conditions for safe performance.

Administrative controls and PPE influence individual behaviour but cannot compensate for flawed design. Accident analyses throughout this book show that failures attributed to 'human error' often originate from weaknesses in the upper three tiers of the hierarchy.

19.3.3 Science-Based Safety Management Must Address Technical, Organisational, and Human Issues (MTO)

19.3.3.1 Why Human Factors Engineering Is a Foundation in Automation and AI

A consistent pattern across domains is that technological change is rarely accompanied by equivalent organisational or human-centred design capability. This imbalance produces unsafe assumptions and hidden vulnerabilities.

'Blaming Human Error' obscures systemic design deficiencies. Across the chapters, explanations such as 'loss of situational awareness' or 'human error' are shown to be downstream symptoms of deeper failures, including weak project scoping, inadequate Management of Change (MoC), missing HF analysis early in design, inconsistent interfaces, alarm flooding, cognitive overload, and insufficient preparation for automation edge cases. Operator actions cannot be understood without examining the system that shaped them.

19.3.3.1.1 Human Factors Are More Critical with Automation and AI

As Bainbridge (1983) observed, the more a system is automated, the more crucial human intervention becomes at precisely the moments when information is degraded, and uncertainty is highest. AI-enabled systems amplify this challenge because edge cases, model brittleness, and opaque decision processes place additional burdens on operators. Systematic safety management must therefore integrate the following:

- Safety-II and learning from normal work
- Human-centred design (ISO 9241-210 (2019))
- Meaningful Human Control
- Human–AI teaming requirements
- Safety-critical task analysis
- Alarm management standards
- Situation awareness and workload modelling
- Scenario-based evaluation (e.g., CRIOP)

These elements collectively ensure that design, operation, and learning remain aligned with human capabilities.

19.3.3.2 The Readiness Challenges of MTO

Across many chapters, a recurring theme is the rapid deployment of advanced automation without corresponding investment in human-centred engineering or organisational capability. This creates a readiness gap, manifested as mismatches between system assumptions and operational realities in:

- Technological readiness (TRL)
- Human readiness (HRL)
- Organisational readiness (ORL)

Without deliberate integration of TRL, HRL, and ORL, systems become mismatched with human and organisational capability. The result is brittle automation, limited transparency, unrealistic operational assumptions, and inadequate preparation for anomalies and edge cases. Across the chapters, insufficient HRL and ORL repeatedly appear as root causes behind technological failures. These gaps manifest as mismatches between how we imagine work versus how work is actually performed.

Human Readiness Level (HRL) assesses how effectively HF, user needs, and human-system interaction are incorporated into a system. According to ANSI/HFES 400-2021, it ranges from basic awareness (HRL 1) to validated operational performance (HRL 9), emphasising usability, situational awareness, and training.

The Technology Readiness Level (TRL) indicates the maturity of a technology, from initial concept (TRL 1) to full operational deployment (TRL 9). It is frequently utilised in engineering and innovation to monitor technical feasibility.

The Organisational Readiness Level (ORL) assesses an organisation's preparedness to adopt, support, and sustain a new system or change. It ranges from initial awareness (ORL 1) to strategic renewal (ORL 9), including leadership engagement, cultural alignment, and system learning.

Approaches such as Action Research (Greenwood and Levin, 2006) and HOP (Dekker & Conklin, 2014) support the development of readiness through just culture and systems thinking, thereby creating a foundation for learning and change.

19.3.4 Synthesis of Challenges

The challenges identified across industries can be synthesised into three overarching requirements for safe and sustainable automation.

First, front-end safety engineering must ensure that systems are conceptually sound, technically robust, and grounded in HF evidence before deployment.

Second, operational safety management must account for human limitations and strengths, ensuring that operators have sufficient time, information, and workload support to maintain control during normal and degraded conditions.

Third, organisations must embed systematic learning processes that integrate work-as-imagined (WAI) with WAD, transforming operational insights and lessons from accidents into improvements in design, procedures, and training.

Together, these three pillars form a coherent Safety-by-Design model: design shapes operational possibilities, operations generate improvement opportunities, and learning feeds back into design. This lifecycle must be explicitly engineered and reinforced if organisations are to maintain control as automation and complexity increase. This lifecycle should be verified and validated (e.g., by the CRIOP method).

19.4 FRONT-END SAFETY ENGINEERING

Front-end engineering shapes safety outcomes. Decisions made during the conceptual and early design stages set constraints that are difficult to alter later. The book's case studies consistently reveal that the most serious failures in Chernobyl, Boeing 737 MAX, and USS John McCain stemmed from design assumptions rather than operational actions at the sharp end. A strong concept and design process form the first pillar of Safety by Design. Eliminating hazards is often the best first step. No amount of training or discipline can replace design choices that ignore human cognitive and perceptual abilities. In addition, early design and safety inclusion reduce the later costs of changes.

Safety, efficiency, and usability begin with defining scope and requirements that accurately mirror possibilities and constraints – including technological options in the context of human strengths and limitations, task demands, environmental conditions, and anticipated challenges.

19.4.1 System Scoping, Problem Definition, and Human-Centred Design

Before implementing any automation or AI component, the system must be specified appropriately. Throughout the book, failures in early scoping repeatedly resulted in unclear human–automation task boundaries, misaligned expectations, excessive cognitive demands, and unrealistic training assumptions.

Clear scoping should define the purpose and operational design domain (ODD), which tasks automation performs versus which tasks humans must retain, the information needed for Situational Awareness, expected anomalies and edge cases, the alarm philosophy and failure mode behaviour, required training and competencies, and the demands of night work or remote supervision.

Norwegian regulations provide strong formal support for human-centred design. Facilities Regulation §10 requires that installations, systems, and equipment be designed in the most robust and simple manner possible so that the possibility of human error is limited. Facilities Regulation §9 emphasises the qualification of new technologies and methods, aligning closely with the Technology Readiness Level (TRL) and, implicitly, we argue, with the Human Readiness Level (HRL).

Management Regulation (MR) §18 incorporates the principles of ISO 11064, mandating human-centred design of control rooms and operator environments. MR §19 emphasises the appropriate handling and use of data, a requirement that has become central in AI-based systems where training data, model drift, and representativeness directly affect system behaviour.

The EU AI Act enhances these requirements by mandating that high-risk AI systems incorporate mechanisms enabling operators to oversee, understand, and interrupt automated processes (EU AI Act Art. 14). These obligations reflect a shared principle across safety-critical areas: systems must be designed from the outset to uphold MHC. The primary human-centred design standards are ISO 9241-210, which establishes user-centred design principles, and ISO 11064 (2000), applicable to control facilities. These approaches are based on Human Factors Engineering (HFE) methods, as explained in Stanton et al. (2013).

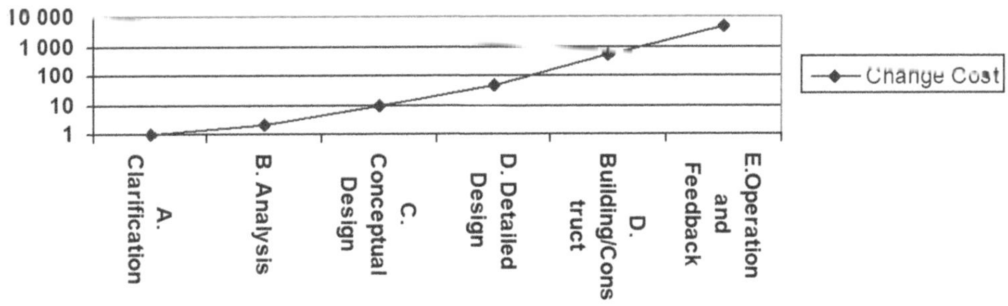

FIGURE 19.3 The cost of change is dependent on the phase (clarification through operation).

Source: Boehm (1976), Szymberski (1997), Samset (2001), Behm (2005), and Driscoll et al. (2008).

19.4.1.1 Reducing Costs by Front-End Design

The cost of changes rises considerably from one phase of the design process to the next. Experience across various industries (software, construction industry) indicates that the cost of a change grows exponentially as the project advances through its stages (Boehm, 1976; Szymberski, 1997; Samset, 2001; Behm, 2005; Driscoll et al., 2008), see Figure 19.3.

Both the cost-benefit considerations and the importance of designing for safety (Behm et al., 2014) are supported by accident analysis. This highlights the need to conduct systematic HFE analysis early and to validate it. By performing a CRIOP validation at the outset, we can proactively eliminate risks and improve safety, ensuring hazards are identified and addressed before they become costly and difficult to manage. To illustrate this development of change cost, the cost of the same change could be as follows:

- 10–100 USD in the analysis phase
- 100–1,000 USD in the design phase
- 1,000–10,000 USD in the build phase
- 10K–100K USD in operations phase (K = 1,000)

CRIOP supports early front end activities through design checklists and scenario evaluations.

19.4.2 Management of Change as a Safety Barrier

Digital systems constantly evolve. Software updates, new AI models, sensor designs, or vendor changes can markedly alter system behaviour. MoC helps ensure that modifications are assessed for safety impacts, including workload, alarm levels, HMI consistency, and operator training requirements. When MoC is weak or disregarded, new error traps can emerge.

Several accident analyses have shown that undocumented or insufficiently reviewed changes contributed to increased workload, loss of situational awareness, and an inability to maintain control. Appropriate MoC is a fundamental element of Safety by Design.

19.4.3 Situational Awareness and Design for 'Situation at a Glance'

Situational awareness (SA) must be engineered into systems; it cannot be left to operator vigilance alone. Poor SA has been a significant contributing factor in several accidents, where cluttered displays, alarm overload, and inconsistent interfaces disrupted the perceptual basis for rapid decision-making. Designers must ensure that interfaces support all three SA levels defined by Endsley (1995) – perception, comprehension, and projection – so that operators can anticipate how situations will evolve. This is especially important in remote operations where physical cues are absent.

19.4.3.1 Importance of HMI Standardisation

One common weakness identified is the inconsistent HMI design across vendor systems, installations, and organisational units. IEC 63303 addresses this issue by outlining requirements for HMI design, including task analysis (TA), consistency, interaction patterns, terminology, and information display.

The standard also clarifies the need for a situational awareness flow, ensuring that operators can perceive and interpret critical information 'at a glance', a concept consistent with high-performance HMI design (Hollifield et al., 2008) and ecological design principles (Mulder et al.). IEC 63303 functions both as a usability guideline and a safety standard; its adoption can reduce the likelihood of human error and enhances operational efficiency.

19.4.4 Cybersecurity by Design

As systems become more interconnected, cybersecurity and safety are increasingly intertwined. The Hydro (2019), Amedia (2021), Nortura (2021), and Maersk (2017) cyberattacks show how digital breaches can jeopardise operational control. A key standard is NOG 104 (2016), emphasising important guidelines, risk assessments, and the need for proper network segmentation and monitoring.

Security often relies on human knowledge and awareness; however, humans make mistakes. Consequently, the concept of security by design has gained significant attention. The National Institute of Standards and Technology (NIST) in the United States promotes security by design through NIST SP 800-160 (2022) and NIST SP 800-53 (2020). The Cybersecurity and Infrastructure Security Agency (CISA) has also published Secure by Design guidance for software developers.

19.4.5 CRIOP as a Design Verification Tool

CRIOP complements human-centred design by providing a list of best-practice standards and a practical, scenario-based framework for validating design assumptions before systems enter operation (Aas et al., 2009; Johnsen et al., 2025). CRIOP systematically evaluates HMI layout, alarm design, workload, communication, training needs, teamwork, scenario management, and compliance with HF standards. The method reveals discrepancies between WAI and WAD, exposing cognitive bottlenecks, error traps, or interface weaknesses that might remain unnoticed in document-based reviews.

By involving operators, engineers, and HF specialists in scenario walkthroughs, CRIOP supports early verification of design quality, ensuring that systems uphold MHC under both normal and degraded conditions. Furthermore, the participation of management, HF experts, safety specialists, and users in CRIOP workshops encourages organisational learning and development. A brief overview of CRIOP issues is included in the Appendix.

19.5 OPERATIONAL SAFETY MANAGEMENT

In the following section, we have summarised key operational issues to sustain safety management.

19.5.1 Psychological Safety

Psychological safety is one of the most vital yet often neglected parts of systematic safety management. When people fear blame, legal consequences, or damage to their reputation, they stop reporting anomalies, concerns, or near misses. They hide workarounds rather than share them. They avoid raising questions about automation behaviour, unclear responsibilities, or poor working conditions. This silence removes early warning signals that could prevent incidents.

A psychologically safe reporting culture promotes several crucial benefits. It encourages honest reflection on WAD versus WAI and fosters transparent communication about automation surprises. Such a culture enables early identification of new error traps and rapid learning from deviations. It also boosts operators' trust in management, results in more accurate MoC assessments, and enhances the effectiveness of CRIOP workshops.

A poor reporting culture was identified as an issue in the Boeing 737 Max accident (Hopkins, 2025) and has been a problem in Norway's oil and gas industry, as shown in several VG articles (2025).

Psychological safety transforms reporting from a compliance task into an effective safety tool, enabling operators to identify weak signals early before they escalate. Regulations can require reporting systems, but only a positive culture can ensure they are genuinely used. Therefore, psychological safety is a fundamental prerequisite for meaningful human control.

19.5.2 Supply Chains and Responsibility

Safety cannot be outsourced; the 'See-to' responsibility obliges operators and contractors to work together to ensure barrier integrity, follow safety and HF standards, deliver proper training, maintain good ergonomics, and promote an open reporting culture alongside psychological safety to ensure issues are raised and addressed. These issues are also outlined in regulations and industry guidelines, such as the International Association of Oil & Gas Producers, IOGP guidelines (2017). IOGP emphasises the importance of early identification of inherent hazards and risks, conducting systematic verification and monitoring to ensure that engineered controls operate as intended, and performing robust subcontractor management.

19.5.3 Human Oversight in Automated Operations

Human oversight and SA must be maintained, and operators must understand what automation is doing, why it is doing it, anticipate its next actions, and be able to intervene during safety-critical operations.

Experiences from digitalisation and accident investigations show that human oversight fails when interfaces become cluttered, information is fragmented, alarms are overwhelming, cognitive load rises, and fatigue occurs.

Maintaining situational awareness relies on well-designed automation that adheres to HFE standards, adequate time, appropriate workload management, reliable cues, clear interfaces, and effective team communication.

Loss of SA in operations is not considered an 'operator failure'; it is a design failure (Miranda, 2019). Error traps and workarounds point to design mismatches; they reveal hidden system conditions and constraints. They do not indicate operator carelessness but highlight areas where MHC is vulnerable. Night work and fatigue affect SA, vigilance, reaction time, coordination, and the ability to interpret automation behaviour.

CRIOP is vital for validating operations. It checks key issues such as cues supporting SA; compares WAI with WAD; assesses workload; tests alarm behaviour; reviews communication; evaluates the operating room design under all conditions, including remote; and identifies role conflicts. CRIOP is particularly important for remote operations, where low workload, 'being out of the loop', loss of operational practice, boredom, and the absence of physical cues can be challenging.

19.5.4 A Taxonomy for AI Hazard Analysis

As AI more frequently becomes part of safety-critical applications, initial deployments reveal common hazard patterns. These patterns mirror familiar safety mechanisms – automation surprises, mode confusion, data drift, and fragile automation – that appear differently in machine-learning-based systems. The EU AI Act and national regulators (Bergh & Teigen, 2024) are increasingly acknowledging these mechanisms, emphasising oversight, transparency, and human–AI collaboration.

Article 1 of the EU AI Act explicitly states that one of its purposes is to 'promote the uptake of human-centric and trustworthy artificial intelligence'. Key areas emphasised in the Act – AI oversight, AI design, AI maintenance, and AI testing – align closely with the failure modes described by Cummings (2024) in her taxonomy for AI hazard analysis.

Cummings (2024) identifies four broad categories of AI-related hazards, here labelled A to D. Importantly, real-world accidents across different domains demonstrate how each failure mode can result in significant operational or safety repercussions.

19.5.4.1 Inadequate Oversight

A lack of organisational, regulatory, or human supervisory control over AI systems can allow unsafe deployments or unnoticed system degradation. *Example*: The well-known fatal Uber autonomous vehicle crash in 2018 occurred when the AI system misclassified a pedestrian, and the safety driver – who was supposed to provide oversight – was not effectively monitoring the vehicle. A weak oversight model allowed the hazard to materialise, and there were no adequate mechanisms to address operator complacency with automation. Both factors indicate weaknesses in safety culture and oversight.

19.5.4.2 Inadequate AI Design

Flawed design choices – such as poor sensor fusion, insufficient handling of real-world complexity, or unclear interaction logic – can cause AI systems to fail in operational settings.

The Uber case from 2018 is also an example, along with the Boeing 737 MAX MCAS accidents, which were partly due to a design that relied on a single sensor and lacked adequate safeguards and intuitive feedback, illustrating how poor design of an automated decision system can have catastrophic consequences.

19.5.4.3 Inadequate AI Maintenance

Failure to update, recalibrate, or retrain AI models can lead to model drift, decreased accuracy, and inappropriate responses to new environments. *Example*: The ML algorithms commonly used in manufacturing are not robust, and it has been shown that accuracy declines sharply with increasing drift across all tested algorithms.

19.5.4.4 Inadequate AI Testing

Testing methods that rely too heavily on simulation, fail to validate edge cases, or assume idealised operator behaviour risk missing critical real-world hazards. *Example*: The Tesla Autopilot crashes involving stationary emergency vehicles repeatedly demonstrated that real-world edge cases – such as lateral hazards or stopped objects in a lane – were not adequately covered in testing protocols.

These examples demonstrate that AI hazards are not abstract concerns; they are recurring patterns observed whenever AI systems are introduced into safety-critical work without sufficient organisational scaffolding. As Bergh and Teigen (2024) note, regulatory expectations for AI oversight are increasing because past accidents have revealed the consequences of deploying AI systems that are not thoroughly designed, maintained, or validated for real-world conditions.

Building a solid taxonomy for AI hazard analysis is therefore vital – not only to meet the requirements of the EU AI Act but also to ensure that future autonomous or decision-support systems learn from past mistakes. A structured, cross-sector learning approach will help avoid repeating known hazards and aid in creating reliable, human-centric AI for high-risk operations.

The previous sections described the structural and operational foundations for safe automation. However, even the best design cannot anticipate all future conditions. Sustained safety, therefore, depends on the organisation's capacity to detect weak signals, learn from operations, and transform those insights into design and procedural improvements.

19.6 LEARNING FROM OPERATIONS AND ACCIDENTS

Sustained safety depends on an organisation's ability to detect weak signals, learn from normal work, and convert those insights into improved system design and procedures. Learning systems must include the entire socio-technical context – technology, operations, and HF – so that MHC is reinforced over time.

A vital aspect of this is learning from WAD versus Work-as-imagined, as outlined by the Federation of Norwegian Industries (2025). Normal work uncovers adaptive strategies and early warning signs. These are crucial inputs for redesigning the working environment and procedures; they can also inform interface redesigns, update protocols, adjust alarm thresholds, and enhance training.

Regulatory and organisational learning also depends on a framework that facilitates learning. Organisations must conduct regular HF assessments that include psychological safety (Thun & Tinmannsvik, 2021), transparent reporting, Just Culture, and cross-disciplinary reviews.

19.6.1 Aligning Work-as-Imagined with Work-as-Done

Work-as-Imagined (WAI) refers to how tasks, processes, and workflows are expected to be performed in accordance with procedures, policies, or management assumptions. It is often based on ideal conditions and theoretical models.

WAD describes how tasks are performed in practice, accounting for constraints such as the quality of existing systems, time pressure, resource limitations, variability, training, and human adaptation. If there are large differences between WAI and WAD, the combined WAD can lead to accidents, as suggested by Figure 19.4.

We aim to support individuals undertaking the actual work in a safe, supportive environment. Good practice in HFE and verification/validation (by CRIOP) helps minimise discrepancies between WAI and WAD. This is achieved by focusing on improved project scoping, system design, usability, and human capabilities. Some strategies and HFE practices to align WAI with WAD and avoid accidents are as follows:

- **Observational Studies and Field Research** – Conduct ethnographic studies, direct observations, and task analyses to understand real workflows, especially in risky areas.
- **User-Centred Design** – Design tools, interfaces, and processes based on user needs and behaviours rather than idealised expectations, using ISO 9241-210/ISO 11064 as a guide.
- **Functional Analysis, Task Analysis, and Cognitive Load Assessment** – Perform functional analysis, and TA (or Safety-Critical Task Analysis) to evaluate workload, attention demands, and decision-making processes to design systems that support human performance (e.g., through function allocation and assessment of the degree of automation).

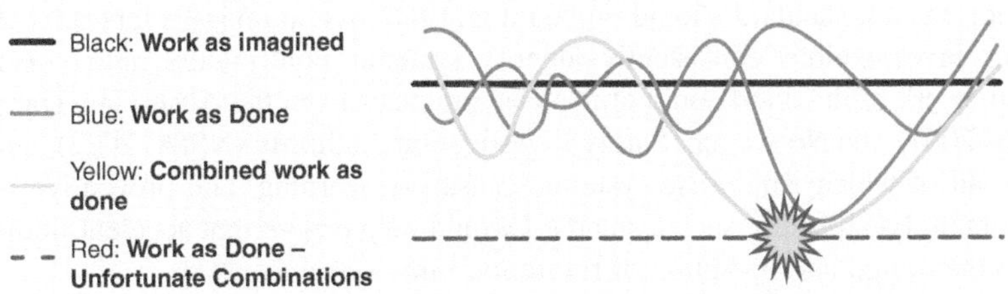

FIGURE 19.4 Gap between work-as-imagined and work-as-done, and its influence on safety.

Source: Hollnagel (2017).

- **Scenario Analysis, Test, Simulation, and Prototyping** – Analyse, explore, and test new systems in controlled but realistic environments before full implementation.
- **Training and Adaptive Systems** – Provide training that reflects real-world variability and design adaptive procedures that allow for flexibility without compromising safety.
- **Feedback Loops and Exploration of Scenarios** – Implement mechanisms for workers to report inefficiencies or procedural deviations to improve systems continuously.

By integrating HFE principles, organisations can align expectations with reality, improve safety, efficiency, and worker well-being while maintaining system effectiveness.

19.6.2 Learning from Accidents

Despite decades of accident investigations, several industries, including the maritime sector, continue to reproduce well-known failure patterns. HF issues such as fatigue, cognitive overload, alarm flooding, and poor interface design appear across multiple cases, yet findings are seldom translated into specific design requirements for new vessels. The learning process remains incomplete.

Examples such as *KNM Helge Ingstad*, *City of Rotterdam*, *USS John S. McCain*, and *PSV Sjøborg* show that poor interface design, inconsistent system layouts, alarm overload, and inadequate support for situational awareness create predictable cognitive traps for operators. However, these system-level issues are rarely highlighted in accident reports or developed into actionable design requirements for future vessels.

The accident statistics and the clear impact of HF highlight the need for stronger organisational learning. Many seafarers report inadequate rest, heavy workloads, and poor bridge ergonomics – conditions that directly threaten safety – yet current accident investigations, inspection regimes, and follow-up mechanisms fail to identify or address these issues. For instance, the Norwegian National Audit Office (2023) concluded that the Norwegian Maritime Authority's inspection system is not sufficiently designed to spot poor working and living conditions. The result is a learning system that observes symptoms but does not tackle their root causes.

Preventing recurring failures requires that accident investigations directly inform early-stage design decisions, including interface layout, alarm management strategies, workload-reducing workflows, and mechanisms that support team situational awareness. Accident analysis must move beyond attributing blame to operators. A systematic approach should include documenting events and actors, analysing breakdowns in situational awareness, evaluating interface and alarm design, reviewing management-of-change processes, examining workload and staffing, and assessing whether human-factors engineering methods – such as TA – were integrated into the design and operational procedures. Such an approach is illustrated by Winge et al. (2023) and several cases discussed in this book.

There is also a definite need to establish an international standard for accident analysis that avoids assigning blame and instead promotes learning and improvement across the entire socio-technical system. Such a standard should embed a non-punitive, human-factors-oriented approach and ensure that investigations consistently identify systemic contributors rather than individual failings. A framework of this kind could utilise best practices from the Australian Transport Safety Bureau (ATSB, 2007), the Norwegian Safety Investigation Authority (NSIA, 2022), and the Safety Forum (2019), all of which emphasise systemic causation, learning, and preventive action rather than assigning fault. Establishing such a standard would help ensure that accident findings systematically inform the design and operation of future maritime systems.

Human-centred design methods, cognitive task analysis (CTA), and scenario-based evaluations such as CRIOP offer the structured frameworks needed to translate accident findings into tangible design improvements. CRIOP, in particular, acts as an investigative toolset capable of identifying design omissions, missing standards in human-factors engineering, inadequate alarm philosophies, and disparities between 'work-as-imagined' and 'work-as-done'. By adopting these approaches, the industry can turn accidents into opportunities for learning rather than repeating them. Without closing this learning loop, maritime systems risk continuing to be designed around technology rather than the humans who must operate it – ensuring that today's accidents become tomorrow's as well.

19.7 RECOMMENDATIONS

For industry, the evidence in this book highlights the importance of systematically integrating HF methods throughout the entire design and operational lifecycle.

This includes mandating TA, usability evaluation, system testing, and scenario-based verification during early design; aligning HRL and TRL evaluations; adopting unified HMI standards, such as IEC 63303; strengthening alarm management in accordance with EEMUA 191 (2024); and ensuring that CRIOP is used not only during design but also as part of MoC. Organisations should also develop competencies in situational awareness, design, and simulator-based training, while maintaining continuous monitoring of MHC in automated operations.

Regulators, for their part, should mandate HF-based verification and validation, update certification pathways for autonomous and remote systems, and make HF-oriented accident investigations mandatory.

This could be supported by an international standard for accident analysis based on best practices such as ATSB (2007), NSIA (2022) the Norwegian Safety Investigation Authority, and the Safety Forum (2019). They should reinforce the establishment of Just Culture principles and transparent reporting mechanisms, and include AI-specific requirements for oversight, predictability, and explainability within regulatory frameworks. Together, these steps encourage systems that remain safe, understandable, and controllable as complexity increases.

19.8 SUMMARY: A SAFETY-BY-DESIGN FRAMEWORK FOR THE FUTURE

Automation, AI, and remote operations hold significant promise, but realising this potential requires safety and efficiency to evolve together. The chapters in this book show that Safety by Design is not a single technique but an integrated lifecycle philosophy that connects design, operations, and organisational learning. Disciplines once treated separately – HF, automation engineering, training, alarm management, and accident analysis – must now function as a coherent whole.

MHC emerges as a central integrating concept linking the following:

- Design integrity
- Operational readiness
- Organisational learning
- Governance and regulation
- Verification and validation
- Cross-sector sharing of best practices

HF methods (and tools to validate them, such as CRIOP) form the practical backbone of this integration. Safety by Design is no longer optional – it is the foundation of safe and sustainable automation. The next generation of automated, AI-supported, and remote systems must be built on rigorous, systemic safety management, combining technical excellence with a deep respect for human capabilities and limitations of automation.

AI and remote operations will continue to evolve. Still, the Safety-by-Design principles outlined in this chapter offer a robust foundation for ensuring that future systems remain effective, understandable, and under MHC. Safety is not a static property; it emerges from the continuous interaction between people, technology, and organisations.

By adopting strong design practices, operational vigilance, and organisational learning, industries can ensure that increasingly complex systems remain safe, adaptable, trustworthy, and ethically responsible. This is the essence of Safety by Design.

APPENDIX: TOP TEN HUMAN FACTORS ACTIVITIES IN CRIOP (OR ANY CHECKLIST)

The CRIOP method encompasses a comprehensive range of HF questions, standards, and best practices. In practice, the project team, together with users, selects critical HF activities and scenarios of interest to resolve the key HF and safety issues to achieve tangible safety improvements. Drawing on extensive use of CRIOP in automation, remote operations, and safety-critical industries, the following top ten HF activities highlight the essentials to explore in every project – from concept development to commissioning and operation. Instead of viewing them as a simple checklist, it is more effective to see them as a logical sequence:

Start by defining the right problem → understand real work → design around human capabilities → validate through scenario analysis → ensure robustness over time.

CRIOP is not only about compliance. It is also a powerful learning tool that reveals WAD through scenario-based discussions. These scenarios bring engineers, operators, vendors, and managers into the same room to explore real-world conditions, making invisible problems visible, and motivating improvements. This is often the most valued part of CRIOP for participants.

The suggested top ten HF activities to be selected (with CRIOP reference):

1. DEFINE THE PROJECT CLEARLY (G1): THE RIGHT SCOPE, THE RIGHT PROBLEM, THE RIGHT PEOPLE

CRIOP begins by clarifying why the project exists and the key challenges. This includes verifying that the scope covers the full MTO triad – Man, Technology, Organisation – rather than isolated technical components. Early HF involvement helps ensure that Technology Readiness Level (TRL) and Human Readiness Level (HRL; ANSI/HFES 400-2021) are aligned, and that MoC is activated from the start. Early identification of AI challenges, automation brittleness (NAS 2021) and classical Fitts List allocation issues prevents unsafe assumptions later.

2. IDENTIFY AND MITIGATE MAIN RISKS (G4): INCLUDING HUMAN FACTORS RISKS

CRIOP encourages teams to explore past incidents, near misses, and system vulnerabilities. The outcome is a set of realistic hazards and mitigations that address not only process safety but also psychological safety, usability, reporting culture, human performance, cybersecurity, situational awareness, and workload.

3. ESTABLISH A USER-CENTRED DEVELOPMENT PROCESS (G3)

A CRIOP-verified process uses standards such as ISO 9241-210 and ISO 11064 to guide design. This ensures that operators and HF experts participate from the very beginning, helping to shape workflows, interfaces, alarms, and procedures. Doing this early is crucial for ensuring MHC.

4. ANALYSE TASKS AND SYSTEM BEHAVIOUR (G7): TA, SCTA, CTA, PROTOTYPES

TA and SCTA uncover the actual cognitive demands of work, including 'out-of-the-loop' risks associated with automation. CTA exposes workload drivers and information gaps. CRIOP scenario sessions, simulations, and even eye-tracking reveal hidden constraints and error traps. This forms the behavioural foundation for interface design, alarm philosophy, role allocation, and procedure development.

5. DEFINE RESPONSIBILITIES AND DISTRIBUTED WORK (J1)

Many modern systems involve multiple actors across time, functions, and locations (e.g., local operators, remote operators, vendors, contractors). CRIOP maps these distributed responsibilities – including 'see-to' responsibility – to prevent ambiguity, overload, or gaps in control.

6. ANALYSE SITUATIONAL AWARENESS (SA) AND HMI (C3): DESIGN FOR 'SITUATION AT A GLANCE'

SA must be designed, not assumed. CRIOP checks whether information flows support Endsley's SA model (perception–comprehension–projection) and whether HMIs follow best practice (IEC 63303 2024). Scenario analysis is especially valuable here: participants often discover that the actual 'work-as-done' differs from design assumptions, revealing gaps in visibility, timing and sensemaking that affect safety during critical events.

7. VERIFY ALARM STRATEGY AND ALARM USABILITY (C9)

CRIOP tests whether alarm philosophy follows EEMUA 191 (2024) and IEC 62682 (2022), ensures prioritisation, avoids alarm floods, and supports operators in defining abnormal conditions. Real-world evidence, such as that reported by Havtil (2022), indicates that standing alarms and poor configuration are present, degrading performance and requiring reporting. CRIOP provides a structured way to surface and prioritise these issues.

8. ANALYSE INFORMATION AND COMMUNICATION NEEDS (C1)

Effective coordination in safety-critical systems depends on shared situational awareness and reliable exchange of information. CRIOP identifies what information is needed, who needs it, when, and under what conditions – across local and remote actors.

9. TEST USABILITY, ROBUSTNESS, AND MOC DISCIPLINE (G10)

CRIOP stresses that systems must be tested for real-world performance, including degraded modes and hazardous situations (DFUs). It checks usability with real users and verifies that MoC processes prevent 'silent drift' of system behaviour over time.

10. TRAIN SYSTEMATICALLY FOR TASKS AND HAZARDS (T1)

Training must be based on TA and on realistic scenarios involving unwanted events. CRM, risk-based training, and simulator exercises ensure that teams are prepared for the unexpected. CRIOP checks whether training evolves alongside system changes, DFUs, and operational risk.

Together, these ten activities ensure that HF are integrated across the full lifecycle – from project definition to training and operation. They help projects design technology that supports human performance, avoids error traps, and enhances safety, efficiency, and usability. When combined with scenario-based exploration of WAD, CRIOP becomes not just a compliance activity but a practical and motivating tool for learning, collaboration, and system improvement. The success of the Unified Bridge (Bjørneseth, 2021) demonstrates the value of this approach in real-world projects.

REFERENCES

Aas, A. L., Johnsen, S. O., & Skramstad, T. (2009). CRIOP: a human factors verification and validation methodology that works in an industrial setting. In Proceedings of 28th International Conference on Computer Safety, Reliability, and Security (SAFECOMP'09), Hamburg, Germany, September 15–18 (pp. 243–256). Springer, Berlin, Heidelberg.

Amedia (2021). https://www.amedia.no/aktuelt/nyheter/amedia-utsatt-for-et-alvorlig-dataangrep

ANSI/HFES (2021). *Human Readiness Level Scale in the System Development Process* (ANSI/HFES 400-2021).

Antonsen, S., Ramstad, L., & Kongsvik, T. (2007). Unlocking the organization: Action research as a means of improving organizational safety. *Safety Science Monitor*, 11(1).

ATSB (2007). *Australian Transport Safety Bureau Aviation Research and Analysis.*

Bainbridge, L. (1983). Ironies of automation. In *Analysis, Design and Evaluation of Man–Machine Systems* (pp. 129–135). Pergamon.

Behm, M. (2005). Linking construction fatalities to the design for construction safety concept. *Safety Science*, 43(8), 589–611.

Behm, M., Culvenor, J., & Dixon, G. (2014). Development of safe design thinking among engineering students. *Safety Science*, 63.

Bergh, L. I. V., & Teigen, K. S. (2024). AI safety: A regulatory perspective. In Proceedings of the 4th International Conference on Applied Artificial Intelligence (ICAPAI) (pp. 1–6). IEEE.

Bergh, L. I. V., Teigen, K. S., & Dørum, F. (2024). Human performance and automated operations: A regulatory perspective. *Ergonomics*, 67(6), 744–758. https://doi.org/10.1080/00140139.2024.2321457

Bjørneseth, F. B. (2021). Unified bridge–design concepts and results. *Sensemaking in Safety Critical and Complex Situations*, 135–153.

Boehm (1976). Software engineering. *IEEE Transactions on Computers*, C–25(12), 1226–1241. https://doi.org/10.1109/tc.1976.1674590

Bradshaw, J. M., Hoffman, R. R., Woods, D. D., & Johnson, M. (2013).The seven deadly myths of "autonomous systems". *IEEE Intelligent Systems*, 28(3), 54–61.

Bye, R. J., Johnsen, S. O., & Lillehammer, G. (2018). Addressing differences in safety influencing factors – A comparison of offshore and onshore helicopter operations. *Safety*, 4(1), 4.

Cummings, M. L. (2024). A taxonomy for AI hazard analysis. *Journal of Cognitive Engineering and Decision Making*, 18(4), 327–332.

Dekker, S. (2019). *Foundations of Safety Science: A Century of Understanding Accidents and Disasters.* Routledge.

Dekker, S., & Conklin, T. (2014). *Safety Differently.* London: CRC Press.

Driscoll, T. R., Harrison, J. E., Bradley, C., & Newson, R. S. (2008). The role of design issues in work-related fatal injury in Australia. *Journal of Safety Research*, 39(2), 209–214.

Dyreborg, J., Lipscomb, H. J., Nielsen, K., Törner, M., Rasmussen, K., Frydendall, K. B., ··· & Kines, P. (2022). Safety interventions for the prevention of accidents at work: A systematic review. *Campbell Systematic Reviews*, 18(2), e1234.

EEMUA 191: The Engineering Equipment and Materials Users Association (EEMUA) (2024). *Alarm Systems: A Guide to Design, Management and Procurement* (Standard No. 191). Retrieved from https://online.standard.no/nb/eemua-publication-191

Elvik, R. (2021). Democracy, governance, and road safety. *Accident Analysis & Prevention*, 154, 106067.

Endsley, M. R. (1995). Toward a theory of situation awareness in dynamic systems. *Human Factors*, 37(1), 32–64.

Endsley, M. R., (2019). Human Factors & Aviation Safety, Testimony to the United States House of Representatives Hearing on Boeing 737-Max8 Crashes – December 11, 2019.

Federation of Norwegian Industries (2025) *Safety, Leadership, and Learning – A Practical Guide to HOP.*

Fitts, P. M. (Ed.). (1951). *Human Engineering for an Effective Air Navigation and Traffic Control System*, 84pp. National Research Council, Committee on Aviation Psychology, Division of Anthropology and Psychology.

Greenwood, D. J., & Levin, M. (2006). *Introduction to Action Research: Social Research for Social Change.* SAGE Publications.

Hancock, P. A. (2021). Months of monotony–moments of mayhem: Planning for the human role in a transitioning world of work. *Theoretical Issues in Ergonomics Science*, 22(1), 63–82.

Havtil (2022). Ansvar, kompetanse og vedlikehold avalarmhåndteringssystemer i kontrollrom/ (Responsibility, competence and maintenance of alarm handling systems in control rooms). Activity: 992923.

Hollifield, B. R., Oliver, D., Nimmo, I., & Habibi, E. (2008). *The High Performance HMI Handbook: A Comprehensive Guide to Designing, Implementing and Maintaining Effective HMIs for Industrial Plant Operations*. Houston: Pas.

Hollnagel, Erik (2017). *Safety-II in practice: Developing the resilience Potentials*. Routledge.

Hopkins, A. (2025). *Boeing, the 737 MAX Crisis and Aviation Safety: The Perils of Profit-driven Engineering*. CRC Press.

Hydro (2019). https://www.hydro.com/Document/Index?name=General%20cyber-attack%20presentation%20April%2012.pdf&id=28255

IEC - International Electrotechnical Commission 62682: 2022 Management of alarm systems for process industries.

IEC 63303: 2024 - International Electrotechnical Commission Human machine interface for process automation systems.

Insag-7 (1992). *The Chernobyl Accident: Updating of INSAG-1*. IAEA.

IOGP (2017). Report 423 'HSE management – guidelines for working together in a contract environment'.

ISO 11064 Principles for Design of Control Centres, International Organization for Standardization. (2000).

ISO 9241-210 (2019). *Ergonomics of Human-System Interaction Part 210: Human-Centred Design for Interactive Systems*.

Johnsen, S. O., Thieme, C., Aminoff, H., Park, J.; Fernander, M., Boren, M., Kjøl, Å. A., & Kirwan, B. (2025). *CRIOP A Scenario Method for Crisis Intervention and Operability Analysis*. Norway: SINTEF Technology and Society.

Kinnersley, S., & Roelen, A. (2007). The contribution of design to accidents. *Safety Science*, 45(1–2).

Maersk (2017). https://sosintel.co.uk/case-study-maersks-response-to-notpetya-how-cybersecurity-best-practices-mitigated-a-major-cyberattack

Manuele (2005). Risk assessment & hierarchies of control. *Professional Safety*, 50(5), 33.

Miranda, A. T. (2019). Misconceptions of human factors concepts, *Theoretical Issues in Ergonomics Science*, 20(1), 73–83,

MiT (2025). *The GenAI Divide State of Ai in Business 2025*. MiT NANDA Project.

Moura, R., Beer, M., Patelli, E., Lewis, J., & Knoll, F. (2016). Learning from major accidents to improve system design. *Safety Science*, 84, 37–45.

NAS - National Academies of Science (2021) *Human-AI Teaming: State of the Art and Research Needs*.

NIOSH (2024). *Prevention Through Design Program*. https://www.cdc.gov/niosh/research-programs/portfolio/ptd.html

NIST SP 800-160 (2022). Engineering Trustworthy Secure Systems.

NIST SP 800-53 (2020). Security and Privacy Controls for Information Systems and Organizations.

NOG 104 (2016). *Norwegian Oil and Gas*. Guideline No. 104 OLF Retningslinjer 104 ISBR.

Nortura (2021). https://www.linkedin.com/pulse/when-cyberattacks-hit-close-home-case-nortura-peter-ottis

Norwegian National Audit (2023). Sjøfartsdirektoratets arbeid med å fremme gode arbeids- og levevilkår til sjøs Dokument 3:9 (2022–2023).

NSIA (Norwegian Safety Investigation Authority) (2022). Human Factors in the NSIAs Safety Investigations.

NTSB - National Transportation Safety Board (2019). Collision between US Navy Destroyer John S McCain and Tanker Alnic MC. Singapore Strait 5 Miles Northeast of Horsburgh Lighthouse August 21, 2017. Marine Accident Report: NTSB/MAR-19/01 PB2019-100970).

Rasmussen, J. (1997). Risk management in a dynamic society: A modelling problem. *Safety Science*, 27(2–3), 183–213.

Safety Forum (2019). Learning from Incidents/Report from Safety Forum.

Samset, K. (2001). *Prosjektvurdering i tidligfasen. Fokus på konseptet (Front-end Assessment of Projects. Focus on the concept)*, University Textbook. Trondheim: Tapir Academic Press. ISBN: 82-519-1679-8; Side 39 - Figure 4.2.

Stanton, N. A., Salmon, P. M., Rafferty, L. A., Walker, G. H., Baber, C. & Jenkins, D. P. (2013). *Human Factors Methods: A Practical Guide for Engineering and Design*. CRC Press.

Szymberski, R. T. (1997). Construction project safety planning. *Tappi Journal*, 80(11), 69–74.

Thun, S., & Tinmannsvik, R. (2021). PRoIND - Proaktive indikatorer for psykososialt og organisatorisk arbeidsmiljø.

VG (2025). https://www.vg.no/nyheter/i/8plG3Q/vg-avsloerer-varslene-som-forsvinner-paa-melkoeya

Winge, S., Kilskar, S. S., & Johnsen, S. O. (2023). *A Guide for Identifying Human Factors in Accident Investigations*.

Index

Pages in *italics* refer to figures and pages in **bold** refer to tables.

P

R

S

T

U

V

For Product Safety Concerns and Information please contact our
EU representative GPSR@taylorandfrancis.com Taylor & Francis
Verlag GmbH, Kaufingerstraße 24, 80331 München, Germany